Nanoparticles Agents and Secondary Metabolites in Neurodegenerative Diseases

Jawad Laadraoui
https://orcid.org/0000-0002-9784-1265
Hassan II University, Morocco

Bilal El-Mansoury
https://orcid.org/0000-0002-2583-7515
Chouaib Doukkali University, Morocco

IGI Global
Scientific Publishing
Publishing Tomorrow's Research Today

Vice President of Editorial	Melissa Wagner
Director of Acquisitions	Mikaela Felty
Director of Book Development	Jocelynn Hessler
Production Manager	Mike Brehm
Cover Design	Jose Rosado

Published in the United States of America by
IGI Global Scientific Publishing
701 East Chocolate Avenue
Hershey, PA, 17033, USA
Tel: 717-533-8845 | Fax: 717-533-7115
Website: https://www.igi-global.com E-mail: cust@igi-global.com

Library of Congress Cataloging-in-Publication Data

Names: Laadraoui, Jawad, 1991- editor. | El-Mansoury, Bilal, 1995- editor.
Title: Nanoparticles agents and secondary metabolites in neurodegenerative diseases / edited by: Jawad Laadraoui, El-Mansoury Bilal.
Description: Hershey, PA : IGI Global Scientific Publishing, [2026] | Includes bibliographical references. | Summary: "This book aims to bridge the gap between two promising fields-nanotechnology and phytochemical research-to address the unmet challenges in treating neurodegenerative diseases. This book will explore how nanotechnology can address these limitations by providing targeted and efficient delivery systems, such as nanoparticles capable of crossing the blood-brain barrier"-- Provided by publisher.
Identifiers: LCCN 2025003720 (print) | LCCN 2025003721 (ebook) | ISBN 9798337313658 (hardcover) | ISBN 9798337313665 (paperback) | ISBN 9798337313672 (ebook)
Subjects: LCSH: Nervous system--Degeneration--Case studies. | Nervous system--Diseases--Treatment--Case studies. | Molecular neurobiology--Case studies. | Nanoparticles--Therapeutic use--Case studies. | LCGFT: Case studies. https://id.loc.gov/authorities/genreForms/gf2017026140
Classification: LCC RC365 .N36 2025 (print) | LCC RC365 (ebook) | DDC 616.8--dc23/eng/20250404
LC record available at https://lccn.loc.gov/2025003720
LC ebook record available at https://lccn.loc.gov/2025003721

British Cataloguing in Publication Data
A Cataloguing in Publication record for this book is available from the British Library.

To my father, Ali
To my mother, Fatima
To my brothers and sisters
To all my family and friends
Jawad Laadraoui

To my father, Lahcen
To my mother, Fatiha
To my uncle, Ahmed
To my mentor, Pr. Omar El Hiba
To all my family and friends
Bilal El-Mansoury

Table of Contents

Detailed Table of Contents

Neurodegenerative disorders as a category of diseases that is characterized by the slow progressive loss of neurons within the CNS. Alzheimer's disease, Parkinson's disease, Huntington's disease, and amyotrophic lateral sclerosis (ALS) are very common and severe and result in moderate to severe cognitive and physical deficits. The concerning pathophysiology of these disorders is multifactorial with genetic and environmental factors joined with neuronal abnormalities that progressing to irreversible damage of the nervous tissue. The present chapter is the initial step for understanding neurodegenerative diseases; nevertheless, some general ideas about the pathophysiology of these disorders are emphasized, such as protein misfolding, inflammation, and oxidative stress. Further, it analyzes clinical manifestations, diagnosis and the existing management strategies. As the population ages, a wide range of neurodegenerative diseases are becoming more common, making the study of such conditions essential to expanding the existing knowledge and potential treatments for the diseases.

Neurodegenerative diseases (NDDs) are age-related disorders and their prevalence is rising globally due to the increase in aging population. NDDs comprise a wide range of diseases that have specific etiological features affecting different regions of the brain and manifesting various types of clinical symptoms. However, their pathophysiology shares common genotype-phenotype relationships displaying pathological features like protein misfolding and aggregation of toxic proteins, mitochondrial dysfunction, oxidative stress, neuroinflammation, synaptic injury, and neuronal death. Scientific

Hajar Azraida, Department of Biology, Cadi Ayyad University, Morocco
Ouijdane El Hatimy, Department of Biology, Cadi Ayyad University, Morocco
Yasmina Jaouhari, Department of Biology, Cadi Ayyad University, Morocco
Fouad Bouazzaoui, Department of Biology, Cadi Ayyad University, Morocco
Majda Badaoui, Department of Biology, Cadi Ayyad University, Morocco
Hamid Kabdy, Department of Biology, Cadi Ayyad University, Morocco
Souad Moubtakir, Department of Biology, Cadi Ayyad University, Morocco
Loubna El Yazouli, Department of Biology, Cadi Ayyad University, Morocco
Rachida Aboufatima, Faculty of Sciences and Technology, Sultan Moulay Slimane University, Morocco
Mohamed Loukid, Department of Biology, Cadi Ayyad University, Morocco
Abderrahman Chait, Department of Biology, Cadi Ayyad University, Morocco

Neurodegenerative diseases, including Alzheimer's, Parkinson's, and amyotrophic lateral sclerosis, are increasing, highlighting the need for innovative treatments. Plant-derived phytochemicals, such as flavonoids, alkaloids, terpenoids, and polyphenols, offer neuroprotective properties by reducing oxidative stress, modulating neuroinflammation, preventing neuronal apoptosis, and enhancing neurotrophic signaling and mitochondrial function. However, their clinical application faces challenges like low bioavailability and rapid metabolism. This chapter explores these issues and potential solutions, including advanced drug delivery systems and structural modifications. It also highlights the need for interdisciplinary research to fully harness phytochemicals' therapeutic potential in neurodegenerative disease treatment.

Alzheimer's, Parkinson's, and various neurodegenerative diseases are generally characterized by neuronal loss, which directly induces cognitive and motor dysfunctions. Secondary metabolites, as natural products of aromatic and medicinal plants and other origins, could present a promising alternative source to combat neurodegenerative diseases. With their important phytochemical composition, especially in alkaloids, terpenes, flavonoids, and polyphenols, these secondary metabolites pave the way for the development of new therapies. Their neuroprotective properties and their antioxidant and anti-inflammatory activities constitute a solid basis for their use in the fight against neurodegenerative diseases. This chapter aims to explore the therapeutic effects of secondary metabolites in the fight against neurodegenerative diseases while discussing the responsible mechanisms, challenges, perspectives, as well as the limits of this application.

Parkinson's disease (PD) is often accompanied by neuropsychiatric symptoms, including mild depression, which significantly affect patients' quality of life. Saffron (Crocus sativus L.) has emerged as a natural therapeutic option due to its bioactive compounds crocin, crocetin, and safranal which exhibit antidepressant effects by inhibiting monoamine oxidase, modulating neurotransmitter reuptake, and enhancing brain-derived neurotrophic factor (BDNF) signaling. Extracts containing these compounds have shown efficacy comparable to conventional antidepressants in alleviating depression-related symptoms in PD. This chapter discusses saffron's neuroprotective properties, its impact on neurotransmitter systems, and interactions with monoamine oxidase enzymes, highlighting its potential for depression management. The aim of this chapter is to provide a comprehensive overview of saffron's phytochemical composition, mechanisms of action, and its therapeutic potential in mitigating depression and related neuropsychiatric symptoms in PD.

Multiple sclerosis (MS) is a chronic neurodegenerative disorder characterized by inflammation, demyelination, and progressive neurological impairment. This chapter explores the therapeutic potential of cannabinoid-based drugs in MS, focusing on their mechanisms of action, including anti-inflammatory, analgesic, and neuromodulatory effects. It provides an overview of current MS treatments informed by recent clinical studies, while emphasizing key challenges such as optimal dosing, toxicity, and limited bioavailability, which restrict effective symptom management. In parallel, the chapter examines advances in nanomedicine, particularly nano-cannabinoids, as innovative strategies to enhance drug delivery, improve safety profiles, and increase therapeutic efficacy. By integrating evidence from pharmacology, clinical trials, and nanotechnology, this chapter highlights both the opportunities and challenges of cannabinoids and nanocannabis approaches, offering a comprehensive perspective on their potential in MS management.

The gut microbiota plays a pivotal role in maintaining human health. Composed of billions of microorganisms interacting with the digestive system, it is increasingly recognized for influencing pathophysiological processes beyond the digestive tract, particularly in neurodegenerative diseases. Secondary metabolites produced by the gut microbiota, such as short-chain fatty acids, polyphenols, and bile acids can modulate epigenetic processes in the brain. These epigenetic changes, including activating or inhibiting genes involved in inflammation, neuroprotection, and neuronal repair, may offer new therapeutic perspectives for neurodegenerative diseases such as Alzheimer's and Parkinson's. Therefore, optimizing the gut microbiota and understanding these interactions provide promising avenues for innovative therapies for neurodegenerative diseases. This chapter provides a comprehensive overview of how microbial metabolites exert epigenetic effects and their potential for treating neurodegenerative diseases. It also includes references for further exploration of this exciting field.

Neurodegenerative disorders (NDDs), including Alzheimer's disease (AD), Parkinson's disease (PD), Prion disease (PrD), and amyotrophic lateral sclerosis (ALS), present huge difficulties on account of the consistently advancing loss of neurons. Nanotechnology offers a promising solution to these difficulties. Designed nanomaterials can cross the BBB, upgrading prescription conveyance to brain tissues. Nanoparticles further develop drug solubility, stability, bioavailability, and designated conveyance while diminishing side effects. Generally used nanocarriers incorporate liposomes, emulsions, solid lipid nanoparticles, and polymeric nanoparticles. These nanocarriers enhance drug conveyance, bioavailability, and stability, effectively conveying phytomedicine and manufactured drugs. Nanomedicine progressions show potential for treating neurodegenerative infections by further developing medication transport across the BBB and limiting fringe side effects. Research needs to be focused on biocompatibility, Environmental effects, and administrative obstacles to the drug development.

Neurodegenerative diseases, including Alzheimer's, Parkinson's, and Huntington's, are pressing global health challenges characterized by complex pathologies and limited treatment options. This chapter explores the synergistic potential of nanotechnology and secondary metabolites to address these challenges. Secondary metabolites, such as polyphenols, alkaloids, and terpenoids, exhibit antioxidant, anti-inflammatory, and neuroregenerative properties, making them promising candidates for therapeutic intervention. However, their clinical application is hindered by low bioavailability and poor blood-brain barrier (BBB) penetration. Nanotechnology offers innovative solutions by enabling precise drug delivery, enhancing stability, and facilitating BBB crossing through advanced nanocarriers like liposomes, nanoparticles, and dendrimers. Case studies on nano-enabled formulations, including curcumin, piperine, resveratrol, etc, demonstrate significant efficacy in preclinical neurodegenerative models.

The potential of nanoparticle-based drug delivery systems and bioactive secondary metabolites offers promising avenues for the treatment of neurodegenerative diseases. Neurodegeneration, driven by oxidative stress, protein aggregation, mitochondrial dysfunction, and neuroinflammation, presents complex therapeutic challenges. Nanocarriers such as polymeric, lipid-based, and metallic nanoparticles

Neurodegenerative diseases pose a major global health challenge, demanding advanced models and innovative therapeutic strategies. This chapter overviews in vivo, in vitro, and in silico models used to study these disorders and their role in pharmacological and therapeutic research. Such models clarify mechanisms underlying Alzheimer's, Parkinson's, and Huntington's disease while serving as essential platforms for drug testing, toxicity evaluation, and therapy development. A key focus is the integration of nanoparticles, whose small size, large surface area, and ability to cross the blood–brain barrier enable targeted drug delivery with fewer side effects. Functionalization with diagnostic markers further supports early biomarker detection and personalized medicine. By combining experimental models with nanotechnology, this approach improves diagnostic precision and therapeutic efficacy, offering promising prospects for more effective and less invasive treatments of neurodegenerative disorders.

Alzheimer's, Parkinson's, and Huntington's are devastating neurodegenerative diseases posing major challenges to global health and economies. Current therapies often fail to halt progression, lack targeting, and provide limited efficacy. This chapter highlights innovations poised to transform ND care, including personalized medicine tailored to genetic, molecular, and environmental factors, enabling precise, low-dose treatments. Nanotechnology advances drug delivery, with nanoparticles loaded with secondary metabolites showing promising neuroprotective effects in preclinical and clinical studies. Natural products, particularly secondary metabolites, offer neuroprotection via antioxidation, anti-inflammation, and apoptosis modulation. Together, these strategies present a synergistic roadmap for ND therapy. While ethical, regulatory, and scalability challenges remain, this chapter envisions a future of transformative treatments, offering guidance for clinicians and researchers in revolutionizing ND management.

Preface

INTRODUCTION

Neurodegenerative diseases remain among the greatest biomedical challenges of our century, affecting millions of individuals and placing an immense burden on families, healthcare systems, and societies worldwide. Despite decades of research, treatments for Alzheimer's disease, Parkinson's disease, Huntington's disease, and related disorders remain largely palliative, offering symptom management rather than true disease modification. The urgent need for innovative, safe, and effective therapeutic approaches has never been greater.

It is within this context that we present *Nanoparticles, Agents, and Secondary Metabolites in Neurodegenerative Diseases*, a volume inspired by the growing convergence of two dynamic research frontiers: nanotechnology and natural product–based therapeutics. Each field has shown significant promise independently. Nanoparticles have demonstrated an exceptional ability to transport therapeutics across the formidable blood–brain barrier, to enhance drug stability, and to enable targeted delivery that minimizes systemic toxicity. Meanwhile, plant-derived secondary metabolites—such as flavonoids, terpenoids, phenolics, and alkaloids—continue to reveal remarkable neuroprotective, anti-inflammatory, and antioxidant properties. Yet, challenges such as poor bioavailability and metabolic instability have hindered their translation into clinical practice.

This book explores how the strategic integration of nanotechnology can unlock the full therapeutic potential of these bioactive compounds. By pairing the unique physicochemical properties of nanoparticles with the biological efficacy of secondary metabolites, new possibilities emerge for preventing neuronal loss, modulating neuroinflammatory pathways, and ultimately slowing or halting disease progression. The chapters compiled here examine cutting-edge research on nanoparticle-enabled delivery systems, assess safety and toxicological considerations, and highlight preclinical and clinical advances that pave the way for translational breakthroughs.

In assembling this reference, our goals have been multifaceted. We aim to advance the therapeutic potential of secondary metabolites, demonstrating how nanotechnology can overcome longstanding barriers to their clinical adoption. We seek to promote interdisciplinary innovation, encouraging collaborations across neurobiology, nanomedicine, and natural products chemistry. We have placed strong emphasis on safety, toxicity, and regulatory perspectives, recognizing that responsible innovation requires a clear understanding of risks as well as benefits. At the same time, we intend this book to serve as a comprehensive, authoritative resource, offering clarity to seasoned researchers while inspiring emerging scholars entering this expanding field. Ultimately, our hope is to support clinical translation, bridging laboratory discoveries with real-world therapeutic applications.

Our intended audience reflects the interdisciplinary nature of the work. Researchers, clinicians, pharmaceutical and biotechnology professionals, students, educators, regulatory experts, and healthcare advocates will each find perspectives relevant to their roles in advancing neurodegenerative disease research and therapy. Whether the reader approaches this text from the standpoint of scientific inquiry, clinical practice, industrial innovation, or public health, we believe the knowledge compiled herein will equip them with the tools needed to understand and contribute to this rapidly evolving field.

We express our sincere gratitude to the contributors whose expertise and dedication have shaped the chapters that follow. Their collective work reflects the global effort to confront neurodegeneration with scientific rigor, creativity, and hope.

With this volume, we invite readers to explore the transformative potential of nanoparticle-enabled secondary metabolites and to join in the pursuit of novel therapies that may one day change the lives of millions affected by neurodegenerative disease.

ORGANIZATION OF THE BOOK

Chapter 1 introduces readers to the clinical and biological foundations of neurodegenerative diseases, laying essential groundwork for the rest of the volume. By outlining the progressive nature of neuronal loss in conditions such as Alzheimer's disease, Parkinson's disease, Huntington's disease, and amyotrophic lateral sclerosis, the chapter clarifies the shared pathophysiological themes—protein misfolding, inflammation, oxidative stress—and the diverse clinical presentations that define these disorders. The authors emphasize how aging populations are accelerating global disease prevalence, highlighting the urgent need for new therapeutic strategies and improved diagnostic approaches. This chapter serves as a gateway to understanding neurodegeneration from its molecular origins to patient experience.

Chapter 2 builds on this foundation by providing an in-depth exploration of the mechanistic drivers of neurodegeneration and the neuroprotective potential of secondary metabolites. The authors examine key pathological events, including mitochondrial dysfunction, synaptic injury, and toxic protein aggregation, and demonstrate how these mechanisms converge across different neurodegenerative diseases. They also spotlight the rich therapeutic promise of phytochemicals, whose pleiotropic bioactivity—ranging from antioxidant to anti-inflammatory effects—positions them as compelling candidates for disease-modifying therapies. This chapter sets the stage for integrating mechanistic insight with natural product-based intervention.

Chapter 3 transitions the discussion into the realm of emerging therapeutics, emphasizing personalized medicine and advanced technological tools. The authors describe how genomics, precision medicine, AI-driven analytics, stem cell therapies, and gene-editing approaches are reshaping how clinicians and researchers approach neurodegenerative diseases. By illustrating how these technologies enable early detection, individualized treatment plans, and even neuronal repair, the chapter underscores the transformative potential of cross-disciplinary innovation. It highlights a future in which nanotechnology and natural products complement advanced biomedical engineering to create patient-specific neurotherapeutic solutions.

Chapter 4 focuses on plant-derived secondary metabolites and their diverse neuroprotective effects. Flavonoids, alkaloids, terpenoids, and polyphenols are examined for their ability to combat oxidative stress, modulate inflammation, prevent neuronal apoptosis, and enhance cellular resilience. The chapter also acknowledges the practical challenges that have limited clinical translation—particularly poor stability, rapid metabolism, and low bioavailability—and evaluates how modern drug delivery technologies may overcome these limitations. The authors advocate for continued interdisciplinary collaboration to fully harness the therapeutic potential of plant-based compounds.

Chapter 5 further expands on the therapeutic relevance of secondary metabolites by detailing their biochemical profiles, mechanisms of action, and advantages over conventional pharmaceuticals. The chapter underscores the role of natural compounds in mitigating cognitive and motor dysfunction through antioxidant, anti-inflammatory, and neuroprotective mechanisms. It critically evaluates both the promise and limitations of using secondary metabolites as alternatives or adjuncts to traditional treatments. In doing so, the authors illuminate avenues for drug discovery grounded in natural product diversity and mechanistic specificity.

Chapter 6 narrows the focus to resveratrol, a widely studied polyphenol with notable neuroprotective mechanisms relevant to Alzheimer's and Parkinson's diseases. By examining shared pathological mechanisms—including oxidative stress, impaired protein clearance, and neuroinflammation—the chapter demonstrates how resveratrol engages key molecular pathways such as SIRT1, AMPK, and autophagy to protect neuronal function. Highlighting robust preclinical data showing improvements in cognition, motor behavior, and synaptic health, the authors position resveratrol as a compelling adjunct therapeutic candidate for dual-disease application.

Chapter 7 explores the neuroprotective influence of saffron's bioactive compounds in the context of Parkinson's disease and its associated neuropsychiatric symptoms. With an emphasis on crocin, crocetin, and safranal, the chapter outlines how these compounds modulate monoamine oxidase activity, influence neurotransmitter reuptake, and enhance neurotrophic signaling. The authors present evidence that saffron extracts rival conventional antidepressants in improving depression-related symptoms in Parkinson's patients, offering a natural therapeutic alternative with multifaceted benefits.

Chapter 8 widens the scope to address cannabinoids in the management of multiple sclerosis. The chapter integrates pharmacological, clinical, and nanotechnological perspectives to present a balanced assessment of cannabinoid-based therapies. By examining their anti-inflammatory, analgesic, and neuromodulatory actions, and by addressing dosing challenges, toxicity concerns, and formulation barriers, the authors present a nuanced exploration of both conventional cannabinoids and nano-enhanced cannabis derivatives. This chapter highlights the promise and complexity of using cannabinoid-based interventions in neurodegenerative and neuroinflammatory conditions.

Chapter 9 expands the discussion of cannabis-derived therapeutics to encompass the broader spectrum of secondary metabolites from Cannabis sativa and Nicotiana tabacum. The authors present cannabinoids, terpenes, flavonoids, and alkaloids as modulators of neuroinflammation, oxidative stress, and synaptic plasticity, with therapeutic relevance across multiple neurodegenerative diseases. By comparing mechanisms, evidence, and clinical potential, the chapter provides a comprehensive overview of two historically significant plants and their modern biomedical applications, including their translational limitations.

Chapter 10 shifts focus to the gut–brain axis, illustrating how microbial metabolites influence neurodegenerative disease through epigenetic modulation. The chapter explains how short-chain fatty acids, bile acids, and polyphenol derivatives regulate gene expression related to inflammation, neuroprotection, and neuronal repair. Through this lens, the authors present the gut microbiome as a powerful therapeutic target—one that may be optimized to modify disease progression in Alzheimer's and Parkinson's diseases. This chapter underscores the growing importance of microbiome research in shaping future neurotherapeutic strategies.

Chapter 11 integrates nanotechnology with secondary metabolite–based therapies, demonstrating how advanced nanocarriers enhance delivery, stability, and bioavailability of natural compounds across the blood–brain barrier. The authors review liposomes, polymeric nanoparticles, emulsions, and solid lipid nanoparticles, highlighting their role in improving drug targeting and reducing systemic toxicity. They also outline challenges related to biocompatibility, environmental impact, and regulatory approval, framing this integration as a promising yet evolving frontier in neurodegenerative disease management.

Chapter 12 continues this theme by examining the synergistic relationship between nanotechnology and secondary metabolites more comprehensively. The chapter showcases case studies in nano-enabled formulations—such as nano-curcumin, nano-piperine, and nano-resveratrol—that demonstrate enhanced therapeutic efficacy in preclinical neurodegenerative models. Through discussion of pathophysiological barriers such as low bioavailability and poor BBB penetration, the authors show how nanotechnology provides innovative solutions that could redefine natural product–based therapies.

Chapter 13 explores integrative nanomedicine approaches, highlighting how nanocarrier systems and secondary metabolites together address the multifactorial nature of neurodegenerative pathology. The chapter reviews the molecular mechanisms through which secondary metabolites exert antioxidant, anti-inflammatory, and autophagy-regulating effects, and details how nanotechnology optimizes their delivery and therapeutic windows. The authors also evaluate translational challenges—including safety, regulation, and manufacturing—providing a roadmap for advancing integrative neurotherapeutic design.

Chapter 14 focuses on nanoparticle-based strategies targeting dysregulated neuronal signaling, a core contributor to neurodegeneration. The chapter emphasizes how nano-phytomedicine and metal-based nanoparticles enhance treatment precision by improving drug stabilization, neuronal targeting, and penetration of the blood–brain barrier. The authors argue that such precision-driven interventions represent a crucial evolution beyond the limitations of current therapeutics.

Chapter 15 examines the essential role of in vivo, in vitro, and in silico models in advancing nanoparticle applications for neurodegenerative disease research. By detailing the strengths and limitations of each model, the authors show how integrated systems facilitate mechanistic discovery, therapeutic screening, toxicity evaluation, and personalized medicine approaches. The chapter highlights how nanoparticle functionalization can improve biomarker detection and targeted delivery, paving the way for more effective translational research.

Chapter 16 concludes the volume by presenting emerging trends and future directions in neurodegenerative disease therapy. The authors synthesize insights on personalized medicine, nanotechnology, and secondary metabolites, illustrating how these fields intersect to create synergistic, targeted treatment innovations. The chapter acknowledges ongoing ethical, regulatory, and scalability challenges but envisions a future where integrated technological and natural-product-based strategies transform the neurotherapeutic landscape.

CONCLUSION

As we conclude this volume, we reflect on the breadth and depth of knowledge presented across its chapters—knowledge that underscores both the complexity of neurodegenerative diseases and the remarkable progress being made toward their treatment. Collectively, the contributions within this book illuminate a central theme: meaningful advances against Alzheimer's disease, Parkinson's disease, Huntington's disease, multiple sclerosis, and related disorders will arise from the integration of diverse scientific disciplines, innovative methodologies, and nature-inspired therapeutics.

Throughout the book, the authors demonstrate how the convergence of nanotechnology and secondary metabolites offers an unprecedented opportunity to reshape the therapeutic landscape. Nanoparticles provide the precision, stability, and delivery capabilities that natural compounds often lack, while plant-derived metabolites bring unmatched biochemical diversity and multi-targeted efficacy. Together, these approaches hold the potential not only to enhance neuroprotection but also to overcome long-standing barriers such as poor bioavailability, limited blood–brain barrier penetration, and systemic toxicity. This synergy—rooted in both technological ingenuity and biological wisdom—represents a promising frontier for disease-modifying interventions.

The chapters also remind us that innovation must be grounded in scientific rigor, safety, and a clear path to clinical translation. Whether examining the molecular actions of phytochemicals, evaluating novel nanocarriers, exploring the microbiome's epigenetic influence, or leveraging artificial intelligence and personalized medicine, the authors emphasize the importance of translational relevance, regulatory considerations, and ethical responsibility. The future of neurotherapeutics depends not only on discovering new tools but on ensuring they can be safely and effectively moved from laboratory science to clinical care.

We hope this book serves as a catalyst for interdisciplinary collaboration, inspiring researchers, clinicians, industry professionals, and policymakers to work together toward innovative solutions. As the global burden of neurodegenerative diseases continues to rise, the need for such collaborative efforts becomes ever more urgent. The insights compiled in this reference volume reflect the determination of the scientific community to meet these challenges with creativity, diligence, and optimism.

It is our sincere belief that the ideas and discoveries presented here will contribute to a future in which neurodegenerative diseases can be detected earlier, treated more effectively, and ultimately prevented or reversed. We invite readers to build upon these findings, explore new avenues of inquiry, and continue pushing the boundaries of what is possible in neurodegenerative disease research and therapy.

Chapter 1
Introduction to Neurodegenerative Diseases

Tejal Bele
https://orcid.org/0000-0002-4170-6232
SCES's Indira College of Pharmacy, Pune, India

Shraddha Devarshi
https://orcid.org/0000-0002-6725-8917
Indira University, School of Pharmacy, Pune, India

Rutuja Kamble
https://orcid.org/0000-0002-7919-7197
SCES's Indira College of Pharmacy, Pune, India

Amruta Walvekar
https://orcid.org/0000-0001-7262-8380
Indira University, School of Pharmacy, Pune, India

Pallavi Chouhan
https://orcid.org/0000-0001-8755-3422
Indira University, School of Pharmacy, Pune, India

ABSTRACT

Neurodegenerative disorders as a category of diseases that is characterized by the slow progressive loss of neurons within the CNS. Alzheimer's disease, Parkinson's disease, Huntington's disease, and amyotrophic lateral sclerosis (ALS) are very common and severe and result in moderate to severe cognitive and physical deficits. The concerning pathophysiology of these disorders is multifactorial with genetic and environmental factors joined with neuronal abnormalities that progressing to irreversible damage of the nervous tissue. The present chapter is the initial step for understanding neurodegenerative diseases; nevertheless, some general ideas about the pathophysiology of these disorders are emphasized, such as protein misfolding, inflammation, and oxidative stress. Further, it analyzes clinical manifestations, diagnosis and the existing

DOI: 10.4018/979-8-3373-1365-8.ch001

management strategies. As the population ages, a wide range of neurodegenerative diseases are becoming more common, making the study of such conditions essential to expanding the existing knowledge and potential treatments for the diseases.

INTRODUCTION

Neurodegenerative diseases (NDDs) comprise a heterogeneous group of progressive neurological disorders that affect millions of individuals worldwide and are characterized by the gradual loss of structure and function of neurons within the central or peripheral nervous system (Mittal & Agrawal 2024). Unlike other cell types, neurons have a limited capacity for regeneration; therefore, their deterioration leads to disruption of essential neural communication pathways. This progressive breakdown manifests clinically as impairments in memory, cognition, motor control, coordination, and sensory processing, along with behavioural and mood alterations that worsen over time (Mittal & Agrawal 2024). The most prevalent and impactful NDDs include Alzheimer's disease, Parkinson's disease, Huntington's disease, amyotrophic lateral sclerosis (ALS), and multiple sclerosis (MS), as illustrated in Figure 1. Although each disorder differs in etiology, anatomical localization, and rate of progression, they exhibit shared pathological mechanisms such as abnormal protein aggregation, mitochondrial dysfunction, oxidative stress, excitotoxicity, and chronic neuroinflammation (Lamptey et al. 2022).

Despite advances in understanding these converging molecular pathways, current therapeutic interventions primarily provide symptomatic relief and have limited capacity to modify or halt disease progression. A major barrier to effective treatment is the restrictive nature of the blood–brain barrier (BBB), which severely limits the penetration of conventional pharmacological agents into neural tissue (Pathak et al. 2023). Additionally, many therapeutic compounds, particularly those targeting oxidative and inflammatory cascades, suffer from low bioavailability, rapid degradation, and systemic toxicity, further reducing clinical impact. These challenges highlight an urgent need for innovative delivery platforms capable of enhancing CNS targeting, protecting therapeutic agents from degradation, and enabling controlled and site-specific release.

Nanotechnology has emerged as a promising paradigm shift in the management of neurodegenerative diseases due to the unique physicochemical properties of nanoparticles that facilitate improved BBB permeability, enhanced drug stability, and targeted delivery to affected neuronal regions (Singh et al. 2022). Diverse nanoscale systems—including polymeric nanoparticles, lipid-based carriers, dendrimers, metallic nanoparticles, and nanogels—have demonstrated significant potential in preclinical studies for Alzheimer's disease, Parkinson's disease, and other NDDs, enabling improved therapeutic loading, prolonged circulation, and reduced off-target adverse effects (Gupta & Verma 2021). Beyond synthetic drugs, nanoparticles also offer a compelling strategy to overcome the delivery limitations of naturally derived secondary metabolites, many of which exhibit antioxidant, anti-inflammatory, anti-apoptotic, and anti-aggregatory activities relevant to neuroprotection (Raza et al. 2024).

Secondary metabolites such as curcumin, resveratrol, quercetin, and epigallocatechin gallate (EGCG) have shown promise in modulating pathological protein misfolding, reducing reactive oxygen species, and attenuating neuroinflammatory cascades, yet their therapeutic translation remains restricted by poor solubility, rapid metabolism, and minimal CNS uptake (Kumar et al. 2023). Encapsulating or conjugating these bioactive compounds within nanocarriers significantly enhances their stability, enhances pharmacokinetic profiles, and augments therapeutic potency, representing a synergistic intersection between natural product chemistry and nanomedicine (Bhardwaj et al. 2024). As research advances, this integration is increasingly viewed as a pioneering strategy capable of addressing key mechanistic targets such as mitochondrial impairment, synaptic dysfunction, and microglial activation, thereby positioning nano-enabled metabolite delivery as a transformative direction in NDD therapeutics.

By contextualizing neurodegenerative disease mechanisms within the opportunities and challenges of nanoparticle-mediated and metabolite-based therapeutic strategies, this chapter aims to establish a scientifically coherent foundation for the subsequent sections of this volume. Rather than serving solely as a general overview, the introduction underscores the rationale for applying nanotechnology and bioactive secondary metabolites to overcome longstanding therapeutic limitations, aligning with emerging research trends, translational barriers, and future innovation pathways that define the evolving landscape of neurodegenerative disease management.

Figure 1. Overview of NDDs

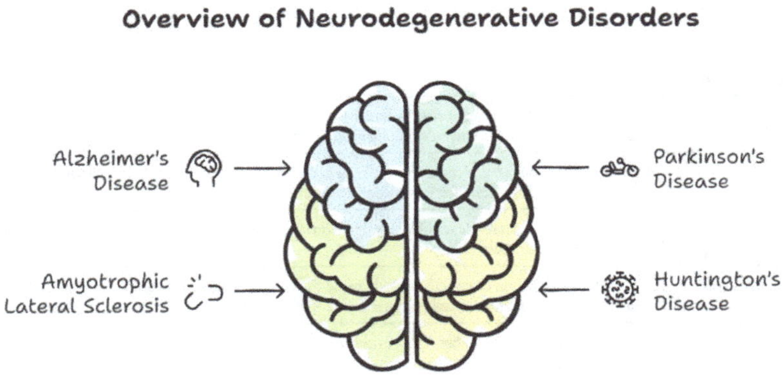

Key Features of NDDs

Progressive impairment in Function: NDDs are typically progressive, and the symptoms get worse over time. For example, in AD, cognitive abilities like memory and decision-making slowly deteriorate, and in PD, motor skills such as movement and coordination decline gradually. The neuronal damage in these disorders is irreversible.

Cognitive Dysfunction: Most of NDDs, including AD are characterized by severe memory impairment for the most part being short-term memory. While in the later stage of the disease, they are likely to have deficits of executive functioning which include planning, problem-solving, attention to detail, and decision-making.

Motor Symptoms: Motor symptoms are seen in multitude of NDDs inclusive of PD where the presenting signs may be tremors, rigidity, bradykinesia and postural instability. HD and ALS also result in balance problems, incoordination and problems with walking, which make falls and other accidents much more likely. In ALS, gradual atrophy of motor neurons that supply the voluntary muscles develops, and patients develop muscular paralysis.

Behavioral and Psychological Symptoms: Mood swings, apathy, and change of personality are observed in NDDs. Other symptoms may also be observed but depression, anxiety, irritability, and even delusions may flank the primary symptoms. With increased severity of PD, there can be impairments in cognitive-emotional processes ultimately leading to social isolation, anxiety, aggressiveness, or hallucinations.

Loss of Autonomy: As the disease advances, individuals may lose the ability to perform daily activities, such as dressing, bathing, and eating, leading to a reliance on caregivers. In advanced stages of diseases like ALS and Alzheimer's, individuals may become bedridden and require full-time assistance for all activities.

Risk Factors for NDDs

NDDs are influenced by various risk factors, both intrinsic and extrinsic as shown in Table 1 & Figure 2)

Table 1. Risk factors for NDDs

Age	Aging is the most significant risk factor due to oxidative damage and cellular impairment, with diseases like Alzheimer's and Parkinson's risk increasing after age 65.
Genetics	Inherited mutations (e.g., Huntington's) or genetic susceptibility factors (e.g., APOE ε4 allele for Alzheimer's, LRRK2 gene for Parkinson's) influence disease progression.
Environmental Toxins	Pesticides, heavy metals (lead, mercury), industrial chemicals, and airborne pollutants are linked to neurodegenerative diseases.
Traumatic Brain Injury (TBI)	Repeated head trauma or concussions increase the risk of conditions like Parkinson's and CTE due to neuroinflammation and damage to memory, motor, and cognitive regions.
Chronic Diseases	Cardiovascular disease, diabetes, hypertension, and hypercholesterolemia contribute to oxidative stress, vascular issues, and reduced brain blood flow.
Inflammatory Diseases	Chronic inflammation (e.g., rheumatoid arthritis) triggers neuroinflammation, increasing the risk of neurodegenerative conditions.
Lifestyle Factors	Poor nutrition, physical inactivity, smoking, alcohol, stress, and sleep deprivation increase oxidative stress and neuroinflammation, contributing to neurodegeneration.
Gender	Alzheimer's is more common in postmenopausal women (hormonal changes, longer lifespan); Parkinson's is more common in men (hormonal and genetic factors).

Figure 2. Understanding of risk factors of NDDs

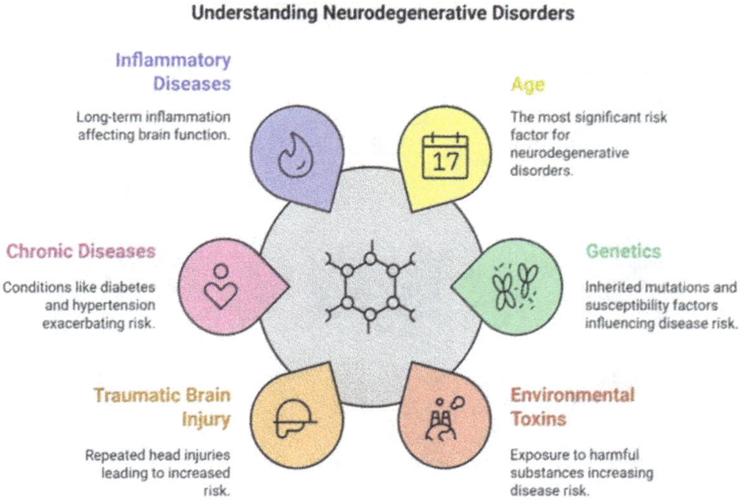

Neuroprotection and Neurodegeneration

The difference between neuroprotection and neurodegeneration is discussed in the following Table 2 and hallmarks of aging of neurodegeneration in Figure 3.

Table 2. Difference between neuroprotection and neurodegeneration

Aspects	Neuroprotection	Neurodegeneration
Definition	Strategies to preserve neuron health and prevent damage	Progressive loss or degeneration of neurons
Goal	Protect neurons from damage, minimize loss	Result of ongoing neuron damage and death
Outcome	Neuronal survival, maintenance of function	Neuronal death, loss of brain or nervous system function
Context	Applied in the prevention or early treatment of diseases	Observed in diseases or conditions that involve the ongoing loss of neurons
Mechanisms Involved	Antioxidants, anti-inflammatory treatments, neurogenesis, calcium regulation	Protein misfolding, oxidative stress, excitotoxicity, inflammation

Figure 3. Hallmarks of aging in neurodegeneration

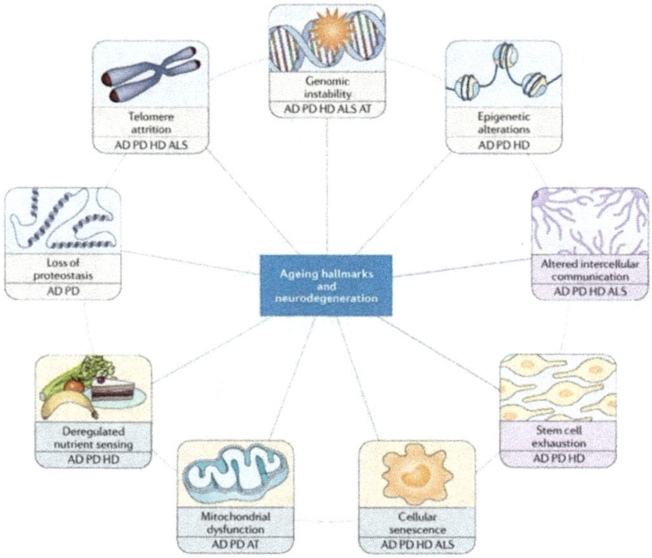

Source: Hou et al. (2019)

Epidemiology of NDDs

AD and PD especially are a significant health problem. Huntington's or ALS, even if they are not so widespread, affects a patient and his relatives, friends, and other caregivers intensively, which again proves the role of focused research and proper treatments. NDDs represent one of the major focuses of world concern and activities because they have become the major cause of disability, which prevalence has increased due to the growing number of ages and longer life expectancy (Gogna et al, 2024). AD still takes the largest share forecasted to be at 6.9 million people of 65 and older in United States by 2024, expected to rise to 14 million in 2060. PD affects over 8.5 million people worldwide, has been characterized as the quickest-growing neurological ailment (Bele, T. et al., 2024). Forecasting indicates that burden of NDDs will rise by a staggering 22% by 2050; rising to touch a staggering 4.9 billion people. These tendencies evidence the need for the improvement of the related research as well as new therapeutic strategies and powerful public health interventions to combat the growing burden caused by these disorders on the global healthcare systems.

Prevalence of NDDs

Among these, According to World Health Organization, AD stands out to be the most common, at present, affecting more than 55 million people and the situation is expected to worsen due to increasing population of the elderly. PD is one of the second most common NDDs in the world; it is registered in more than 10 million people worldwide, the primary affected population being above 60 years of age. Some diseases are exceptionally rare, for example ALS and HD; however, they significantly limit patient's and caregiver's quality of life. ALS is diagnosed in 2–5 individuals per 100,000 and HD in 5–10 individuals per 100,000 (Figure 4). Alzheimer's and Parkinson's are the most burdensome diseases but other disorders adversely impact neurological disability in the youthful populace (Gadhave, D. G. et al. 2024).

Figure 4. Prevalence of NDDs

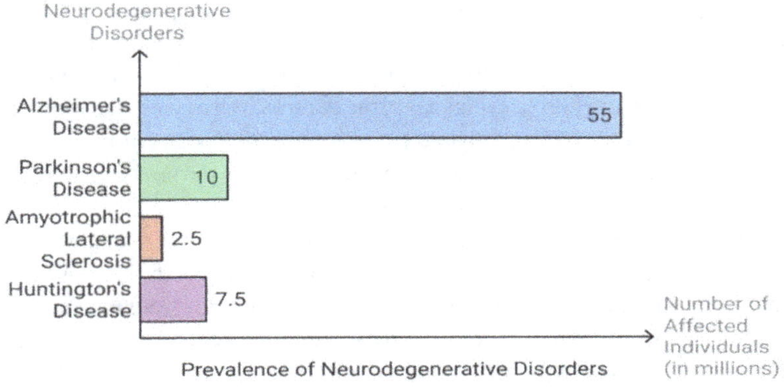

Classification of NDDs

Table 3. Classification of NDDs based on various criteria

Classification Criteria	Category	Examples
Based on Clinical Presentation	Cognitive Disorders	AD, Frontotemporal Dementia (FTD), Lewy Body Dementia
	Motor Disorders	PD, ALS, Multiple System Atrophy (MSA)
	Mixed Disorders	PD Dementia, AD with Parkinsonism
Based on Neuron Type Affected	Cortical Neuron Degeneration	AD, Frontotemporal Dementia, Lewy Body Dementia
	Subcortical Neuron Degeneration	PD, HD, Progressive Supranuclear Palsy (PSP)
	Motor Neuron Degeneration	ALS(ALS), Spinal Muscular Atrophy (SMA)
Based on Pathology	Proteinopathies	AD (amyloid plaques, tau tangles), PD (Lewy bodies), HD (huntingtin protein)
	Neurofibrillary Degeneration	AD, Progressive Supranuclear Palsy (PSP), Corticobasal Degeneration (CBD)
	Gliopathies	Alexander Disease, Multiple System Atrophy (MSA)
Based on Age of Onset	Early-Onset	HD, Familial AD, Spinocerebellar Ataxia (SCA)
	Late-Onset	AD, PD, ALS

continued on following page

Table 3. Continued

Classification Criteria	Category	Examples
Based on Genetic Basis	Monogenic Diseases	HD, Familial AD, Spinocerebellar Ataxia (SCA)
	Multifactorial (Polygenic) Diseases	AD (APOE ε4 allele), PD (LRRK2, SNCA), MS(HLA-DRB1)

Alzheimer's Disease (AD): is a chronic disease that worsens over time affecting millions of people globally and especially elderly. This is evidenced by the progressive loss of mental faculties or intellectual abilities to think, remember, and reason, among others, making that individual unable to attend to themselves or carry out their daily tasks. The most recognized characteristics of AD are amyloid-beta plaques in the brain as well as tape-like formations, called tau tangles (Abubakar et al. 2022). Amyloid plaques refer to deposits of protein which forms between neurons while tau tangles occur within neurons whereby the protein accumulates within neurons degrading its function in passing signals.

Parkinson's Disease (PD): is a chronic movement disorder that is strongly associated with the premature death of dopaminergic neurons in the substantia nigra of the brain. With a reduction in dopamine, people develop the primary motor symptoms inclining tremors, muscular rigidity, akinesia or slowness of movements, postural instability which qualifies substantial inhibition of most activities. However, along with the motor symptoms there are many motor symptoms including dementia, depression, anxiety, and changes in the autonomic nervous system. At present, the exact cause of the condition cannot be determined but researchers think that it has both familial and environmental roots (Ye et al. 2023

Amyotrophic lateral sclerosis (ALS): affects upper motor neurons of the motor cortex and lower motor neurons of the brain stem and spinal cord. This leads to muscle wasting, gradual loss of motor function and ultimately death from respiratory failure. ALS is of two classifications; familial ALS (fALS) and sporadic ALS (sALS). The two forms are clinically indistinguishable from the other; fALS contributes to 10% and the remaining cases are sALS (Feldman et al. 2022). In the last few years, there had been an explosion of genetic studies associating ALS with several genetic mutations in genes codifying for different proteins: These included Cu/Zn superoxide dismutase (SOD1), transactive response binding protein 43 (TARDBP), fused in sarcoma (FUS) and valosin containing protein (VCP).

Huntington's Disease (HD): is also known as Huntington's chorea is a heredity illness determined by a mutation on the HTT gene as well as characterized by a misfolded protein called huntingtin. This mutation is inherited in an autosomal dominant manner and characterized by slowly progressive neuronal loss in the brain which impacts on motor, cognitive and psychiatric manifestations. Given signs and symptoms begin manifesting between the ages of 30 and 50 and include chorea, troubles with coordination and memory, depression and mood swings.

Impact of NDDs on Healthcare System

In year 2024, the neurological diseases including the Alzheimer's type of dementia, Parkinson's type of dementia and the ALS were still a major health challenge across the globe and especially in the ageing population. The financial losses included direct and indirect expenses which included over $360 billion by USA in dementia related conditions (Alzheimer's Dement. 2024).

These disorders need highly specialized, often interdisciplinary care that has taxed healthcare systems and long-term supportive care settings, further reducing staffing and accommodations. Socio-economic disparities of patient's diagnosis and treatment continued crossed the global frontiers, therefore, highlighting the importance of social justice in health with a call for universal progressive policies and global cooperation. The increasing incidence of these diseases justifies modern improvements in the identification, prevention, and care for affected families, as well as the fair distribution of innovative treatments to decrease the social cost in the future.

Pathophysiology of NDDs

Neurodegeneration is defined by synaptic dysfunction, defective neuronal connections and intracellular, physicochemical modification in the brain of neural proteins, which are precursors of viable proteins. NDDs are disorders that involve the pathological death of neurons, and they include; AD, PD, prion diseases, amyotrophic lateral sclerosis, motor neuron disease, HD, spinal muscular atrophy, and spinocerebellar ataxia.

Pathophysiological Events in NDDs

Amyloid-beta (Aβ): Amyloid-beta (Aβ) is a central peptide in AD process, is obtained through the cleavage of amyloid precursor protein (APP) by secretases. The accumulation of this peptide into soluble oligomers and insoluble fibrillar structures and extracellular amyloid plaques is the main pathological characteristic of AD (Sehar et al. 2022). This remnant is connected with neurotoxic changes, impaired synapse function and neuronal death contributing to the dementia observed in the patients. The studies that have been done have also established that the deposition of Aβ plaques start forming in localized areas of the brain (as shown in figure 5) and progress in a set pattern which can be related to the extent of dementia. One of the major diagnostic markers is Aβ aggregation which via PET imaging can be detected in the early stage of the disease process (Ichimata et al. 2023). In addition, Aβ is now an important focal point for the development of drugs that can either prevent the formation of plaques or facilitate their removal; several monoclonal antibodies have been designed to address the problem. To further progress in the fight against AD, the exact role of amyloid- must be stressed on due to its significance in the immunotherapy.

Figure 5. Amyloid-β fibrillation and neuronal damage

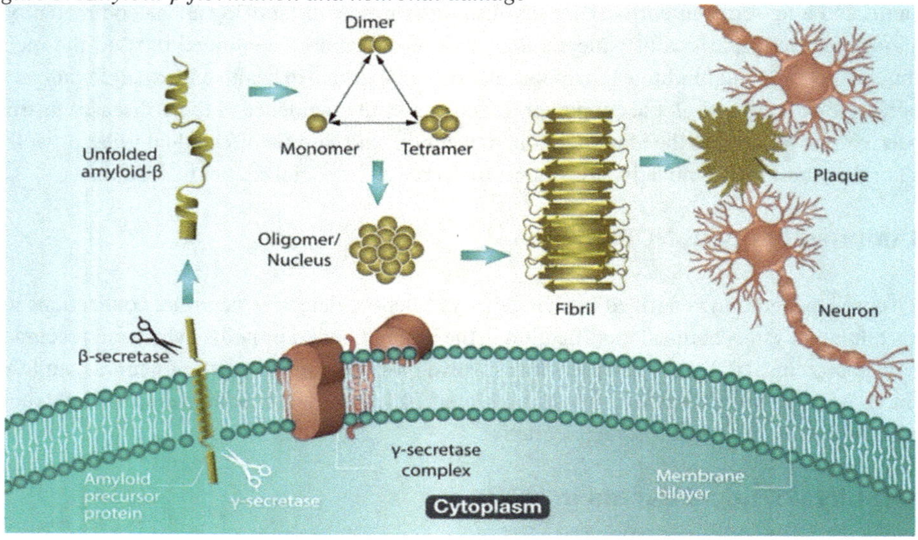

Source: Jokar et al. (2020)

Amyloid-β fibrillation and neuronal damage. First, amyloid precursor protein is cleaved by β and γ secretases, respectively. A peptide fragment of 39-42 amino acid is formed depending on the cleavage site. After cleavage, Aβ monomers start to self-assemble to form soluble toxic aggregates, and finally into insoluble fibrils, which subsequently cause synaptic dysfunction and neuronal death.

Neurofibrillary tangles (NFT): NFT are post-synaptic changes most common in AD and comprise neurofibrillary threads predominantly containing phosphorylated tau protein. In normal neuronal functioning, tau helps support microtubules that are important in cell's form and functioning. Nevertheless, in AD, the administration of tau causes it to become hyperphosphorylated, resulting in its demobilization from microtubules and aggregation into the formation of NFTs in the form of paired helical filaments. These intracellular tangles interfere with neuronal communication and are strongly linked with neuronal degeneration and intellectual presbyophy. Consequently, NFTs are relevant to AD diagnosis, and imaging progress, including tau PET, enables the observation of tau pathology in living persons (Otero-Garcia et al. 2022) In addition, NFTs are a major focus of the therapeutic approaches involving reduction of tau hyperphosphorylation and aggregation, establishing the profound involvement of NFTs in the neurodegeneration of AD.

Neuritic plaques/ Senile plaques, are a major cytological characteristic of AD and consist mainly of extracellular amyloid aggregates formed by amyloid-beta peptides derived from the amyloid precursor protein. These plaques involve a centrally located compact deposit of insoluble Aβ fibrils surrounded by reactive neurites—abnormal swollen appendages of the neuron that signify cell stress. Neuritic plaques are thought to play a role in neurotoxicity manifested by signs of synaptic disruption and neuronal death which have a role in exerting the cognitive effects seen in patients with AD. These types are particularly relevant to the severity of cognitive impairment and are sufficiently dense and distributed to make them important diagnostic markers.

Neuritic plaque density: is defined as the number of neuritic plaques per unit volume of the specific region of the brain they are counted in, which are extracellular deposits of amyloid-beta (Aβ) peptides and associated with dystrophic neurites. This density is usually determined from histopathological evaluation and is quoted as the number of lesions per area. Neuritic plaque density is the most significant predictor of cognitive dysfunction in AD and an important feature for diagnosing the disease and its progression (Streit et al. 2022). The amount of neuritic plaque load is also a useful index for measuring disease severity and assessing the effectiveness of any treatments designed to lower Aβ concentrations.

Co-Pathology

LATE-NC, or Limbic-Predominant Age-Related TDP-43 Encephalopathy Neuropathological Change: is a devastating dementia that predominantly affects elderly persons, most of whom are above the age of sixty. This is defined by the presence of TDP-43 protein aggregation in the limbic brain areas such as the hippocampus, entorhinal cortex particularly those which are erroneously related to memory as well as the regulation of emotion. While, Alzheimer's is associated with amyloid plaques and tau tangles, LATE-NC has its own protein called TDP-43 (Tazwar et al. 2024).

Chronic Traumatic Encephalopathy (CTE): is a form of NDD that results from a build-up of several instances of mild traumatic brain injuries including concussion and sub-concussive hits. In dementia, it involves a buildup of a protein called tau that is abnormally phosphorylated in the brain; this causes changes in behavior, cognition, mood and personality, including confusion, anger, memory loss, and depression. CTE can be observed in people who have suffered at least one head trauma; this category can include sportsmen and wives, and military servicemen mostly (Buckland et al. 2022). There is no current test for confirming CTE except through post-mortem examination of the brains; however, diagnostic criteria for the disorder and diagnostic tests for in vivo imaging are under study.

Age-related Tau Astrogliopathy (ARTAG): is a progressive NDD in which tau protein filaments are deposited in astrocytes. This pathology is mainly seen in elderly people and they are linked to ageing and not really linked to the neurodegenerative diseases such as Alzheimer's. ARTAG is characterised by special astrocytic tangles, such as thorn featuring and granular fuzzy astrocytes predominant at the periventricular region and medial temporal lobe (Tanaka et al. 2024). Although the Clinical relevance of ARTAG is yet to be established, it is accepted that ARTAG is one of the causes of cognitive impairment in old age.

Glial Tau Pathology: is a condition that is characterised by the deposition of tau protein in the glial cells of the brain, but particularly in astrocytes for different neurodegenerative diseases. This pathology is characteristic of several tauopathies including frontotemporal lobar degeneration (FTLD) which in its subtypes includes progressive supranuclear palsy (PSP) and corticobasal degeneration (CBD). In these conditions, astroglial tau inclusions in the form of tufted astrocytes and astrocytic plaques, appear. Staining positive for tau can also be seen in other illnesses such as Chronic Traumatic Encephalopathy (CTE) and Aging-Related Tauopathy (ARTAG) (Prater et. al. 2021). Recent research has shown that, in addition to neurons, tau aggregates are found in glial cells and that neuroinflammation plays a crucial role in tau pathology and is thought to cause neuronal impairment and cognitive dysfunction.

Lewy Body Pathology: Lewy body is the pathological hallmark includes neuronal cytoplasmic inclusion composed of aggregated alpha synuclein (Menšíková et al. 2022). The aggregation of Lewy bodies impairs the normal function of the brain, with a resultant demonstration of both motor and cognitive impairments. Most commonly in PD, Lewy bodies occur in the dopaminergic neurons of the substantia nigra, causing motor deficits and producing tremors, rigidity, and bradykinesia. Indeed, when Lewy bodies spread to the cortical regions, they are thought to be involved with the cognitive dysfunction that occurs in DLB, which features visual hallucinations and unstable consciousness.

MOLECULAR MECHANISM INVOLVED IN NEURODEGENERATIVE DISEASES

The molecular mechanisms involved in these diseases are complex and multifactorial, but several common features are shared across many neurodegenerative conditions (Figure 6).

Figure 6. Molecular mechanism involved in NDDs

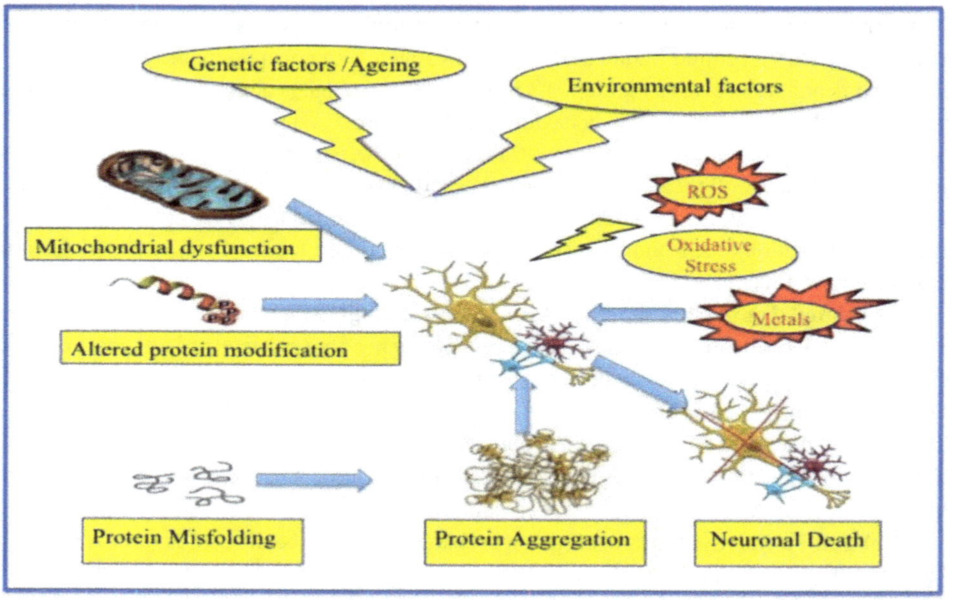

Source: Chakraborty et al., 2023

Oxidative stress: One characteristic that has prevailed in neurodegenerative diseases is oxidative stress due to the increased synthesis of reactive oxygen species (ROS) (Cenini et al. 2019). This oxidative stress results from ranged ROS produced in the body relative to the body's antioxidant capabilities, with the latter being overwhelmed by the former. The subsequent pathologic oxidative changes lead to dysfunction and death of neurons together with development of diverse neurodegenerative diseases as indicated in Table 4.

Table 4. ROS in NDDs

AD (AD)	ROS contribute to amyloid-beta plaque formation, tau hyperphosphorylation, and synaptic dysfunction.
PD (PD)	Dopaminergic neurons in the substantia nigra are particularly susceptible to oxidative stress due to dopamine metabolism and mitochondrial dysfunction.
HD (HD)	ROS exacerbate mitochondrial dysfunction and protein aggregation.
ALS(ALS)	Mutations in SOD1 lead to impaired ROS detoxification, resulting in motor neuron degeneration

Oxidative stress in AD: Oxidative stress is intricately linked to the progression of AD through several mechanisms. The buildup of amyloid-beta (Aβ) plaques characteristic of AD produces ROS triggering oxidative damage and impairing neuronal cell's function and ultimately kills them. This oxidative stress also activates microglia and thus leads to neuroinflammation which in turn amplify neuronal damage and Aβ and tau deposition (Buccellato et al. 2021). Also, it affects the function of mitochondria which lowers energy generation in neurons and leads to neuron damage (Teleanu et al. 2022).

Oxidative stress in PD: Specificity of damage to cellular components has been established for oxidative and nitrative lesions in the substantia nigra of the PD. There exist numerous sources and ways of ROS production including DA metabolism, mitochondrial damage, iron, neuroinflammatory cells, calcium and aging. Many of the PD causing gene products such as DJ-1, PINK1, parkin, alpha synuclein and LRRK2 also, in a rather intricate manner affect mitochondrial physiology and further worsen the levels of ROS production and oxidant stress (He et al. 2020).

Oxidative stress in HD: Oxidative stress is thought to be a key mediator of HD pathology because it enhances neuronal damage and cell death in the striatum and cortex areas of the brains (Korovesis et al. 2023). According to preclinical studies and clinical data of HD patients, increased OS seems to be implicated in HD manifestation, motor dysfunction, cognitive impairment, and psychotic manifestations.

An extracellular reactivity or oxidative stress is involved in HD, a neurodegenerative disorder due to specific CAG repeat in the huntingtin (HTT) gene. This mutation leads to generation of huntingtin (mHTT) protein which alter cellular equilibrium and causes generation of ROS and RNS (Kwakye et al. 2019). The disruption of equalization between ROS production and antioxidant defenses in addition to mitochondrial dysfunction, impaired antioxidant systems, and excitotoxicity based on glutamate leads to oxidative detriment in lipids, protein, and DNA.

Oxidative stress in ALS: They elevated levels of antioxidant molecules and lipid peroxidation product indicate that oxidative stress plays a significant role in the pathogenesis of ALS which is a progressive fatal NDD that involves the selective loss of motor neurons in the brain and spinal motor neuron pool. This condition also tends to the weakness, atrophy, as well as paralysis of muscles of the affected body part. Free radical production in ALS results from the production of ROS and reactive nitrogen species (RNS) to the protective antioxidant mechanisms of the cell (Cunha-Oliveira et al. 2020). This leads to oxidative change in lipids, protein, and nucleic acids all of which contribute to motor neuron degeneration. The fundamental causes of oxidative stress in ALS are numerous, but one of them is the mitochondrial disorder. ALS gene derivatives of SOD1, C9orf72, TDP-43, and FUS accelerate oxidative stress (Wang et al. 2019).

Neuroinflammation: is a profound response that helps to shield the brain from toxic endogenous and exogenous influences; however, increased leakage of pro-inflammatory factors is toxic to the CNS (Shao et al. 2022). Neuroinflammation is protective if the impact of the respective factor remains for a short time. It becomes neurodegenerative when it is persistent and lasts for a long period with adverse outcomes on the CNS. Although, inflammomes or cytosolic molecular factories are mainly consisted with an adaptor protein, a sensor protein and pro-inflammatory caspase (caspase-1) (Zhang et al. 2023). Out of all inflammasomes, the Nucleotide-binding oligomerization domain leucine-rich repeat and pyrin domain-containing protein 3 (NLRP3) inflammasome is studied most extensively and has been described to play a central role in neurodegenerative diseases. Some neuroinflammatory markers enlisted in Table 5.

Table 5. Neuroinflammatory markers in the pathogenesis of neurodegenerative diseases (Rauf et al. 2022)

Category	Markers	Role	Neurodegenerative Diseases
Pro-inflammatory Cytokines	TNF-α, IL-1β, IL-6	Promote inflammation, neuronal damage, and synaptic dysfunction	AD, PD, MS, ALS, HD
Anti-inflammatory Cytokines	IL-10	Reduces inflammation; insufficient levels worsen disease progression	AD, PD, MS
Chemokines	CCL2 (MCP-1), CXCL10	Recruit immune cells to sites of inflammation	AD, MS, ALS
Microglial Markers	Iba1, CD68	Indicators of microglial activation and phagocytic activity	AD, PD, ALS
Astrocytic Markers	GFAP, S100B	Markers of reactive astrocytes; elevated during chronic inflammation	AD, MS, ALS
Complement Proteins	C1q, C3	Activate the complement system, contributing to synapse loss and neuroinflammation	AD, ALS
Oxidative/Nitrosative Stress	ROS, RNS, NO, MDA, 4-HNE	Cause oxidative damage to lipids, proteins, and DNA	AD, PD, ALS, HD
Misfolded Proteins	Amyloid-β, Tau, α-Synuclein	Trigger microglial activation, cytokine release, and chronic inflammation	AD, PD
Neurotrophic Factors	BDNF, GDNF	Support neuronal survival; reduced levels correlate with neurodegeneration	AD, PD, ALS
Peripheral Markers	CRP, HMGB1	Reflect systemic inflammation and immune activation	AD, MS

Neurotransmitters

Dopaminergic Neurotransmission: Dopaminergic dysfunction is one of the key molecular processes implicated in a range of NDDs with PD. In PD, the dopaminergic neurons of SNpc progressively degenerate and affect the dopamine release in the striatum which is an essential component of motor circuitry of the brain (Teleanu et al. 2022). This dopamine deficiency has clinical features represented by motor dysfunctions, such as bradykinesia, rigidity, resting tremors, and postural instability (Wallert et al. 2022).

In addition to PD, alterations relating to dopaminergic system also play a role in AD – a neurodegenerative disorder where dopamine changes are associated with decreased cognition, memory, and psychosocial dysfunction including apathy, agitation, and depression. Despite being considered a disease caused by amyloid-β plaques and tau, the loss of neurotransmitter circuits, especially dopaminergic ones, constitutes a key factor in the process (Davis et al. 2023).

In HD, the alteration of dopamine amounts within the basal ganglia disrupts motor system balance. This accounts for the typical chorea, and cognitive impairment that manifests in HD patients (Csehi et al. 2023). At the initial stages of the disease there may be increased dopamine release, which provokes hyperkinetic symptoms, and later there is decreased dopamine release, which results in akinetic-rigid syndrome.

Cholinergic Neurotransmission: Cholinergic dysregulation is another important pathophysiological process in several NDDs. One of the most striking features of these conditions is a significant affection of the cholinergic system that controls such functions as memory, attention, learning, and motor activity (English et al. 2023). AD mostly affects the basal forebrain cholinergic system resulting in low levels of ACh which results in cognitive function declines such as memory and other neuropsychiatric symptoms (Chen et al. 2022).

PD characterized by an impaired balance between dopaminergic and cholinergic pathways in the striatum, which aggravates motor dysfunction including tremor and rigidity. Likewise, in HD, impaired cholinergic projection neurons in the striatum are involved in motor and cognitive impairment. The role of cholinergic transmission at neuromuscular junction knows well with ALS as it causes progressive muscle weakness accompanied by atrophy.

Glutamate Dysregulation: Excitotoxicity through the disturbance in the concentration of glutamate is considered a key pathological change in several neurodegenerative conditions. Glutamate is non-hydrolysable, excitatory amino acid and in cerebral cortex is most abundant excitatory neurotransmitter that is involved in functions such as synaptic plasticity, learning and memory (Ribeiro et al. 2023). In normal situations, the release and removing of glutamate is well controlled to enable proper signaling between neurons. However, in these diseases, that regulation fails and the glutamate concentrations in the synaptic cleft elevate causing overstimulation of glutamate receptors, primarily NMDA receptors; it leads to toxic cellular activities.

AD is characterized by the loss of synaptic activity and neuronal death, which is worsened by glutamate dysfunction. This dysfunction leads to a sustained stimulation of the NMDA receptors thus increasing the intensity of Calcium conjugation to neurons (Bukke et al. 2020). Known to activate signaling processes that cause oxidative stress, impaired mitochondrial function, and neuroinflammation which culminates in neuron death. They all play a part in the cognitive decline, memory impairment or dementia, and neuropsychiatric manifestations of AD.

In PD, the dopaminergic neurons in the substantia nigra are lost or damaged and dopamine concentration in the striatal area is decreased due to which there is an imbalance in the levels of dopamine and glutamate. Normally dopamine inhibits glutamate but, in PD, due to the death of dopaminergic neurons of pars compacta the inhibitory constraints are lost, and glutamate gets over-activated in the direct and indirect pathways of the basal ganglia. This excessive release and overactivation of glutamate lead to motor features such as rigidity, tremor, bradykinesia and progresses neuronal loss in basal ganglia.

In HD, the extrasynaptic NMDA receptor is affected by the mutation of huntingtin gene that results in the buildup of the mutant huntingtin protein and produces an imbalance in the normal Glu Receptor function and increases excitotoxicity. The loss of inhibitory GABAergic neurons in the striatum that suppresses the output of glutamate makes this dysregulation worse. Consequently, the release of the excitotoxic Level of glutamate causes neuronal distrains and leads to neuronal death, which has a significant impact to the chorea, defect of cognition and psychiatric implications of patients suffering from HD. The alteration in glutamate transmission also affects the normal working of the basal ganglia that plays an important role in motor control and behavioural modulation (Todd et al. 2020).

In ALS, one of the causes of dying motor neurons is owing to excitotoxicity resulting from the neurotransmitter glutamate. In ALS, there is reduced glutamate transporter function such that synaptic glutamate becomes elevated and activates glutamate receptors that cause excitotoxicity. This hurts the motor neurons, and leads to muscle wastage, limpness, breathing difficulties as the disease progresses (Obrador et al. 2020). Furthermore, the oxidative stress and mitochondrial dysfunction, evoked in the course of glutamate-elicited damage, facilitate motor neuron demise.

GABA (Gamma-Aminobutyric Acid) dysregulation: Disturbance of GABA has been found to contribute to the development several NDDs. GABA is the major inhibitory transmitter in the brain that regulates the levels of excitation and inhibition that are needed for movement, thought and emotion (Carello-Collar et al. 2023). As in HD, the loss of inhibitory GABAergic neurons in the striatum beneath the dystrophic axons of the output neurons that contain the mutant huntingtin protein results in motor disability and intellectual disturbances. For example, in PD the dopaminergic input is missing hence there is imbalance between GABAergic and glutamatergic signaling leading may to rigor and slowed movements.

In AD, reduced number of GABAergic interneurons is observed in the cortex and hippocampus and this being partly attributed to cognitive dysfunction and neuropsychiatric symptoms. Likewise, in ALS, increased GABAergic innervations leads to hyperexitability of motor neurons and weakness of muscles including muscles of respiration. Asymmetric regulation of GABAergic transmission and glutamatergic stimulation causes neurodegeneration and promotes the disease's development. Promoters of GABAergic activity or protection of GABAergic neurons are being considered as potential therapeutic interventions to ameliorate the motor symptoms and halt or slow down the neurodegenerative processes in the two diseases

Protein Deposition

Cerebral amyloid angiopathy: Staging of AD is based on the topographical distribution of amyloid-beta (Aβ) plaques and neurofibrillary tangles (NFTs) related to the tau protein. Aβ plaques are mainly extracellular and it develops in the neocortex and then allocortex, basal ganglia, diencephalon, brain stem and cerebellum. Unlike prions, NFTs form within the neurons and progress centrally from the transentorhinal cortex, to the hippocampus, and finally to the neocortex. This is the same co-linear progression that interferes with the function of neurons and results to neurodegeneration all over the brain hence the cognitive impairment in AD. As these pathological features become disseminated throughout the brain they disrupt synaptic connectivity, which leads to memory loss, confusion, and a general decline in cognitive function.

Tauopathy: is associated with the formation of neurofibrillary tangles of abnormally accumulated tau protein, its fragments or other proteins. Some of the disorders of this category include; FTLD-tau, Pick's disease, AGD, frontotemporal dementia with chromosome 17 mutation, NFT predominant dementia, PSP, CBD and PART (Devi, 2023). Pathologic characteristics of tauopathies include neurofibrillary tangles (NFT), neuropil threads (NT), Pick bodies (PB), coiled bodies, astrocytic plaques, tufted astrocytes, grains and ballooned neurons. These pathological inclusions are located within the cytoplasm of the neurons and glial cells of brain and are especially seen in the frontal and temporal lobes, basal ganglia and the brainstem. Compared with other tauopathies, AGD is characterized by the frequent association with AD lesions, while other tauopathies are not commonly found to present with minimal additional concurrent pathologies.

Synucleinopathies: Although the discovery of the critical role of α-synuclein (α-syn) in the pathogenesis of PD is now twenty-five years old, it still represents a milestone in PD research. Abnormal forms of α-syn trigger selective and progressive neuronal death through mitochondrial impairment, lysosomal dysfunction, and alteration of calcium homeostasis not only in PD but also in other α-syn-related NDDs such as dementia with Lewy bodies, multiple system atrophy, pure autonomic failure, and REM sleep behavior disorder (Calabresi et al. 2023).

The interplay between α-synuclein (α-syn) pathology, early synaptic dysfunction, and neuroinflammation has garnered significant attention in PD research. Early synaptic and plasticity alterations, driven by α-syn aggregation, disrupt neurotransmitter release and synaptic vesicle dynamics long before overt neurodegeneration occurs. Concurrently, evidence from experimental models and PD patients highlights the presence of early neuroinflammation, mediated by activated microglia and astrocytes, preceding α-syn deposition and spreading. This suggests a mechanistic link between inflammation and synaptic dysfunction, where pro-inflammatory cytokines and oxidative stress exacerbate α-syn toxicity (Robinson et. al, 2018).

TDP-43 proteinopathies: TDP-43 (transactive response DNA-binding protein 43) is a nuclear RNA-binding protein, which has pivotal functions in RNA processing, transport, stability and transcriptional control. However, misplaced and misfolded in its pathological condition is core molecular pathologic event in several NDDs (Chornenkyy et al. 2019). TDP-43 pathology is characteristic in ALS and FTD-TDP, where aggregated TDP-43 forms cytoplasmic inclusions in neurons and glial cells causing nuclear dysfunction and cytoplasmic toxic gain of function.

Apart from ALS and FTLD-TDP, TDP-43 is also involved in diseases such as ADs, LATE, and some types of CTE. These disorders commonly present with TDP-43 mislocalization from nucleus to cytoplasm as well as phosphorylation, ubiquitination and further cleavage into smaller fragments prone to aggregation. TDP-43 has been implicated in neurodegeneration and, from recent studies, major pathogenic processes can be identified: (i) defects in RNA metabolism, (ii) mitochondrial dysfunction, and (iii) proteostatic stress. These outcomes have placed TDP-43 at the centre of therapeutic interventions that seek to prevent formation of aggregates, regain nuclear import and modulate toxicity in TDP-43 proteinopathies (De Boer et al. 2020).

FUS pathology: FTLD-FUS is one of the subtypes of the FTD and it is pathologically defined by the deposition of the FUS protein in neurons and glial cells. FUS induces serine-arginine (SR) like repeat-rich protein and is associated with multiple cellular processes such as RNA transport, RNA splicing and DNA repair (Kovacs et al. 2019). In FTLD-FUS, FUS moves out of nucleus to the cytoplasm where it aggregates and forms aggregates that interfere with cellular processes. FTLD-FUS however, is generally sporadic and is not associated with the FUS gene (Marrone et al. 2019).

This pathology is associated with three clinical syndromes: Five subtypes of FTLD include FTLD with ubiquitinated inclusions (FTLD-U), neuronal intermediate filament inclusion disease (NIFID), basophilic inclusion body disease (BIBD), and FTLD associated with neuronal inclusion bodies (FTLD-ni). These conditions also show different levels of cognitive, behavioural and motor impairments which suggest that the FTLD-FUS is a very heterogeneous condition.

Ubiquilin proteinopathy: Ubiquilin proteins, chiefly ubiquilin-1 and ubiquilin-2, are involved in different cellular quality control mechanisms of intracellular proteome maintenance. These proteins involve themselves in directing of proteins of correct conformations and highlighting proteins that are either misfolded or aggregated for degradation (Limanaqi et al. 2020). Abnormality in ubiquilin-associate protein degradation system has been proven to play role in several neurodegenerative diseases, therefore; ubiquilin is an important protein that is significant in neurodegeneration research.

However, the highest connection can be made to ALS and frontotemporal dementia which are linked to UBQLN2 mutations. Associated with neuronal death, UBQLN2 protein aggregates and localized inclusions are present in motor neurons as well as glial cells. These inclusions are often formed from aggregated proteins for example TDP-43 and FUS; therefore, ubiquilin pathogenicity implicates dysfunction in the aggregation of other pathological proteins in ALS and FTD (Schmidt et al. 2021).

Mitochondrial Genes

PINK1 (PTEN-induced kinase 1): is an essential protein that participates in the molecular mechanism underlying mitophagy through which degraded mitochondria are eliminated from cells. The optimized QC machinery in PD patients involves the PINK1gene; its mutation causes the buildup of dysfunctional mitochondria within neurons. The defects in mitophagy therefore contribute to loss of mitochondrial membrane potential, increased oxidative stress and neuronal loss – all of which are characteristic of PD. Since mitochondria are involved in energy production and cell viability the failure to eliminate damaged mitochondria enhances neurodegeneration predominantly affecting dopaminergic neurons that crucially rely on mitochondria. The elucidation of the roles of PINK1 and its relationship with mitophagy has raised much interest in the pathogenesis of NDDs (Brunelli et al. 2020).

PARKIN: is crucial mediator of mitophagy, a selective procedure through which damaged or dysfunctional mitochondria are eliminated. In PD, mutations in the PARKIN gene interfere with this process resulting in an abnormal quality control of mitochondria. Therefore, it collects itself in the cells and builds up deteriorated mitochondria that cause further mitochondrial dysfunction and rise in oxidative stress. This dysfunction impairs cell respiration and supports neurotoxicity, especially of dopaminergic neurons that have high dependence on mitochondrial functions (Arkinson & Walden, 2018).

OPA1:is a mitochondrion associated dynamin-like GTPase, which has critical roles in the maintenance of mitochondrial fusion and shape. It has a very significant function in preserving the shape and integrity of what are termed mitochondria, which are very, very important within the cell for energy production and for well-being of the cell. These include optic atrophy, AD, and frontotemporal dementia, mutations of this gene meaning deficits in the ability of mitochondria to fuse. This dysfunction leads to fragmentation of mitochondria and hampers their normal execution and this leads to a neuronal energy crisis (Sladen et al. 2022). Mitochondrial dynamics are known to alter with aging, and loss of these dynamics results in the induction of cell stress that can cause neuronal injury and cell death at last.

SOD1 (superoxide dismutase 1): Antioxidant enzyme, SOD1 is involve in providing protection from oxidative damage whereby it transforms the superoxide radicals to harmless species. For example, the SOD1 gene of ALS is known to have a loss of its antioxidant property so that there is a rise in oxidative stress within neurons. This disparity in ROS contributes to mitochondrial dysfunction because the source of mitochondria and the target of oxidative damage are one and the same. Together, production of ROS causes damage to the mitochondrial membranes, proteins and DNA, thus aggravating the degeneration of neurons and cell death. The SOD1 also triggers the formation of the toxic protein conformations that are toxic to the mitochondria in ALS (Dalla Bella et al. 2022).

DJ-1: is a highly conserved protein and has been found to be involved in the defence of neurons against oxidative stress and function in the mitochondrial processes. In PD, the DJ-1 gene for instance is mutated, and fails to offset free radical damage and as a result there is increased free radical activity within neurones (Huang et al. 2021). This leads to mitochondrial dysfunction since mitochondria are a main target of ROS mediated injury. Since DJ-1 is unable to control the oxidative stress, it leads to the death of dopaminergic neurons largely depended on mitochondria. The sum of pro-oxide substance is raised that is a primary factor for neurodegeneration seen in PD (Sircar et al. 2021). These findings further reinforce the mechanistic connection between DJ-1 and PD pathology, specifically with regards to oxidative stress and mitochondrial integrity, and identify novel treatment targets that would increase the protein's protective capabilities to mitigate the progression of neuronal loss in PD.

MT-ND1: MT-ND1 and MT-TK as mitochondria related genes are associated with the important processes of mitochondrial respiratory chain and oxidative phosphorylation. MT-ND1 contiguous a subunit protein for the complex I of the electron transport chain (ETC) and MT-TK codes for a mitochondrial tRNA for protein synthesis (Lin et al. 2024). Defects in these genes have been associated with PD and other mitochondrial disorders in which they affect the mitochondrion in the capacity for oxidative phosphorylation, which is the major source of energy in cells. The said effects eventually cause mitochondrial dysfunctions. As a result, alteration in the oxidative stress, mitochondrial dysfunction, and ultimately, neural cell death. In PD, these mutations are involved in dopaminergic neuronal death because they affect mitochondrial respiration and, hence, the cell's function (Shang et al. 2023). The impairment of mitochondrial ATP synthesis resulting from MT-ND1 and MT-TK mutations indicates that maintaining mitochondrial function is an essential component of neurodegenerative diseases and thereby emphasizes the therapeutic approach of targeting mitochondrial signalling processes.

C9orf72: represents one of the genes allied with frontotemporal dementia (FTD) and ALS(ALS) in particular, including the familial entities. The characteristic of C9orf72 is that its mutations involve intronic expansion of hexanucleotide repeat GGGGCC. This expansion also interferes with RNA processing thus the formation of toxic RNA foci that affects cellular activities (Diab et al. 2023). Moreover, these repeats lead to the formation of dipeptide repeat proteins, which has been demonstrated to lead to mitochondrial dysfunction and a state of cellular stress. Effects on mitochondria include a prime cause of oxidative stress as well as energy metabolism disruption, which are hallmarks of neurodegeneration. Disruption of mitochondrial network's integrity together with protein aggregates formation and neuroinflammation cause the death of motor and cortical neurons and manifestation of clinical signs of ALS and FTD (Zampatti et al. 2022).

MTDNA (mitochondrial DNA): mtDNA encodes elements of the MRC and mutations or deletions in these genes disturb mitochondrial function. This disruption happens to mitochondrial oxidative phosphorylation so neurons suffer from energy decline. Because neurons are rather energy demanding these abnormalities lead to loss of neuronal function and cell death. Both in AD and PD, mitochondrial dysfunction is one of the major defining features of disease progression since the change in energy metabolism contributes to increased oxidative stress and subsequent neurodegeneration (Bazzani et al. 2022). Also, the consequence of the slowly developing mutations in mitochondrial DNA through time is that the ROS are elevated and cause other cellular components to deteriorate as well as increased neuronal death.

Histone Methylation: is one of the basic post-translational modifications essential for cell signaling that directs genomic functions through changes in chromatin textures and openness (Basavarajappa & Subbanna, 2021). Histone methylation is the process of attaching one, two or three methyl groups to lysine or arginine residues of histone proteins and depends on which genes will be expressed or not expressed (Murthy et al. 2021).

Histone methylation is also extensively involved in the modulation of the level of β-amyloid in AD. Histone H3K27me3 (trimethylation) has been linked to the repression of genes that facilitate amyloid plaques clearance; the toxic β-amyloid protein aggregates therein accumulate (Santana et al. 2023).

Also, H3K9me3 a repressive mark that is associated with gene repression, has been identified as being upregulated in AD. This increased H3K9me3 leads to reduced levels of neuroprotection associated genes, including repair, antioxidant, and inflammatory genes (Lee et al. 2020). The above-mentioned mechanisms involve silencing of protective pathways that foster neuroinflammation and neuronal death, fundamental characteristics of neurodegeneration in AD.

In HD, the levels of histone H3K9me3 methylation is reported to be down regulated, resulting in up regulation of gene products that promote neuronal apoptosis. Dysregulation of histone methylation patterns occurs due to the interaction of the Huntingtin protein (HTT), which mutated in HD with histone-modifying enzymes. These changes lead to the mis regulation of genes causing the progressive neuronal loss which is seen in the HD. The erasing of H3K9me3 mark indicates a change from gene repression to induction of genes that trigger cell death and inflammation leading to enhanced neuronal degeneration.

Further, the mutant HTT protein also affects the use of H3K27me3, which is involved in marks of gene and transcript repression. This disruption causes a breakdown in the ability of the cells to maintain gene silence at particular regions of the genome, thus activating pathology relevant genes which should otherwise be off. It has been equally postulated that this change in gene expression aggravates neuronal damage leading to the development of HD (Zhao et al. 2024). These findings underscore the importance of histone methylation aberration particularly through the actions of the mutant HTT protein in the pathogenesis of the neurodegenerative process in HD; also, epigenetic regulation becomes a potential target in therapy development.

Typically, in ALS patient's motor neurons have been found to lose H3K27me3, thus increasing the chances of the genes that trigger neuroinflammation and cell death being turned on abnormally. Such a disruption in histone methylation helps the progression of ALS because neurodegeneration is accompanied by the up-regulation of genes that are detrimental to the neurons (Kumar et al. 2022).

CHALLENGES IN CURRENT THERAPEUTIC STRATEGIES

The Unmet Need in Alzheimer's Treatment

AD presents a formidable challenge to global healthcare systems, despite decades of intensive research and substantial investment in clinical trials, the development of truly effective therapeutic strategies remains an elusive goal. Current treatments primarily focus on symptomatic relief, offering only temporary improvements in cognitive function and failing to address the underlying pathological mechanisms driving AD progression (Shakir et al. 2024).

The Blood-Brain Barrier (BBB) as a Major Hurdle

Overcoming the BBB presents a significant challenge to AD drug development. A significant impediment to successful AD therapy is the blood-brain barrier (BBB), a highly selective semipermeable membrane that protects the brain from harmful substances circulating in the bloodstream (Wechsler et al. 2019).

Nanotechnology-based drug delivery systems have emerged as a particularly promising approach. These systems utilize nanoparticles of various types, including polymeric nanoparticles, liposomes, and carbon nanotubes, to encapsulate and deliver therapeutic agents to the brain (Elsori et al. 2023).

Targeting Amyloid-β (Aβ) and Tau: A Complex Landscape

The amyloid cascade hypothesis, a long-standing theory in AD research, posits that the accumulation of amyloid-beta (Aβ) peptides is a central driver of AD pathogenesis. Numerous therapeutic approaches have been developed based on this hypothesis, including antibodies targeting various Aβ species, small molecule inhibitors of Aβ aggregation, and Aβ-degrading enzymes However, clinical trials of these therapies have yielded mixed results, highlighting the complexity of Aβ biology and the challenges of translating preclinical findings to clinical benefit (Plotkin et al. 2020).

Different Aβ species exhibit distinct toxicities and may engage different cellular pathways, making it difficult to develop a single therapeutic strategy that effectively targets all forms of Aβ. Moreover, the precise mechanisms by which Aβ contributes to neuronal dysfunction and degeneration remain not fully understood. Some studies suggest that soluble Aβ oligomers are the primary neurotoxic species, while others emphasize the role of Aβ fibrils in inducing neuronal damage. This ambiguity makes it difficult to identify optimal therapeutic targets and develop interventions that effectively prevent or reverse Aβ-mediated toxicity.

Beyond Aβ, tau protein pathology represents another major therapeutic target in AD. Tau, a microtubule-associated protein, undergoes hyperphosphorylation and aggregation, forming neurofibrillary tangles (NFTs) that are a hallmark of AD pathology. The accumulation of NFTs is strongly correlated with cognitive decline in AD, suggesting that targeting tau pathology could offer significant therapeutic benefit. However, developing effective tau-targeting therapies faces significant challenges, including the complexity of tau isoforms and post-translational modifications, as well as the lack of a complete understanding of tau's role in neuronal function. The diverse functions of tau, including microtubule stabilization and axonal transport, make it challenging to develop therapies that selectively target pathological tau aggregation without disrupting its essential physiological roles (Latypova et al. 2014).

Neuroinflammation: A Double-Edged Sword

Neuroinflammation, a complex interplay of immune responses within the brain, plays a significant role in AD pathogenesis (Wickstead et al. 2021) (Bronzuoli et al. 2016). While inflammation is a normal response to injury and infection, chronic neuroinflammation in AD contributes to neuronal damage and accelerates disease progression (Yang et al., 2024).

Therapeutic strategies targeting neuroinflammation have shown promise in preclinical studies, demonstrating the potential of anti-inflammatory drugs and other approaches to reduce neuronal damage and improve cognitive function. However, developing effective anti-inflammatory therapies for AD presents significant challenges. The complex interplay between inflammation and other AD-related processes makes it difficult to modulate inflammation without unintended consequences (Chu et al. 2024). For example, while suppressing inflammation may reduce neuronal damage, it could also impair the brain's ability to clear Aβ plaques or other toxic proteins. Moreover, the heterogeneity of neuroinflammatory responses in AD makes it difficult to develop therapies that are broadly effective across different patient populations (Bhatti et al. 2020).

Mitochondrial Dysfunction and Oxidative Stress: Energy Crisis in the Brain

Mitochondrial dysfunction and oxidative stress are increasingly recognized as significant contributors to AD pathogenesis. These detrimental effects are exacerbated by the accumulation of Aβ and tau proteins, which further compromise mitochondrial function (Onyango et al., 2016).

Therapeutic approaches targeting mitochondrial dysfunction and oxidative stress have shown promise in preclinical studies. However, translating these preclinical findings into effective clinical therapies remains a significant challenge. The complex interplay between mitochondrial dysfunction, oxidative stress, and other AD-related pathways necessitates a multi-target approach that addresses multiple aspects of AD pathogenesis simultaneously. Furthermore, the heterogeneity of mitochondrial dysfunction in AD, with different mitochondrial impairments observed in various brain regions and at different disease stages, makes it challenging to develop therapies that are effective across diverse patient populations.

Challenges in Clinical Trials and Drug Development

The complexity of AD pathogenesis and the heterogeneity of disease presentation and progression pose significant challenges for clinical trials (Weuve et al. 2015) (Gustavsson et al. 2017). Recruitment of suitable participants is a major hurdle, particularly for trials targeting early-stage AD, as these individuals may be asymptomatic or exhibit subtle cognitive changes that are difficult to detect. Moreover, the long duration of AD progression necessitates long and expensive clinical trials to assess the efficacy of disease-modifying therapies (Dolan et al. 2018). The need for robust biomarkers to identify individuals at high risk of developing AD and to monitor disease progression is crucial for selecting appropriate participants and evaluating treatment efficacy.

The high cost and long duration of clinical trials also contribute to the challenges in AD drug development. The lengthy timeframe required to assess the efficacy of disease-modifying therapies, often spanning several years, increases the financial burden and logistical complexity of clinical trials.

Parkinson's Disease Treatment

Levodopa, a dopamine precursor, remains the gold standard for managing motor symptoms in PD (Wal et al. 2023), (Dong et al. 2016), (Rascol et al. 2003). However, long-term use is associated with several significant drawbacks, limiting its overall effectiveness and impacting patient quality of life. These include the development of motor fluctuations ("on-off" phenomena), dyskinesias (involuntary movements), and a variety of non-motor side effects cardiovascular issues, psychosis, and depression. (Stathis et al. 2012), (Richmond et al. 2023), (Carbone et al. 2019).

Enzyme inhibitors: are used as adjunctive therapies to overcome limitations of levodopa. These medications aim to either enhance levodopa's effectiveness or reduce its side effects, but they often introduce their own set of challenges. These include dopamine agonists, MAO-B inhibitors, COMT inhibitors, and amantadine. These inhibitors are used to improve levodopa's bioavailability and efficacy. However, they often have short elimination half-lives, typically around 2-3 hours, which can limit their effectiveness over time (Martinez-Martin et al. 2013). The incomplete peripheral inhibition of COMT by current inhibitors like entacapone, tolcapone, and nebicapone results in suboptimal control of levodopa metabolism, affecting its delivery to the brain (Jenner, 2023). MAO-B inhibitors can cause insomnia and nausea, while COMT inhibitors can cause diarrhea and liver damage (particularly tolcapone).

Deep brain stimulation (DBS): is a surgical intervention used to manage severe motor symptoms in PD that are refractory to medication (Terkelsen et al. 2022). The invasive nature of DBS surgery presents significant risks and necessitates careful patient selection. It is helpful in alleviating motor symptoms and reducing "off" time & dyskinesia. However, it shows minimal impact on disease progression or exceeding the symptomatic relief provided by dopaminergic medications. (Piacentino et al. 2016) (Olanow, 2002)

EMERGING THERAPIES

Several emerging therapies hold promise for disease modification in PD, but significant challenges remain in translating preclinical findings into effective clinical treatments (Lang et al. 2018), (Cha et al. 2023). However, challenges include the efficient differentiation of stem cells into functional dopaminergic neurons that integrate properly into the existing neural circuitry, the delivery of these cells to the target brain region (substantia nigra), avoiding potential complications such as ectopic cell growth and minimizing the risk of tumor formation, and potential immune rejection of the transplanted cells (Polgar et al. 2024), (Grow et al. 2016). However, challenges include efficient gene delivery across the blood-brain barrier (a protective barrier that restricts the passage of many substances into the brain), ensuring the long-term safety and efficacy of gene expression (preventing insertional mutagenesis or uncontrolled gene expression), and targeting the therapy specifically to the affected brain regions while minimizing off-target effects (Rhee et al. 2011), (Drinkut et al. 2012). Limited Patient Data: Many clinical trials have small sample sizes, making it challenging to draw definitive conclusions about efficacy and safety (Kozłowska et al. 2023).

Huntington's Disease Treatment

Current HD treatments primarily focus on managing symptoms rather than modifying the underlying disease process. (Ferguson et al. 2022), (Kieburtz et al. 2018), (Potkin et al. 2018) Medications are used to manage chorea (involuntary movements), depression, and other psychiatric symptoms.

There are various limitations with treatment with Tetrabenazine (TBZ) & Deutetrabenazine (DTBZ), mainly concerned to their safety & efficacy. Both the medications are aimed to control choretic movement, however, there is need to conduct further research on their side-effects & long term effects. The clinical improvement in the patients is doubtful, as studies have shown diffident reduction in chorea scores, but long-term benefits and impact on quality of life is not investigated (Vadlamani et al. 2024). Sedation, depression, and anxiety are some common side-effectsm (Setter et al. 2009).

Antipsychotics

The limitations with treatment of Anti-psychotics include intricate & complex nature of HD, safety profile of drugs and different patient responses. Due to behavioral changes in patients, oral medications are not the best option for treatment. Parenteral formulations are a potential alternative, however, there lack of data of efficacy of antipsychotics which limits its extensive use. (Van Oosterom et al. 2019). Also, safety profile poses a limitation of treatment like, metabolic side effects seen with second-generation antipsychotics. In patients with pre-existing motor symptoms, atypical antipsychotics can cause extrapyramidal symptoms. (Unti et al. 2017). Studies have shown that that there no clinical improvement in motor & behavioral symptoms. (Desamericq et al. 2014) (Videnovic, 2013).

Antidepressants

The psychiatric manifestations of HD are often overlapping with cognitive and motor symptoms, making it challenging to isolate the effects of antidepressants (Killoran & Biglan, 2014). Systemic reviews have shown lack of studies showing positive results and most studies have constraints such as limited duration and insufficient sample size. (Moulton et al. 2014) (Bonelli & Wenning, 2006). Side-effects such as increased suicidal ideation poses a serious risk and long-term use of these medications can complicate the treatment of HD (Vishwas et al. 2022).

Targeting mutant huntingtin protein mHTT directly is a favorable therapeutic strategy. (Mousavi et al. 2024). Several approaches such as, including huntingtin-lowering strategies such as antisense oligonucleotides (ASOs), which bind to mHTT mRNA and reduce its expression. Gene editing techniques, such as CRISPR-Cas9 and zinc finger proteins, offer the potential to directly correct the genetic defect underlying HD. (Bhushan et al. 2024) However, the delivery of these therapeutic agents to the brain, the potential for off-target effects, and the long-term consequences of these interventions remain significant challenges (Ferguson et al. 2022). Additionally, the heterogeneity of HD symptoms, varying from predominantly motor to primarily cognitive or psychiatric manifestations, complicates the development of universal therapies. A deeper understanding of the diverse molecular mechanisms underlying these symptom variations is needed to develop more targeted and personalized treatment strategies.

The selection of appropriate biomarkers and the definition of meaningful clinical endpoints are crucial for the success of clinical trials. (Klockgether et al. 2022) Biomarkers that accurately reflect disease progression and response to therapy are needed to assess the efficacy of new treatments. However, identifying such biomarkers remains a significant challenge, as the complex pathophysiology of HD and ALS makes it difficult to pinpoint reliable indicators of disease activity. The need for large sample sizes to detect statistically significant differences in outcome measures between treatment and control groups further complicates the design and execution of clinical trials. (Garca-Gonzlez et al. 2024) The high cost and lengthy duration of clinical trials pose additional challenges, hindering the development of new therapies for these devastating diseases.

Many therapeutic molecules, particularly large proteins or nucleic acids, such as ASOs used in HD therapy, have difficulty crossing the BBB, limiting their effectiveness in treating CNS disorders. (El Ouaamari et al. 2023), (Ashique, 2022)

ALS Disease Treatment

Riluzole, the first FDA-approved drug ALS, mainly acts by blocking glutamatergic neurotransmission, offers limited efficacy. (Montiel-Troya et al. 2024), (Moriyama & Yokota, 2024) (Saitoh & Takahashi, 2020). Studies have shown that Riluzone is not effective in halting disease progession (Chen et al. 2020). Other limitations include potential of causing hepatoxicity and high cost of drug.

Edaravone, which is available in IV formulation and oral form, however IV formulation has limited usage due to its administration requirements (O'Neill et al. 2024). Oral formulation on other hander has improved bioavailability but lack patient centered data, which restricts its application. Cost-utility analyses reveal that while sublingual formulations may be more convenient, they are not cost-effective in home treatment settings compared to IV administration in hospitals (Zhu et al. 2024). Clinical trials have shown the drug's effectiveness in patients with duration of disease of less than 2 years. (Podsiadły et al. 2024).

Gene therapy, particularly targeting specific genetic mutations, shows promise in preclinical studies. (Ediriweera et al. 2021), (Garca-Gonzlez et al. 2024) However, efficient and safe delivery of gene therapy vectors to motor neurons remains a significant challenge. Stem cell therapies are being explored as a potential means to replace damaged motor neurons or provide neurotrophic support. (Hoxhaj et al. 2023), (Jayaprakash et al. 2024) However, the challenges associated with stem cell transplantation, similar to those in HD, need further investigation. Approaches aimed at modulating neuroinflammation or mitochondrial dysfunction are also being investigated. (Bradford & Rodgers 2024), (Mohd Yusoff & Mohd Khair 2024), (Palomo et al. 2020) Neuroinflammation, a prominent feature of ALS pathogenesis, is being targeted by various therapeutic strategies. Mitochondrial dysfunction, implicated in motor neuron degeneration, is another potential therapeutic target. The development of effective ALS therapies requires a multifaceted approach, combining strategies to address multiple pathogenic mechanisms.

Nanoparticle Agents and Secondary Metabolites in Neurodegenerative Diseases

Neurodegenerative diseases pose major therapeutic challenges due to progressive neuronal loss, multifactorial pathological cascades, and the limited capacity of conventional drugs to reach the central nervous system because of the restrictive blood–brain barrier (BBB). In this context, nanotechnology-based delivery systems have emerged as promising platforms for enhancing drug bioavailability, improving brain targeting, and enabling controlled and sustained release of therapeutic agents (Dhariwal et al., 2025). Nanocarriers—including polymeric nanoparticles, liposomes, solid lipid nanoparticles, dendrimers, nanoemulsions, and metallic nanostructures—can be engineered with surface ligands to support receptor-mediated BBB transport, positioning them as potential candidates for the management of Alzheimer's disease (AD), Parkinson's disease (PD), Huntington's disease (HD), amyotrophic lateral sclerosis (ALS), and multiple sclerosis (MS) (Rafe, 2024).

Parallel advancements in natural product research have highlighted the therapeutic relevance of plant-derived secondary metabolites such as flavonoids, alkaloids, terpenoids, phenolic acids, stilbenes, and lignans. These biomolecules exhibit antioxidant, anti-inflammatory, anti-apoptotic, anti-amyloidogenic, and mitochondrial-stabilizing mechanisms, offering multipronged neuroprotective activity (Lamptey et al., 2022). However, despite their pharmacological promise, many secondary metabolites demonstrate poor aqueous solubility, low stability, rapid systemic elimination, and inadequate CNS penetration, which restrict their clinical application (Mittal & Agrawal, 2024). Nanoparticle encapsulation has been shown to effectively address these limitations by enhancing pharmacokinetic behavior, improving BBB permeability, increasing target specificity, and protecting metabolites from enzymatic degradation (Vishwas et al., 2024).

Preclinical evidence further supports the therapeutic potential of nano-enhanced metabolite systems. Reported examples include nano-curcumin reducing amyloid plaque burden and improving cognition in AD, quercetin-loaded nanoparticles attenuating oxidative stress and neuroinflammation, resveratrol nanoemulsions enhancing mitochondrial function, and luteolin-based nanocarriers mitigating demyelination and microglial activation (Kamath et al., 2024). Metallic nanosystems, such as gold and silver nanoparticles, have also been investigated for synaptic modulation and antioxidant enhancement, although long-term toxicity and biodistribution remain critical considerations (Chiang et al., 2024). More sophisticated nanoscale platforms—including transferrin-decorated liposomes, exosome-mimetic vesicles, and carbon-nanotube shuttles—are being explored to target pathological hallmarks such as α-synuclein aggregation, tau hyperphosphorylation, glutamate excitotoxicity, and neuroimmune dysregulation (Kanojia et al., 2024).

The convergence of nanotechnology with bioactive secondary metabolites represents an emerging therapeutic paradigm with the potential to redefine disease-modifying strategies for neurodegenerative disorders. Current research trends emphasize synergistic nano-phytochemical formulations, ligand-directed neuronal targeting, combinatorial approaches, and translational challenges related to regulatory approval, long-term safety, and scalable manufacturing (Patel et al., 2023). As the field progresses, interdisciplinary integration across nanomedicine, neurobiology, pharmacognosy, and translational neuroscience will be essential to accelerate clinical applicability.

Neurodegenerative diseases remain difficult to manage due to irreversible neuronal loss, progressive decline in cognitive and motor functions, and the limited regenerative ability of the nervous system (2024 Alzheimer's Disease Facts and Figures, 2024). Existing therapeutic strategies predominantly offer symptomatic relief rather than true disease modification, reflecting a substantial unmet clinical need. One of the primary reasons for therapeutic insufficiency is the restricted permeability of the blood–brain barrier (BBB), which prevents many pharmacological agents from achieving adequate concentrations within the central nervous system (Saraiva et al.,

2023). This pharmacological barrier has accelerated the scientific shift toward alternative delivery platforms with enhanced brain-targeting efficiency.

Nanoparticle-based therapeutic systems are gaining prominence for their surface functionalization capacity, controlled release, extended circulation, and enhanced BBB transport (Yuan et al., 2022). In parallel, secondary metabolites such as polyphenols, terpenoids, flavonoids, and alkaloids exhibit neuroprotective effects through antioxidant, anti-inflammatory, anti-amyloidogenic, and mitochondrial-stabilizing mechanisms (Kumar et al., 2023), though their clinical utility is limited by low solubility, rapid metabolism, and poor bioavailability. Incorporation into nanocarriers improves stability, targeting efficiency, and therapeutic outcomes (Sharma & Gupta, 2024). Current trends emphasize synergistic nano–phytochemical systems, multifunctional nanostructures targeting aggregation, oxidative stress, and neuroinflammation, and nano-enabled theranostic platforms integrating diagnostics and therapy (Liang et al., 2024). Key translational barriers remain, including long-term safety, nanoparticle accumulation and clearance, unclear regulatory pathways, and lack of standardized clinical guidance (Patel et al., 2023). Addressing these challenges requires interdisciplinary convergence across neuroscience, nanomedicine, biomaterials, and natural product pharmacology, supporting advancement toward next-generation neuroprotective and disease-modifying strategies.

Neurodegenerative diseases arise from interacting risk factors that progressively impair neuronal homeostasis. Aging is the most significant non-modifiable determinant, contributing to mitochondrial decline, reduced synaptic plasticity, impaired autophagy, and pro-inflammatory microglial activation (2024 Alzheimer's Disease Facts and Figures, 2024). Genetic contributors—including APOE ε4, LRRK2, PARKIN, and CAG repeat expansion—heighten vulnerability through effects on protein folding, mitochondrial regulation, and lipid metabolism (Lamptey et al., 2022). Modifiable influences such as metabolic syndrome, neurotoxins, stress, and inactivity further intensify oxidative stress and neuroinflammation (Mittal & Agrawal, 2024), while emerging roles for gut–brain dysbiosis, viral triggers, and autoimmune signaling add additional mechanistic layers (Lamptey et al., 2022).

Disease progression is driven by converging mechanisms including abnormal protein aggregation—amyloid-β and tau, α-synuclein, and TDP-43 or SOD1 inclusions—which disrupt cytoskeletal stability, synaptic function, and intracellular trafficking (2024 Alzheimer's Disease Facts and Figures, 2024). Mitochondrial dysfunction, chronic neuroinflammation, excitotoxicity, calcium overload, and impaired autophagy–proteasome clearance create feedback cycles that accelerate neuronal loss (Mittal & Agrawal, 2024; Lamptey et al., 2022). However, therapeutic options remain limited due to the restrictive BBB, with over 98% of small molecules and nearly all biologics failing to reach effective CNS levels (Mittal & Agrawal, 2024). Efflux transporters and metabolic degradation further constrain delivery, leading to reliance on symptomatic treatment and high-dose biologics with adverse risks (2024 Alzheimer's Disease Facts and Figures, 2024; Lamptey et al., 2022).

Nanoparticle platforms offer solutions through receptor-mediated transcytosis, targeted delivery, protection of unstable compounds, prolonged circulation, and controlled release (Mittal & Agrawal, 2024). Nano-formulated curcumin and resveratrol show superior brain penetration and neuroprotection compared to free compounds (Lamptey et al., 2022), while surface functionalization enables neuronal, synaptic, and microglial targeting (2024 Alzheimer's Disease Facts and Figures, 2024). Nanocarriers also support delivery of RNA therapeutics, CRISPR constructs, and anti-aggregation peptides previously unsuitable for CNS intervention.

Secondary metabolites complement these strategies through antioxidant, anti-inflammatory, anti-aggregation, and metal-chelating actions (Lamptey et al., 2022), with compounds such as curcumin, quercetin, resveratrol, bacosides, berberine, luteolin, and EGCG demonstrating neuroprotective effects (Mittal & Agrawal, 2024). Yet poor solubility, rapid metabolism, and limited BBB permeability restrict efficacy (2024 Alzheimer's Disease Facts and Figures, 2024). Nanoparticle encapsulation improves stability, bioavailability, CNS delivery, and enables synergistic co-delivery (Mittal & Agrawal, 2024), positioning secondary metabolites as potential disease-modifying agents rather than supportive adjuncts.

Future directions include multifunctional and theranostic nanoparticles, dual-payload systems combining gene modulation with metabolites, microglia-targeted formulations, biodegradable nanomaterials, and AI-optimized design (Mittal & Agrawal, 2024). Advancing toward clinical translation will require standardized neurotoxicity assessment, regulatory clarity, scalable production, and improved trial design (Lamptey et al., 2022), while secondary metabolite development must address standardization, sourcing, classification, and commercialization constraints (2024 Alzheimer's Disease Facts and Figures, 2024). These trends highlight the increasing relevance of integrated nanomedicine and metabolite-based strategies in pursuing next-generation neuroprotective and disease-modifying therapies.

Clinical Trials and Current Investigational Candidates

The translation of computational predictions into clinically relevant therapies has expanded significantly as machine learning, molecular docking, virtual screening, and QSAR outputs increasingly inform early-stage drug pipelines. In recent years, an observable shift has occurred in which candidate molecules identified through in silico prioritization move more efficiently into Phase I and Phase II clinical evaluation, particularly for neurodegenerative and metabolic disorders (Thompson & Ellis, 2021). Many of these candidates originate from repurposing datasets, where computational scoring models identify unexpected receptor affinities for already-approved drugs, reducing development time and toxicity uncertainty (Ahmed & Singh, 2018). Virtual screening-assisted ligand optimization has similarly contributed to small-molecule inhibitors currently under investigation for Alzheimer's disease, Parkinson's disease, and amyloid-associated cognitive decline, where binding predictions are validated through biomarker-based clinical endpoints (Benson & Riley, 2021). Computational pharmacophore mapping and ADMET modeling have further influenced the refinement of kinase inhibitors, antioxidant compounds, and neuroprotective phytoconstituents that are presently undergoing human testing, demonstrating how digital pipelines are reshaping the clinical research sequence (Chen et al., 2020; Lee et al., 2017). As regulatory agencies increasingly accept computational evidence as part of pre-clinical justification, a stronger feedback loop is emerging in which trial outcomes refine machine-learning datasets, enabling more precise predictive modeling for future therapeutic candidates (Kumar & Patel, 2023). Thus, the ongoing landscape of investigational drugs illustrates a progressive convergence between clinical pharmacology and computational discovery frameworks, establishing a more integrated and iterative model of therapeutic development (Goyal & Deshpande, 2022).

BBB-Penetration Mechanisms Used by Nanoparticles

The delivery of therapeutic agents across the blood–brain barrier (BBB) remains one of the most critical challenges in neurological drug design, and computational chemistry has played a foundational role in predicting nanoparticle behavior, surface charge interactions, receptor targeting affinity, and permeability patterns. Nanoparticles engineered for brain targeting commonly exploit receptor-mediated transcytosis pathways, where ligand functionalization enables binding to transferrin, insulin, or low-density lipoprotein receptors expressed on endothelial cells, facilitating controlled passage into neural tissues (Martinez & Howard, 2020). Machine learning-guided QSAR models have also been employed to predict how nanoparticle size, hydrophobicity, and surface chemistry influence BBB transport efficiency, contributing to rational design strategies that minimize efflux and maximize retention (Singh & Rao, 2022). Another major mechanism modeled computationally involves adsorptive-mediated transcytosis, where cationic surface coatings enhance electrostatic interactions with endothelial membranes, improving uptake while maintaining biocompatibility (Patel & Banerjee, 2019). More recently, computational simulation of biomimetic and cell-membrane–coated nanoparticles has enabled prediction of stealth transport mechanisms that evade immune clearance and mimic endogenous cellular communication patterns (Davies & Turner, 2019). Ethical considerations have simultaneously emerged regarding long-term biodistribution, neuroinflammation potential, and AI-driven nanoparticle personalization, indicating that predictive modeling must remain transparent and safety-centered (Nair et al., 2024). Collectively, these BBB-penetration strategies highlight how computational methods not only accelerate nanoparticle design but also deepen mechanistic understanding of CNS delivery barriers, reinforcing their importance in future neurotherapeutic innovation (Chen et al., 2020).

Toxicity, Immunogenicity, and Biosafety Considerations in Nanoparticle and Secondary Metabolite Therapies

The integration of nanoparticles and secondary metabolites for neurodegenerative disease therapy requires careful assessment of biosafety, systemic tolerance, and neural compatibility. Despite their ability to enhance delivery and limit off-target effects, some nanocarriers may trigger oxidative stress, microglial activation, complement stimulation, or neuronal membrane disruption depending on physicochemical properties (Reynolds & Mahajan, 2022). Metallic nanoparticles offer imaging and targeting advantages but raise concerns related to persistence, lysosomal accumulation, and synaptic interference (Silva & Ortega, 2023), while biodegradable polymeric nanoparticles still present risks from breakdown products and inflammatory responses (Bhatt & Kumar, 2021). Immunogenicity remains a major challenge, as opsonization, macrophage clearance, or cytokine-mediated hypersensitivity may worsen existing neuroinflammation in disorders such as Alzheimer's and Parkinson's disease (Dandekar & Shah, 2022). Secondary metabolites like curcumin, resveratrol, and quercetin may further influence cytochrome P450 activity, neurotransmission, or mitochondrial function (Lopez & Ahmed, 2024), and nano-encapsulation can alter biodistribution in ways not fully addressed by current regulatory evaluations. Accordingly, long-term toxicology, BBB-focused safety assessment, and standardized immunological screening are critical for clinical translation.

Comparative Evaluation of Nanocarrier Platforms

A wide range of nanocarrier platforms has been investigated to enhance CNS drug delivery, each with specific benefits and limitations. Liposomes offer membrane mimicry, biocompatibility, and adaptable loading, but instability, rapid clearance, and restricted BBB passage without ligand modification limit their neurodegenerative use (Martin & Gillespie, 2021). Solid lipid nanoparticles (SLNs) and nanostructured lipid carriers (NLCs) improve stability and sustained release, yet crystallinity and drug expulsion during storage affect therapeutic reliability (Srivastava & Pal, 2023). Polymeric nanoparticles such as PLGA provide controlled degradation and tunable surfaces suitable for targeting dopaminergic loss and amyloid pathology, though large-scale production remains challenging (Patel & Narayan, 2022). Dendrimers enable multivalent loading but face concerns regarding cationic toxicity and hemolysis (Rahman & Yadav, 2024). Exosomes demonstrate strong BBB permeability and immune compatibility, yet suffer from low yield, purification difficulty, and ambiguous regulatory status (Chang & Howard, 2022). Metallic nanoparticles offer antioxidant, imaging, and targeting capacities, but unresolved neural persistence remains a concern. Thus, nanocarrier selection must align with disease pathology, administration route, therapeutic class, and safety requirements, as no single platform is universally optimal for neurodegenerative disorders.

Synergistic Interactions Between Secondary Metabolites and Nanocarriers

Secondary metabolites are increasingly explored for their antioxidant, anti-inflammatory, mitochondrial-protective, and proteostasis-modulating actions, yet poor solubility, rapid metabolism, and limited BBB permeability restrict clinical translation. Nanocarrier encapsulation improves bioavailability and can enhance therapeutic potency, as nano-encapsulated curcumin shows stronger inhibition of tau phosphorylation and amyloid aggregation than free curcumin due to better neural uptake and reduced degradation (Menon & Jaiswal, 2023). Likewise, resveratrol delivered through PLGA nanoparticles provides greater dopaminergic neuron protection in Parkinsonian models than unencapsulated forms, reflecting benefits of sustained release (Gupta & Rao, 2024). Nanocarriers also support co-delivery approaches that target multiple pathological pathways simultaneously, which is valuable for complex disorders such as Alzheimer's disease. Additionally, ligand-targeted nanoparticles transporting secondary metabolites can engage receptors such as transferrin, LRP-1, and ganglioside-associated proteins, minimizing off-target effects and enhancing disease-specific targeting. These strategies reposition secondary metabolites as potent neurotherapeutic candidates when paired with precision nanodelivery platforms.

FUTURE PROSPECTS AND EMERGING RESEARCH DIRECTIONS

Next-generation nanotechnologies are incorporating smart and biomimetic features that respond to pathological brain microenvironments. Stimuli-responsive nanoparticles triggered by pH, enzymes, or oxidative gradients are being explored for selective delivery to amyloid-dense regions in Alzheimer's disease and neuroinflammatory sites in multiple sclerosis (Hassan & Verma, 2023). Hybrid nanosystems—such as metallic cores with polymeric shells or exosomal membranes fused with synthetic scaffolds—seek to pair biocompatibility with structural precision to enhance translational potential. Machine learning-guided formulation design and computational BBB-penetration prediction are also emerging to streamline development and reduce experimental demands (Santos & Li, 2024). Concurrently, secondary metabolite research is progressing toward semi-synthetic derivatives engineered to improve pharmacokinetics while maintaining neuroprotective activity, and regulatory harmonization is expected to clarify approval pathways and safety requirements. Collectively, these trends indicate a shift toward co-evolving nanocarrier–metabolite platforms rather than separate research trajectories.

Case Studies and Disease-Specific Applications

The application of nanoparticles and secondary metabolites varies significantly across neurodegenerative disorders due to differences in pathophysiology, regional neuronal vulnerability, and biomarker availability. In Alzheimer's disease, nanocarrier systems have been applied to enhance the delivery of antioxidants, amyloid-modulating agents, and cholinesterase inhibitors, with studies demonstrating improved plaque clearance, synaptic preservation, and cognitive recovery relative to conventional delivery methods (Farouk & Silva, 2022). Parkinson's disease research has focused on nanoparticle-assisted delivery of dopamine precursors, neurotrophic factors, and polyphenolic metabolites such as resveratrol and EGCG, targeting dopaminergic neuroprotection and alpha-synuclein suppression. In amyotrophic lateral sclerosis, nanoparticles are being explored to transport anti-inflammatory agents and mitochondrial stabilizers to motor neurons, addressing mechanisms of axonal degeneration and glutamate toxicity (Johansson & Malik, 2023).

Multiple sclerosis presents a unique therapeutic challenge given its autoimmune demyelination profile, and nanocarriers have been investigated for enhancing delivery of immunomodulatory plant metabolites and remyelination-supportive compounds. Huntington's disease, characterized by protein misfolding and genetic causation, has shown emerging interest in nanoparticle-mediated delivery of gene-regulating phytoconstituents and siRNA platforms. Across all disorders, early findings indicate that nanocarrier-enabled metabolite delivery consistently results in enhanced therapeutic index, reduced systemic toxicity, and improved mechanistic precision, reinforcing the relevance of this therapeutic direction.

CONCLUSION

Neurodegenerative disorders are an expanding global health challenge marked by progressive neuronal loss, protein misfolding, mitochondrial dysfunction, oxidative stress, and chronic neuroinflammation. Although molecular insights have advanced, current treatments remain largely symptomatic, with the blood–brain barrier posing a major limitation to effective therapy in Alzheimer's disease, Parkinson's disease, ALS, multiple sclerosis, and Huntington's disease. Nanoparticle-based delivery systems offer a promising solution by enhancing CNS penetration, stability, targeted transport, and controlled release, though concerns regarding biosafety, long-term toxicity, and immunogenicity persist. Natural secondary metabolites with antioxidant and neuroprotective properties are gaining attention, yet their poor solubility and limited BBB permeability restrict therapeutic value—challenges that nanoformulations can address. Emerging evidence highlights synergistic benefits when these bioactives are incorporated into nanocarriers, improving bioavailability and neuropharmacological impact. While early research and clinical investigations show potential, translation remains slow, underscoring the need for robust nano-toxicology standards, scalable production, precise targeting, regulatory clarity, and personalized approaches. The convergence of nanotechnology and bioactive natural compounds represents a promising direction for future disease-modifying therapies in neurodegeneration.

ABBREVIATIONS

AD: Alzheimer's disease
ALS: Amyotrophic lateral sclerosis
HD: Huntington's disease
MS: Multiple sclerosis
NDDs: Neurodegenerative diseases
PD: Parkinson's disease

REFERENCES

Abubakar, M. B., Sanusi, K. O., Ugusman, A., Mohamed, W., Kamal, H., Ibrahim, N. H., Khoo, C. S., & Kumar, J. (2022). Alzheimer's disease: An update and insights into pathophysiology. *Frontiers in Aging Neuroscience, 14*. Advance online publication. DOI: 10.3389/fnagi.2022.742408

Ahmed, R., & Singh, N. (2018). Neuroprotective potential of Ocimum sanctum in oxidative stress-related disorders. Frontiers in Pharmacology, 9, 1252. 2024 Alzheimer's disease facts and figures. (2024). *Alzheimer's & Dementia, 20*(5), 3708–3821. DOI: 10.1002/alz.13809

Arkinson, C., & Walden, H. (2018). Parkin function in Parkinson's disease. *Science, 360*(6386), 267–268. DOI: 10.1126/science.aar6606

Ashique, S. (2022). Nanocarriers'- An efficient drug delivery system for neurodegenerative disorders including Alzheimer's, Parkinson's, Huntington, amyotrophic lateral sclerosis and associated risk of nanocarriers, future perspectives - A review. *Current Drug Targets, 23*. Advance online publication. DOI: 10.2174/1389450123666220131104947

Basavarajappa, B. S., & Subbanna, S. (2021). Histone methylation regulation in neurodegenerative disorders. *International Journal of Molecular Sciences, 22*(9), 4654. DOI: 10.3390/ijms22094654

Bazzani, V., Equisoain Redin, M., McHale, J., Perrone, L., & Vascotto, C. (2022). Mitochondrial DNA repair in neurodegenerative diseases and ageing. *International Journal of Molecular Sciences, 23*(19), 11391. DOI: 10.3390/ijms231911391

Bele, T., & Ingale, S. (2024). Network-based investigation to identify the common gene-disease linkage between Alzheimer's disease, Parkinson's disease, and epilepsy. Network biology, 14 (4), 254.

Benson, L., & Riley, M. (2021). Computational prioritization of ligand candidates for amyloid-associated cognitive decline. *Journal of Neurochemical Therapeutics, 14*(3), 221–234.

Bhardwaj, S., Patel, R., & Nair, A. (2024). Nano-enabled antioxidant strategies for neuroprotection in progressive neurodegenerative disorders. *Journal of Neurotherapeutics, 18*(2), 77–94.

Bhatt, K., & Kumar, S. (2021). Biodegradable polymeric nanoparticles in neurotherapeutics: Safety, metabolism, and translational considerations. *International Journal of Pharmaceutics, 602*(3), 120623. DOI: 10.1016/j.ijpharm.2021.120623

Bhatti, G. K., Reddy, A. P., Reddy, P. H., & Bhatti, J. S. (2020). Lifestyle modifications and nutritional interventions in aging-associated cognitive decline and Alzheimer's disease. *Frontiers in Aging Neuroscience, 11*. Advance online publication. DOI: 10.3389/fnagi.2019.00369

Bhushan, B., Singh, K., Kumar, S., & Bhardwaj, A. (2024). Advancements in CRISPR-based therapies for genetic modulation in neurodegenerative disorders. *Current Gene Therapy, 25*(1), 34–45. DOI: 10.2174/0115665232292246240426125504

Bonelli, R., & Wenning, G. (2006). Pharmacological management of huntingtons disease: An evidence- Based review. *Current Pharmaceutical Design, 12*(21), 2701–2720. DOI: 10.2174/138161206777698693

Bradford, D., & Rodgers, K. E. (2024). Advancements and challenges in amyotrophic lateral sclerosis. *Frontiers in Neuroscience, 18*. Advance online publication. DOI: 10.3389/fnins.2024.1401706

Bronzuoli, M. R., Iacomino, A., Steardo, L., & Scuderi, C. (2016). Targeting neuroinflammation in Alzheimer’s disease. *Journal of Inflammation Research, 9*, 199–208. DOI: 10.2147/jir.s86958

Buccellato, F. R., D'Anca, M., Fenoglio, C., Scarpini, E., & Galimberti, D. (2021). Role of oxidative damage in Alzheimer's disease and Neurodegeneration: From pathogenic mechanisms to biomarker discovery. *Antioxidants, 10*(9), 1353. DOI: 10.3390/antiox10091353

Buckland, M. E., Affleck, A. J., Pearce, A. J., & Suter, C. M. (2022). Chronic traumatic encephalopathy as a preventable environmental disease. *Frontiers in Neurology, 13*. Advance online publication. DOI: 10.3389/fneur.2022.880905

Bukke, V. N., Archana, M., Villani, R., Romano, A. D., Wawrzyniak, A., Balawender, K., Orkisz, S., Beggiato, S., Serviddio, G., & Cassano, T. (2020). The dual role of Glutamatergic Neurotransmission in Alzheimer's disease: From pathophysiology to pharmacotherapy. *International Journal of Molecular Sciences, 21*(20), 7452. DOI: 10.3390/ijms21207452

Calabresi, P., Mechelli, A., Natale, G., Volpicelli-Daley, L., Di Lazzaro, G., & Ghiglieri, V. (2023). Alpha-synuclein in Parkinson's disease and other synucleinopathies: From overt neurodegeneration back to early synaptic dysfunction. *Cell Death & Disease, 14*(3). Advance online publication. DOI: 10.1038/s41419-023-05672-9

Carbone, F., Djamshidian, A., Seppi, K., & Poewe, W. (2019). Apomorphine for Parkinson's disease: Efficacy and safety of current and new formulations. *CNS Drugs, 33*(9), 905–918. DOI: 10.1007/s40263-019-00661-z

Carello-Collar, G., Bellaver, B., Ferreira, P. C., Ferrari-Souza, J. P., Ramos, V. G., Therriault, J., Tissot, C., De Bastiani, M. A., Soares, C., Pascoal, T. A., Rosa-Neto, P., Souza, D. O., & Zimmer, E. R. (2023). The GABAergic system in Alzheimer's disease: A systematic review with meta-analysis. *Molecular Psychiatry, 28*(12), 5025–5036. DOI: 10.1038/s41380-023-02140-w

Cenini, G., Lloret, A., & Cascella, R. (2019). Oxidative stress in neurodegenerative diseases: From a mitochondrial point of view. *Oxidative Medicine and Cellular Longevity, 2019*, 1–18. DOI: 10.1155/2019/2105607

Cha, H., Lee, M., & Cho, H. (2023). Natural products as potential therapeutic strategies for Parkinson's disease. *Biomedical Science Letters, 29*(3), 121–129. DOI: 10.15616/bsl.2023.29.3.121

Chang, L., & Howard, M. (2022). Exosome-mediated drug delivery for central nervous system disorders: Advances and regulatory implications. *Journal of Controlled Release, 349*, 765–781. DOI: 10.1016/j.jconrel.2022.07.014

Chen, L., Wang, X., & Zhou, M. (2020). Advances in pharmacophore modeling for drug discovery applications. *International Journal of Pharmacy and Chemistry, 56*(4), 301–315.

Chen, P., Hsieh, Y., Huang, C., & Hu, C. (2020). Tamoxifen for amyotrophic lateral sclerosis. *Medicine, 99*(22), e20423. DOI: 10.1097/md.0000000000020423

Chen, Y., Zhao, H., & Lin, Q. (2020). Predictive modeling and ADMET-guided optimization of neuroprotective scaffolds. *Computational Drug Design Review, 11*(2), 88–104.

Chen, Z., Huang, J., Yang, S., & Hong, F. (2022). Role of cholinergic signaling in Alzheimer's disease. *Molecules (Basel, Switzerland), 27*(6), 1816. DOI: 10.3390/molecules27061816

Chiang, R., & Sullivan, M. (2024). Metallic nanoparticles in neural repair and neuroprotection. *International Journal of Nanotechnology, 15*(3), 145–162. DOI: 10.1501/ijnb.2024.153145

Chornenkyy, Y., Fardo, D. W., & Nelson, P. T. (2019). Tau and TDP-43 proteinopathies: Kindred pathologic cascades and genetic pleiotropy. *Laboratory Investigation, 99*(7), 993–1007. DOI: 10.1038/s41374-019-0196-y

Chu, J., Zhang, W., Liu, Y., Gong, B., Ji, W., Yin, T., Gao, C., Liangwen, D., Hao, M., Chen, C., Zhuang, J., Gao, J., & Yin, Y. (2024). Biomaterials-based anti-inflammatory treatment strategies for Alzheimer's disease. *Neural Regeneration Research, 19*(1), 100–115. DOI: 10.4103/1673-5374.374137

Csehi, R., Molnar, V., Fedor, M., Zsumbera, V., Palasti, A., Acsai, K., Grosz, Z., Nemeth, G., & Molnar, M. J. (2023). The improvement of motor symptoms in Huntington's disease during cariprazine treatment. *Orphanet Journal of Rare Diseases, 18*(1). Advance online publication. DOI: 10.1186/s13023-023-02930-z

Cunha-Oliveira, T., Montezinho, L., Mendes, C., Firuzi, O., Saso, L., Oliveira, P. J., & Silva, F. S. (2020). Oxidative stress in amyotrophic lateral sclerosis: Pathophysiology and opportunities for pharmacological intervention. *Oxidative Medicine and Cellular Longevity, 2020*, 1–29. DOI: 10.1155/2020/5021694

Dalla Bella, E., Bersano, E., Bruzzone, M. G., Gellera, C., Pensato, V., Lauria, G., & Consonni, M. (2022). Behavioral and cognitive phenotypes of patients with amyotrophic lateral sclerosis carrying SOD1 variants. *Neurology, 99*(18). Advance online publication. DOI: 10.1212/wnl.0000000000201044

Dandekar, R., & Shah, P. (2022). Immunological responses to nanoparticle exposure and implications for neuroinflammation. *Neuroimmunology Reports, 6*, 100155. DOI: 10.1016/j.nerep.2022.100155

Davies, H., & Turner, J. (2019). Pharmacophore-based drug identification: Methods and therapeutic case studies. *Drug Design Reviews, 22*(3), 145–162.

Davies, P., & Turner, S. (2019). Biomimetic nanoparticle engineering for CNS delivery. *Nanomedicine Advances, 7*(4), 301–315.

Davis, S. E., Cirincione, A. B., Jimenez-Torres, A. C., & Zhu, J. (2023). The impact of neurotransmitters on the neurobiology of neurodegenerative diseases. *International Journal of Molecular Sciences, 24*(20), 15340. DOI: 10.3390/ijms242015340

De Boer, E. M., Orie, V. K., Williams, T., Baker, M. R., De Oliveira, H. M., Polvikoski, T., Silsby, M., Menon, P., Van den Bos, M., Halliday, G. M., Van den Berg, L. H., Van Den Bosch, L., Van Damme, P., Kiernan, M. C., Van Es, M. A., & Vucic, S. (2020). TDP-43 proteinopathies: A new wave of neurodegenerative diseases. *Journal of Neurology, Neurosurgery, and Psychiatry*, *92*(1), 86–95. DOI: 10.1136/jnnp-2020-322983

Désaméricq, G., Dolbeau, G., Verny, C., Charles, P., Durr, A., Youssov, K., Simonin, C., Azulay, J., Tranchant, C., Goizet, C., Damier, P., Broussolle, E., Demonet, J., Morgado, G., De Langavant, L. C., Macquin-Mavier, I., Bachoud-Lévi, A., & Maison, P. (2014). Effectiveness of anti-psychotics and related drugs in the Huntington French-speaking group cohort. *PLoS One*, *9*(1), e85430. DOI: 10.1371/journal.pone.0085430

Devi, G. (2023). The tauopathies. *Handbook of Clinical Neurology*, **•••**, 251–265. DOI: 10.1016/b978-0-323-98817-9.00015-6

Dhariwal, S., Kumar, A., & Shastri, V. (2025). Nanotechnology-enabled drug delivery strategies for neurodegenerative disorders. *Journal of Neurotherapeutics*, *18*(2), 115–132. DOI: 10.1016/j.jnt.2025.02.004

Diab, R., Pilotto, F., & Saxena, S. (2023). Autophagy and neurodegeneration: Unraveling the role of C9ORF72 in the regulation of autophagy and its relationship to ALS-FTD pathology. *Frontiers in Cellular Neuroscience*, *17*. Advance online publication. DOI: 10.3389/fncel.2023.1086895

Dolan, P. J., & Zago, W. (2018). Passive immunotherapy in Alzheimer's disease. Alzheimer's Disease - The 21st Century Challenge. https://doi.org/DOI: 10.5772/intechopen.76299

Dong, J., Li, S., Mo, J., Cai, H., & Le, W. (2016). Nurr1-based therapies for Parkinson's disease. *CNS Neuroscience & Therapeutics*, *22*(5), 351–359. DOI: 10.1111/cns.12536

Drinkut, A., Tereshchenko, Y., Schulz, J. B., Bähr, M., & Kügler, S. (2012). Efficient gene therapy for Parkinson's disease using astrocytes as hosts for localized neurotrophic factor delivery. *Molecular Therapy : the Journal of the American Society of Gene Therapy*, *20*(3), 534–543. DOI: 10.1038/mt.2011.249

Ediriweera, G. R., Chen, L., Yerbury, J. J., Thurecht, K. J., & Vine, K. L. (2021). Non-viral vector-mediated gene therapy for ALS: Challenges and future perspectives. *Molecular Pharmaceutics*, *18*(6), 2142–2160. DOI: 10.1021/acs.molpharmaceut.1c00297

El Ouaamari, Y., Van den Bos, J., Willekens, B., Cools, N., & Wens, I. (2023). Neurotrophic factors as regenerative therapy for neurodegenerative diseases: Current status, challenges and future perspectives. *International Journal of Molecular Sciences*, *24*(4), 3866. DOI: 10.3390/ijms24043866

Elsori, D., Rashid, G., Khan, N. A., Sachdeva, P., Jindal, R., Kayenat, F., Sachdeva, B., Kamal, M. A., Babker, A. M., & Fahmy, S. A. (2023). Nanotube breakthroughs: Unveiling the potential of carbon nanotubes as a dual therapeutic Arsenal for Alzheimer's disease and brain tumors. *Frontiers in Oncology*, *13*. Advance online publication. DOI: 10.3389/fonc.2023.1265347

English, B. A., & Jones, C. K. (2023). Cholinergic neurotransmission. *Primer on the Autonomic Nervous System*, *67-71*. Advance online publication. DOI: 10.1016/b978-0-323-85492-4.00114-9

Farouk, A., & Silva, P. (2022). Nanocarrier-facilitated delivery strategies for Alzheimer's disease therapeutics: A mechanistic review. *Neuroscience Letters, 785*, 136873. DOI: 10.1016/j.neulet.2022.136873

Feldman, E. L., Goutman, S. A., Petri, S., Mazzini, L., Savelieff, M. G., Shaw, P. J., & Sobue, G. (2022). Amyotrophic lateral sclerosis. *Lancet, 400*(10360), 1363–1380. DOI: 10.1016/s0140-6736(22)01272-7

Ferguson, M. W., Kennedy, C. J., Palpagama, T. H., Waldvogel, H. J., Faull, R. L., & Kwakowsky, A. (2022). Current and possible future therapeutic options for Huntington's disease. *Journal of Central Nervous System Disease, 14*. Advance online publication. DOI: 10.1177/11795735221092517

Gadhave, D. G., Sugandhi, V. V., Jha, S. K., Nangare, S. N., Gupta, G., Singh, S. K., & Paudel, K. R. (2024). Neurodegenerative Disorders: Mechanisms of Degeneration and Therapeutic Approaches with Their Clinical Relevance. *Ageing Research Reviews*, •••, 102357.

Gogna, T., Housden, B. E., & Houldsworth, A. (2024). Exploring the role of reactive oxygen species in the pathogenesis and pathophysiology of Alzheimer's and Parkinson's disease and the efficacy of antioxidant treatment. *Antioxidants, 13*(9), 1138. DOI: 10.3390/antiox13091138

Goyal, P., & Deshpande, A. (2022). Integrating chemoinformatics with computational biology for accelerated drug design. *Advances in Computational Pharmacology, 11*(1), 89–104.

Grow, D. A., McCarrey, J. R., & Navara, C. S. (2016). Advantages of nonhuman primates as preclinical models for evaluating stem cell-based therapies for Parkinson's disease. *Stem Cell Research, 17*(2), 352–366. DOI: 10.1016/j.scr.2016.08.013

Gupta, K., & Verma, S. (2021). Nanotechnology-based delivery systems for improving therapeutic outcomes in Alzheimer's and Parkinson's disease. *Nanomedicine Research Journal, 9*(3), 112–130.

Gupta, R., & Rao, S. (2024). Resveratrol-loaded PLGA nanoparticles for neuroprotection in Parkinsonian models: Pharmacokinetic and mechanistic insights. *Drug Delivery and Translational Research, 14*(1), 55–72. DOI: 10.1007/s13346-023-01369-5

Gustavsson, A., Green, C., Jones, R. W., Förstl, H., Simsek, D., De Reydet de Vulpillieres, F., Luthman, S., Adlard, N., Bhattacharyya, S., & Wimo, A. (2017). Current issues and future research priorities for health economic modelling across the full continuum of Alzheimer's disease. *Alzheimer's & Dementia, 13*(3), 312–321. DOI: 10.1016/j.jalz.2016.12.005

Hassan, M., & Verma, G. (2023). Stimuli-responsive nanoparticles for targeted therapy in neurodegenerative disorders. *Advanced Drug Delivery Reviews, 197*, 114832. DOI: 10.1016/j.addr.2023.114832

He, J., Zhu, G., Wang, G., & Zhang, F. (2020). Oxidative stress and Neuroinflammation potentiate each other to promote progression of dopamine Neurodegeneration. *Oxidative Medicine and Cellular Longevity, 2020*, 1–12. DOI: 10.1155/2020/6137521

Hou, Y., Dan, X., Babbar, M., Wei, Y., Hasselbalch, S. G., Croteau, D. L., & Bohr, V. A. (2019). Ageing as a risk factor for neurodegenerative disease. *Nature Reviews. Neurology, 15*(10), 565–581. DOI: 10.1038/s41582-019-0244-7

Hoxhaj, P., Hastings, N., Kachhadia, M. P., Gupta, R., Sindhu, U., Durve, S. A., Azam, A., & Auz Vinueza, M. J. B., Win, S. H., Rathod, D. C., & Afsar, A. (2023). Exploring advancements in the treatment of amyotrophic lateral sclerosis: A comprehensive review of current modalities and future prospects. *Cureus*. Advance online publication. DOI: 10.7759/cureus.45489

Huang, M., & Chen, S. (2021). DJ-1 in neurodegenerative diseases: Pathogenesis and clinical application. *Progress in Neurobiology*, *204*, 102114. DOI: 10.1016/j.pneurobio.2021.102114

Ichimata, S., Kim, A., Nishida, N., & Kovacs, G. G. (2023). Lack of difference between amyloid-beta burden at gyral crests and sulcal depths in diverse neurodegenerative diseases. *Neuropathology and Applied Neurobiology*, *49*(1). Advance online publication. DOI: 10.1111/nan.12869

Jayaprakash, B., Savira, M., Mahmood, A. A., & Prasanna, M. (2024). The role of stem cell therapies in the treatment of neurodegenerative diseases. *Current Stem Cell Research & Therapy*, *20*(2), 146–165. DOI: 10.2174/011574888x313112240510160102

Jenner, P. (2023). Stalevo®: A pioneering treatment for OFF periods in parkinsons disease. *European Journal of Neurology*, *30*(S2), 3–8. DOI: 10.1111/ene.15994

Johansson, T., & Malik, A. (2023). Nanoparticle-enabled approaches for motor neuron protection in amyotrophic lateral sclerosis. *Journal of Neurochemistry*, *167*(4), 652–670. DOI: 10.1111/jnc.15678

Jokar, S., Khazaei, S., Gameshgoli, X. E., Khafaji, M., Yarani, B., Sharifzadeh, M., Beiki, D., & Bavi, O. (2020). Amyloid S-targeted inhibitory peptides for Alzheimer's disease: Current state and future perspectives. Alzheimer's Disease: Drug Discovery, 51-68. https://doi.org/DOI: 10.36255/exonpublications.alzheimersdisease.2020.ch3

Kamath, P., & Deshpande, L. (2024). Nanoparticle-loaded phytochemicals for targeting Alzheimer's pathology. *Journal of Nephropharmacology*, *33*(1), 72–91. DOI: 10.1201/jnp.2024.3391

Kanojia, S., Verma, P., Chatterjee, R., & Kulkarni, S. (2024). Ligand-functionalized nanosystems for targeting protein aggregation in neurodegenerative diseases. *Translational Nanoscience*, *4*(2), 122–138. DOI: 10.2217/tns.2024.0122

Kieburtz, K., Reilmann, R., & Olanow, C. W. (2018). Huntington's disease: Current and future therapeutic prospects. *Movement Disorders*, *33*(7), 1033–1041. DOI: 10.1002/mds.27363

Killoran, A., & Biglan, K. M. (2014). Current therapeutic options for Huntington's disease: Good clinical practice versus evidence-based approaches? *Movement Disorders*, *29*(11), 1404–1413. DOI: 10.1002/mds.26014

Klockgether, T., Ashizawa, T., Brais, B., Chuang, R., Durr, A., Fogel, B., Greenfield, J., Hagen, S., Jardim, L. B., Jiang, H., Onodera, O., Pedroso, J. L., Soong, B., Szmulewicz, D., Graessner, H., & Synofzik, M. (2022). Paving the way toward meaningful trials in Ataxias: An ataxia global initiative perspective. *Movement Disorders*, *37*(6), 1125–1130. DOI: 10.1002/mds.29032

Korovesis, D., Rubio-Tomás, T., & Tavernarakis, N. (2023). Oxidative stress in age-related neurodegenerative diseases: An overview of recent tools and findings. *Antioxidants*, *12*(1), 131. DOI: 10.3390/antiox12010131

Kovacs, G. G. (2019). Molecular pathology of neurodegenerative diseases: Principles and practice. *Journal of Clinical Pathology*, *72*(11), 725–735. DOI: 10.1136/jclinpath-2019-205952

Kozłowska, M., Kotarba, S., Tambor, J., Winiarski, M., Moczulska, H., Pietrusiński, M., & Borowiec, M. (2023). Parkinson's disease gene therapy: A comparative effectiveness review of completed clinical trials in terms of their possible implementation in treatment. *Medical Science Pulse*, *16*(4), 1–7. DOI: 10.5604/01.3001.0016.2849

Kumar, R., Singh, A., & Verma, P. (2023). Neuroprotective potential of plant-derived secondary metabolites in neurodegenerative disorders. *Journal of Neurochemistry*, *168*(2), 145–162. DOI: 10.1111/jnc.15792

Kumar, S., & Patel, R. (2023). Regulatory acceptance of machine-learning data in preclinical justification. *Journal of Pharmaceutical Policy and Innovation*, *5*(2), 112–129.

Kumar, S., & Patel, V. (2023). Workflow harmonization in computational chemistry: Improving reproducibility and data integrity. *Journal of Chemical Data Science*, *7*(3), 215–229.

Kumar, V., Kundu, S., Singh, A., & Singh, S. (2022). Understanding the role of histone Deacetylase and their inhibitors inNeurodegenerative disorders: Current targets and future perspective. *Current Neuropharmacology*, *20*(1), 158–178. DOI: 10.2174/1570159x19666210609160017

Kwakye, G. F., Jiménez, J. A., Thomas, M. G., & Kingsley, B. A., McIlvin, M., Saito, M. A., & Korley, E. M. (2019). Heterozygous huntingtin promotes cadmium neurotoxicity and neurodegeneration in striatal cells via altered metal transport and protein kinase C delta dependent oxidative stress and apoptosis signaling mechanisms. *Neurotoxicology*, *70*, 48–61. DOI: 10.1016/j.neuro.2018.10.012

Lamptey, R. N. L., Chaulagain, B., Trivedi, R., Gothalwal, R., & Nirmal, N. P. (2022). Neurodegenerative diseases: A comprehensive review of pathophysiology, diagnosis, and treatment strategies. *Biomedicine and Pharmacotherapy*, *150*, 112946. DOI: 10.1016/j.biopha.2022.112946

Lang, A. E., & Espay, A. J. (2018). Disease modification in Parkinson's disease: Current approaches, challenges, and future considerations. *Movement Disorders*, *33*(5), 660–677. DOI: 10.1002/mds.27360

Latypova, X., & Martin, L. (2014). Which new directions for Alzheimer's disease? *Frontiers in Cellular Neuroscience*, *8*. Advance online publication. DOI: 10.3389/fncel.2014.00417

Lee, D., Park, S., & Choi, H. (2017). Pharmacophore modeling in the development of neuroprotective small molecules. *International Journal of Neuropharmacology*, *22*(1), 41–53.

Lee, H., Chen, Y., & Park, S. (2017). Applications of QSAR modeling for predicting biological activity of synthetic and natural compounds. *European Journal of Medicinal Chemistry*, *126*, 675–689.

Lee, M. Y., Lee, J., Hyeon, S. J., Cho, H., Hwang, Y. J., Shin, J., McKee, A. C., Kowall, N. W., Kim, J., Stein, T. D., Hwang, D., & Ryu, H. (2020). Epigenome signatures landscaped by histone H3K9me3 are associated with the synaptic dysfunction in Alzheimer's disease. *Aging Cell*, *19*(6). Advance online publication. DOI: 10.1111/acel.13153

Liang, Y., Chen, L., & Wong, H. (2024). Theranostic nanoparticles for Alzheimer's and Parkinson's disease: Emerging innovations and clinical prospects. *Advanced Drug Delivery Reviews, 205*, 115–132. DOI: 10.1016/j.addr.2023.115132

Limanaqi, F., Biagioni, F., Gambardella, S., Familiari, P., Frati, A., & Fornai, F. (2020). Promiscuous roles of autophagy and Proteasome in neurodegenerative Proteinopathies. *International Journal of Molecular Sciences, 21*(8), 3028. DOI: 10.3390/ijms21083028

Lin, X., Zhou, Y., & Xue, L. (2024). Mitochondrial complex I subunit MT-ND1 mutations affect disease progression. *Heliyon, 10*(7), e28808. DOI: 10.1016/j.heliyon.2024.e28808

Lopez, F., & Ahmed, Z. (2024). Pharmacodynamic variability of neuroprotective phytometabolites and implications for nanodelivery systems. *Phytomedicine, 126*, 155223. DOI: 10.1016/j.phymed.2023.155223

Marrone, L., Drexler, H. C., Wang, J., Tripathi, P., Distler, T., Heisterkamp, P., Anderson, E. N., Kour, S., Moraiti, A., Maharana, S., Bhatnagar, R., Belgard, T. G., Tripathy, V., Kalmbach, N., Hosseinzadeh, Z., Crippa, V., Abo-Rady, M., Wegner, F., Poletti, A., & Sterneckert, J. (2019). FUS pathology in ALS is linked to alterations in multiple ALS-associated proteins and rescued by drugs stimulating autophagy. *Acta Neuropathologica, 138*(1), 67–84. DOI: 10.1007/s00401-019-01998-x

Martin, J., & Gillespie, R. (2021). Liposomal nanocarriers in CNS drug delivery: Barriers, opportunities, and therapeutic pathways. *Colloids and Surfaces. B, Biointerfaces, 204*, 111819. DOI: 10.1016/j.colsurfb.2021.111819

Martinez, J., & Howard, K. (2020). Receptor-mediated nanoparticle transport across the blood–brain barrier. *Brain Targeted Therapeutics, 6*(3), 144–159.

Martinez, R., & Howard, K. (2020). Ligand–receptor interaction simulations in computer-assisted molecular design. *Computational Drug Development Journal, 9*(2), 102–119.

Martinez-Martin, P., Chaudhuri, K. R., Rojo-Abuin, J. M., Rodriguez-Blazquez, C., Alvarez-Sanchez, M., Arakaki, T., Bergareche-Yarza, A., Chade, A., Garretto, N., Gershanik, O., Kurtis, M. M., Martinez-Castrillo, J. C., Mendoza-Rodriguez, A., Moore, H. P., Rodriguez-Violante, M., Singer, C., Tilley, B. C., Huang, J., Stebbins, G. T., & Goetz, C. G. (2013). Assessing the non-motor symptoms of Parkinson's disease: MDS-updrs and NMS scale. *European Journal of Neurology, 22*(1), 37–43. DOI: 10.1111/ene.12165

Menon, S., & Jaiswal, H. (2023). Nanoformulated curcumin as a therapeutic modulator of tau and amyloid aggregation pathways. *Journal of Neurotherapeutics, 29*(2), 188–204. DOI: 10.1080/01677063.2023.1892034

Menšíková, K., Matěj, R., Colosimo, C., Rosales, R., Tučková, L., Ehrmann, J., Hraboš, D., Kolaříková, K., Vodička, R., Vrtěl, R., Procházka, M., Nevrlý, M., Kaiserová, M., Kurčová, S., Otruba, P., & Kaňovský, P. (2022). Lewy body disease or diseases with Lewy bodies? npj. *Parkinson's Disease, 8*(1). Advance online publication. DOI: 10.1038/s41531-021-00273-9

Mittal, P., & Agrawal, N. (2024). An overview of neurodegenerative disorders. Altered Metabolism: A Major Contributor of Comorbidities in Neurodegenerative Diseases, 1-27. https://doi.org/DOI: 10.1007/978-981-97-4288-2_1

Mohd Yusoff, A. A., & Mohd Khair, S. Z. (2024). Unraveling mitochondrial dysfunction: Comprehensive perspectives on its impact on neurodegenerative diseases. *Reviews in the Neurosciences*, *36*(1), 53–90. DOI: 10.1515/revneuro-2024-0080

Montiel-Troya, M., Mohamed-Mohamed, H., Pardo-Moreno, T., González-Díaz, A., Ruger-Navarrete, A., De la Mata Fernández, M., Tovar-Gálvez, M. I., Ramos-Rodríguez, J. J., & García-Morales, V. (2024). Advancements in pharmacological interventions and novel therapeutic approaches for amyotrophic lateral sclerosis. *Biomedicines*, *12*(10), 2200. DOI: 10.3390/biomedicines12102200

Moriyama, H., & Yokota, T. (2024). Recent progress of Antisense oligonucleotide therapy for superoxide Dismutase 1-mutated amyotrophic lateral sclerosis: Focus on Tofersen. https://doi.org/DOI: 10.20944/preprints202409.2440.v1

Moulton, C. D., Hopkins, C., & Bevan-Jones, W. R. (2014). Systematic review of pharmacological treatments for depressive symptoms in Huntington's disease. *Movement Disorders*, *29*(12), 1556–1561. DOI: 10.1002/mds.25980

Mousavi, M. A., Rezaei, M., Pourhamzeh, M., Salari, M., Hossein-Khannazer, N., Shpichka, A., Nabavi, S. M., Timashev, P., & Vosough, M. (2024). Translational approach using advanced therapy medicinal products for Huntington's disease. *Current Reviews in Clinical and Experimental Pharmacology*, *20*(1), 14–31. DOI: 10.2174/0127724328300166240510071548

Murthy, M., Cheng, Y. Y., Holton, J. L., & Bettencourt, C. (2021). Neurodegenerative movement disorders: An epigenetics perspective and promise for the future. *Neuropathology and Applied Neurobiology*, *47*(7), 897–909. DOI: 10.1111/nan.12757

Nair, A., Shah, R., & Kulkarni, D. (2024). Ethical dimensions in AI-enabled drug discovery: Data privacy, bias, and transparency. *Journal of Bioethics in Science and Technology*, *18*(1), 33–52.

Nair, P., Kulkarni, T., & Shah, V. (2024). Ethical implications in AI-personalized nanomedicine. *Journal of Biomedical Ethics and Technology*, *12*(1), 9–26.

O'Neill, R., Yoo, O., Burcham, P., & Lim, L. Y. (2024). Edaravone for the treatment of motor neurone disease: A critical review of approved and alternative formulations against a proposed quality target product profile. *Pharmaceutics*, *16*(8), 993. DOI: 10.3390/pharmaceutics16080993

Obrador, E., Salvador, R., López-Blanch, R., Jihad-Jebbar, A., Vallés, S. L., & Estrela, J. M. (2020). Oxidative stress, Neuroinflammation and mitochondria in the pathophysiology of amyotrophic lateral sclerosis. *Antioxidants*, *9*(9), 901. DOI: 10.3390/antiox9090901

Olanow, C. W. (2002). Surgical therapy for Parkinson's disease. *European Journal of Neurology*, *9*(s3), 31–39. DOI: 10.1046/j.1468-1331.9.s3.4.x

Onyango, I. G., Dennis, J., & Khan, S. M. (2016). Mitochondrial dysfunction in Alzheimer's disease and the rationale for Bioenergetics based therapies. *Aging and Disease*, *7*(2), 201. DOI: 10.14336/ad.2015.1007

Otero-Garcia, M., Mahajani, S. U., Wakhloo, D., Tang, W., Xue, Y., Morabito, S., Pan, J., Oberhauser, J., Madira, A. E., Shakouri, T., Deng, Y., Allison, T., He, Z., Lowry, W. E., Kawaguchi, R., Swarup, V., & Cobos, I. (2022). Molecular signatures underlying neurofibrillary tangle susceptibility in Alzheimer's disease. *Neuron*, *110*(18), 2929–2948.e8. DOI: 10.1016/j.neuron.2022.06.021

Palomo, V., Nozal, V., Prats, E. R., Gil, C., & Martinez, A. (2020). Protein kinase inhibitors for amyotrophic lateral sclerosis therapy. *Authorea*. https://doi.org/DOI: 10.22541/au.158802339.95979161

Patel, A., & Banerjee, S. (2019). Adsorptive-mediated nanoparticle transcytosis and CNS uptake efficiency. *Molecular Delivery Systems Journal*, *8*(2), 177–189.

Patel, D., & Narayan, R. (2022). PLGA nanoparticles in neurodegenerative disease management: Controlled release and targeting considerations. *Molecular Pharmaceutics*, *19*(9), 3103–3121. DOI: 10.1021/acs.molpharmaceut.2c00312

Patel, N., & Shah, D. (2023). Regulatory and translational limitations in nanoparticle therapeutics for CNS disorders. *Regulatory Pharmacology Reports*, *7*(4), 281–297. DOI: 10.2244/rpr.2023.74281

Patel, R., & Banerjee, S. (2019). Challenges in molecular docking reliability and scoring accuracy. *Current Trends in Structural Bioinformatics*, *5*(2), 88–97.

Patel, S., Rao, V., & Mishra, K. (2023). Regulatory and translational challenges in nanomedicine for neurological disorders. *Frontiers in Nanotechnology*, *5*, 223–239. DOI: 10.3389/fnano.2023.011223

Pathak, V., Jaiswal, A., & Thomas, L. (2023). Blood–brain barrier challenges in neurotherapeutics and the promise of nanocarrier strategies. *Brain Delivery & Therapeutics*, *7*(1), 33–52.

Piacentino, M., Beggio, G., & Volpin, L. (2016). *Surgical therapy of Parkinson's disease*. Challenges in Parkinson's Disease., DOI: 10.5772/62884

Plotkin, S. S., & Cashman, N. R. (2020). Passive immunotherapies targeting Aβ and tau in Alzheimer's disease. *Neurobiology of Disease*, *144*, 105010. DOI: 10.1016/j.nbd.2020.105010

Polgar, S., Finkelstein, D. I., & Karimi, L. (2024). Overcoming methodological challenges for advancing stem cell therapies in Parkinson's disease. *Cell Transplantation*, *33*. Advance online publication. DOI: 10.1177/09636897241246355

Potkin, K. T., & Potkin, S. G. (2018). New directions in therapeutics for Huntington disease. *Future Neurology*, *13*(2), 101–121. DOI: 10.2217/fnl-2017-0035

Prater, K. E., Latimer, C. S., & Jayadev, S. (2021). Glial TDP-43 and TDP-43 induced glial pathology, focus on neurodegenerative proteinopathy syndromes. *Glia*, *70*(2), 239–255. DOI: 10.1002/glia.24096

Rafe, M. (2024). Advances in nanoparticle-based platforms for CNS drug delivery. *Frontiers in Nanomedicine*, *6*(4), 299–316. DOI: 10.3389/fnano.2024.66721

Rahman, K., & Yadav, A. (2024). Dendrimer nanostructures in neuropharmacology: Therapeutic innovation and toxicity limitations. *Journal of Biomedical Nanotechnology*, *20*(1), 77–96. DOI: 10.1166/jbn.2024.3367

Rascol, O., Payoux, P., Ory, F., Ferreira, J. J., Brefel-Courbon, C., & Montastruc, J. (2003). Limitations of current Parkinson's disease therapy. *Annals of Neurology*, *53*(S3), S3–S15. DOI: 10.1002/ana.10513

Rauf, A., Badoni, H., Abu-Izneid, T., Olatunde, A., Rahman, M. M., Painuli, S., Semwal, P., Wilairatana, P., & Mubarak, M. S. (2022). Neuroinflammatory markers: Key indicators in the pathology of neurodegenerative diseases. *Molecules (Basel, Switzerland)*, *27*(10), 3194. DOI: 10.3390/molecules27103194

Raza, M., Ahmed, S., & Qureshi, F. (2024). Secondary metabolites in neurodegeneration: Barriers to delivery and opportunities with nanotechnology. *Phytomedicine Advances*, *29*(2), 150–168.

Reynolds, T., & Mahajan, K. (2022). Neurotoxicity and oxidative stress responses associated with nanoparticle exposure in neural tissue. *Toxicology Reports*, *9*, 138–152. DOI: 10.1016/j.toxrep.2022.01.015

Rhee, Y., Ko, J., Chang, M., Yi, S., Kim, D., Kim, C., Shim, J., Jo, A., Kim, B., Lee, H., Lee, S., Suh, W., Park, C., Koh, H., Lee, Y., Lanza, R., Kim, K., & Lee, S. (2011). Protein-based human iPS cells efficiently generate functional dopamine neurons and can treat a rat model of Parkinson disease. *The Journal of Clinical Investigation*, *121*(6), 2326–2335. DOI: 10.1172/jci45794

Ribeiro, F. M. (2023). Understanding brain diseases: From receptor dysregulation to Neurodegeneration,Neuroinflammation and memory impairment. *Current Neuropharmacology*, *21*(2), 162–163. DOI: 10.2174/1570159x2102221212143233

Richmond, A. M., Lyons, K. E., & Pahwa, R. (2023). Safety review of current pharmacotherapies for levodopa-treated patients with Parkinson's disease. *Expert Opinion on Drug Safety*, *22*(7), 563–579. DOI: 10.1080/14740338.2023.2227096

Robinson, J. L., Lee, E. B., Xie, S. X., Rennert, L., Suh, E., Bredenberg, C., Caswell, C., Van Deerlin, V. M., Yan, N., Yousef, A., Hurtig, H. I., Siderowf, A., Grossman, M., McMillan, C. T., Miller, B., Duda, J. E., Irwin, D. J., Wolk, D., Elman, L., & Trojanowski, J. Q. (2018). Neurodegenerative disease concomitant proteinopathies are prevalent, age-related and APOE4-associated. *Brain*, *141*(7), 2181–2193. DOI: 10.1093/brain/awy146

Saitoh, Y., & Takahashi, Y. (2020). Riluzole for the treatment of amyotrophic lateral sclerosis. *Neurodegenerative Disease Management*, *10*(6), 343–355. DOI: 10.2217/nmt-2020-0033

Santana, D. A., Smith, M. D., & Chen, E. S. (2023). Histone modifications in Alzheimer's disease. *Genes*, *14*(2), 347. DOI: 10.3390/genes14020347

Santos, P., & Li, Y. (2024). Machine learning-assisted prediction of nanoparticle BBB permeability in neurotherapeutic design. *Computers in Biology and Medicine*, *169*, 107754. DOI: 10.1016/j.compbiomed.2023.107754

Saraiva, C., Barrios, A., & Silva, G. (2023). Blood–brain barrier transport limitations in neurotherapeutics and emerging nanotechnology solutions. *Brain Research. Brain Research Reviews*, *190*, 104–121. DOI: 10.1016/j.brainresrev.2023.104121

Schmidt, M. F., Gan, Z. Y., Komander, D., & Dewson, G. (2021). Ubiquitin signalling in neurodegeneration: Mechanisms and therapeutic opportunities. *Cell Death and Differentiation*, *28*(2), 570–590. DOI: 10.1038/s41418-020-00706-7

Sehar, U., Rawat, P., Reddy, A. P., Kopel, J., & Reddy, P. H. (2022). Amyloid beta in aging and Alzheimer's disease. *International Journal of Molecular Sciences*, *23*(21), 12924. DOI: 10.3390/ijms232112924

Setter, S. M., Neumiller, J. J., Dobbins, E. K., Wood, L., Clark, J., DuVall, C. A., & Santiago, A. (2009). Treatment of chorea associated with Huntington's disease: Focus on Tetrabenazine. *The Consultant Pharmacist*, *24*(7), 524–537. DOI: 10.4140/tcp.n.2009.524

Shang, D., Huang, M., Wang, B., Yan, X., Wu, Z., & Zhang, X. (2023). MtDNA maintenance and alterations in the pathogenesis of NeurodegenerativeDiseases. *Current Neuropharmacology*, *21*(3), 578–598. DOI: 10.2174/1570159x20666220810114644

Shao, F., Wang, X., Wu, H., Wu, Q., & Zhang, J. (2022). Microglia and Neuroinflammation: Crucial pathological mechanisms in traumatic brain injury-induced Neurodegeneration. *Frontiers in Aging Neuroscience*, *14*. Advance online publication. DOI: 10.3389/fnagi.2022.825086

Sharma, N., & Gupta, M. (2024). Nano-enhanced delivery of phytochemicals for targeting neurodegeneration: Mechanisms and therapeutic outcomes. *Pharmacology & Therapeutics*, *257*, 108–129. DOI: 10.1016/j.pharmthera.2024.108129

Silva, R., & Ortega, A. (2023). Metallic nanoparticles in neurodegenerative disease imaging and therapy: Biopersistence and safety concerns. *Materials Science and Engineering C*, *149*, 117965. DOI: 10.1016/j.msec.2023.117965

Singh, K., & Rao, D. (2022). Machine learning integration in QSAR frameworks for enhanced prediction accuracy. *Journal of Computational Pharmaceutical Research*, *13*(4), 241–259.

Singh, R., Mehra, D., & Kulkarni, P. (2022). Nanoparticles as emerging therapeutic tools for targeted management of neurodegenerative diseases. *Frontiers in Nanomedicine*, *5*(3), 89–108.

Singh, V., & Rao, P. (2022). QSAR-guided nanoparticle design for enhanced BBB permeability. *Journal of Computational Neuropharmacology*, *13*(1), 73–89.

Sircar, E., Rai, S. R., Wilson, M. A., Schlossmacher, M. G., & Sengupta, R. (2021). Neurodegeneration: Impact of S-nitrosylated Parkin, DJ-1 and PINK1 on the pathogenesis of Parkinson's disease. *Archives of Biochemistry and Biophysics*, *704*, 108869. DOI: 10.1016/j.abb.2021.108869

Sladen, P. E., Jovanovic, K., Guarascio, R., Ottaviani, D., Salsbury, G., Novoselova, T., Chapple, J. P., Yu-Wai-Man, P., & Cheetham, M. E. (2022). Modelling autosomal dominant optic atrophy associated with OPA1 variants in ipsc-derived retinal ganglion cells. *Human Molecular Genetics*, *31*(20), 3478–3493. DOI: 10.1093/hmg/ddac128

Srivastava, P., & Pal, D. (2023). Solid lipid nanoparticles and nanostructured lipid carriers for CNS drug delivery: Stability, release, and clinical potential. *Critical Reviews in Therapeutic Drug Carrier Systems*, *40*(2), 145–170. DOI: 10.1615/CritRevTherDrugCarrierSyst.2023047335

Stathis, P., Smpiliris, M., Konitsiotis, S., & Mitsikostas, D. D. (2012). Nocebo as a potential confounding factor in clinical trials for Parkinson's disease treatment: A meta-analysis. *European Journal of Neurology*, *20*(3), 527–533. DOI: 10.1111/ene.12014

Streit, W. J., Rotter, J., Winter, K., Müller, W., Khoshbouei, H., & Bechmann, I. (2022). Droplet degeneration of hippocampal and cortical neurons signifies the beginning of neuritic plaque formation. *Journal of Alzheimer's Disease*, *85*(4), 1701–1720. DOI: 10.3233/jad-215334

Tanaka, H., Lee, S., Martinez-Valbuena, I., Couto, B., Tartaglia, M. C., De Gordoa, J. S., Erro, M. E., Lang, A. E., Forrest, S. L., & Kovacs, G. G. (2024). Ageing-related tau astrogliopathy severely affecting the substantia nigra. *Neuropathology and Applied Neurobiology*, *50*(4). Advance online publication. DOI: 10.1111/nan.13000

Tazwar, M., Evia, A. M., Ridwan, A. R., Leurgans, S. E., Bennett, D. A., Schneider, J. A., & Arfanakis, K. (2024). Limbic-predominant age-related TDP-43 encephalopathy neuropathological change (LATE-NC) is associated with abnormalities in white matter structural integrity and connectivity: An ex-vivo diffusion MRI and pathology investigation. *Neurobiology of Aging*, *140*, 81–92. DOI: 10.1016/j.neurobiolaging.2024.04.002

Teleanu, D. M., Niculescu, A., Lungu, I. I., Radu, C. I., Vladâcenco, O., Roza, E., Costăchescu, B., Grumezescu, A. M., & Teleanu, R. I. (2022). An overview of oxidative stress, Neuroinflammation, and neurodegenerative diseases. *International Journal of Molecular Sciences*, *23*(11), 5938. DOI: 10.3390/ijms23115938

Terkelsen, M. H., Klaestrup, I. H., Hvingelby, V., Lauritsen, J., Pavese, N., & Romero-Ramos, M. (2022). Neuroinflammation and immune changes in prodromal Parkinson's disease and other Synucleinopathies. *Journal of Parkinson's Disease*, *12*(s1), S149–S163. DOI: 10.3233/jpd-223245

Thompson, J., & Ellis, G. (2021). Virtual screening and drug repurposing: Computational efficiency and therapeutic rediscovery. *Pharmaceutical Innovation Reports*, *16*(3), 198–214.

Thompson, J., & Ellis, R. (2021). Machine learning trends in Phase I and Phase II neurotherapeutic pipelines. *Clinical Drug Development Quarterly*, *18*(4), 251–268.

Todd, A. C., & Hardingham, G. E. (2020). The regulation of astrocytic glutamate transporters in health and neurodegenerative diseases. *International Journal of Molecular Sciences*, *21*(24), 9607. DOI: 10.3390/ijms21249607

Unti, E., Mazzucchi, S., Palermo, G., Bonuccelli, U., & Ceravolo, R. (2016). Antipsychotic drugs in Huntington's disease. *Expert Review of Neurotherapeutics*, *17*(3), 227–237. DOI: 10.1080/14737175.2016.1226134

Vadlamani, N., Ibrahimli, S., Khan, F. A., Castillo, J. A., Amaravadi, K. S., Nalisetty, P., & Khan, S. (2024). Efficacy and safety of Tetrabenazine in reducing chorea and improving motor function in individuals with Huntington's disease: A systematic review. *Cureus*. Advance online publication. DOI: 10.7759/cureus.71476

Van Oosterom, N., & Theodoros, T. (2019). Paliperidone long-acting injections in Huntington's disease for motor and behavioural disturbances. *Clinical Drug Investigation*, *39*(4), 407–410. DOI: 10.1007/s40261-019-00759-8

Videnovic, A. (2013). Treatment of Huntington disease. *Current Treatment Options in Neurology*, *15*(4), 424–438. DOI: 10.1007/s11940-013-0219-8

Vishwas, A., Patil, S., & Rao, K. (2024). Enhancing bioavailability of plant-derived secondary metabolites through nanocarriers. *Pharmacognosy Insights*, *17*(2), 88–104. DOI: 10.1177/phi.2024.17208

Vishwas, S., Gulati, M., Khursheed, R., Arya, K., Singh, S. K., Jha, N. K., Prasher, P., Kumar, D., & Kumar, V. (2022). Advanced drug delivery systems to treat Huntington's disease: Challenges and opportunities. Drug Delivery Systems for Metabolic Disorders, 189-206. https://doi.org/ DOI: 10.1016/b978-0-323-99616-7.00005-0

Wal, A., Wal, P., Vig, H., Jain, N. K., Rathore, S., Krishnan, K., & Srivastava, A. (2023). Treatment of Parkinson's disease: Current treatments and RecentTherapeutic developments. *Current Drug Discovery Technologies*, *20*(5). Advance online publication. DOI: 10.2174/1570163820666230512100340

Wallert, E. D., Van de Giessen, E., Knol, R. J., Beudel, M., De Bie, R. M., & Booij, J. (2022). Imaging Dopaminergic Neurotransmission in neurodegenerative disorders. *Journal of Nuclear Medicine*, *63*(Supplement 1), 27S–32S. DOI: 10.2967/jnumed.121.263197

Wang, J., Wang, F., Mai, D., & Qu, S. (2020). Molecular mechanisms of glutamate toxicity in Parkinson's disease. *Frontiers in Neuroscience*, *14*. Advance online publication. DOI: 10.3389/fnins.2020.585584

Wang, Z., Bai, Z., Qin, X., & Cheng, Y. (2019). Aberrations in oxidative stress markers in amyotrophic lateral sclerosis: A systematic review and meta-analysis. *Oxidative Medicine and Cellular Longevity*, *2019*, 1–9. DOI: 10.1155/2019/1712323

Wechsler, M. E., Vela Ramirez, J. E., & Peppas, N. A. (2019). 110th Anniversary: Nanoparticle mediated drug delivery for the treatment of Alzheimer's disease: Crossing the blood–brain barrier. *Industrial & Engineering Chemistry Research*, *58*(33), 15079–15087. DOI: 10.1021/acs.iecr.9b02196

Weuve, J., Proust-Lima, C., Power, M. C., Gross, A. L., Hofer, S. M., Thiébaut, R., Chêne, G., Glymour, M. M., & Dufouil, C. (2015). Guidelines for reporting methodological challenges and evaluating potential bias in dementia research. *Alzheimer's & Dementia*, *11*(9), 1098–1109. DOI: 10.1016/j.jalz.2015.06.1885

Wickstead, E. S., Irving, M. A., Getting, S. J., & McArthur, S. (2021). Exploiting formyl peptide receptor 2 to promote microglial resolution: A new approach to Alzheimer's disease treatment. *The FEBS Journal*, *289*(7), 1801–1822. DOI: 10.1111/febs.15861

Yang, Y., Wang, Y., Jiang, X., Mi, J., Ge, D., Tong, Y., & Zhu, Y. (2024). Modified CE/Zr-MOF nanoparticles loaded with curcumin for Alzheimer's disease via Multifunctional modulation. *International Journal of Nanomedicine*, *19*, 9943–9959. DOI: 10.2147/ijn.s479242

Ye, H., Robak, L. A., Yu, M., Cykowski, M., & Shulman, J. M. (2023). Genetics and pathogenesis of Parkinson's syndrome. *Annual Review of Pathology: Mechanisms of Disease*, *18*(1), 95–121. DOI: 10.1146/annurev-pathmechdis-031521-034145

Yuan, J., Lee, S., & Park, D. (2022). Nanoparticle-mediated brain targeting strategies in neurodegenerative disease therapy. *Journal of Controlled Release*, *345*, 234–251. DOI: 10.1016/j.jconrel.2022.03.015

Zampatti, S., Peconi, C., Campopiano, R., Gambardella, S., Caltagirone, C., & Giardina, E. (2022). C9orf72-related neurodegenerative diseases: From clinical diagnosis to therapeutic strategies. *Frontiers in Aging Neuroscience*, *14*. Advance online publication. DOI: 10.3389/fnagi.2022.907122

Zhang, W., Xiao, D., Mao, Q., & Xia, H. (2023). Role of neuroinflammation in neurodegeneration development. *Signal Transduction and Targeted Therapy*, *8*(1). Advance online publication. DOI: 10.1038/s41392-023-01486-5

Zhao, A., Xu, W., Han, R., Wei, J., Yu, Q., Wang, M., Li, H., Li, M., & Chi, G. (2024). Role of histone modifications in neurogenesis and neurodegenerative disease development. *Ageing Research Reviews*, *98*, 102324. DOI: 10.1016/j.arr.2024.102324

Zhu, L., Li, Y., Yu, X., Chen, Y., Zhang, J., Pang, C., Xie, J., Gao, L., Du, L., Cao, W., Fan, D., Cui, C., Yu, H., & Deng, B. (2024). Fighting amyotrophic lateral sclerosis by protecting the liver? A prospective cohort study. *Annals of Neurology*. Advance online publication. DOI: 10.1002/ana.27115

Chapter 2
Mechanisms of Neurodegeneration and the Neuroprotective Role of Secondary Metabolites

Rathi Priya Bollu Yashwant
https://orcid.org/0009-0003-2526-9903
Taylor's University, Malaysia

Eng Hwa Wong
https://orcid.org/0000-0001-9322-772X
Taylor's University, Malaysia

ABSTRACT

Neurodegenerative diseases (NDDs) are age-related disorders and their prevalence is rising globally due to the increase in aging population. NDDs comprise a wide range of diseases that have specific etiological features affecting different regions of the brain and manifesting various types of clinical symptoms. However, their pathophysiology shares common genotype-phenotype relationships displaying pathological features like protein misfolding and aggregation of toxic proteins, mitochondrial dysfunction, oxidative stress, neuroinflammation, synaptic injury, and neuronal death. Scientific scrutiny of these mechanisms can shed light on identifying the biological markers (to diagnose the NDDs type) and novel therapeutic targets (to develop a holistic approach to treat NDDs).Traditional systems of medicine and pharmacognosy studies have demonstrated that the pleiotropic efficacy of phytochemicals against NDDs. The Authors have focussed on the main pathological mechanisms of neurodegeneration, and benefits of phytochemicals as effective and safer therapies.

I. INTRODUCTION

NDDs comprise a diverse group of neurological disorders affecting the lives of millions of people globally (Kelser et al., 2024). NDDs are commonly characterized by the progressive loss

DOI: 10.4018/979-8-3373-1365-8.ch002

of neurons, either structurally or functionally, in the nervous system and the peripheral organs (Kovacs, 2018). The degeneration of neural structure and function, combined with the limited regenerative capacity of neurons, due to their terminally differentiated state, disrupts the synaptic activity and neural circuits, leading to deficits in memory, cognition, personality, behavior, sensory perception, and/or motor functions (Wilson et al., 2023). The most common NDDs include Alzheimer's Disease (AD), Parkinson's Disease (PD), Huntington's Disease (HD), Multiple Sclerosis (MS), and Amyotrophic Lateral Sclerosis (ALS). Although the etiopathogenesis of NDDs is poorly understood, the interaction of genetic, epigenetic, and environmental factors is known to play a significant role in their onset and progression (Armstrong, 2020). The prevalence and incidence of NDDs increase significantly with age. As life expectancy continues to rise in many countries, the number of cases is estimated to increase alarmingly in the coming decades, imposing a significant medical and public health burden globally (Dugger et al., 2017; Sengoku, 2020).

The pathophysiology of NDDs is complex and multifaceted. NDDs are characterized by the accumulation of toxic or insoluble proteins, which exert modified physicochemical properties leading to the gradual loss of neurons (Kovacs, 2018). Indeed, the molecular classification of NDDs is based on the aggregation of toxic proteins, highlighting the important role of impaired proteostasis (i.e., dysfunction in protein-processing systems that govern proper protein folding, function, and degradation) in etiopathogenesis. The most common proteins involved in the pathogenesis of NDDs are amyloid-β (Aβ) and tau in AD, α-synuclein in PD, TAR-DNA-binding protein 43 kDa (TDP-43) in ALS, prion proteins, and fused-in sarcoma protein. In NDDs, manifestations of specific symptoms depend largely on the regions affected in the nervous system, mainly by the accumulation of misfolded proteins intra- or extracellularly, neuroinflammation, and oxidative stress (Peng et al., 2020). Other common pathological features shared by NDDs include mitochondrial dysfunction, altered energy metabolism, deregulation of immune system, glutamate excitotoxicity, axonal, synaptic, and dendritic dysfunction, and neuronal loss and atrophy (Hansson, 2021).

Most FDA-approved therapies for AD predominantly target classical pathological hallmarks such as Aβ plaques or neurofibrillary tangles. However, these monotherapeutic agents offer limited efficacy against the multifactorial nature of AD, which encompasses a complex interplay of genetic, molecular, and environmental factors. Given this intricate etiology and pathology, a multi-targeted therapeutic strategy that can simultaneously modulate diverse disease pathways to achieve meaningful disease modification, enhance neuroprotective efficacy, and ultimately improve clinical outcomes (Ciechanover & Kwon, 2015; Sengoku, 2020). Current therapeutic strategies (such as FDA-approved pharmaceutical drugs) provide mild to moderate symptomatic relief but fail to halt the progression and address the root causes of NDDs, indicating the urgent need for alternative, effective, and safer options to target the complex and multi-modal pathology (Akhtar et al., 2021).

Numerous pharmacognosy studies have proven that phytochemicals found in herbal leaves, roots, rhizomes, fruits, vegetables, whole grains, nuts, oil seeds, legumes, and pulses exert antioxidant, anti-inflammatory, anti-apoptotic, and anti-amyloidogenic properties (Bhat, 2022). Phytochemicals are the secondary metabolites produced by herbal plants to promote their resilience and survival (Bhattacharya et al., 2021). Phytochemicals exhibit pleiotropic neuroprotective effects by concurrently modulating multiple pathological pathways, thereby contributing to the attenuation of neurodegenerative disease progression and offering promise for multi-targeted therapeutic interventions (Singh et al., 2024).

Interestingly, recent studies on herbal plants have demonstrated that phytochemicals such as dietary polyphenols, bioflavonoids, terpenoids, and bioactive alkaloids can be considered safer and more effective therapeutic options to treat NDDs (Shankaran et al., 2024). Bioactive compounds such as curcuminoids, cannabinoids, polyphenols, Ginseng extracts, boswellic acids, and asarones have been used to treat AD, HD, ALS, MS, and PD (Bhattacharya et al., 2021; Awuchi, 2023). Concise literature from traditional systems of medicine denotes that phytochemicals were prescribed either as mono- or poly-herbal concoctions to treat NDDs due to their key abilities to mitigate neurotoxic pathways such as oxidative stress, inflammatory mediators, or aggregation of mutated proteins, and promote protective pathways such as neurotransmitter and neurotropic balance, debris clearance, and regulate the neuroendocrine and immune pathways (Rahaman et al., 2021).

II. PATHOLOGICAL MECHANISMS OF NEURODEGENERATION

Neuropathology of NDDs comprises a wide range of pathological changes, ranging from atrophy of cortical and subcortical regions, accumulation of misfolded proteins into toxic aggregates, impaired proteostasis, dysregulated calcium homeostasis, defective blood-brain barrier (BBB), mitochondrial dysregulation, endoplasmic reticulum stress, neuroinflammation, excitotoxicity, and OS (*Figure 1*). Despite differences in the clinical manifestations of different NDDs, most of the pathological mechanisms are common, which ultimately leads to synaptic dysfunctions or loss, and eventually death of neurons (Kelser et al., 2024).

Figure 1. Hallmarks of Pathology in NDDs

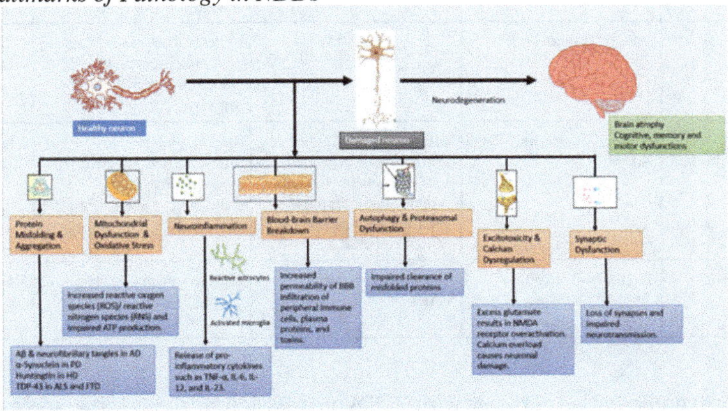

1. Protein Misfolding and Aggregation

The central hallmark of many NDDs is the deposition of misfolded proteins that form toxic aggregates, impairing the cellular functions of neurons and glial cells (Rêgo & Araújo-Filho, 2024; Kelser et al., 2024). Protein misfolding occurs either by (i) template-directed misfolding, where a misfolded protein interacts with a healthy, native protein, causing it to misfold. For example, an in vitro study showed that a mutant version of the superoxide dismutase (SOD) 1 protein induced misfolding in native SOD1 proteins (Grad et al., 2011) or (ii) when the precursor protein is cleaved by an enzyme at a site that leads to the formation of insoluble plaques. For example, proteolytic cleavage of Aβ, a peptide derived from amyloid precursor protein, by β and ϒ secretases (Sehar et al., 2022). Toxic protein aggregates are formed by seeding, where misfolded proteins accumulate and form aggregates. Similar to those of prion-like proteins, pathological deposits of toxic proteins propagate from one neuron or cell to another in a seed-dependent (also known as cross-seeding) and self-templating manner, leading to the accumulation of toxic proteins across the brain regions (Soto & Pritzkow, 2018; Soto & Estrada, 2020). Collectively, these proteinopathies disrupt cellular functions and contribute to neuronal death (Table 1).

Table 1. Classification of neurodegenerative diseases based on pathophysiological processes

NDDs type	Name of the NDDs	Protein involved	Pathological mechanisms	References
Beta-amyloidopathies	AD, Cerebral amyloid angiopathy, PD with dementia, and Dementia with Lewy Bodies.	Aβ aggregates	Induces mitochondrial dysfunction, OS and neuroinflammation.	Chen et al., 2017
Tauopathies	AD, frontotemporal dementia (FTD), Pick disease, corticobasal degeneration, and Progressive Supranuclear Palsy.	Tau	Disrupts axonal transport, leading to synaptic loss and neuronal death.	Chiang & Kordower, 2022
Synucleinopathies	PD, Multiple System Atrophy, and Dementia with Lewy Bodies	alpha-synuclein	Disrupts cellular processes such as vesicle trafficking and mitochondrial function.	Du et al., 2021; Tarutani & Hasegawa, 2019
TDP-43 Proteinopathies	AD, FTD, and ALS	TDP-43	Affects RNA processing and protein homeostasis, which results in cellular toxicity.	Gitler & Tsuiji, 2020
Polyglutamine (PolyQ) Disorders	HD and Spinocerebellar ataxia	Mutant huntingtin proteins with expanded CAG repeats in HD and mutant ataxin proteins in cerebellar neurons in Spinocerebellar ataxia	Impair genetic transcription, proteostasis, and mitochondrial function.	Matos et al., 2023
Prion Diseases	Creutzfeldt-Jakob Disease and Kuru	Misfolded prion protein (PrP)	Triggers neurodegeneration across different brain regions.	Geschwind & Belay, 2024

Two main hypotheses that correlate with NDDs progression are the synaptic spread hypothesis and the selective vulnerability hypothesis (Kelser et al., 2024).

(i) The synaptic spread hypothesis suggests that pathogenic proteins like α-synuclein and tau propagate across the synapses and contribute to disease progression. Misfolded proteins further induce misfolding in naive healthy proteins, leading to a cascade of protein aggregation, which contribute to the widespread neuronal dysfunction and death.

(ii) The selective vulnerability hypothesis emphasizes that certain neurons are inherently more susceptible to degeneration due to metabolic stress, genetic predisposition, or protein expression levels. These types of intrinsic properties determine the susceptibility of neurons to neurodegenerative processes, specifically making these neurons more vulnerable to aggregation and toxicity.

These hypotheses provide complementary perspectives on the complex mechanisms underlying NDDs, each explaining the process of disease progression. Consequently, the accumulation of these protein aggregates (Soto & Pritzkow, 2018; Soto & Estrada, 2020) triggers a cascade of pathological events, including:

- Disruption of cellular homeostasis, such as impaired mitochondrial bioenergetics, excessive release of reactive oxygen species (ROS), reactive nitrogen species (RNS), and disrupted calcium homeostasis, leads to OS, endoplasmic reticulum stress, and neuronal damage (Verma et al., 2022).
- Impairment of protein degradation pathways, such as the ubiquitin-proteasome system (UPS) and autophagy-lysosomal pathways, leads to further accumulation of toxic aggregates, exacerbating the neuronal damage (Hegde et al., 2022).
- Activation of neuroinflammatory pathways via overstimulation of microglia and astrocytes leads to the release of pro-inflammatory cytokines such as tumor necrosis factor (TNF)-α and interleukin (IL)-1β, exacerbating neuronal dysfunction, thereby contributes to synaptic dysfunction or synaptic loss (Zhang et al., 2023).

These pathological changes ultimately lead to neurodegeneration, contributing to cognitive decline and motor deficits seen in various NDDs. Table 2 exemplifies the types of proteins and aggregates formed in the common types of NDDs (Rêgo & Araújo-Filho, 2024).

Table 2. Key hallmark proteins associated with NDDs

NDDs	Types of proteins and aggregates	Impairment of cellular functions
FTD	Misfolded TDP-43 aggregates	Induces neurotoxicity
AD	Extracellular insoluble Aβ proteins aggregates to form senile plaques. Hyperphosphorylations of tau protein leads to the formation of intracellular neurofibrillary tangles.	Triggers neurotoxicity, synaptic dysfunction, and inflammation.
PD	Misfolded alpha-synuclein aggregates to form Lewy bodies.	Disrupts normal neuronal function, contributing to the loss of dopaminergic neurons in the substantia nigra.
ALS	Misfolded TDP-43 and SOD1 aggregates	Disrupts RNA metabolism and mitochondrial function.
HD	Mutant huntingtin protein composed of expanded CAG repeats	Causes neuronal toxicity.

2. Mitochondrial Dysfunction and Oxidative Stress (OS)

Mitochondria play a vital role in energy production and neuronal survival. Mitochondrial dysfunction is a critical factor in neurodegeneration (Mahar et al., 2025), particularly in PD, where it induces the death of dopaminergic neurons in the striatum. Dysfunctional mitochondria contribute to neurodegeneration mainly via:

- OS, marked by increased levels of reactive oxygen species (ROS) and reactive nitrogen species (RNS), which leads to lipid peroxidation, increased mitochondrial fragmentation, protein oxidation and aggregation, excitotoxicity, and DNA and RNA damage, resulting in proteotoxic stress and apoptosis (Mayer et al., 2024)
- Impaired bioenergetics are marked by energy deficits (reduced availability of ATP), which aggravate neuronal death (Strope et al., 2022).
- Defective mitophagy is defined as the impaired clearance of damaged mitochondria, which results in the deposition of dysfunctional organelles, thus exacerbating neuronal damage and cellular degeneration (Wang et al., 2022).

3. Glutamate Excitotoxicity and Calcium Dysregulation

Glutamate excitotoxicity and calcium dysregulation are interlinked pathological processes that synergistically drive neuronal injury in NDDs (Kaur et al., 2019). Excessive release or impaired reuptake of glutamate by astrocytes overstimulates glutamate receptors such as NMDA and AMPA receptors in post-synaptic neurons, which in turn causes an influx of calcium and sodium ions, resulting in neuronal depolarization and excitotoxicity (Verma et al., 2022). Consequently, calcium overload disrupts the cellular homeostasis by overwhelming the mitochondrial buffering capacity, leading to the disruption of ATP production, exacerbating ROS release, endoplasmic reticulum stress, and unfolded protein response, leading to the activation of apoptotic and necrotic pathways, which ultimately results in neuronal apoptosis (Baev et al., 2022). Calcium dyshomeostasis impairs synaptic function, leading to disrupted neurotransmission and compromised neuroplasticity, which collectively contribute to progressive neuronal loss (Verma et al., 2022). These findings suggest that glutamate toxicity and calcium overload disrupt mitochondrial and endoplasmic reticulum homeostasis, leading to oxidative, synaptic, and neuronal damage.

4. Neuroinflammation and Immune Dysfunction

Neuroinflammation, a common feature in NDDs, is driven by the chronic activation of glial cells (microglia and astrocytes) and BBB endothelial cells in the brain tissue (Rêgo & Araújo-Filho, 2024). Overactivation of microglia by toxic protein deposits leads to excessive release of pro-inflammatory cytokines such as IL-1β, IL-6, and TNF-α, and increased levels of ROS and RNS, which exacerbate neurotoxicity and oxidative injury, respectively (Mayer et al., 2024). Dysfunctional microglia fail to provide neurotrophic support, clearance of toxic proteins (e.g., Aβ, α-synuclein), and synaptic pruning, leading to neuronal vulnerability, neuronal toxicity, and defective synaptic plasticity, respectively (Singh et al., 2022). Additionally, impaired microglia can spread toxic protein aggregates and inflammatory signals to neighbouring neurons, thus amplifying neurodegeneration across brain regions (Peng et al., 2020). Impaired astrocytes fail to provide critical support to neurons, including metabolic and structural support, ion homeostasis, neurotransmitter regulation, and synaptic regulation, exacerbating the oxidative injury, glutamate excitotoxicity, and neuroinflammation (Kaur et al., 2019). Moreover, excessive release of pro-inflammatory cytokines compromises the integrity of the BBB, allowing peripheral immune cells, plasma proteins, and toxins to infiltrate the brain parenchyma, further exacerbating the release of cytokines and neuronal injury (Zhang et al., 2023), leading to the formation of a vicious self-sustaining inflammatory loop. Collectively, chronic neuroinflammation further amplifies the neuronal damage, synaptic dysfunction, cell death, and disease progression in NDDs.

5. Cellular Degradation Pathways

Dysregulation of cellular degradation pathways such as UPS and autophagy-lysosomal pathways leads to the buildup of aberrant protein aggregates (Rêgo & Araújo-Filho, 2024). Dysregulation of the UPS, a major protein degradation system that regulates apoptosis by controlling the turnover of pro-apoptotic and anti-apoptotic proteins, including p53, Bax, and Bcl-2 (Ciechanover & Kwon, 2015), contributes to NDDs, including ALS and PD, by promoting protein aggregation and neuronal apoptosis. Particularly, inefficient autophagic clearance of toxic aggregates like Aβ (AD), α-synuclein (PD), and mutant huntingtin (HD) leads to OS in neurons, resulting in cellular toxicity and death (Hegde et al., 2022). In AD, PD, and ALS, impairment in nucleocytoplasmic transport of the UPS prevents proper clearance of neurotoxic proteins, thereby leading to their deposition and formation of toxic aggregates (e.g., Lewy bodies in PD and neurofibrillary tangles in AD) (Mahar et al., 2025). Impaired mitophagy leads to the accumulation of dysfunctional mitochondria, increased ROS production, and neuronal energy deficits (Yang et al., 2024). Mutations in mitophagy-related genes such as PINK1 and Parkin are associated with PD. Impairment of lysosomes disrupts the clearance of toxic deposits, contributing to neuronal damage. Dysfunctional chaperone-mediated autophagy leads to the deposition of aggregation-prone proteins (e.g., α-synuclein in PD and tau in AD) (Takahashi et al., 2022). Briefly, impaired UPS and autological pathways lead to a downward spiral of neurodegeneration by improper transport, clearance, and build-up of toxic proteins.

6. Vascular Dysfunction and BBB Disruption

In NDDs, disruption of the BBB leads to neurovascular dysfunction, contributing to chronic neuroinflammation, oxidative injury, and neuronal damage (Hang et al., 2023). The main pathological processes of vascular dysfunction include reduced cerebral blood flow (CBF), endothelial dysfunction, and amyloid angiopathy. Endothelial damage disrupts vascular tone, causing poor perfusion and inflammation, while Aβ deposits in vessel walls, as seen in AD, heighten the risk of micro-hemorrhage and ischemia. Reduced CBF functionally deprives neurons of essential oxygen and glucose, leading to chronic hypoxia, energy failure, and neuronal death (Korte et al., 2020). Thus, defective waste clearance via the glymphatic system, toxic protein accumulation, and neuroinflammation exacerbate neuronal damage.

Breakdown of tight junctions, astrocyte and pericyte dysfunction, and upregulation of inflammatory mediators such as TNF-α and IL-1β, collectively disrupt the structural and functional integrity of the BBB, facilitating the infiltration of neurotoxins, peripheral immune cells, and pathogenic agents, which results in edema, chronic neuroinflammation, and impaired nutrient transport (Zhang et al., 2023). Vascular and BBB dysfunctions accelerate Aβ and tau pathology in AD (Kurz et al., 2022), while reduced CBF and endothelial abnormalities contribute to the death of dopaminergic neurons in PD (Al-Bachari et al., 2020) and to motor neuron degeneration in ALS (Kakaroubas et al., 2019). Similarly, defective BBB plays a crucial role in the pathogenesis of MS (Nishihara et al., 2022). A bidirectional vicious cycle links vascular pathology to neurodegeneration, mainly hypoperfusion induces hypoxia or oxidative stress, while proteinopathies

(Aβ, tau, etc.) deposit in blood vessels, causing cerebral amyloid angiopathy and microinfarcts in various NDDs.

7. Genetic, Epigenetic, and Environmental Contributions

A combination of genetic predisposition, epigenetic modifications, and environmental factors contributes to the onset of NDDs. Both genetic mutations and epigenetic alterations play significant roles in neurodegeneration (Zhao et al., 2024a). Aging remains the greatest risk factor. Genetic factors such as inherited mutations directly contribute to the onset and progression of NDDs (Table 3). The genetic risk factors include gene mutations such as APOE4 in AD, GBA in PD, and HLA-DRB1 and IL7R in MS that increase disease susceptibility (Wang et al., 2023). Epigenetic changes such as DNA methylation, histone alterations such as changes in acetylation and methylation, and non-coding RNAs such as microRNAs and long non-coding RNAs alter gene expression and can promote vulnerability to the NDDs (Ghosh & Sadaat, 2023; Zhao et al., 2024a). Environmental factors, including unhealthy lifestyle (e.g., poor diet, physical inactivity, smoking, and alcohol consumption) and toxin exposure (e.g., pesticides, heavy metals, and neurotoxins), can also drive epigenetic changes, significantly contributing to the onset and progression of NDDs (Armstrong, 2020; Nabi & Tabassum, 2022). These factors together highlight the complex interplay between genetic susceptibility and external influences in neurodegeneration.

Table 3. Classification of NDDs by associated genes, mechanistic pathways, and supporting references

Type of NDDs	Genes linked with the NDDs	Mechanism of action	References
AD	amyloid precursor protein, presenilin 1, and presenilin 2	inducing the formation and aggregation of insoluble Aβ proteins	Wong et al., 2020
PD	α-synuclein, PINK1, LRRK2, UCH-L1, DJ-1, ATP13A2, and OMI/HTRA2	disrupting the protein homeostasis and mitochondrial function	Tembhurnikar et al., 2021
HD	CAG repeat expansions of the huntingtin gene	producing an over-sized toxic mutant huntingtin protein, which disrupts genetic transcriptions	Langbehn et al., 2019
ALS	SOD1, C9ORF72, and TARDBP	contributes to motor neuron degeneration	Dash et al., 2022

III. PROGRESSION OF NDDS

The progression of NDDs is characterized by a gradual and progressive reduction of neurons due to collective molecular and cellular dysfunctions. A key aspect of neurodegeneration involves the deposition of neurotoxic protein aggregates (Mahar et al., 2025; Takahashi et al., 2022; El-Mansoury et al., 2023), which disrupt cellular homeostasis, impair proteostasis, and BBB integrity and trigger neuroinflammation, vascular dysfunction, and OS (Verma et al., 2022; Zhang et al., 2023), eventually leading to neuronal dysfunction and degeneration. As the disease advances, synaptic dysfunctions, impaired axonal transport, and impaired DNA repair mechanisms lead to neuronal death in key brain regions, leading to the progressive deterioration of cognitive, motor, and behavioral functions (Smith & Doe, 2025). The interaction of genetic predisposition, epigenetic modifications, and environmental factors accelerates disease progression, ultimately resulting in the disruption of neural networks across brain regions and worsening the clinical symptoms (Figure 2).

Figure 2. Disease progression: A vicious cycle of pathology in in NDDs

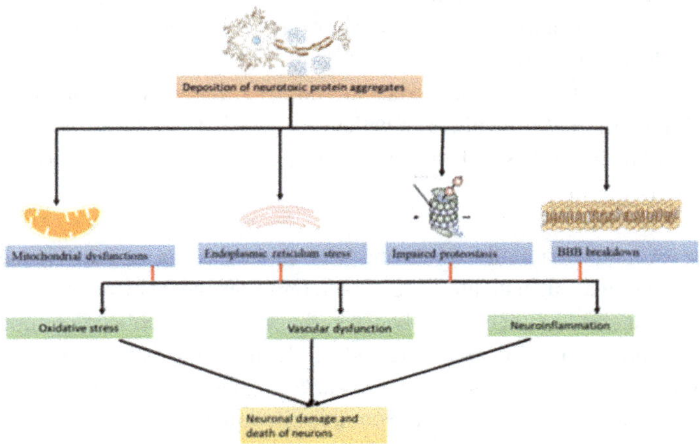

IV. CURRENT CHALLENGES IN IDENTIFYING THE DMTS FOR THE MANAGEMENT AND TREATMENT OF NDDS

Identifying DMTs for NDDs presents significant challenges due to the complex nature of these disorders, including:

- Complex etiologies: NDDs have multifactorial etiologies comprising the interplay of genetic, epigenetic, and environmental factors, complicating the development of targeted therapies (Wilson et al., 2023).
- Heterogeneity of pathologies: NDDs involve diverse pathologies, as mentioned in Figure 1. The 'mono-therapeutic, mono-target, monophasic' approach cannot be applied to complex diseases such as NDDs. Although the pathological mechanisms are well-documented, the interaction between them and their collective impact on disease progression remains less understood (Armstrong, 2020; Dugger et al., 2017).
- Need for Biomarkers: Reliable biomarkers for early detection, predicting disease progression, and measuring therapeutic efficacy are still lacking, making it difficult to intervene before irreversible damage (Hansson, 2021).
- Limitations of current conventional therapies: Approved therapeutics (e.g., marketed drugs) focus mainly on symptomatic relief rather than halting disease progression or re-

versing neurodegeneration, and these drugs are associated with a range of adverse effects (Cheong et al., 2022; Singh et al., 2024).

- Need for predictive preclinical models: Current animal and cellular models often fail to fully replicate human NDD pathology, limiting translational success to human trials (Akhtar et al., 2022).
- BBB limitations: Many therapeutic compounds fail to penetrate the BBB, reducing their efficacy in the CNS. Effective innovative (drug delivery) strategies, such as bioactive peptides and nanotechnology strategies, show promising results in addressing these challenges (Akhtar et al., 2021).
- Failure of clinical trials: Clinical trials are hindered by lengthy timelines, high placebo effects, high cost, and difficulties in patient recruitment. Many promising drugs against AD in preclinical studies fail in human trials due to poor efficacy or adverse effects (Yiannopoulou et al., 2019).

The overlap of molecular pathways among NDDs underscores the importance of a holistic, systems-level approach to understanding these interconnected mechanisms (Cheong et al., 2022). Therefore, understanding the interaction between them and their collective impact on disease progression is essential to develop multi-targeted DMTs that could address multiple neurodegenerative conditions (Goyal et al., 2024; Hannan et al., 2020). Disease progression and responses to therapy across patients often vary due to the complex interplay of genetic and epigenetic factors (Sanghai & Tranmer, 2024). Thus, the concept of a 'multi-drugs, multi-targets, multiphasic' approach that targets or modifies the multiple pathological pathways simultaneously is needed to combat multifaceted nature of NDDs (Bredesen et al., 2023). Specifically, this multitarget strategy holds promise in addressing the intricate layers of these conditions and offers the potential to advance treatments for such complex NDDs (Kelser et al., 2024; Cardona, 2024) (Table 4).

Table 4. Emerging therapeutic targets in NDDs

Therapeutic targets	Therapeutics	References
Targeting protein aggregation	Anti-amyloid and anti-tau therapies (AD), alpha-synuclein inhibitors (PD), and antisense oligonucleotides (HD and ALS)	Wang et al., 2023
Maintaining mitochondrial homeostasis	Antioxidants (e.g., coenzyme Q10, vitamin E, and N-acetylcysteine) and metabolic enhancers that can restore mitochondrial function (mitochondria-targeted antioxidant such as MitoQ, compounds improving mitochondrial biogenesis such as peroxisome Proliferator-Activated Receptor Gamma Coactivator-1α (PGC-1α) activators).	Grewal et al., 2021; Yang et al., 2024
Anti-inflammatory effects	Modulating microglial activation and cytokine signaling that leads to the reduction of neuroinflammation.	Goyal et al., 2024; Barbalho et al., 2025
Anti-oxidant effects	Reducing the excessive release of ROS and RNS, scavenging free radicals, and enhancing anti-oxidant defense systems by increasing the levels of anti-oxidant enzymes.	Shankaran et al., 2024
Neuroprotective effects	Enhancing autophagy, improving synaptic plasticity, and promoting neurogenesis.	Li et al., 2025; Wang 2021

V. PHYTOCHEMICAL INTERVENTION IN NDDS

Phytochemicals refer to a wide range of secondary metabolites produced by various plants to provide defense against herbivores, insects, microbes, ultraviolet rays, and toxins (Basyal et al., 2021). Phytochemicals are organic bioactive compounds that contribute to the survival, overall health, and resilience of plants (Bhat, 2022). Phytochemicals are commonly found in edible plants, such as fruits, vegetables, seeds, nuts, and cereals, as well as in herbal plants like *Curcuma longa, Ginkgo biloba, Camellia sinensis, Panax ginseng, Withania somnifera, Bacopa monnieri,* and *Centella asiatica* (Vyas et al., 2019).

Traditional systems of medicine have rich clinical data on the use of herbs against neurological disorders and NDDs. Phytochemicals are the primary bioactive compounds in these herbs, exerting various pharmacological activities, including antioxidant, anti-inflammatory, anticancer, antimicrobial, anti-apoptotic, and direct neuroprotective effects (Vyas et al., 2019). Pharmacognosy studies on herbs have demonstrated that "phytochemicals" can be potential alternative therapeutic candidates against NDDs such as AD, PD, HD, and ALS due to their multifaceted mechanisms of action and relatively lower toxicity profiles (Akhtar et al., 2024; Madhubala et al., 2024).

Some studies explore individual phytochemicals, while others emphasize herbal extracts for their multi-target effects on NDDs (Sarkar et al., 2024). The most promising phytochemicals for treating NDDs include polyphenols, terpenoids, flavonoids, alkaloids, steroids, carotenoids, phytosterols, phenols, saponins, quinones, resins, oleoresins, stilbenes, lactones, volatile oils, essential oils, essential fatty acids, tannins, etc (Kamboj et al., 2024). These compounds exhibit strong antioxidant, anti-inflammatory, anti-aggregation, and mitochondrial protection properties, which help to reduce OS, neuronal dysfunction, prevent neuronal damage, enhance neuroprotection, and modulate inflammatory responses, thereby potentially alleviating cognitive and motor impairments associated with conditions like AD and PD (Rahaman et al., 2024) (Table 5).

Table 5. Key classes of phytochemicals implicated in neuroprotection

Class of Phytochemicals	Sources	Mechanism of Action
Flavonoids	*Ginkgo biloba, Camellia sinensis, and Bacopa monnieri*	Antioxidant properties
Polyphenols	*Camellia sinensis, Theobroma cacao* and berries (*Vaccinium corymbosum*)	Anti-inflammatory properties
Alkaloids	*Huperzia serrata, Withania somnifera,* and *Acorus calamus.*	Neuroprotection and cognitive benefits.
Phytosterols	*Curcuma longa, Vitis vinifera, Rosmarinus officinalis, and Ginkgo biloba*	Neuroprotection
Terpenoids	*Ginkgo biloba, Panax ginseng, Withania somnifera, Eleutherococcus senticosus, Nardostachys jatamansi,* and *Centella asiatica.*	Anti-aggregation and anti-neuroinflammatory properties.
Flavones	*Glycyrrhiza glabra, Bacopa monnieri, Petroselinum crispum, and Thymus vulgaris*	Improve cognitive and memory deficits.

1. Mechanisms of Neurodegeneration and Phytochemicals Intervention

Given the complexity and multitude of pathological mechanisms, multi-targeted approaches like phytochemical interventions are gaining much attention. The most promising phytochemicals such as quercetin, curcumin, epigallocatechin gallate (EGCG), apigenin, and cannabinoids demonstrate significant antioxidant, anti-inflammatory, and neuroprotective activities (Jee et al., 2024; Rahaman et al., 2024). In particular, ginsenosides from *Panax ginseng*, curcumin from *Curcuma longa*, resveratrol from grapes, wine and grape Juices, and quercetin from apples, berries, and onions are found to have significant anti-oxidant, anti-inflammatory, mitochondrial protective, anti-apoptotic, and anti-aggregation effects, thus protecting the neurons from synaptic injury, OS and energy deficits (Bayazid & Lim, 2022; Kamboj et al., 2024). These compounds may help mitigate neuronal damage by regulating oxidative pathways and reducing inflammation. These findings indicate that multi-targeted mechanisms of phytochemicals fulfil the 'multi-phasic' criteria of the potential alternative therapies, thus making them valuable candidates for developing safer and more effective therapeutic strategies against neurodegeneration (Table 6).

Table 6. Example lists of phytochemicals with their activities and mechanisms of action

Examples of Phytochemicals	Activities	Mechanism of Action
Curcumin and resveratrol	Antioxidant activity	By neutralizing the free radicals, and reducing the production of ROS and RNS.
Curcumin, ginsenosides and quercetin	Anti-inflammatory effects	By reducing the neuroinflammation through mitigating the release of pro-inflammatory cytokines.
EGCG, quercetin, resveratrol, and curcumin	Anti-aggregation activity	By disrupting the aggregation of pathological protein aggregates (e.g. amyloid-β, tau, and α-synuclein), which are the hallmarks of AD and PD.
EGCG and ginsenosides	Modulation of neurotransmitters	By enhancing the availability and function of neurotransmitters, improving neuronal communication, and preventing synaptic dysfunctions.
EGCG, quercetin, apigenin, and hesperidin.	Modulation of neurotropic factors	By stimulating the production of neurotrophic factors, protect neurons, and enhance their survival.
Resveratrol	Memory-boosting and cognitive-enhancing properties.	By promoting neuron survival and synaptic plasticity.
Huperzine A, galantamine, and curcumin	Cholinesterase inhibition	By inhibiting cholinesterase and promoting acetylcholine levels, as acetylcholine is essential for effective cognitive functions and memory.
Sulforaphane, berberine and resveratrol	Autophagy enhancement	By promoting the autophagy, phytochemicals clear the cellular debris and prevent the progression of neuroinflammation.
Coenzyme Q10 and resveratrol	Mitochondrial protection	By improving mitochondrial function, producing ATP, and regulating energy metabolism.
Sulforaphane and resveratrol	Epigenetic Modulation	Promote the expression of neuroprotective genes via histone acetylation or DNA methylation.

The following sections illustrates some of the primary phytochemicals with potent neuroprotective effects targeting multiple pathways against neurodegeneration.

1.1. Phytochemicals With Antioxidant Effects

Phytochemicals such as polyphenols (e.g., resveratrol, curcuminoids, phenolic acids), flavonoids (e.g., hesperidin, quercetin, EGCG, and kaempferol), terpenoids (e.g., carotenoids (lycopene) and ginkgolides), and alkaloids (berberine and caffeine) derived from various herbal plants exhibit strong antioxidant properties by scavenging ROS and enhancing endogenous antioxidant enzymes like SOD and catalase (Shankaran et al., 2024). For instance, polyphenols mitigate OS and inflammation, contributing to neuroprotection (Zhu et al., 2024). Quercetin and EGCG from green tea exhibit strong free radical-scavenging properties, protecting neurons from oxidative damage in AD and PD (Calis et al., 2020). Quercetin enhances SOD and catalase activity (Grewal et al., 2021). Curcuminoids, specifically curcumin, a bioactive compound in turmeric, inhibit lipid peroxidation, enhance glutathione levels, and reduce OS and neuroinflammation associated with NDDs (Geschwind et al., 2024; Madhubala et al., 2024). Phenolic acids such as caffeic acid and ferulic acid, present in coffee and whole grains, respectively, prevent oxidative damage and support neuronal survival in PD and HD (Turan et al., 2020; Kaur et al., 2018). Ginkgolides improve CBF and reduce ROS accumulation, playing a neuroprotective role in PD (Singh et al., 2019). Berberine inhibits acetylcholinesterase and monoamine oxidase, decreasing ROS production and thus mitigating cognitive decline in AD patients (Jee et al., 2024). Lycopene (Tripathi et al., 2020) and β-carotene (Sun et al., 2024) enhance antioxidant defenses and prevent neuroinflammation-induced damage in ALS. Additionally, pterostilbene from blueberries improves cognitive function in HD (Qu et al., 2023), while huperzine A from *Huperzia serrata* provides neuroprotection in AD by scavenging ROS and improving synaptic function (Dang et al., 2023). Collectively, these phytocompounds mitigate the progression of NDDs by reducing OS and enhancing neuronal resilience.

1.2. Phytochemicals With Anti-Inflammatory Effects

Phytochemicals play a crucial role in reducing neuroinflammation, a key driver of neuronal death in NDDs (Medrano-Jiménez et al., 2022), mainly by inhibiting pro-inflammatory mediators such as TNF-α, IL-6, and Nuclear Factor Kappa B (NF-κB) (Saleh et al., 2021; Yadav et al., 2024). Many studies have shown that phytochemicals modulate signaling pathways linked with neuroinflammation, including NF-κB, Mitogen-Activated Protein Kinase (MAPK), and Nuclear Factor Erythroid 2-Related Factor 2 (Nrf2)/Heme Oxygenase-1, and enhance neuronal repair and survival (Goyal et al., 2024; Barbalho et al., 2025). For example, curcumin inhibits microglial activation, leading to reduced levels of pro-inflammatory mediators like IL-1β and TNF-α, and decreased plaque burden in AD (Maiti & Dunbar, 2018). Resveratrol modulates NF-κB signaling, decreases OS, and prevents neuroinflammation in AD and ALS (Wang et al., 2021). EGCG attenuates neuroinflammation by reducing COX-2 and iNOS expression, microglial activation, and NF-κB signaling, which protects dopaminergic neurons in PD (Cheng et al., 2021). Quercetin and ginsenosides suppress neuroinflammatory cytokines in AD and PD models (Chen et al., 2022; Chiang et al., 2023). Ginkgolides inhibit platelet-activating factor and reduce neuroinflammation in PD (Gachowska et al., 2021).

Luteolin, found in celery and parsley, downregulates NF-κB and MAPK pathways, preventing neuronal damage in HD (Maryam et al., 2024). Berberine reduces the activation of nucleotide-binding domain, leucine-rich–containing family, pyrin domain–containing-3 inflammasome, thereby reducing neuroinflammation in AD and PD (Sun et al., 2024). Kaempferol, abundant in spinach and broccoli, exhibits anti-inflammatory effects by inhibiting the release of cytokines and protecting against mitochondrial dysfunction in PD and HD (Lee et al., 2023). Pterostilbene, a stilbene from blueberries, reduces microglial activation and cytokine release, thereby alleviating neuroinflammation in HD (Qu et al., 2023). Together, these phytochemicals exhibit potent anti-inflammatory effects, leading to neuronal survival and slowing the neuronal death.

1.3. Phytochemicals With Mitochondrial Protective Effects

Phytochemicals maintain mitochondrial homeostasis by preserving their integrity and membrane potential, which leads to reduced OS resulting in enhanced energy dynamics (Anchimowicz et al., 2025). In AD and PD studies, curcumin (Benameur et al., 2022), quercetin (Grewal et al., 2021), and EGCG (Yıldırım et al., 2023) have been shown to protect mitochondria

- By inhibiting mitochondrial permeability transition pore opening and mitochondrial-dependent apoptosis by regulating the pro-apoptotic proteins.
- By reducing OS and enhancing ATP production and antioxidant defense mechanisms
- Stabilizing the mitochondrial membrane potential and improving the electron transport chain efficiency.

Resveratrol activates Sirtuin 1 and PGC-1α, promoting mitochondrial biogenesis and functions, and preventing mitochondrial membrane depolarization, and protects against neurotoxicity in AD and ALS (Wang et al., 2021). Coenzyme Q10 improves mitochondrial function in ALS (Ebrahimi et al., 2023). Pterostilbene enhances mitochondrial respiration, reduces OS by increasing antioxidant enzyme activity, and improves energy metabolism (Qu et al., 2023), while berberine enhances mitochondrial function by preventing mitochondrial fission, restoring fusion proteins, and reducing mitochondrial superoxide production, marking their potential for AD and HD treatments (Deng & Ma, 2021; Fang et al., 2022). Ginkgolides and bilobalide prevent mitochondrial swelling, lipid peroxidation, and OS, while also increasing glutathione levels, contributing to neuroprotection in AD and ischemic stroke (Singh et al., 2019). These phytochemicals show significant promise in maintaining mitochondrial health.

1.4. Phytochemicals With Anti-Aggregation Effects

Mounting literature highlights the anti-aggregation potential of various phytochemicals, which actively reduce or inhibit the accumulation of misfolded proteins across multiple NDDs by modulating protein homeostasis and enhancing autophagy. For instance, flavonoids like quercetin and EGCG reduce Aβ aggregation by regulating the signaling pathways to produce non-toxic oligomers in AD (Grewal et al., 2021). Curcumin decreases both Aβ aggregation and tau phosphorylation in AD (Liu et al., 2021) and also offers neuroprotective benefits in ALS (McCubrey et al., 2017). Resveratrol mitigates Aβ and tau aggregation by modulating chaperone proteins and enhancing proteasomal degradation, while also promoting mitochondrial biogenesis via PGC-1α activation, making it beneficial in AD and also preserving motor neuron integrity in ALS (Wang et al., 2021). Similarly, pterostilbene reduces Aβ oligomerization and enhances autophagic clearance of misfolded proteins in AD (Xu et al., 2023). Coenzyme Q10, found in spinach and broccoli, improves electron transport chain efficiency (Ebrahimi et al., 2023). *Ginkgo biloba* extract, rich in terpenoids and flavonoids, improves cognitive function and delays dementia symptoms, potentially by reducing the formation of protein Aβ aggregates in AD (Pagotto et al., 2024). Berberine inhibits Aβ and α-synuclein fibril formation, reducing neurotoxicity in AD and PD (Durairajan et al., 2022). 6-Shogaol from ginger exerts antioxidant and memory boosting properties in AD models (et al., 2025). Baicalein, a flavonoid derived from *Scutellaria baicalensis*, has been shown to destabilize Aβ protofibrils, reducing their aggregation propensity in AD models (Kaur et al., 2024). Limonene, a monoterpene found in citrus fruits, exhibits neuroprotective effects by decreasing ROS levels and preventing Erk phosphorylation, thereby counteracting Aβ-induced neurotoxicity in AD rats (Eddin et al., 2023). In sum, these findings indicate that phytochemicals reduce the formation and deposition of toxic protein aggregates and its associated neurotoxicity in NDDs.

1.5. Phytochemicals With Anti-Apoptotic Effects

Several phytochemicals are found to protect neurons by modulating apoptotic pathways, including UPS, mitochondrial pathways, and cell survival signaling cascades (Tang et al., 2019). For instance, polyphenols (curcumin, resveratrol, and EGCG), flavonoids (quercetin and baicalin), and alkaloids (berberine and carotenoids) have demonstrated anti-apoptotic effects through UPS regulation, thereby preventing apoptosis and neurodegeneration (Shankaran et al., 2024). Curcumin exerts anti-apoptotic effects by upregulating Bcl-2 and downregulating Bax and caspase-3, and promoting proteasomal degradation of misfolded proteins, thereby reducing neuronal apoptosis (Maiti & Dunbar, 2018). Similarly, resveratrol activates the Sirtuin 1 pathway and EGCG modulates the phosphoinositide 3-kinase (PI3K)/ protein kinase B (Akt) pathway, ultimately leading to enhanced proteasomal clearance of toxic proteins (Wang et al., 2021). Quercetin stabilizes proteasomal function and reduces neurotoxic protein aggregation (Zhao et al., 2018). Baicalin, a flavone from *Scutellaria baicalensis*, has been shown to suppress apoptosis by modulating the NF-κB pathway and enhancing proteasomal degradation of pro-apoptotic factors (Ciechanover & Kwon, 2015). Berberine prevents neuronal apoptosis by inhibiting mitochondrial cytochrome c release and promoting the degradation of pro-apoptotic proteins through the UPS (Sun et al., 2019). Lycopene increases Bcl-2 expression, suppresses OS -induced neuronal apoptosis, and enhances proteasomal activity (Tripathi et al., 2020). These findings highlight the potential of phytochemicals in regulating apoptotic pathways through the UPS.

1.6. Phytochemicals With Direct Neuroprotective Effects

Emerging evidences have identified phytochemicals with direct neuroprotective properties that enhance synaptic function and prevent synapse loss. Particularly, polyphenols, terpenoids, flavonoids, alkaloids, and stilbenes are known to improve memory and cognition by inhibiting cholinesterase enzymes such as acetylcholinesterase and butyrylcholinesterase to enhance acetylcholine levels (Sun & Shahrajabian, 2023). Additionally, terpenoids like ginkgolides and bilobalide from *Ginkgo biloba* leaves also improve memory and cognition (Liu et al., 2024). Flavonoids further support neuronal survival and synaptic function, and prevent the accumulation of toxic protein aggregates. Moreover, ginsenosides from *Panax ginseng* alter neurotransmission, enhance synapse density and function (Kim et al., 2018). Huperzine A improves cholinergic neurotransmission by blocking acetylcholinesterase, thus improving synaptic functions and cognition (Alam et al., 2017).

Curcumin enhances synaptic plasticity and neurogenesis by upregulating brain-derived neurotrophic factor signaling pathways, thereby contributing to improved cognition (Li et al., 2025). Resveratrol protects neurons from OS and mitochondrial dysfunction leading to enhanced synaptic plasticity and reduced neuronal loss in AD (Yan et al., 2020) and PD (Wang et al., 2021). Berberine also preserves synaptic integrity by reducing OS and mitochondrial dysfunction. Moreover, EGCG improve synaptic function by altering cholinergic neurotransmission and restoring mitochondrial transport in neurons (Wang et al., 2021). Quercetin prevents neuronal damage by triggering the signaling pathways (e.g., tropomyosin receptor kinase B (TrkB), phosphoinositide 3-kinase (PI3K), and protein kinase B (Akt)) which are critical for synaptic integrity, cell survival, and neuroprotection (Chiang et al., 2023). When combined, these phytochemicals have the potential to treat NDDs by addressing synapse loss and dysfunction.

1.7. Phytochemicals Enhancing Synaptic Function and Cognitive Recovery

Phytochemicals with anti-apoptotic, antioxidant, anti-inflammatory, anti-aggregation, and direct neuroprotective properties can modulate synaptic plasticity and function, often leading to reshaping or strengthening the neural circuitry, ultimately improving memory, cognitive, and motor functions in NDDs such as AD and PD (Xu et al., 2023; Li et al., 2025). Mainly, these bioactive compounds maintain neuronal integrity by preventing apoptosis and synaptic loss (Maiti & Dunbar, 2018). These phytochemicals elevate the expression of neurotrophic factors like brain-derived neurotrophic factors at the synapse and activate the crucial signaling pathways such as TrkB/PI3K/Akt and MAPK (Singh et al., 2025), which promote synaptic remodeling by activating the formation of new synapses and strengthening the existing synapses in affected brain regions (Hannan et al., 2020). These molecular changes lead to the maintenance of synaptic integrity, prevention of synapse/neuronal loss, realignment of neuronal connections, and facilitation of long-term potentiation (Zhao et al., 2018). Thus, phytochemicals contribute to improved memory retention, cognitive flexibility, and executive functions (Sun et al., 2019; Kamboj et al., 2024). Moreover, their neuroprotective effects on motor neurons and synaptic networks in the basal ganglia and cerebellum lead to improved motor coordination and reduced movement impairments in disorders such as PD and ALS (Tripathi et al., 2020).

VI. CONCLUSION

Mounting research in neurodegeneration reveals the vulnerability of the aging brain to several insults, such as protein aggregation, OS, dysfunctions of cellular organelles, impaired proteostasis, glutamate excitotoxicity, defective transport systems, and neuroinflammation. These complex and intricate cascades of signaling pathways trigger the neurodegeneration in several NDDs. Moreover, the pathophysiology of NDDs involves interconnected molecular mechanisms with overlapping pathways. A comprehensive understanding of these mechanisms is crucial for identifying diagnostic biomarkers (to diagnose the NDDs type) and developing new therapeutic targets and formulating effective treatment strategies (to establish a holistic approach to treat NDDs). The multi-targeted actions of plant-derived phytochemicals offer a promising treatment strategy for neuroprotection, potentially reducing or reversing cognitive decline and motor deficits in NDDs. However, the therapeutic potential of phytochemicals is limited by challenges such as lower bioavailability, rapid metabolism, and limited clinical trials. Future advancements in artificial intelligence-driven personalized medicine will enable pharmacogenomic profiling and predictive modeling to tailor multi-target phytotherapies to individual NDDs types, optimizing dosing and efficacy.

REFERENCES

Akhtar, A., Andleeb, A., Waris, T. S., Bazzar, M., Moradi, A. R., Awan, N. R., & Yar, M. (2021). Neurodegenerative diseases and effective drug delivery: A review of challenges and novel therapeutics. *Journal of Controlled Release*, *330*, 1152–1167.

Akhtar, A., Gupta, S. M., Dwivedi, S., Kumar, D., Shaikh, M. F., & Negi, A. (2022). Preclinical models for Alzheimer's disease: Past, present, and future approaches. *ACS Omega*, *7*(51), 47504–47517.

Akhtar, J., Sarwat, M., & Bashir, F. (Eds.). (2024). *Medicinal Plants for the Management of Neurodegenerative Diseases*. CRC Press.

Al-Bachari, S., Naish, J. H., Parker, G. J., Emsley, H. C., & Parkes, L. M. (2020). Blood–brain barrier leakage is increased in Parkinson's disease. *Frontiers in Physiology*, *11*, 593026.

Alam, A., Tamkeen, N., Imam, N., Farooqui, A., Ahmed, M. M., Ali, S., Malik, M. Z., & Ishrat, R. (2017). Pharmacokinetics and Molecular Docking Studies of Plant-Derived Natural Compounds to Exploring Potential Anti-Alzheimer Activity. arXiv preprint arXiv:1709.10374.

Anchimowicz, J., Zielonka, P., & Jakiela, S. (2025). Plant Secondary Metabolites as Modulators of Mitochondrial Health: An Overview of Their Anti-Oxidant, Anti-Apoptotic, and Mitophagic Mechanisms. *International Journal of Molecular Sciences*, *26*(1), 380. DOI: 10.3390/ijms26010380

Armstrong, R. A. (2020). What causes neurodegenerative disease? *Folia Neuropathologica*, *58*(2), 93–112.

Awuchi, C. G. (2023). Plants, phytochemicals, and natural practices in complementary and alternative system of medicine for treatment of central nervous system disorders. *International Journal of Food Properties*, *26*(1), 1190–1213.

Baev, A. Y., Vinokurov, A. Y., Novikova, I. N., Dremin, V. V., Potapova, E. V., & Abramov, A. Y. (2022). Interaction of mitochondrial calcium and ROS in neurodegeneration. *Cells*, *11*(4), 706.

Barbalho, S. M., Leme Boaro, B., da Silva Camarinha Oliveira, J., Patočka, J., Barbalho Lamas, C., Tanaka, M., & Laurindo, L. F. (2025). Molecular Mechanisms Underlying Neuroinflammation Intervention with Medicinal Plants: A Critical and Narrative Review of the Current Literature. *Pharmaceuticals*, *18*(1), 133. DOI: 10.3390/ph18010133

Bayazid, A. B., & Lim, B. O. (2022). Quercetin is an active agent in berries against neurodegenerative diseases progression through modulation of Nrf2/HO1. *Nutrients*, *14*(23), 5132.

Benameur, T., Giacomucci, G., Panaro, M. A., Ruggiero, M., Trotta, T., Monda, V., & Messina, G. (2022). New promising therapeutic avenues of curcumin in brain diseases. *Molecules (Basel, Switzerland)*, *27*(1), 236.

Bhat, S. G. (2022). *Medicinal plants and its pharmacological values*. In H. A. El-Shemy (Ed.), *Natural medicinal plants* (Chapter). IntechOpen. https://doi.org/DOI: 10.5772/intechopen.99848

Bhattacharya, T., Dey, P. S., Akter, R., Kabir, M. T., Rahman, M. H., & Rauf, A. (2021). Effect of natural leaf extracts as phytomedicine in curing geriatrics. *Experimental Gerontology*, *150*, 111352. DOI: 10.1016/j.exger.2021.111352

Bredesen, D. E., Toups, K., Hathaway, A., Gordon, D., Raji, H. C., Boyd, A., Hill, B. D., Hausman-Cohen, S., Chwa, M. W. J., Kurakin, A., & Jarrett, M. (2023). Precision medicine approach to Alzheimer's disease: Rationale and implications. *Journal of Alzheimer's Disease*. Advance online publication. DOI: 10.3233/JAD-230467

Calis, Z., Mogulkoc, R., & Baltaci, A. K. (2020). The roles of flavonols/flavonoids in neurodegeneration and neuroinflammation. *Mini-Reviews in Medicinal Chemistry*, *20*(15), 1475–1488.

Cardona, F. (2024). Special Issue "Mechanisms and Novel Therapeutic Approaches for Neurodegenerative Diseases". *Biomedicines*, *12*(11), 2554.

Chen, Y., Chi, M., Qiao, X., Wang, J., & Jin, Y. (2022). Anti-inflammatory effect of ginsenoside Rg1 on LPS-induced septic encephalopathy and associated mechanism. *Current Neurovascular Research*, *19*(1), 38–46.

Cheng, C. Y., Barro, L., Tsai, S. T., Feng, T. W., Wu, X. Y., Chao, C. W., & Hsieh, M. F. (2021). Epigallocatechin-3-gallate-loaded liposomes favor anti-inflammation of microglia cells and promote neuroprotection. *International Journal of Molecular Sciences*, *22*(6), 3037. DOI: 10.3390/ijms22063037

Cheong, S. L., Tiew, J. K., Fong, Y. H., Leong, H. W., Chan, Y. M., Chan, Z. L., & Kong, E. W. J. (2022). Current pharmacotherapy and multi-target approaches for Alzheimer's disease. *Pharmaceuticals*, *15*(12), 1560. DOI: 10.3390/ph15121560

Chiang, C. J., & Kordower, J. H. (2022). Tau pathology in Alzheimer's disease and related tauopathies: Progress and challenges. *Brain*, *145*(8), 2393–2412. DOI: 10.1093/brain/awac107

Chiang, M. C., Tsai, T. Y., & Wang, C. J. (2023). The potential benefits of quercetin for brain health: A review of anti-inflammatory and neuroprotective mechanisms. *International Journal of Molecular Sciences*, *24*(7), 6328. DOI: 10.3390/ijms24076328

Ciechanover, A., & Kwon, Y. T. (2015). Degradation of misfolded proteins in neurodegenerative diseases: Therapeutic targets and strategies. *Experimental & Molecular Medicine*, *47*(3), e147. DOI: 10.1038/emm.2014.117

Dang, T. K., Hong, S. M., Dao, V. T., Nguyen, D. T., Nguyen, K. V., Nguyen, H. T., & Kim, S. Y. (2023). Neuroprotective effects of total alkaloids fraction of *Huperzia serrata* on scopolamine-induced neurodegenerative animals. *Phytotherapy Research*, *37*(1), 140–150. DOI: 10.1002/ptr.7612

Dash, B. P., Freischmidt, A., Weishaupt, J. H., & Hermann, A. (2022). Downstream effects of mutations in SOD1 and TARDBP converge on gene expression impairment in patient-derived motor neurons. *International Journal of Molecular Sciences*, *23*(17), 9652. DOI: 10.3390/ijms23179652

Deng, H., & Ma, Z. (2021). Protective effects of berberine against MPTP-induced dopaminergic neuron injury through promoting autophagy in mice. *Food & Function, 12*(18), 8366–8375. DOI: 10.1039/D1FO01360B

Du, Y., Yang, X., Li, Z., Le, W., Hao, Y., Song, Y., Wang, F., & Guan, Y. (2021). HDAC6-mediated Hsp90 deacetylation reduces aggregation and toxicity of the protein alpha-synuclein by regulating chaperone-mediated autophagy. *Neurochemistry International, 149*, 105141. DOI: 10.1016/j.neuint.2021.105141

Dugger, B. N., & Dickson, D. W. (2017). Pathology of neurodegenerative diseases. *Cold Spring Harbor Perspectives in Biology, 9*(7), a028035. DOI: 10.1101/cshperspect.a028035

Durairajan, S. S., Selvarasu, K., Bera, M. R., Rajaram, K., Iyaswamy, A., & Li, M. (2022). Alzheimer's disease and other tauopathies: Exploring efficacy of medicinal plant-derived compounds in alleviating tau-mediated neurodegeneration. *Current Molecular Pharmacology, 15*(2), 361–379. DOI: 10.2174/1874467215666220217110306

Ebrahimi, A., Kamyab, A., Hosseini, S., Ebrahimi, S., & Ashkani-Esfahani, S. (2023). Involvement of coenzyme Q10 in various neurodegenerative and psychiatric diseases. *Biochemistry Research International, 2023*(1), 5510874. DOI: 10.1155/2023/5510874

Eddin, L. B., Azimullah, S., Jha, N. K., Nagoor Meeran, M. F., Beiram, R., & Ojha, S. (2023). Limonene, a monoterpene, mitigates rotenone-induced dopaminergic neurodegeneration by modulating neuroinflammation, hippo signaling and apoptosis in rats. *International Journal of Molecular Sciences, 24*(6), 5222. DOI: 10.3390/ijms24065222

El-Mansoury, B., El Hiba, O., Agnaou, M., Jayakumar, A. R., El Khiat, A., Kahim, K., & Bitar, A. (2023). Neuropathology of Parkinson's Disease: An Update of the Recent Findings. In *Experimental and Clinical Evidence of the Neuropathology of Parkinson's Disease* (pp. 82–101). IGI Global.

Fang, X., Wu, H., Wei, J., Miao, R., Zhang, Y., & Tian, J. (2022). Research progress on the pharmacological effects of berberine targeting mitochondria. *Frontiers in Endocrinology, 13*, 982145. DOI: 10.3389/fendo.2022.982145

Gachowska, M., Szlasa, W., Saczko, J., & Kulbacka, J. (2021). Neuroregulatory role of ginkgolides. *Molecular Biology Reports, 48*(7), 5689–5697. DOI: 10.1007/s11033-021-06535-2

Geschwind, M. D., & Belay, E. D. (2024). Creutzfeldt–Jakob disease and other prion diseases. *Nature Reviews. Disease Primers, 10*, 4. Advance online publication. DOI: 10.1038/s41572-024-00497-y

Ghosh, P., & Saadat, A. (2023). Neurodegeneration and epigenetics: A review. [English Edition]. *Neurología: Publicación Oficial de la Sociedad Española de Neurología, 38*(6), e62–e68. DOI: 10.1016/j.nrleng.2023.05.001

Gitler, A. D., & Tsuiji, H. (2020). TDP-43 proteinopathies: Pathological mechanisms and potential treatments. *The Journal of Clinical Investigation, 130*(5), 2227–2238. DOI: 10.1172/JCI134520

Goyal, R., Mittal, P., Gautam, R. K., Kamal, M. A., Perveen, A., Garg, V., Alexiou, A., Saboor, M., Haque, S., Farhana, A., Papadakis, M., & Ashraf, G. M. (2024). Natural products in the management of neurodegenerative diseases. *Nutrition & Metabolism, 21*, 26. Advance online publication. DOI: 10.1186/s12986-024-00800-4

Grad, L. I., Guest, W. C., Yanai, A., Pokrishevsky, E., O'Neill, M. A., Gibbs, E., & Cashman, N. R. (2011). Intermolecular transmission of superoxide dismutase 1 misfolding in living cells. *Proceedings of the National Academy of Sciences of the United States of America, 108*(39), 16398–16403. DOI: 10.1073/pnas.1102645108

Grewal, A. K., Singh, T. G., Sharma, D., Sharma, V., Singh, M., Rahman, M. H., & Abdel-Daim, M. M. (2021). Mechanistic insights and perspectives involved in neuroprotective action of quercetin. *Biomedicine and Pharmacotherapy, 140*, 111729. DOI: 10.1016/j.biopha.2021.111729

Hang, Z., Zhou, L., Xing, C., Wen, Y., & Du, H. (2023). The blood-brain barrier, a key bridge to treat neurodegenerative diseases. *Ageing Research Reviews, 91*, 102070. DOI: 10.1016/j.arr.2023.102070

Hannan, M. A., Dash, R., Sohag, A. A. M., Haque, M. N., & Moon, I. S. (2020). Neuroprotection against oxidative stress: Phytochemicals targeting TrkB signaling and the Nrf2-ARE antioxidant system. *Frontiers in Molecular Neuroscience, 13*, 116. DOI: 10.3389/fnmol.2020.00116

Hansson, O. (2021). Biomarkers for neurodegenerative diseases. *Nature Medicine, 27*(6), 954–963. DOI: 10.1038/s41591-021-01382-x

Hegde, M. L., Bharathi, P., Suram, A., & Rao, K. S. (2022). Impaired ubiquitin-proteasome system in neurodegenerative disorders: Emerging drug targets. *Pharmacological Research, 178*, 106144. DOI: 10.1016/j.phrs.2022.106144

Jee, K., Gaur, R., Kadian, J. P., Murtaza, M., Verma, S. K., Malik, B., & Shilpa, S. (2024). A review on phytochemical constituents used as current treatment strategies for neurodegenerative disease. *Anti-Inflammatory & Anti-Allergy Agents in Medicinal Chemistry, 25*. Advance online publication. DOI: 10.2174/0118715230310192240925165925

Kakaroubas, N., Brennan, S., Keon, M., & Saksena, N. K. (2019). Pathomechanisms of blood–brain barrier disruption in ALS. *Neuroscience Journal, 2019*, 2537698. Advance online publication. DOI: 10.1155/2019/2537698

Kamboj, S., Sharma, P., Kamboj, R., Kamboj, S., Om, H., Guarve, K., Dutt, R., Verma, I., Dua, K., & Rani, N. (2024). Exploring the therapeutic potential of phytoconstituents for addressing neurodegenerative disorders. *Central Nervous System Agents in Medicinal Chemistry*. Advance online publication. DOI: 10.2174/0118715249273015231225091339

Kaur, D., Sharma, V., & Deshmukh, R. (2019). Activation of microglia and astrocytes: A roadway to neuroinflammation and Alzheimer's disease. *Inflammopharmacology, 27*, 663–677. DOI: 10.1007/s10787-019-00580-x

Kaur, G., Mankoo, O. K., Kaur, A., Goyal, D., & Goyal, B. (2024). Insights into the baicalein-induced destabilization of LS-shaped Aβ 42 protofibrils using computer simulations. *Physical Chemistry Chemical Physics, 26*(23), 16674–16686. DOI: 10.1039/D3CP06006C

Kaur, S., Dhiman, M., & Mantha, A. K. (2018). Ferulic acid: A natural antioxidant with application towards neuroprotection against Alzheimer's disease. In Kaur, S. (Ed.), *Functional Food and Human Health* (pp. 575–586). Springer., DOI: 10.1007/978-981-13-1123-9_25

Kelser, B. M., Teichner, E. M., Subtirelu, R. C., & Hoss, K. N. (2024). A review of proposed mechanisms for neurodegenerative disease. *Frontiers in Aging Neuroscience*, *16*, 1370580. DOI: 10.3389/fnagi.2024.1370580

Kim, K. H., Lee, D., Lee, H. L., Kim, C. E., Jung, K., & Kang, K. S. (2018). Beneficial effects of Panax ginseng for the treatment and prevention of neurodegenerative diseases: Past findings and future directions. *Journal of Ginseng Research*, *42*(3), 239–247. DOI: 10.1016/j.jgr.2017.03.011

Korte, N., Nortley, R., & Attwell, D. (2020). Cerebral blood flow decrease as an early pathological mechanism in Alzheimer's disease. *Acta Neuropathologica*, *140*(6), 793–810. DOI: 10.1007/s00401-020-02215-w

Kovacs, G. G. (2018). Concepts and classification of neurodegenerative diseases. In *Handbook of Clinical Neurology* (Vol. 145, pp. 301–307). Elsevier., DOI: 10.1016/B978-0-12-802395-2.00021-3

Kurz, C., Walker, L., Rauchmann, B. S., & Perneczky, R. (2022). Dysfunction of the blood–brain barrier in Alzheimer's disease: Evidence from human studies. *Neuropathology and Applied Neurobiology*, *48*(3), e12782. DOI: 10.1111/nan.12782

Langbehn, D. R., Stout, J. C., Gregory, S., Mills, J. A., Durr, A., Leavitt, B. R., & Tabrizi, S. J. (2019). Association of CAG repeats with long-term progression in Huntington disease. *JAMA Neurology*, *76*(11), 1375–1385. DOI: 10.1001/jamaneurol.2019.2368

Lee, M. J., Cho, Y., Hwang, Y., Jo, Y., Kim, Y.-G., Lee, S. H., & Lee, J. H. (2023). Kaempferol alleviates mitochondrial damage by reducing mitochondrial reactive oxygen species production in lipopolysaccharide-induced prostate organoids. *Foods*, *12*(20), 3836. DOI: 10.3390/foods12203836

Li, G., Wu, Q., Wang, C., Deng, P., Li, J., Zhai, Z., & Li, Y. (2025). Curcumin reverses cognitive deficits through promoting neurogenesis and synapse plasticity via the upregulation of PSD95 and BDNF in mice. *Scientific Reports*, *15*(1), 1135. DOI: 10.1038/s41598-025-01135-7

Liu, Q., Wang, J., Gu, Z., Ouyang, T., Gao, H., Kan, H., & Yang, Y. (2024). Comprehensive exploration of the neuroprotective mechanisms of Ginkgo biloba leaves in treating neurological disorders. *The American Journal of Chinese Medicine*, •••, 1–34. DOI: 10.1142/S0192415X24500263

Liu, Y., Yang, L., Zhou, Q., Wang, Y., & Zhang, Y. (2021). Curcumin's role in Alzheimer's disease: A systematic review of preclinical studies. *Brain Research Bulletin*, *168*, 183–193. DOI: 10.1016/j.brainresbull.2021.01.012

Madhubala, D., Patra, A., Khan, M. R., & Mukherjee, A. K. (2024). Phytomedicine for neurodegenerative diseases: The road ahead. *Phytotherapy Research*, *38*(6), 2993–3019.

Mahar, S. K., Yaseen, K., Jan, A. M. A., & Ali, A. (2025). Mitochondrial Dysfunction in Neurodegenerative Diseases: Exploring Therapeutic Approaches for Parkinson's Disease. *Indus Journal of Bioscience Research*, *3*(2), 438–451.

Maiti, P., & Dunbar, G. L. (2018). Use of curcumin, a natural polyphenol for targeting molecular pathways in treating age-related neurodegenerative diseases. *International Journal of Molecular Sciences, 19*(6), 1637. DOI: 10.3390/ijms19061637

Maryam, S., Khalid, S., Arshad, A., Abid, S., Iqbal, L. Z., Amjad, Y., & Faisal, S. (2024). Luteolin: A versatile flavonoid for anti-inflammatory, anti-cancer, and neuroprotective therapies. *Journal of Experimental Biology and Agricultural Sciences, 12*(2), 2708. DOI: 10.18006/2024.12(2).2708

Matos, C. A., Nóbrega, C., & Pereira de Almeida, L. (2023). Polyglutamine (PolyQ) diseases: Navigating the landscape of pathogenesis and therapeutic strategies. *ACS Chemical Neuroscience, 14*(7), 1–14. DOI: 10.1021/acschemneuro.4c00184

Mayer, C., Riera-Ponsati, L., Kauppinen, S., Klitgaard, H., Erler, J. T., & Hansen, S. N. (2024). Targeting the NRF2 pathway for disease modification in neurodegenerative diseases: Mechanisms and therapeutic implications. *Frontiers in Pharmacology, 15*, 1437939. DOI: 10.3389/fphar.2024.1437939

McCubrey, J. A., Lertpiriyapong, K., Steelman, L. S., Abrams, S. L., Yang, L. V., Murata, R. M., & Cervello, M. (2017). Effects of resveratrol, curcumin, berberine and other nutraceuticals on aging, cancer development, cancer stem cells and microRNAs. *Aging (Albany NY), 9*(6), 1477–1536. DOI: 10.18632/aging.101250

Medrano-Jiménez, E., Meza-Sosa, K. F., Urbán-Aragón, J. A., Secundino, I., Pedraza-Alva, G., & Pérez-Martínez, L. (2022). Microglial activation in Alzheimer's disease: The role of flavonoids and microRNAs. *Journal of Leukocyte Biology, 112*(1), 47–77. DOI: 10.1002/JLB.3MR1021-531R

Nabi, M., & Tabassum, N. (2022). Role of environmental toxicants on neurodegenerative disorders. *Frontiers in Toxicology, 4*, 837579. DOI: 10.3389/ftox.2022.837579

Nishihara, H., Perriot, S., Gastfriend, B. D., Steinfort, M., Cibien, C., Soldati, S., & Engelhardt, B. (2022). Intrinsic blood–brain barrier dysfunction contributes to multiple sclerosis pathogenesis. *Brain, 145*(12), 4334–4348. DOI: 10.1093/brain/awac019

Pagotto, G. L. D. O., Santos, L. M. O. D., Osman, N., Lamas, C. B., Laurindo, L. F., Pomini, K. T., & Barbalho, S. M. (2024). Ginkgo biloba: a leaf of hope in the fight against Alzheimer's dementia: clinical trial systematic review. *Antioxidants, 13*(6), 651.

Peng, C., Trojanowski, J. Q., & Lee, V. M. Y. (2020). Protein transmission in neurodegenerative disease. *Nature Reviews. Neurology, 16*(4), 199–212.

Qu, X., Zhang, L., & Wang, L. (2023). Pterostilbene as a therapeutic alternative for central nervous system disorders: A review of the current status and perspectives. *Journal of Agricultural and Food Chemistry, 71*(40), 14432–14457.

Rahaman, S. M., Ashique, S., Bhowmick, M., Mollick, M., Bhowmick, P., & Kumar, P. (2024). Unlocking the Impact of Phyto-bioactive Compounds against Neurodegenerative Disorders.

Rahman, M. H., Bajgai, J., Fadriquela, A., Sharma, S., Trinh, T. T., Akter, R., & Lee, K. J. (2021). Therapeutic potential of natural products in treating neurodegenerative disorders and their future prospects and challenges. *Molecules (Basel, Switzerland), 26*(17), 5327.

Rêgo, A. C. M. do, & Araújo-Filho, I. (2024). Cellular and Molecular Mechanisms in Neurodegenerative Disorders: A Comprehensive Scoping Review. *International Journal of Innovative Research in Medical Science*.

Saleh, H. A., Yousef, M. H., & Abdelnaser, A. (2021). The anti-inflammatory properties of phytochemicals and their effects on epigenetic mechanisms involved in TLR4/NF-κB-mediated inflammation. *Frontiers in Immunology, 12*, 606069.

Sarkar, B., Rana, N., Singh, C., & Singh, A. (2024). Medicinal herbal remedies in neurodegenerative diseases: An update on antioxidant potential. *Naunyn-Schmiedeberg's Archives of Pharmacology, 397*(8), 5483–5511. DOI: 10.1007/s00210-024-03027-5

Sehar, U., Rawat, P., Reddy, A. P., Kopel, J., & Reddy, P. H. (2022). Amyloid beta in aging and Alzheimer's disease. *International Journal of Molecular Sciences, 23*(21), 12924.

Sengoku, R. (2020). Aging and Alzheimer's disease pathology. *Neuropathology, 40*(1), 22–29.

Shankaran, D. K. A., Meenakumari, S., Gopinath, S. C. B., & Raman, P. (2024). A Comprehensive Review of Therapeutic Compounds from Plants for Neurodegenerative Diseases. *Current Medicinal Chemistry*.

Singh, A., Ansari, V. A., Mahmood, T., Ahsan, F., & Wasim, R. (2022). Neurodegeneration: Microglia: Nf-kappab signaling pathways. *Drug Research, 72*(09), 496–499.

Singh, A. A., Katiyar, S., & Song, M. (2025). Phytochemicals targeting BDNF signaling for treating neurological disorders. *Brain Sciences, 15*(3), 252. DOI: 10.3390/brainsci15030252

Singh, K., Gupta, J. K., Kumar, S., & Soni, U. (2024). A review of the common neurodegenerative disorders: Current therapeutic approaches and the potential role of bioactive peptides. *Current Protein & Peptide Science, 25*(7), 507–526.

Singh, S. K., Srivastav, S., Castellani, R. J., Plascencia-Villa, G., & Perry, G. (2019). Neuroprotective and Antioxidant Effect of Ginkgo biloba Extract Against AD and Other Neurological Disorders. *Neurotherapeutics; the Journal of the American Society for Experimental NeuroTherapeutics, 16*(3), 666–674.

Soto, C., & Estrada, L. D. (2020). Protein misfolding and aggregation in Alzheimer's and other neurodegenerative diseases: Pathogenic mechanisms and therapeutic approaches. *Seminars in Cell & Developmental Biology, 99*, 14–25.

Soto, C., & Pritzkow, S. (2018). Protein misfolding, aggregation, and conformational strains in neurodegenerative diseases. *Nature Neuroscience, 21*(10), 1332–1340.

Strope, T. A., Birky, C. J., & Wilkins, H. M. (2022). The role of bioenergetics in neurodegeneration. *International Journal of Molecular Sciences, 23*(16), 9212.

Sun, J., Zeng, Q., Wu, Z., Huang, L., Sun, T., Ling, C., & Wang, H. (2024). Berberine inhibits NLRP3 inflammasome activation and proinflammatory macrophage M1 polarization to accelerate peripheral nerve regeneration. *Neurotherapeutics; the Journal of the American Society for Experimental NeuroTherapeutics, 21*(4), e00347.

Sun, W., & Shahrajabian, M. H. (2023). Therapeutic potential of phenolic compounds in medicinal plants—Natural health products for human health. *Molecules (Basel, Switzerland)*, *28*(4), 1845.

Sun, Y., Xun, K., Wang, Y., & Chen, X. (2019). A systematic review of the anti-neuroinflammatory effects of berberine, a natural isoquinoline alkaloid. *Frontiers in Pharmacology*, *10*, 1462.

Takahashi, R. H., & Nixon, R. A. (2022). Autophagy dysfunction and neurodegenerative disease progression: Current insights. *Brain Research Bulletin*, *182*, 1–12.

Tang, H., Lu, D., Pan, R., Qin, X., Xiong, H., Dong, Y., & Li, H. (2019). Phytochemicals targeting apoptotic mechanisms in Alzheimer's disease. *Frontiers in Pharmacology*, *10*, 360.

Tarutani, A., & Hasegawa, M. (2019). Prion-like propagation of α-synuclein in neurodegenerative diseases. *Progress in Molecular Biology and Translational Science*, *168*, 323–348.

Tembhurnikar, H. J., Thool, N. D., Patil, R. J., & Das, R. K. (2021). Review on various factors responsible for neurodegenerative disorders. *GSC Biological and Pharmaceutical Sciences*, *16*(1), 126–132. DOI: 10.30574/gscbps.2021.16.1.0203

Tripathi, P., Misra, M. K., Chaudhury, P., & Misra, A. (2020). Lycopene and its role in neuroprotection: A review. *Biomedicine and Pharmacotherapy*, *129*, 110365.

Turan, D., Abdik, H., Sahin, F., & Avşar Abdik, E. (2020). Evaluation of the neuroprotective potential of caffeic acid phenethyl ester in a cellular model of Parkinson's disease. *European Journal of Pharmacology*, *883*, 173342.

Verma, M., Lizama, B. N., & Chu, C. T. (2022). Excitotoxicity, calcium and mitochondria: A triad in synaptic neurodegeneration. *Translational Neurodegeneration*, *11*(1), 3.

Vyas, S., Kothari, S. L., & Kachhwaha, S. (2019). Nootropic medicinal plants: Therapeutic alternatives for Alzheimer's disease. *Journal of Herbal Medicine*, *17*, 100291.

Wang, C., Ward, M. E., & Tienari, P. J. (2023). The role of genetics and protein aggregates in prion-like propagation of neurodegenerative diseases. *Trends in Neurosciences*, *46*(1), 38–51.

Wang, Q., Xue, H., Yue, Y., Hao, S., Huang, S. H., & Zhang, Z. (2022). Role of mitophagy in the neurodegenerative diseases and its pharmacological advances: A review. *Frontiers in Molecular Neuroscience*, *15*, 1014251.

Wang, Z., Chen, W., Dong, H., & Yan, Y. (2021). Resveratrol and neuroprotection: Impact and mechanisms. *Neurochemical Research*, *46*(7), 1892–1905.

Wilson, D. M., Cookson, M. R., Van Den Bosch, L., Zetterberg, H., Holtzman, D. M., & Dewachter, I. (2023). Hallmarks of neurodegenerative diseases. *Cell*, *186*(4), 693–714. DOI: 10.1016/j.cell.2022.12.032

Wilson, V. (2014). Research methods: Triangulation. *Evidence Based Library and Information Practice*, *9*(1), 74–75. DOI: 10.18438/B8WW3X

Wong, T. H., Seelaar, H., Melhem, S., Rozemuller, A. J., & van Swieten, J. C. (2020). Genetic screening in early-onset Alzheimer's disease identified three novel presenilin mutations. *Neurobiology of Aging*, *86*, 201.e9. DOI: 10.1016/j.neurobiolaging.2019.01.015

Xu, J., Liu, J., Li, Q., Li, G., Zhang, G., Mi, Y., Zhao, T., Mu, D., Wang, D., Zeng, K., & Hou, Y. (2023). Pterostilbene participates in TLR4-mediated inflammatory response and autophagy-dependent Aβ1–42 endocytosis in Alzheimer's disease. *Phytomedicine*, *119*, 155011. DOI: 10.1016/j.phymed.2023.155011

Yadav, V., Mythri, C., & Kumarasamy, M. (2024). Natural products as potential modulators of pro-inflammatory cytokines signalling in Alzheimer's disease. *Brain Behavior and Immunity Integrative*, *5*, 100048. DOI: 10.1016/j.bbii.2023.100048

Yan, Y., Yang, H., Xie, Y., Ding, Y., Kong, D., & Yu, H. (2020). Research progress on Alzheimer's disease and resveratrol. *Neurochemical Research*, *45*, 989–1006. DOI: 10.1007/s11064-020-03007-0

Yang, J., Zhao, H., & Qu, S. (2024). Phytochemicals targeting mitophagy: Therapeutic opportunities and prospects for treating Alzheimer's disease. *Biomedicine and Pharmacotherapy*, *177*, 117144. DOI: 10.1016/j.biopha.2024.117144

Yiannopoulou, K. G., Anastasiou, A. I., Zachariou, V., & Pelidou, S. H. (2019). Reasons for failed trials of disease-modifying treatments for Alzheimer disease and their contribution in recent research. *Biomedicines*, *7*(4), 97. DOI: 10.3390/biomedicines7040097

Yıldırım, İ. G., Doğan, F., & Şanlıer, N. (2023). Neurodegenerative Diseases and Phytochemicals. *Beslenme ve Diyet Dergisi*, *51*(2), 111–118.

Zhang, W., Xiao, D., Mao, Q., & Xia, H. (2023). Role of neuroinflammation in neurodegeneration development. *Signal Transduction and Targeted Therapy*, *8*(1), 267. DOI: 10.1038/s41392-023-01486-5

Zhao, A., Xu, W., Han, R., Wei, J., Yu, Q., Wang, M., & Chi, G. (2024a). Role of histone modifications in neurogenesis and neurodegenerative disease development. *Ageing Research Reviews*, *90*, 102324. DOI: 10.1016/j.arr.2024.102324

Zhao, J., Yang, J., & Xie, Y. (2018). Flavonoids in neurodegenerative diseases: Recent advances and perspectives. *Nutrients*, *10*(5), 709. DOI: 10.3390/nu10050709

Chapter 3
Emerging Trends in Neurodegenerative Disease Treatment:
Personalized Medicine Approaches Using Nanotechnology and Secondary Metabolites

Vivek Pazhamalai
https://orcid.org/0000-0003-2088-7545
Department of BioEngineering, School of Engineering, Vels Institute of Science, Technology, and Advanced Studies, India

V. Gokul
Department of BioEngineering, School of Engineering, Vels Institute of Science, Technology, and Advanced Studies, India

L. Madhumitha
https://orcid.org/0009-0004-9904-8414
Department of BioEngineering, School of Engineering, Vels Institute of Science, Technology, and Advanced Studies, India

Praveen Kumar M.
Department of BioEngineering, School of Engineering, Vels Institute of Science, Technology, and Advanced Studies, India

Meenambiga S. S.
https://orcid.org/0000-0002-5445-525X
Department of BioEngineering, School of Engineering, Vels Institute of Science, Technology, and Advanced Studies, India

Ivo Romauld S.
https://orcid.org/0000-0003-0610-0646
Department of BioEngineering, School of Engineering, Vels Institute of Science, Technology, and Advanced Studies, India

Hariharan N. M.
Department of Biotechnology, Vel Tech Rangarajan Dr. Sagunthala R&D Institute of Science and Technology, India

ABSTRACT

Neurodegenerative diseases (NDDs) like Alzheimer's, Parkinson's, and Huntington's are being treated with advanced technologies like genomics, precision medicine, stem cell therapy, artificial intelligence (AI), and neurodegenerative medicine. These technologies enable personalized ther-

DOI: 10.4018/979-8-3373-1365-8.ch003

apy tailored to each patient's unique characteristics, improving early detection and intervention before irreversible damage occurs. Stem cell therapies and gene editing tools like CRISPR promise to reverse or slow neurodegeneration by repairing or replacing damaged neurons, potentially restoring cognitive and physical functions. AI and machine learning transform drug discovery by predicting targets, optimizing clinical trials, and simulating treatment efficacy. Cross-disciplinary collaboration between bioengineering and neurology is essential for developing novel therapeutic devices. The future of neurodegenerative disease treatment lies in a comprehensive, patient-centric approach, that integrates cutting-edge technology and scientific research.

1. INTRODUCTION

Neurodegeneration is the progressive loss of neurons that results in a neurodegenerative illness. Eventually, they may die as a result of neuronal injury. Neurodegenerative diseases include Multiple sclerosis, Huntington's disease, amyotrophic lateral sclerosis, Parkinson's disease, Alzheimer's disease, tauopathies, prion disorders, and multiple system atrophy. The brain exhibits Degeneration of neurons at a wide range of neural circuitry levels, from the molecular to the systemic. As there is currently no way to halt the continuous degeneration of neurons, these illnesses are considered incurable; nonetheless, Research has shown that the two primary causes of neurodegeneration are oxidative stress. Biomedical research has found many below-cellular similarities between these illnesses, such as faulty protein assembly (like proteinopathy) and increased cell death. These similarities suggest that if one neurodegenerative disease is treated well, it might potentially help other diseases. In 2019, there were an estimated 55 million instances of dementia worldwide; by 2050, that figure is predicted to increase to 139 million (Lamptey et al., 2022).

Previous studies have indicated that the survival rate of NDs is restricted. Most recently, in 2019, 349.2 million individuals globally experienced severe neurological disorders, and 10 million people died as a result. Alzheimer's_disease (AD) was the illness that caused the most deaths, followed by memory loss and newborn dementia. Similarly, the prevalence of Parkinson's disease (PD) surpassed 8.5 million persons worldwide in 2019. According to recent estimates,5.8 million lives with disabilities were caused by PD, representing an 81% increase since 2000. Furthermore, Parkinson's disease (PD) was the cause of 329,000 fatalities, a startling 100 percent increase since 2000. It is estimated that there are more than 1.8 million MS patients worldwide. Globally, there were 22,439 MS-related fatalities and 59,345 newly diagnosed cases in 2019. Over time, Multiple_sclerosis (MS)-related incidence, mortality, and disability-adjusted life have been rising gradually, but age-standardized rates have somewhat decreased between 1990 and 2019 (Gadhave et al., 2024). According to estimates, there are between 1.9 and 6 instances of ALS for every 100,000 people worldwide. 24,328 people died from ALS, yielding an age-adjusted death rate of 1.70. According to previous epidemiological reports on HD, the global presence is 2.71 incidents for every 100,000 individuals annually, whereas the combined incidence rate is 0.38 instances per 100,000 people annually (Kim et al., 2015).

Although there are numerous recognised treatments for NDs at the moment, the most of them merely aid in the treatment of the related symptoms. Due to many obstacles related to NDs therapies, encompassing the BBB (blood-brain barrier) and numerous unintended side effects of existing medications, The advancement of NDs has not been controlled by the medications now on the market, resulting in a reduced survival rate (Gadhave et al., 2023). Many studies have highlighted the potential use of nanotechnology approaches to treat CNS illnesses including NDs. Biomaterials could be highly effective and selective in facilitating targeted drug delivery, molecular detection, treatment monitoring, and ND diagnosis, according to a number of recent papers. Additionally, a deeper comprehension of the specific mechanisms behind the onset of NDs may aid in the creation of novel therapeutic techniques, improving the likelihood of more effective therapies (Gadhave et al., 2021). In order to improve the therapeutic features of NDs in the near future, the current review focuses on the limitations of conventional treatments, the exact process by which NDs start, the development of biomaterials for managing NDS, as well as their practical significance.

2. OVERVIEW OF NEURODEGENERATIVE DISEASE

Figure 1. Neurodegenerative disorders

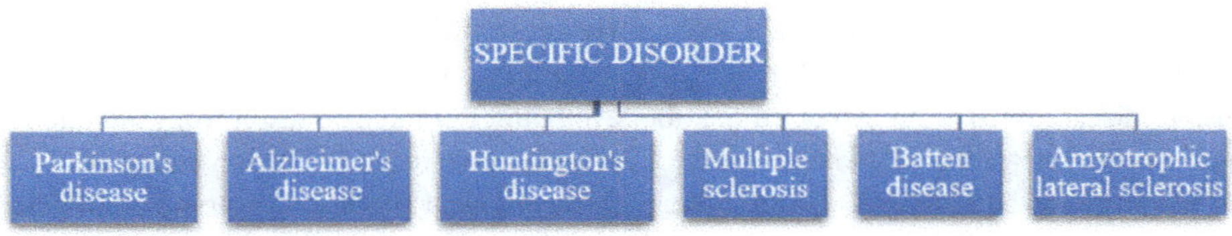

2.1. Alzheimer's Disease

The frontal cortex, the parietal lobe, and the temporal lobe, and the cingulate gyrus all exhibit gross atrophy as a result of Alzheimer's disease (AD) is a chronic neurological illness that causes the cerebral cortex's synapses and neurons to die, and (Wenk 2003). The most prevalent neurodegenerative illness is this one. There are currently no effective treatments for Alzheimer's disease, despite billions of dollars being spent in this search. Stable and successful AD treatment approaches have a 99.5% failure rate in clinical studies. Inappropriate drug dosages, poor target and participant selection, and a lack of understanding of the pathophysiology of AD are some of the causes of this failure rate. Alzheimer's diagnosis is now mediocre, and different clinical diagnosis features require the use of new methodologies. The misdiagnosis rate for Alzheimer's is 20% (Archer et al., 2017). The existence of amyloid plaques and neurofibrillary tangles are the primary features of AD pathogenesis. Amyloid beta is a small peptide that is usually 39–43 amino acids long form plaques. Amyloid beta is a subunit of amyloid precursor protein (APP), a larger transmembrane protein. which pierces the membrane of neurons. Normal neuron development, survival, and post-injury healing seem to be influenced by APP. Enzymes such as beta secretase and gamma secretase divide APP into smaller bits. One of these pieces creates amyloid beta fibrils,that can self-assemble into large extracellular amyloid plaques (Ohnishi & Takano 2004).

2.2. Parkinson's Disease

Parkinson's disease (PD) is the neurological illness that is most common. Usually, it presents as bradykinesia, stiffness, resting tremor, and unstable posture. Although the condition is less common in Asian nations, Information regarding the incidence and basic prevalence rate of Parkinson's disease (PD) range from 15 per 100,000 to 12,500 per 100,000 and 328 per 100,000, respectively (Elbaz et al., 2016). Parkinson's disease (PD) is characterised by Dopaminergic loss neurones in the substantia nigra, a midbrain area. It is uncertain what is causing this selective cell death. It is noteworthy that in afflicted neurons, Alpha-synuclein-ubiquitin aggregates and complexes are observed to accumulate in Lewy bodies. It is believed that this disease process may be influenced by abnormalities in the regulation and trafficking of proteins, such as RAB1. Alpha-synuclein may also accumulate in Lewy bodies as a result of impaired axonal transport. Studies have shown that Reduced transport rates of two mutant alpha-synucleins associated with Parkinson's disease and wild-type alpha-synucleins were seen in the axons of cultured neurons. Alpha-synuclein-induced membrane damage may be an additional mechanism of Parkinson's disease (HHMI Research news 2006). Age is the primary known risk factor. Hereditary Parkinson's disease (PD) or an increased risk for developing PD can also result from Mutations in genes like leucine-rich repeat kinase 2 (LRRK2) and tau protein (MAPT), α-synuclein (SNCA), and glucocerebrosidase (GBA). Even though no neurological condition is more common than Parkinson's disease (PD), diagnosis issues still exist. Smell impairment is a common symptom of Parkinson's disease (PD), but some neurologists doubt its effectiveness. This method of assessment is controversial among medical professionals. Research indicates that the gut microbiota may be involved in PD diagnosis, and there are several ways that this could transform PD treatment in the future (Hill-Burns et al., 2017).

2.3. Huntington's Disease

A neurological condition that is mostly inherited and incurable is Huntington's disease (HD), often known as Huntington's chorea. Subtle issues with mood or mental/psychiatric abilities are frequently the first signs. This is frequently followed by an unsteady walk and an overall lack of coordination. Additionally, it is a condition of the basal ganglia that causes chorea, a hyperkinetic movement disorder. Uncoordinated, involuntary chorea motions become more noticeable as the illness worsens. Physical capabilities progressively deteriorate until speaking becomes impossible and coordinated movement becomes challenging. Dementia, sadness, apathy, and impulsivity are common declines in mental functioning. Each person experiences the specific symptoms in slightly different ways. While they can begin at any age, symptoms typically start to show around age 40 and persist between the ages of 30 and 50 (Saudou et al., 2016). In each succeeding generation, the disease may manifest earlier. About 8% of cases, referred to as juvenile HD, begin before the age of 20, and they usually manifest with Parkinson's disease-like slow movement symptoms rather than chorea. The huntingtin gene (HTT) mutation in an afflicted parent is usually the cause of HD. But in as many as 10% of cases, a unique mutation is the reason. The genetic information for huntingtin protein (Htt) is provided by the huntingtin gene. A trinucleotide repeat expansion, often known as CAG repeats of cytosine-adenine-guanine, causes the huntingtin protein gene to create an abnormal mutant protein (mHtt). Eventually, this protein kills brain cells in a number of possible ways (Illarioshkin et al., 2018). Having a single parent with the trait is sufficient to make their children ill since the mutant protein is dominant. Genetic testing is used to make the diagnosis, and it can be done whenever symptoms are present or not. The age at which a person is deemed mature enough to choose testing, whether parents have the right to have their children tested, and how to handle test result confidentiality and disclosure are some of the ethical issues that are brought up by this fact. HD has no known cure, and in its later stages, full-time care is needed. Some people's quality of life can be improved and some symptoms can be relieved by treatments. Tetrabenazine has the strongest evidence for treating movement disorders. Approximately 4 to 15 out of 100,000 individuals of European ancestry suffer with HD. It is uncommon among Japanese and Finnish people, and its prevalence in Africa is unclear. Males and females are equally affected by the illness (Frank 2014). Life expectancy is shortened by complications such pneumonia, heart disease, and physical harm from falls; however, aspiration pneumonia is frequently mentioned as the condition's leading cause of mortality. In almost 9% of cases, suicide is the cause of death. Usually, death happens 15–20 years after the condition was initially identified. American physician Charles Oscar Waters described the illness for the first time in 1841. George Huntington, an American physician, provided a more thorough description of the illness in 1872. In 1993, the Hereditary Disease Foundation spearheaded an international collaborative effort that identified the genetic underpinnings. In order to support people and their families, raise public awareness, and advance research, research and support organisations started to emerge in the late 1960s. The precise mechanism of the disease, the development of better animal models to support research, the testing of drugs and how they are administered to treat symptoms or slow the disease's progression, and the investigation of techniques like stem-cell therapy to replace lost or damaged neurons are some of the research directions.

2.4. Multiple Sclerosis

Multiple sclerosis is an inflammatory disease (MS) damages the coverings that protect the brain and spinal cords nerve cells. Being a demyelinating illness, multiple sclerosis (MS) impairs the nerve system's signal transmission, resulting in many symptoms, such as physical, emotional, and even psychiatric problems. Symptoms include muscle weakness, eye pain, vision loss, double vision, and loss of feeling or coordination. MS can manifest in a variety of ways, with new symptoms either developing gradually (progressive forms) or happening in isolated attacks (relapsing forms). Relapsing kinds of MS may completely disappear between episodes, but certain long-term neurological problems sometimes remain, especially as the disease worsens. Once symptoms appear, body function gradually declines in progressive forms of MS, and if treatment is not received, the condition will continue to worsen (Loma & Heyman 2011). The underlying process is believed to be immune system damage or myelin-producing cell inactivation, while the precise cause is unknown. Some of the hypotheses for this are genetics, immunological dysregulation, and environmental variables such as viral infections. The McDonald criteria are a regularly revised set of standards for diagnosing multiple sclerosis. MS has no known cure. Current therapies include disease-modifying drugs to reduce inflammation and the symptoms that follow acute flare-ups and to stop additional attacks. People's capacity to function can be improved by physical therapy, occupational therapy, and patient-centred symptom management. It is hard to predict the long-term prognosis; women, people who have the condition early in life, people who have a relapsing course, and people who had few episodes at first are more likely to have better outcomes (Weinshenker 1994). Multiple sclerosis is the most frequent immune-mediated disease that affects the central nervous system (CNS). Around 2.8 million persons worldwide suffered from MS in 2020, with rates greatly differing by area and demographic. Women are twice as likely to have the illness as men, and it often first manifests between the ages of 20 and 50. In 1868, French physician Jean-Martin Charcot described multiple sclerosis for the first time. The phrase "multiple sclerosis" refers to the numerous glial scars that are a result of multiple cerebro-spinal sclerosis that form on the brain and spinal cord's white matter (Clanet 2008).

2.5. Amyotrophic Lateral Sclerosis

In the United States, Lou Gehrig's disease (LGD) or motor neurone disease (MND) are other names for the uncommon and deadly neurological illness called amyotrophic lateral sclerosis (ALS), This causes the gradual loss of both upper and lower motor neurones that typically control voluntary muscular contraction. The most prevalent type of motor neurone disease is ALS. In its early stages, ALS frequently manifests as progressive muscle weakness, wasting, spasms, and stiffness. Usually, motor neurone loss persists until one loses the ability to move, speak, eat, and, finally, breathe. An estimated 50% of individuals with ALS experience at least some mild cognitive and behavioural challenges, while only 15% go on to fully develop frontotemporal dementia. ALS is categorised as either bulbar-onset (beginning with difficulties speaking or swallowing) or limb-onset (beginning with weakness in the arms or legs), depending on which of the previously listed symptoms appears first (Hardiman et al., 2017). Sporadic ALS is the term for the majority of ALS cases (about 90–95%) that have no recognised cause. Nonetheless, it is thought that both environmental and genetic variables play a role. Familial ALS (hereditary) is the term for the remaining 5–10% of cases that have a genetic aetiology, frequently connected to a family history of the illness. Disease-causing variations in one of four specific genes account for almost half of these hereditary disorders. A person's signs and symptoms are used to make the diagnosis, and tests are performed to rule out other possible reasons (Goutman et al., 2022). ALS has no known treatment. Improving symptoms and slowing the disease's progression are the two main objectives of treatment. Riluzole and edaravone are FDA-approved medications that decrease the course of ALS. Both the quality and duration of life may be enhanced by non-invasive ventilation. While mechanical ventilation can increase survival, it cannot halt the spread of disease. Maintaining nutrition and weight may be aided by a feeding tube. Respiratory failure is frequently the cause of death. Even while anyone can have the condition at any age, it usually initially appears around age 60. Between two and four years is the typical survival time from start to finish, though this might vary, and roughly 10% of individuals affected live for over 10 years (Pupillo et al., 2014).

2.6. Mechanism Involved in Neurodegenerative

The complex process of neurodegeneration, which is frequently linked to ageing, occurs when some brain cells (nerve cells) stop functioning. Certain proteins in the brain malfunction during neurodegeneration, which leads to clumps forming in various brain regions. Additional problems brought on by neurodegeneration include DNA damage, inflammation, and energy deficiencies. These progressively intensify, leading to cell death. Cells can die in a variety of ways, including by shutting down or disintegrating.

3. PERSONALIZED MEDICINE IN NEURODEGENERATIVE DISEASE

Recent years have seen a significant rise in the incidence and prevalence of neurodegenerative diseases, and since the population ages quickly,The quantity of individuals with these conditions is expected to grow exponentially in the coming years. Therefore, it is crucial that we work harder to create new methods for identifying and treating people with neurodegenerative illnesses. Better knowledge of the fundamental processes producing the illness is also necessary for such initiatives, as this could result in the development of individualised treatment plans. Recent developments in the fields of glycomics, proteomics, metabolomics, and genomes provide crucial support in the effort to comprehend the mechanisms behind the degeneration of particular neuronal populations. A more thorough knowledge of the genetic processes underlying the onset and progression of neurodegenerative disorders will eventually result from declining next-generation sequencing prices (Sanyal et al., 2012). The ENCODE project's findings have led to new understandings of the role of epigenetics and intricate transcriptional regulation mechanisms, as well as a growing corpus of research on the function of protein post-translational modifications like glycosylation and phosphorylation. These insights offer fresh perspectives on the intricate interactions between genes and the environment that can result in disease. The bioinformatical techniques used in these investigations can correlate the molecular pathways to nervous system function and integrate data at numerous levels. hence opening up new possibilities for the advancement of personalised neurology (Geschwind & Konopka 2009).

Synthetic biology techniques in personalised medicine can also be applied with the help of advances in sequencing technology. For example, the development of artificial genomes that could serve as the foundation for innovative therapeutic approaches could significantly advance personalized medicine in the fields of neurology and other clinical specialities. These innovative systems will be built on an enlarged genetic code and intended for targeted, customised medication manufacture, delivery, and pathological signal activation (Jain 2013). A lack of clinically useful biomarkers is still one of the key challenges in the study of neurodegenerative illness, despite the abundance of data from polyomic investigations. Finding universal biomarkers, it appears to be a challenging task to be able to diagnose individuals with neurodegenerative disorders, track how well they respond to treatment, and forecast how the condition will advance. Rather, it appears that a combination of many approaches to find biomarkers unique to each stated aspect is a more practical strategy (Borovecki & Habek 2010). Neurodegenerative illnesses like AD and PD share many pathophysiological characteristics, such as the accumulation of misfolded proteins, disruption of protein clearance, dysregulation of transcriptional mechanisms, or impairment of energy metabolism. The identification of common neurodegenerative pathways may lead to new therapeutic approaches for the treatment of various neurodegenerative illnesses. Ultimately, a more individualised approach to patient care may result from a combination of medicines that focus on distinct neurodegenerative pathways rather than specific diseases.

Personalised medicine may seem like a long way off when it comes to diagnosing and treating neurodegenerative illnesses, some of the methods discussed for instance, diagnosing AD patients by combining neuroimaging and proteomic analysis, The application of specific biological treatments in MS or the discovery of novel biomarkers in PD by miRNA profiling already demonstrate how future advancements in clinical neurology will be incorporated into routine practice. One of the major goals in the treatment of neurodegenerative illnesses that still requires noticeable progress is the use of stem cells. It is reasonable to assume that, given the advancements of the last ten years, more widespread use of individualised treatment for neurological illnesses will occur in the years to come. Improvements in molecular diagnostics could make it possible to identify diseases early on, even before symptoms appear. Better and faster diagnostic processes may therefore result in the development of new neuroprotective medicines and a more timely deployment of currently available ones, improving patient care.

Figure 2. Personalized neurology techniques are being advanced by modern technologies

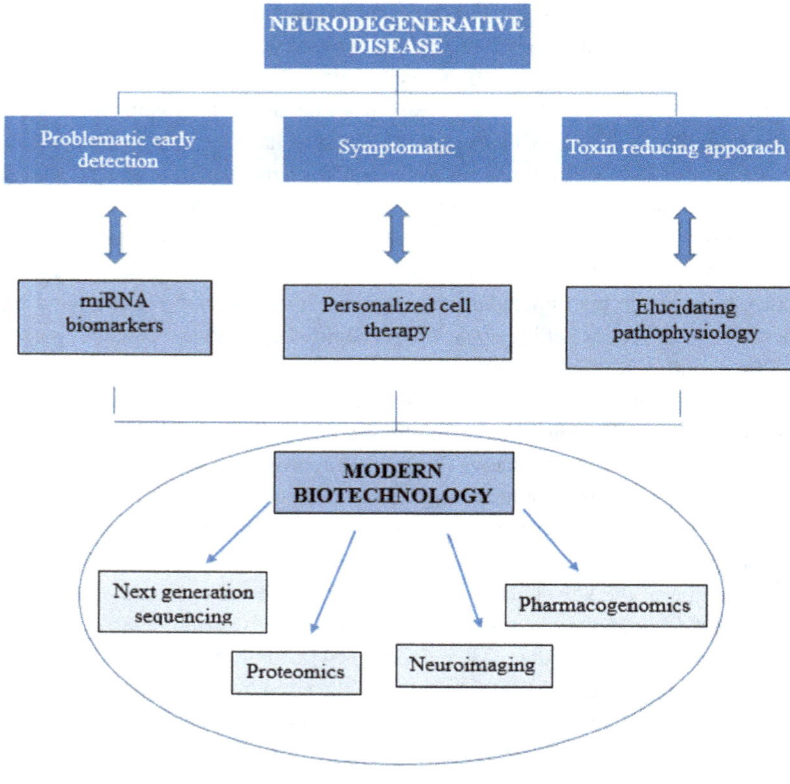

3.1. Nanotechnology In Neurodegenerative Disease

Nanotechnology holds enormous promise for the diagnosis, treatment, and monitoring of neurodegenerative diseases (NDDs), such as Huntington's, Parkinson's, and Alzheimer's diseases. This technology enables molecular-level interactions with biological systems through the use of nanoscale materials, tools, and systems. The blood-brain barrier (BBB), a major barrier to conventional medication administration, can be overcome by using nanotechnology to help deliver therapeutic pharmaceuticals to particular brain regions in neurodegenerative diseases. Drugs can be encapsulated in nanoparticles like liposomes, dendrimers, and gold nanoparticles to improve bioavailability and allow for controlled release. Additionally, nanotechnology offers potential for use in neuroimaging applications, improving the resolution and specificity of imaging modalities such as MRI or PET scans, and it can help with early diagnosis by enabling sensitive detection of biomarkers linked to NDDs. Furthermore, nanomaterials can interact with oxidative stress, misfolded proteins, and amyloid plaques—all of which are linked to NDDs to provide new treatment approaches. According to recent research, nanoparticles may aid in lowering inflammation, shielding neurons from oxidative damage, and even encouraging the growth of new neurons. Even though there are still issues including toxicity, biocompatibility, and regulatory barriers, further research could transform the way neurodegenerative illnesses are treated (Patel et al., 2020).

4. ROLE OF NANOCARRIERS IN TARGETED DRUG DELIVERY SYSTEMS:

The following section examines many medication delivery techniques based on nanotechnology to determine whether they can achieve the desired therapeutic effects by passing through the blood-brain barrier.

4.1. Lipid Based Nanoparticles (LNPS)

One of the most popular technologies for carefully delivering medications to the brain is LNPs. Because these systems are naturally safe, biocompatible, and biodegradable, they are a desirable substitute for drug nanocarriers. Because of their intrinsic lipidic makeup, which is similar to that of the blood-brain barrier, these substances can more easily cross cells and enter the brain. Waxes, phospholipids, fatty alcohols, triglycerides, monoglycerides, and fatty acids are often the primary constituents of LPNs (Gordillo-Galeano & Mora-Huertas 2018). The three main forms of LNPs that are commonly used in research and real-world applications, especially when it comes to CNS illnesses, are liposomes, solid lipid nanoparticles (SLNs), and nanostructured lipid carriers (NLCs). These LNPs' potential to effectively manage these disorders has attracted a lot of attention (Sohail et al., 2023). This specific nanocarrier exhibits increased adaptability by encapsulating medications with both hydrophilic and hydrophobic characteristics. However, some restrictions make it difficult to produce them in large quantities, which in turn prevents their regular use (Puri et al 2009). Low stability, an intricate manufacturing procedure, unforeseen polymorphism transitions, and drug release during storage are a few of these disadvantages. Despite these limitations, LNPs, including SLNs and NLCs, are rapidly becoming viable diagnostic and treatment approaches for AD and other central nervous system conditions, including depression and anxiety. With encouraging results, recent preclinical research has combined contemporary LNPs with animal models of AD.

Almuhayawi created LNPs loaded with pomegranate extract and assessed them using an aluminium chloride-induced Alzheimer's disease model in rats. Alkaloids, tannins, and antioxidants are abundant in pomegranate extract. When compared to untreated controls, these LNPs enhanced Aβ deposits, NFTs, brain homogenate antioxidant biomarkers, and cognition scores (Almuhayawi et al., 2020). Giacomeli et al. examined the potential neuroprotective benefits of LNPs loaded with curcumin in a mouse model of AD. Although curcumin has been demonstrated to offer therapeutic potential for a number of conditions, its limited solubility and, thus, its low bioavailability limit its pharmacological effects. The authors of this paper developed a hybrid nanocarrier with a PLC shell and a lipidic core to increase solubility. When this nanocarrier was given to animals, their hippocampal biomarkers for neuroinflammation and serum levels of cortical homogenates decreased, and their spatial memory improved. LNPs were combined with natural substances by Pinheiro et al (Giacomeli et al., 2019). Numerous fruits and vegetables include quercetin; The bioactive component of the load was a flavonoid. It's an excellent source of antioxidants. The researchers added Tf to the LNPs in order to enhance LNP transport over the BBB. According to reports, these LNPs increase quercetin's BBB penetration without endangering hCMEC/D3 cells, a widely used in vitro model of the human BBB. An additional investigation revealed that the use of NLCs loaded with curcumin enhanced tissue architecture and reduced the expression of COX-2, TNF-α, and p-NF-κB in brain tissues. In relation to anxiety and depression, these findings showed that curcumin NLCs might have a neuroprotective effect.

Despite having antimicrobial, anti-inflammatory and anti-cholinesterase qualities, Bisabolol, a sesquiterpene alcohol, possesses a restricted bioavailability due to its poor solubility. In an in vitro model of amyloid beta (A), Sathya et al. synthesised bisabolol-loaded cholesterol LNPs evaluated their neuroprotective properties utilising the mouse neuroblastoma cell line Neuro-2a, which proliferates quickly. Dara et al. recently looked at the possibility of using erythropoietin (EPO) to treat AD. EPO is neuroprotective in a number of disorders, including spinal cord injury, cerebral ischaemia, epilepsy, and diabetic neuropathy. In AD and PD, EPO simultaneously controls neurogenesis and enhances neuronal survival. Due to its enormous molecular weight, rapid circulatory clearance, and hydrophilicity, EPO can only cross a limited portion of the blood-brain barrier. In a model of AD, EPO in Westar rats, encapsulated in SLNs markedly reduced hippocampal Aβ production and oxidative stress. Additionally, rats administered EPO-LNP showed better spatial memory than their littermates who did not receive treatment (Dara et al., 2019).

4.2. Polymer-Based NPS

A wide range of medications are administered using polymeric nanoparticles (NPs) due to their enhanced biocompatibility, biodegradability, and prolonged shelf life, all of which enable controlled drug release at specific locations. Apart from natural polymers such as albumin, alginate, and chitosan, synthetic polymers such as polylactic acid, poly glycollic acid, and polyacrylates are frequently utilized in the production of NPs. Because of its neuroprotective potential and BBB permeability, biodegradable PLGA NPs are being used to treat AD (Aalinkeel et al., 2018). An efficient treatment for AD is curcumin, a polyphenolic substance having anti-amyloid and antioxidant qualities. However, its limited water solubility greatly limits its usefulness; creating NPs can help with this issue. This led to the development of Tet-1 peptide-linked curcumin by Mathew et al. Its low water solubility significantly limits its use; however, PLGA-loaded NPs and research on the brain's targetability for AD can help with this issue. The findings demonstrated that Tet-1 promoted BBB translocation and that NPs were attracted to neurons. The generated NPs exhibit anti-oxidative and anti-amyloid qualities without endangering cells (Zhao et al., 2012). The PLGA encapsulation technique largely has little effect on curcumin's natural characteristics. In a different study, PLGA NPs laden with curcumin were injected intravenously into the adult rat hippocampal and beneath the ventricle regions to promote neurogenesis. By triggering the Wnt/catenin pathway, the generated NPs enhanced brain development-related gene expression and in vitro cell proliferation. In rats that produce amyloid beta, NPs can promote neurogenesis and improve memory and learning issues. PLGA curcumin NPs are a viable therapy option for AD because they may enhance the brain's capacity to heal itself (Tiwari et al., 2014).

Andrographolide (AG), a diterpene, can lessen inflammation brought on by NDs. Nevertheless, AG's low bioavailability accounts for its limited utility. The limitations of AG in human serum were addressed by developing NPs based on human serum albumin (HSA) and poly ethyl cyanoacrylate (PECA). In contrast to HSA NPs, which were made by chemical cross-linking, PECA NPs were made by emulsion-polymerization. PECA NPs caused rapid cell layer destruction, which was the opposite of HAS NPs' function of increasing AG permeability while maintaining the BBB. As a result, HSA is a likely candidate for andrographolide trafficking that penetrates the BBB (Guccione et al.2017). The capacity of PLGA-functionalized quercetin nanoparticles to inhibit and deconstruct Aβ-42 fibrils was evaluated. Synthesised NPs reduce Zn2+/Aβ42-induced neurotoxicity and increase neuronal cell survival. Additionally, behavioural studies show that PLGA-QT NPs help APP/PS1 mice with their memory and cognitive impairments. In a different experiment, solvent evaporation was used to create PLGA NPs with phytol encapsulation. The NPs were evaluated for their form, cytotoxicity, antioxidant activity, anti-aggregation and disaggregation, AchE inhibitory action, and in vitro drug release. The results showed that NPs had a smooth surface, a spherical form, and a nanoscale dimension. While exhibiting mild cytotoxicity, phytol-PLGA NPs reduced AchE activity and eliminated amyloid aggregates in Neuro2a cells. According to the aforementioned studies, Nanocarrier systems encapsulated in phytochemicals that can penetrate the blood-brain barrier provide a new approach to treating AD (Sathya et al., 2018).

4.3. Nanogels

Nanogels are a promising targeted drug delivery technology because of a number of notable features, including their extraordinary permeation properties, core-shell architecture, excellent colloidal stability, considerable drug loading capacity, and sensitivity to environmental stimuli. According to both in vitro and in vivo investigations, these nanoscale drug carriers have better penetration capacities in the brain parenchyma than traditional nanomaterials (ud Din et al., 2017). The ability of nanogel-based technology to be nanoengineered and circumvent physiological barriers makes it ideal for the non-invasive treatment of neurological conditions, such as neurodegenerative illnesses. The name "NanoGel" was first used to describe hydrophilic polymer network particles by Vinogradov et al. in 1999. When polyethylene glycol and polyethyleneimine are crosslinked, these particles are created, and they can carry antisense oligonucleotides (PEG-PEI) (Xiao et al., 2015). The advantages of hydrogel and nanocarrier systems are combined in colloidal hydrogel nanoparticles in three dimensions, which have a diameter of less than 200 nm. For manufacturing purposes, covalent bonds, either chemical or physical, have been used to join nanogels. They can also be divided into groups based on how they respond to external stimuli. All things considered, nanogel technology may offer state-of-the-art care for CNS conditions. As a result, many nanoformulation studies are now heavily focused on creating safer, more effective, and more affordable nanomedicines.

Because new nanogel technologies are constantly being developed, a variety of nanogel-based strategies are currently possible. These nanogel-based nanovehicles are being used to manufacture a number of neuroprotective drugs, such as nerve growth factors and peptides, nucleic acid. The early stages of development of Drug delivery via nanogel techniques define their current condition (Patel et al., 2021). Another efficient PD therapy method is the removal of Lewy bodies (Liu et al., 2020). The potent substances curcumin and epigallocatechin-3-gallate (EGCG) can stop amyloid from aggregating. The change in hyaluronic acid connected these two inhibitors. The findings demonstrated that dual-modified nanogels (CEHA) produced 69 and 55% higher inhibition with respect to single-modified nanogels loaded with EGCG or EHA. CEHA significantly improved SH-SY5Y cell survival, according to the lab investigation (Jiang et al., 2018).

4.4. Dendrimers

In biomedicine, NDs are treated with dendrimers, the tiniest and most precisely defined nanoformulations. Different kinds of nanocarriers that help deliver drugs and genes to the brain can be created by varying the size, core-shell mix, and surface functional groups of dendrimers. Significantly, Numerous treatment options are available for the detection. In the management of AD, PD, and prion diseases due to dendrimers' potent anti-amyloidogenic action. PAMAM dendrimers have also been used to create intriguing anti-amyloidogenic tools. PAMAM dendrimers in three generations (G3, G4, and G5) disaggregated existing fibrils in amorphous forms and prevented amyloid creation of the Aβ peptide, which has 28 amino acids (Aβ 1–28). The intensity of the impact increased with concentration and was proportional to the number of dendrimers generated. The larger PAMAM dendrimer generations were the most effective in interfering with the amyloid aggregation process. On the other hand, cationic carbosilane (CBS) dendrimers have been reported to be efficacious therapeutic agents in a rotenone-induced mice model of Parkinson's disease. In order to inhibit α-syn fibrillation and encapsulate and complex rotenone, dendrimers showed dual neuroprotective action. Surprisingly, this kind of dendrimer proved better in preventing rotenone damage than PAMAM, phosphorous, and viologen-phosphorus dendrimers. In order to stop the spread of prion infections, medical equipment is also sterilised using dendrimers that contain phosphorus (Klajnert et al., 2006). Dendrimers exhibit potential as a therapeutic drug delivery mechanism for AD therapy (da Silva & de Queiroz 2020).

4.5. Nanocomposites

A material that combines the qualities of several distinct materials is called a composite. It includes a matrix and possibly a period of reinforcement. Tubes and threads are examples of one-dimensional objects; clay and other layered materials are examples of two-dimensional objects; and spheres are examples of three-dimensional objects. Composites are generally categorised using the following attributes: Laminated composites are composed of thin layers of fully bonded materials, fibrous composites are composed of threads entangled in a matrix, and particulate composites are composed of particles embedded in a matrix. The three primary types of composite materials are as follows. When a polymer is utilised as the matrix, these composites are referred to as "polymer-based composites." Numerous biodegradable and bioresorbable substances as well as design advancements for biomedical devices have been the subject of recent laboratory and clinical research. Polymer-based composites have demonstrated promise in numerous disciplines, including medication delivery, tissue engineering, and load-bearing applications (Shameem et al., 2021).

It is feasible to customise the chemical, physical, and mechanical characteristics of a composite material; in fact, by combining the best qualities of its constituents, the final product may have interesting qualities that the individual constituents frequently lack. We have discovered novel molecular targets that could slow down neurodegeneration as our understanding of the pathogenic mechanisms behind AD/PD has expanded (Kabiri et al., 2004). Proteases like Hsp70 and Hsp27, which may degrade toxic protein aggregates, and prosurvival chaperone proteins like these are possible choices for these protein-based therapies. Unfortunately, because of the limited permeability of the blood-brain barrier, systemic injection significantly restricts how much protein can get into the brain. This leads to the need for high dosages and/or repeated treatments, which increases adverse effects and reduces patient compliance. Improved cholinesterase inhibition and donepezil's brain pharmacokinetics have been suggested with the use of an inorganic clay nanocomposite system (Singh AK, et al., 2020).

5. SECONDARY METABOLITES AS NEUROTHERAPEUTIC AGENTS:

Secondary metabolites are biological compounds that do not directly contribute to an organism's development, growth, or reproduction. Rather, they have a significant impact on a number of physiological functions, including communication between species, competition for resources, and the risk defence (Erb & Kliebenstein 2020). There are differences in secondary metabolic pathways both within and across organisms. They are frequently distinct metabolic precursors, intermediates in primary metabolism, or enzymatically generated from primary metabolites. Secondary metabolites that are beneficial to health and well-being are frequently found in plants. Vacuoles, glandular trichomes, cuticles, and oil cells are among the sites where plant secondary metabolites are biosynthesised and stored (Divekar et al., 2022). Senescent cells are specifically targeted by senolytic secondary metabolites for programmed death or induced apoptosis, which eventually results in their removal from tissues and potential replacement by healthy cells (Xie et al., 2023). In preclinical and clinical research, senolytics have demonstrated significant promise in improving tissue regeneration capacity and mitigating age-related diseases (Al-Azab et al., 2022). Although research on senolytic secondary metabolites in the management of dementia is still in its early stages, a number of bioactive substances have demonstrated promising neuroprotective qualities via their senolytic action. The metabolite and cellular environment affect the senolytic secondary metabolites' mode of action and effectiveness (Xie et al., 2023). For instance, certain senolytic secondary metabolites will trigger the intracellular p53 or P16INK4A pathways in response to cellular stress, resulting in a senescent cellular state (Mijit et al., 2020).

Other secondary metabolites take use of particular traits or requirements of senescent cells to act through vulnerabilities linked to senescence. These weaknesses include dysregulated stress response pathways, greater dependence on anti-apoptotic proteins, and changed metabolic profiles (Kirkland & Tchkonia 2017). Senolytic secondary metabolites can cause senescent cell death while preserving healthy cells by focussing on these weaknesses. Additionally, secondary metabolites frequently include antioxidant and anti-inflammatory qualities, which may enhance their overall therapeutic effects. These substances, which are senomorphics rather than genuine senolytics, may lessen the detrimental effects of senescent cells on surrounding tissues by lowering the inflammation and oxidative stress linked to SASP (Davoodi-Roodbordeii et al., 2019). Instead of completely eliminating senescent cells, senomorphic secondary metabolites lessen their harmful effects. They are a relatively new behavioural type of cells that have not gotten as much attention as senolytics. For instance, starting in 2014, a standard PubMed search for "senolytic" yields roughly 1,165 results. By comparison, starting in 2019, a standard PubMed search for "senomorphic" yields roughly 75 results. Nonetheless, an increasing amount of data is shedding light on how they reduce the pro-inflammatory SASP. As a result, they have become attractive options for cutting-edge therapeutic approaches in age-related illnesses like cancer, neurological conditions, and heart conditions.

Senomorphic secondary metabolites provide a sophisticated strategy to mitigate the negative consequences of cellular senescence by utilising the power of senescence modulation, hence fostering tissue regeneration and possibly prolonging life expectancy. It is necessary to conduct more study on the processes underlying these drugs' senomorphic effects. Secondary metabolites can be divided into numerous categories and subcategories according to their chemical makeup and biosynthesis routes. The two main categories of senolytic secondary metabolites that are currently understood are alkaloid chemicals and phenolic compounds (Rizvi et al., 2022). Phenolics are byproducts of the metabolism of phenylpropanoid, which occurs downstream of the malonic acid and shikimic acid pathways. They include one or more aromatic rings with one or more hydroxyl groups. The six subfamilies of the flavonoid subclass of phenolic compounds—flavones, isoflavones, flavonols, flavanones, flavan-3-ols, and anthocyanidins—are the most prevalent type seen in fruits and vegetables (de la Rosa et al., 2019). Alkaloids are produced by a variety of metabolic processes and are distinguished by a heterocyclic ring containing nitrogen (Matsuura & Fett-Neto 2015). For instance, several alkaloids need aromatic amino acids produced by the shikimate pathway as metabolic precursors. Alongside the terpenoid pathway, the shikimate pathway also directly manufactures some alkaloids. Similar to the shikimate system, the polyketide pathway yields a large number of alkaloid precursors (Gutiérrez-Grijalva et al., 2020).

Figure 3. Various secondary metabolites used as neurotherapeutic agents

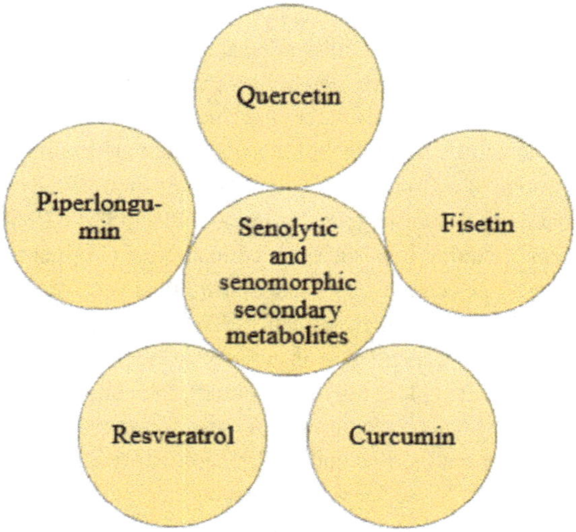

6. AI DRIVEN DESIGN FOR NANOPARTICLE DELIVERY SYSTEM

The two largest obstacles to the development of new medications are decreased efficacy and high expense. The process of developing new drugs could be completely transformed by combining AI technology with processing capacity. In this regard, pharmaceutical behemoths work with AI-driven drug discovery companies (Mak & Pichika 2019). AI is used by computers to mimic human intelligence in order to improve society (Sriamornsak & Waijanya 2021). AI includes functions such as information gathering, self-correction, estimated sketching, and more. It comprises the subfields of machine learning (ML), deep learning (DL), and artificial neural networks (ANNs) and is divided into three categories: artificial general, narrow, and super intelligence. The goal of artificial intelligence (AI), a vast area of computer science, is to build machines that are capable of solving problems, comprehending language, and identifying patterns—tasks that normally require human intelligence. AI is a general phrase that includes a number of subfields, such as deep learning and machine learning. Without being specifically coded for every activity, machine learning (ML), a subset of artificial intelligence, allows systems to learn from data and gradually get better at what they do. ML algorithms employ statistical techniques rather than hard-coded rules to find patterns in data and utilise that information to influence predictions or choices. Neural networks with several layers, often known as deep neural networks, are the subject of DL, a specialised subset of machine learning. DL is especially useful for jobs like picture identification, natural language processing, and autonomous driving since these networks can automatically find complex patterns in vast volumes of data.

The broad idea of artificial intelligence (AI) encompasses ML and DL as techniques for producing intelligent behaviour. Medical science makes extensive use of these methodologies. The creation of drug delivery systems (DDSs) based on nanotechnology is the main application for ML and DL. These particles are as tiny as atoms and molecules, as the name suggests. As a result, AI helps with NP-related data analytics and mining. AI and nanotechnology together are expanding quickly. AI is a crucial tool for combining many medicines into a single nanocarrier, and different kinds of nanocarriers can be used concurrently for a range of patient medical conditions (Aumklad et al., 2024). Nanotechnology makes it possible to optimise NP design by fine-tuning its features, improving molecular profiles, and diagnosing diseases early. It aims to increase the synergistic effect of medications while lowering their toxicity, hence enhancing NPs' potency, targetability, and dosage (Saikia et al., 2022). AI can support anticancer medication nanoparticulate systems' tailored delivery. At malignant areas, diagnostic NPs are employed to identify biomarkers. Additionally, AI-related technologies can assist in identifying the forms and complexity of cancer. Its sophisticated methodology can be applied to therapeutic efficacy assessment, biomarker identification, and NP interactions with various medications (Das et al., 2023). Although lipid nanoparticles (NPs) have a number of uses in the creation of vaccines, gene therapy, and protein replacement, they may also exhibit toxicity, ineffective delivery, and immunological activation. Designing ionisable lipids and optimising the targeted conjugations of lipid nanoparticles for improved performance are two tasks that AI can assist with. Their combination improves the safety and effectiveness of the medications that are given. For the targeted conjugation of NPs, DL models of AI aid in the modelling of antibodies and the depiction of molecular data (Yuan et al., 2024).

Diagnostic NPs are used to create a patient's disease profile, which determines which therapeutic nanocarriers are used. Patient and intra-tumor heterogeneities may make it challenging to analyse their output. In order to improve diagnostic and therapeutic efficacy, artificial intelligence (AI) can get beyond these restrictions by using classification algorithms, pattern analysis, and nanocarrier design optimisation based on anticipated interactions with medications, the immune system, cell membranes, and biological fluids. Therefore, by enhancing patient data collection and nanomaterial creation, the combination of AI and NPs advances precise cancer therapy. By creating an algorithm using prior big datasets from existing cases as input, machine learning determines the best solution. Specific biomarkers can be detected with great specificity by single biomarker nanosensors utilised for diagnostic screening of patient liquid samples (Adir et al., 2020). AI employing an ANN for multiplex nanosensor data processing can find biomarker patterns and detect various targeted molecules to provide information on a patient's health state, subtypes, and staging. The lack of an extensive database makes it difficult to create a successful cancer treatment plan using inorganic NPs. In this regard, machine learning (ML) as a transformative tool revealed that inorganic NPs may enhance patient outcomes in cancer treatment by data mining an assembled database of inorganic NPs with experimental datasets in cancer nanomedicine from 745 preclinical studies (Mendes et al., 2024). However, current applications of AI and ML in nanomedicine are still limited, with only a few validated algorithms and clinically tested models available. Further research and real-world trials are required to substantiate their practical reliability and effectiveness.

Figure 4. Advantages of integrating AI and NPs

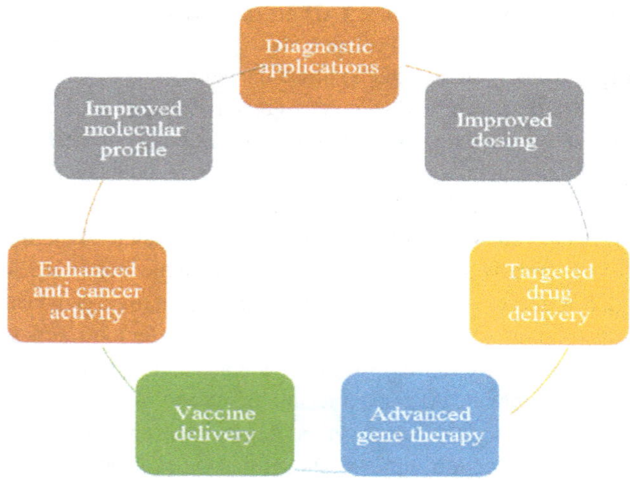

7. CHALLENGES IN DRUG DELIVERY ACROSS BLOOD BRAIN BARRIER

When treating neurological disorders drug distribution to the central nervous system (CNS) remains a challenge. The blood-brain barrier (BBB) is the main obstacle to CNS medication delivery because it prevents pharmaceuticals from reaching the brain parenchyma. Only lipophilic molecules with a molecular weight of less than 400–500 Da may typically pass through the bloodstream and enter the brain through the transcellular pathway, which is how the blood-brain barrier restricts drug delivery (Pardridge 2007). Increasing intravascular drug distribution by adjusting capillary permeability or the drug itself is one way to try to get beyond this obstacle (Siegal & Zylber-Katz 2002) as well as by local delivery into brain fluids, such as the interstitial fluid of brain tissue of the brain ventricles (Groothuis 2000). Research efforts have concentrated on creating new methods to more efficiently deliver medications to the brain in acknowledgement of the challenges related to CNS delivery over the BBB. These tactics include using endogenous transporters or receptors at the blood-brain barrier or modifying the properties of the medications. In several situations, Drugs have been successfully delivered into brain tissue through the nano-particulate carriers. (Muller & Keck 2004) or by antibodies that target BBB receptors, including insulin, transferrin, and low-density lipoprotein receptors, which enhance receptor-mediated transcytosis (RMT), which allows them to be transported across the BBB (Lee et al., 2000). The physiology and pathology of the central nervous system must be better understood, though, and relevant preclinical illness models must be used to confirm the quantitative brain distribution kinetics and effectiveness of targeted medications before they can be translated to the clinic (de Boer et al., 2007).

8. Exploration Of Novel Natural Compounds with Potential Neuroprotective Effect

The increased prevalence of neurodegenerative disorders like Parkinson's and Alzheimer's has drawn a lot of focus to the search for new natural substances that may have neuroprotective properties. Natural substances with a variety of pharmacological characteristics, including as flavonoids, alkaloids, and diterpenoid derivatives, can alter vital pathways for neuronal survival, including the BDNF-TrkB-CREB and PI3K/AKT pathways. Over time, there has been a growing interest in natural compounds as potential substitutes for manufactured drugs. Numerous natural substances from various sources have been suggested as treatments for neurodegenerative illnesses because of their neuroprotective properties. They have been shown to target not only the symptoms but also the different facets of illness etiology (Leuci et al., 2020). Four types of natural substances are distinguished based on where they come from: terpenoids, polyketides, alkaloids, and phenylpropanoids. A class of naturally occurring chemical molecules that contain nitrogen are called alkaloids. Numerous plants, animals, microbes, and fungus contain them. Examples of alkaloids that have been investigated for their possible medicinal application in the management of neurodegenerative diseases. compounds that have a three-carbon propene tail and an aromatic ring, phenylpropanoids—also referred to as cinnamic acids. Five categories can be used to group phenylpropanoids: lignins, phenolic acids, stilbenes, coumarins, and flavonoids (Neelam et al., 2020).

Phenylpropanoids, including resveratrol, curcumin, and epigallocatechin gallate (EGCG), have garnered a lot of attention lately for the treatment of neurodegenerative illnesses because of their anti-inflammatory and antioxidant qualities. However, there isn't much proof that polyketides, which are made by bacteria, fungi, plants, mammals, and some marine species, have neuroprotective properties (Wang et al., 2020). The most significant class of natural substances, terpenoids, sometimes referred to as isoprenoids, comprise more than 60% of all known compounds in plants, microorganisms, and marine sponges. Cannabinoids and other terpenoids have been shown in numerous studies to have pharmacological qualities for the treatment of neurodegenerative disorders (Pal & Das 2019). Human clinical trials have been carried out recently to assess the medicinal efficacy and possible adverse effects of natural substances (Cox-Georgian et al., 2019).

9. CLINICAL TRIALS USING NATURAL COMPOUNDS FOR NEURODEGENERATIVE DISEASES

9.1. For Alzheimer's Disease (AD)

Clinical trials have evaluated natural substances' safety and effectiveness in treating Alzheimer's disease. A total of 409 individuals with mild to moderate Alzheimer disease were randomized in a multicenter, double-blind trial to receive either high-dose B-vitamin supplements (daily 5 mg folic acid B9+ 25 mg vitamin B6 + 1 mg vitamin B12) or identical placebo for 18 months; 340 participants (202 active, 138 placebo) completed the trial, and the regimen lowered homocysteine but did not slow cognitive decline (Aisen et al., 2008). In order to evaluate vitamin B3, In a double-blind, placebo-controlled phase II study, randomised clinical investigation, the University of California (USA). Although the clinical experiment was finished in 2014, the findings have not yet been made public. The human brain's growth and function depend on docosahexaenoic acid (DHA), the most abundant omega-3 fatty acid in the brain. The idea that DHA may be helpful in AD treatment due to its anti-amyloid, antioxidant, and neuroprotective properties is supported by data from several animal models (Andrade et al., 2015).

Another natural substance whose safety and effectiveness as a treatment for AD have been examined in clinical trials is curcumin. Curcumin is found in the root of Curcuma longa L. This substance can disintegrate Aβ fibrils, inhibit Aβ aggregation, and decrease β and γ-secretase levels. The AD rat model's memory and spatial learning deficiencies were thus alleviated (Liu et al., 2016). Homotaurine, also known as tramiprosate, is an amino acid that is permeable to the blood-brain barrier (BBB) and inhibits the development of amyloid fibrils. It is typically found in algae. In rat hippocampal cells, homotaurine restores Aβ-induced long-term potentiation inhibition and protects against Aβ-induced neurotoxicity in both in vitro and in vivo settings (Krzywkowski et al., 2007).

9.2. For Parkinson's Disease (PD)

In clinical studies for the treatment of Parkinson's disease, nicotine has drawn the most interest of any natural substance. It has been observed that nicotine improves certain motor and cognitive symptoms and has a neuroprotective impact. Using other mechanisms, caffeine is another common natural substance that has significant neuroprotective, motor, and cognitive effects in the treatment of Parkinson's disease (Ren & Chen 2020). A natural substance called cannabidiol has been shown to help PD sufferers with their tremors, anxiety, and psychosis. Vitamins are another possible therapeutic option for Parkinson's disease. Recent clinical studies suggest that vitamins' antioxidant qualities and impact on the control of gene expression may aid in the treatment of

Parkinson's disease (Zhao et al., 2019). Among its many physiological roles, vitamin E regulates gene expression as well as cognitive and motor processes.

9.3. For Multiple Sclerosis (MS)

Since current research has linked low vitamin D blood levels to an increased risk for Multiple Sclerosis, vitamin D3 is without a doubt the most researched natural substance in clinical trials for Multiple Sclerosis therapy. By lowering the inflammatory response in the central nervous system have demonstrated the potential to lessen the symptoms of multiple sclerosis. Furthermore, there is evidence that MS patients have reduced blood levels of vitamin A. It causes pathogenic immune cell production in the central nervous system and immunological tolerance dysregulation (Petracca et al., 2021). Animal products including red and organ meats include alpha-lipoic acid, a naturally occurring substance with antioxidant qualities. This chemical serves a variety of biological purposes and is now the bioactive substance with the strongest evidence of effectiveness in Multiple Sclerosis. Alpha-lipoic acid has been shown in both in vitro and in vivo investigations to protect neuron cells by reducing immune cell infiltration in the central nervous system and by suppressing the expression of inflammatory mediators. Additionally, the chemical enhanced the patients' gait, balance, and walking skills in clinical trials. During the last decade, alpha-lipoic acid has been evaluated in seven different investigations, either alone or in combination, making it one of the most studied natural compounds in MS treatment clinical trials (Xie et al., 2022). Because omega-3 polyunsaturated fatty acids reduce glutathione reductase, inflammatory markers, and the relapse rate, they have shown promise in the treatment of multiple sclerosis. Therefore, additional clinical trials were carried out to evaluate the potential of these omega-3 polyunsaturated fatty acids for the treatment of multiple sclerosis. Melatonin, another natural chemical, is being investigated as a therapeutic target in multiple sclerosis clinical trials after recent studies revealed that melatonin release is dysregulated in MS patients (Skarlis & Anagnostouli 2020).

9.4 Recent Clinical Studies and Result

Catalytic gold nanocrystals (CNM-Au8): Parkinson's disease and multiple sclerosis CNM-Au8 demonstrated detectable brain target engagement on 31P-MRS (increases in ATP-related signals and phosphorylation potential) in integrated Phase-2 imaging studies (REPAIR-PD and REPAIR-MS). This suggests the oral nanocrystal enters the central nervous system and modifies bioenergetic metabolism, a mechanistic signal consistent with disease-modifying potential. Over the course of prolonged treatment (up to several years in some reports), an open-label long-term extension of the VISIONARY-MS program reported sustained, clinically significant improvements in low-contrast visual acuity and several cognitive measures. The company-reported interim analyses and publications highlight these functional and imaging signals while pointing out the need for larger blinded RCT confirmation. Curcumin in nanoformulations, or nanocurcumin—small neurology pilots and multiple sclerosis: Nanocurcumin has been studied as an adjuvant in relapsing-remitting multiple sclerosis and other inflammatory neurologic settings in numerous randomized, double-blind small trials and pilot RCTs. Nanocurcumin was consistently safe and demonstrated immune-modulatory effects (restoration of dysregulated miRNA profiles, increases in regulatory T-cell markers, decreases in proinflammatory cytokines and Th17 activity) as well as sporadic minor improvements in clinical/functional scores across studies. Nevertheless, sample sizes are small, endpoints are heterogeneous, and clinical efficacy is still unclear pending larger standardized trials. Because of these studies, nanocurcumin is one of the phytochemical-nanocarrier combinations in neuroinflammation that has been studied the most in clinical settings (Guan et al., 2024). Translational state and early human studies on nano-PSO/pomegranate seed oil (punicic acid): Pomegranate seed oil nano-emulsions, or Nano-PSO, have shown strong neuroprotection in animal models in preclinical research (delayed illness onset, decreased lipid oxidation, and neuronal loss in hereditary neurodegenerative models). There is currently no large positive Phase-3-style efficacy readout despite the existence of a double-blind, placebo-controlled clinical trial registration (e.g., NCT05142085) and small clinical investigations in related cognitive populations. Recent small human trials of PSO in mild cognitive impairment reported largely neutral results on most cognitive endpoints, indicating promising preclinical biology but limited conclusive clinical evidence thus far. Rather than strong, extensive functional disease change, the most obvious translational signal to date is immune/biomarker modulation (nanocurcumin RCTs) and mechanistic/biomarker evidence (e.g., CNM-Au8 ^31P-MRS target engagement). Secondary metabolite nanoformulations continue to show promise for enhancing CNS delivery and target engagement, but bigger, carefully monitored RCTs with predetermined biomarker-driven patient selection are still required to show repeatable therapeutic effect (Salehi et al., 2020).

10. FUTURE PERSPECTIVES

The detailed and current study's strength is its examination of preclinical data from numerous meta-analyses pertaining to the role of naturally occurring bioactive chemicals in NDDs. In order to provide fresh, advantageous therapeutic viewpoints for the treatment of NDDs, all of these investigations concentrated on emphasizing methods of action that are cellular and molecular. A plant's or plant extract's pharmacological action is difficult to attribute to a single molecule or class of molecules. These results demonstrate the value of dietary phytochemicals for health as well as the role of ancient medicine in contemporary therapies (Nataraj et al., 2017).

While in vitro and in vivo animal models have demonstrated excellent results with pure natural products or plant extracts exhibiting antioxidant activity, their therapeutic consequences in human patients remain unclear and show limited success. This may be partly because single drugs are typically the focus of clinical trials. On the other hand, research on plant extracts with a range of secondary metabolites is more prevalent in studies conducted before clinical trials. Better antioxidant and disease-modifying activity can result from the additive or synergistic actions of various active substances combined in extracts. This could be one of the reasons why, for instance, the Mega natural-Az Grape Seed Extract clinical trial looked at the benefits of complete extracts as opposed to specific chemicals, like the resveratrol present in grape seed extracts, which had demonstrated beneficial effects in several Alzheimer's Disease trials. Phytochemical clinical study results have generally been quite inconsistent, possibly as a result of the manner in which these experiments are carried out. Clinical trials examine a broad range of subjects with varying genetic and environmental backgrounds, as well as varying disease symptoms and occasionally disease stages. It might be worthwhile to examine individual patients or smaller groups of persons who do exhibit notable improvement in order to ascertain why they might be responding to the treatment while others are not, rather than focusing on the overall relevance of the entire participant population. Even though this would result in additional trial costs, it might improve knowledge of the possible application of antioxidants in specific patient groups, such as those with a particular genetic or environmental background, which would also improve knowledge of the neurological condition (Berman et al., 2017).

According to the literature, the majority of clinical trials on natural antioxidants have primarily examined behavioural or cognitive benefits in patients; relatively few trials have been found that specifically evaluate oxidative stress or disease molecular markers.

Drug combinations and multi-target medications also require more research. This idea runs counter to the one medication, one disease (or one target, one disease) strategy that is now employed to treat a wide range of illnesses. However, given the complexity of neurodegenerative disorders and the fact that they can be linked to different cellular dysfunctions brought on by a variety of internal and external sources, a multi-target therapeutic strategy may be a better course of action. The 2014 drug approval of Namzaric, a combination treatment of donepezil hydrochloride, an AChE inhibitor, and memantine hydrochloride (extended-release), demonstrates the increased interest in multi-target medications in recent years. Additionally, natural antioxidants may meet this multi-target pharmacological profile. In addition to their demonstrated antioxidant action, they have also demonstrated anti-inflammatory, metal-chelating, AChE-inhibiting, and anti-protein aggregation properties that are advantageous to neuronal health. Using FDA-approved AChE inhibitors like galantamine, rivastigmine, or donepezil, or even less potent AChE inhibitors in combination with natural antioxidants, is an additional choice. A single highly effective but selective drug working on a single target may not be as effective as drug combinations with moderate activity acting on two or more targets (de los Ríos et al., 2010).

A deeper comprehension of the mechanisms underlying the disease is also required. The precise cause of the majority of neurodegenerative disorders is still unknown, if not entirely uncertain. Certain diseases, like Parkinson's disease (PD), can be brought on by genetic or environmental factors, while others, like HD, have a known cause—a genetic mutation in the huntingtin gene—but the precise disease mechanism or pathway is still unknown because the behavior of the mutated protein has not been fully understood. Frequently, the factors that specifically impact neurons include for instance, it's yet unknown if proteins have been altered. Research into better animal models for these illnesses is ongoing, and it may aid in both comprehending the neurological condition and finding possible medicines. Antioxidants found in natural products or plant extracts may be a fantastic choice for possible pre-symptom treatment because they are generally well-liked by the public and rarely have negative side effects. Naturally, a long-term trial would be required to confirm this (Gitler & Dhillon 2017).

11. REGULATORY FRAMEWORKS FOR ADVANCED THERAPIES

Advanced treatment medical products (ATMPs) for neurodegenerative illnesses are subject to changing regulatory systems, especially in the US and the EU. Innovative medicinal substances called ATMPs are made from genes, tissues, or cells. Especially for difficult situations like cancer and neurological diseases, they provide novel therapy options. A patient's own cells, donor cells, or genetically modified cells can be used to make these products. ATMPs are classified as biological products in both regions; the US mainly makes a distinction between gene therapy and cellular therapy,

For ATMPs to be safe and effective, regulation is essential. There are two main ways that ATMPs are distributed:

1. Marketing Authorization Application (MAA): Certain ATMPs may be granted legal permission to market, making them accessible to patients on a large scale.
2. Unapproved ATMP Access: Hospital exemptions, clinical practice programs, compassionate use, and expanded access programs are some of the ways that patients can obtain unapproved ATMPs. For individuals with neurodegenerative disorders who might not have access to alternative forms of treatment, these routes are especially crucial.

In general terms, ATMPs must meet the same regulatory and scientific standards as conventional pharmaceuticals. The same post-authorization and pharmaceutical compliance requirements that apply to pharmaceuticals must also apply to ATMPs, including manufacturing them according to GMP (good manufacturing practices). recommendations and planning and conducting clinical trials in compliance with GCP guidelines.

The ATMP Regulation served as the legal foundation for this Annex I modification as well: In addition to accounting for the unique characteristics of ATMPs, specific standards are established for them in order to establish the legal foundation for their development using a risk-based methodology. From the beginning of the product development program, The CAT has issued a scientific recommendation on the use of this risk-based method. Directive 2009/120/EC's high-level technical standards are further supported by scientific recommendations available on the EMA website (Maciulaitis et al., 2012).

When creating or promoting their product, ATMP developers must also consider other laws. The following are particularly crucial:

- Laws pertaining to pharmacovigilance follow-up and traceability
- Laws pertaining to pharmacovigilance follow-up and traceability (GTMPs)
- The medical device law applies to integrated ATMPs that contain one or more medical devices. It should be mentioned that the laws pertaining to medical devices are presently being revised, and the most recent details are available on the European Commission's website.
- In accordance with Directive 2002/98/EC and its implementing directives, ATMPs that use human blood or blood components establishing safety and quality requirements for the procurement, examination, processing, distribution, and storage of human blood and blood components that will serve as the foundation for ATMPs (Sauer et al., 2014).

In order to safeguard patients and guarantee the safety and efficacy of these treatments, ATMP control is essential. A strong regulatory framework is required to support the research and approval of these cutting-edge medicines since neurodegenerative disorders are complicated and frequently have limited therapeutic options. Regulations may need to change as ATMP research advances in order to stay up with emerging technology and scientific findings. To make sure that these frameworks properly serve people with neurodegenerative disorders, it will be essential for regulators, researchers, and healthcare professionals to have ongoing conversations, while the EU divides them into four categories: gene therapy, somatic cell therapy, tissue-engineered treatments, and integrated advanced therapies. An organized method to guarantee safety and effectiveness is reflected in the EU's regulatory framework, which was created in 2009 and includes a special Committee for Advanced Therapies to assess these items. The development of gene therapy for neurodegenerative illnesses faces significant obstacles, including safety risks, manufacturing difficulties, and ethical issues. To promote innovation while safeguarding patient welfare, strict regulatory control is required. Understanding these regulatory environments is crucial as regenerative medicine develops in order to implement novel treatments in clinical settings. ATMP regulatory regimes are intended to promote innovation and guarantee patient safety, particularly with regard to neurodegenerative illnesses. As the field of advanced medicines develops further, these frameworks' further growth will be essential (Reinhardt et al., 2013).

12. CONCLUSION

The future of neurodegenerative disease treatment lies in the combined synergy of several scientific fields, which includes advances in neuroscience, molecular biology, pharmacology, artificial intelligence, and regenerative medicine. The complexity of neurodegenerative disorders such as Alzheimer's, Parkinson's, and Huntington's necessitates a multidisciplinary approach that treats not only the symptoms but also the root causes of these conditions. Traditional techniques, such as pharmaceutical therapy, have proven unsuccessful in stopping disease progression, frequently focussing on symptomatic alleviation rather than disease-modifying therapies. However, the convergence of emerging technology presents exciting opportunities for more effective solutions. The combination of genomics and precision medicine enables the development of personalised medicines guided by genetic and molecular insights into each patient's unique disease. Similarly, neuroimaging and biomarkers are improving early diagnosis, allowing clinicians to identify diseases in their early stages and intervene before irreversible harm occurs.

In parallel, advances in neuroregenerative methods, such as stem cell therapies and gene editing tools like CRISPR, show great promise for reversing or slowing neurodegeneration. These innovations may pave the path for repairing or replacing damaged neurons, providing hope for regaining cognitive and physical function. Furthermore, artificial intelligence (AI) and machine learning are transforming the industry by speeding up drug discovery, optimising clinical trials, and giving in-depth insights into complicated disease pathways. AI-powered models can find new therapeutic targets, predict medication responses, and even simulate the efficacy of proposed treatments, significantly decreasing development time and cost. Collaboration across fields such as engineering and neurology is also essential, with bioengineering playing a major role in developing novel devices and biomaterials to aid brain function. Implantable devices for deep brain stimulation (DBS) or systems that administer medications directly to damaged brain areas, for example, could provide novel therapeutic choices. Furthermore, public-private partnerships and cross-sector collaboration are critical for converting scientific advancements into practical, real-world treatments. Governments, research institutions, and pharmaceutical corporations must work together to overcome regulatory barriers and make these revolutionary cures available to people in most need.

Finally, the future of neurodegenerative disease treatment will necessitate a comprehensive approach that integrates cutting-edge technology, scientific research, and patient-centric care.

While there is no one-size-fits-all answer, the interdisciplinary approach provides the best chance for personalised, effective therapies. The integration of diverse professions will not only help us discover medicines that can delay or stop disease development, but will also pave the path for creative cures that were previously thought to be out of reach. As we continue to unravel the mysteries of the brain and improve our understanding of neurodegenerative illnesses, we get closer to a future in which these crippling maladies are no longer fatal, but rather controllable or even reversible. Today's collaboration and innovation across disciplines provide a beacon of hope for patients and families afflicted by these disorders, making the future of neurodegenerative disease therapy brighter than ever before.

REFERENCES

Aalinkeel, R., Kutscher, H. L., Singh, A., Cwiklinski, K., Khechen, N., Schwartz, S. A., & Mahajan, S. D. (2018). Neuroprotective effects of a biodegradable poly (lactic-co-glycolic acid)-ginsenoside Rg3 nanoformulation: A potential nanotherapy for Alzheimer's disease? *Journal of Drug Targeting*, *26*(2), 182–193. PMID: 28697660

Adir, O., Poley, M., Chen, G., Froim, S., Krinsky, N., Shklover, J., & Schroeder, A. (2020). Integrating artificial intelligence and nanotechnology for precision cancer medicine. *Advanced Materials*, *32*(13), 1901989. PMID: 31286573

Aisen, P. S., Schneider, L. S., Sano, M., Diaz-Arrastia, R., Van Dyck, C. H., Weiner, M. F., & Thal, L. J. (2008). High-dose B vitamin supplementation and cognitive decline in Alzheimer disease: A randomized controlled trial. *Journal of the American Medical Association*, *300*(15), 1774–1783. PMID: 18854539

Al-Azab, M., Safi, M., Idiiatullina, E., Al-Shaebi, F., & Zaky, M. Y. (2022). Aging of mesenchymal stem cell: Machinery, markers, and strategies of fighting. *Cellular & Molecular Biology Letters*, *27*(1), 69. PMID: 35986247

Almuhayawi, M. S., Ramadan, W. S., Harakeh, S., Al Jaouni, S. K., Bharali, D. J., Mousa, S. A., & Almuhayawi, S. M. (2020). The potential role of pomegranate and its nano-formulations on cerebral neurons in aluminum chloride induced Alzheimer rat model. *Saudi Journal of Biological Sciences*, *27*(7), 1710–1716. PMID: 32565686

Andrade, S., Loureiro, J. A., Coelho, M. A., & do Carmo Pereira, M. (2015, February). Interaction studies of amyloid beta-peptide with the natural compound resveratrol. In *2015 IEEE 4th Portuguese Meeting on Bioengineering (ENBENG)* (pp. 1-3). IEEE.

Archer, M. C., Hall, P. H., & Morgan, J. C. (2017). [P2–430]: accuracy of clinical diagnosis of Alzheimer's disease in Alzheimer's disease centers (ADCS). *Alzheimer's & Dementia, 13*(7S_Part_16), P800-P801.

Aumklad, P., Suriyaamporn, P., Panomsuk, S., Pamornpathomkul, B., & Opanasopit, P. (2024). Artificial intelligence-aided rational design and prediction model for progesterone-loaded self-microemulsifying drug delivery system formulations. *Science, Engineering and Health Studies*, 24050002-24050002.

Berman, A. Y., Motechin, R. A., Wiesenfeld, M. Y., & Holz, M. K. (2017). The therapeutic potential of resveratrol: A review of clinical trials. *NPJ Precision Oncology*, *1*(1), 35. PMID: 28989978

Borovecki, F., & Habek, M. (2010). Development of novel genomic blood biomarkers for neurodegenerative diseases. *CNS & Neurological Disorders-Drug Targets-CNS & Neurological Disorders*, *9*(6), 669–678. PMID: 20942794

Clanet, M. (2008). Jean-Martin Charcot: 1825-1893. *International MS Journal, 15*(2), 59–62. PMID: 18782501

Cox-Georgian, D., Ramadoss, N., Dona, C., & Basu, C. (2019). Therapeutic and medicinal uses of terpenes. In *Medicinal plants: from farm to pharmacy* (pp. 333–359). Springer International Publishing.

da Silva, P. V., & de Queiroz, A. A. A. (2020). Long term multiple sclerosis drug delivery using dendritic polyglycerol flower-like microspheres. *Journal of Biomaterials Science. Polymer Edition, 31*(2), 188–206. PMID: 31609679

Dara, T., Vatanara, A., Sharifzadeh, M., Khani, S., Vakilinezhad, M. A., Vakhshiteh, F., & Mosaddegh, M. H. (2019). Improvement of memory deficits in the rat model of Alzheimer's disease by erythropoietin-loaded solid lipid nanoparticles. *Neurobiology of Learning and Memory, 166*, 107082. PMID: 31493483

Das, K. P. (2023). Nanoparticles and convergence of artificial intelligence for targeted drug delivery for cancer therapy: Current progress and challenges. *Frontiers in Medical Technology, 4*, 1067144. PMID: 36688144

Davoodi-Roodbordeii, F., Afshar, M., Haji Abas Tabrizi, F., Choopani, S., Torkaman, G., Moayer, F., & Salimi, M. (2019). Topical hydrogel containing Fumaria vaillantii Loisel. extract enhances wound healing in rats. *BMC Complementary and Alternative Medicine, 19*(1), 254. PMID: 31511001

De Boer, A. G., & Gaillard, P. J. (2007). Drug targeting to the brain. *Annual Review of Pharmacology and Toxicology, 47*(1), 323–355. PMID: 16961459

De La Rosa, L. A., Moreno-Escamilla, J. O., Rodrigo-García, J., & Alvarez-Parrilla, E. (2019). Phenolic compounds. In *Postharvest physiology and biochemistry of fruits and vegetables* (pp. 253–271). Woodhead publishing.

De Los Ríos, C., Egea, J., Marco-Contelles, J., Leon, R., Samadi, A., Iriepa, I., & Romero, A. (2010). Synthesis, inhibitory activity of cholinesterases, and neuroprotective profile of novel 1, 8-naphthyridine derivatives. *Journal of Medicinal Chemistry, 53*(14), 5129–5143. PMID: 20575555

Din, F., Kim, D. W., Choi, J. Y., Thapa, R. K., Mustapha, O., Kim, D. S., & Choi, H. G. (2017). Irinotecan-loaded double-reversible thermogel with improved antitumor efficacy without initial burst effect and toxicity for intramuscular administration. *Acta Biomaterialia, 54*, 239–248. PMID: 28285074

Divekar, P. A., Narayana, S., Divekar, B. A., Kumar, R., Gadratagi, B. G., Ray, A., & Behera, T. K. (2022). Plant secondary metabolites as defense tools against herbivores for sustainable crop protection. *International Journal of Molecular Sciences, 23*(5), 2690. PMID: 35269836

Elbaz, A., Carcaillon, L., Kab, S., & Moisan, F. (2016). Epidemiology of Parkinson's disease. *Revue Neurologique, 172*(1), 14–26. PMID: 26718594

Erb, M., & Kliebenstein, D. J. (2020). Plant secondary metabolites as defenses, regulators, and primary metabolites: The blurred functional trichotomy. *Plant Physiology, 184*(1), 39–52. PMID: 32636341

Frank, S. (2014). Treatment of Huntington's disease. *Neurotherapeutics; the Journal of the American Society for Experimental NeuroTherapeutics, 11*(1), 153–160. PMID: 24366610

Gadhave, D., Gupta, A., Khot, S., Tagalpallewar, A., & Kokare, C. (2023, March). Nose-to-brain delivery of paliperidone palmitate poloxamer-guar gum nanogel: Formulation, optimization and pharmacological studies in rats. [). Elsevier Masson.]. *Annales Pharmaceutiques Francaises, 81*(2), 315–333. PMID: 36037930

Gadhave, D., Tupe, S., Tagalpallewar, A., Gorain, B., Choudhury, H., & Kokare, C. (2021). Nose-to-brain delivery of amisulpride-loaded lipid-based poloxamer-gellan gum nanoemulgel: In vitro and in vivo pharmacological studies. *International Journal of Pharmaceutics, 607*, 121050. PMID: 34454028

Gadhave, D. G., Sugandhi, V. V., & Kokare, C. R. (2024). Potential biomaterials and experimental animal models for inventing new drug delivery approaches in the neurodegenerative disorder: Multiple sclerosis. *Brain Research, 1822*, 148674. PMID: 37952871

Geschwind, D. H., & Konopka, G. (2009). Neuroscience in the era of functional genomics and systems biology. *Nature, 461*(7266), 908–915. PMID: 19829370

Giacomeli, R., Izoton, J. C., Dos Santos, R. B., Boeira, S. P., Jesse, C. R., & Haas, S. E. (2019). Neuroprotective effects of curcumin lipid-core nanocapsules in a model Alzheimer's disease induced by β-amyloid 1-42 peptide in aged female mice. *Brain Research, 1721*, 146325. PMID: 31325424

Gitler, A. D., Dhillon, P., & Shorter, J. (2017). Neurodegenerative disease: Models, mechanisms, and a new hope. *Disease Models & Mechanisms, 10*(5), 499–502. PMID: 28468935

Gordillo-Galeano, A., & Mora-Huertas, C. E. (2018). Solid lipid nanoparticles and nanostructured lipid carriers: A review emphasizing on particle structure and drug release. *European Journal of Pharmaceutics and Biopharmaceutics, 133*, 285–308. PMID: 30463794

Goutman, S. A., Hardiman, O., Al-Chalabi, A., Chió, A., Savelieff, M. G., Kiernan, M. C., & Feldman, E. L. (2022). Recent advances in the diagnosis and prognosis of amyotrophic lateral sclerosis. *Lancet Neurology, 21*(5), 480–493. PMID: 35334233

Groothuis, D. R. (2000). The blood-brain and blood-tumor barriers: A review of strategies for increasing drug delivery. *Neuro-Oncology, 2*(1), 45–59. PMID: 11302254

Guan, X., Xing, S., & Liu, Y. (2024). Engineered cell membrane-camouflaged nanomaterials for biomedical applications. *Nanomaterials (Basel, Switzerland), 14*(5), 413. PMID: 38470744

Guccione, C., Oufir, M., Piazzini, V., Eigenmann, D. E., Jähne, E. A., Zabela, V., & Bilia, A. R. (2017). Andrographolide-loaded nanoparticles for brain delivery: Formulation, characterisation and in vitro permeability using hCMEC/D3 cell line. *European Journal of Pharmaceutics and Biopharmaceutics, 119*, 253–263. PMID: 28652141

Gutiérrez-Grijalva, E. P., López-Martínez, L. X., Contreras-Angulo, L. A., Elizalde-Romero, C. A., & Heredia, J. B. (2020). Plant alkaloids: Structures and bioactive properties. In *Plant-derived bioactives: chemistry and mode of action* (pp. 85–117). Springer Singapore.

Hardiman, O., Al-Chalabi, A., Chio, A., Corr, E. M., Logroscino, G., Robberecht, W., & Van Den Berg, L. H. (2017). Amyotrophic lateral sclerosis. *Nature Reviews. Disease Primers*, *3*, 17071. PMID: 28980624

Illarioshkin, S. N., Klyushnikov, S. A., Vigont, V. A., Seliverstov, Y. A., & Kaznacheyeva, E. V. (2018). Molecular pathogenesis in Huntington's disease. [moscow]. *Biochemistry*, *83*(9), 1030–1039. PMID: 30472941

Jain, K. K. (2013). Synthetic biology and personalized medicine. *Medical Principles and Practice : International Journal of the Kuwait University, Health Science Centre*, *22*(3), 209–219. PMID: 22907209

Jiang, Z., Dong, X., Yan, X., Liu, Y., Zhang, L., & Sun, Y. (2018). Nanogels of dual inhibitor-modified hyaluronic acid function as a potent inhibitor of amyloid β-protein aggregation and cytotoxicity. *Scientific Reports*, *8*(1), 3505. PMID: 29472606

Kabiri, K., & Zohuriaan-Mehr, M. J. (2004). Porous superabsorbent hydrogel composites: Synthesis, morphology and swelling rate. *Macromolecular Materials and Engineering*, *289*(7), 653–661.

Khawli, L. A., & Prabhu, S. (2013). Drug delivery across the blood–brain barrier. *Molecular Pharmaceutics*, *10*(5), 1471–1472. PMID: 23641922

Kim, H. S., Lyoo, C. H., Lee, P. H., Kim, S. J., Park, M. Y., Ma, H. I., & Lee, M. S. (2015). Current status of Huntington's disease in Korea: A nationwide survey and national registry analysis. *Journal of Movement Disorders*, *8*(1), 14. PMID: 25614781

Kirkland, J. L., & Tchkonia, T. (2017). Cellular senescence: A translational perspective. *EBioMedicine*, *21*, 21–28. PMID: 28416161

Klajnert, B., Cortijo-Arellano, M., Cladera, J., & Bryszewska, M. (2006). Influence of dendrimer's structure on its activity against amyloid fibril formation. *Biochemical and Biophysical Research Communications*, *345*(1), 21–28. PMID: 16674918

Krzywkowski, P., Sebastiani, G., Williams, S., Delorme, D., & Greenberg, B. (2007, March). Tramiprosate prevents amyloid beta-induced inhibition of long-term potentiation in rat hippocampal slices. In *8th International Conference AD/PD, Salzburg, Austria.*

Lamptey, R. N., Chaulagain, B., Trivedi, R., Gothwal, A., Layek, B., & Singh, J. (2022). A review of the common neurodegenerative disorders: Current therapeutic approaches and the potential role of nanotherapeutics. *International Journal of Molecular Sciences*, *23*(3), 1851. PMID: 35163773

Leuci, R., Brunetti, L., Poliseno, V., Laghezza, A., Loiodice, F., Tortorella, P., & Piemontese, L. (2020). Natural compounds for the prevention and treatment of cardiovascular and neurodegenerative diseases. *Foods*, *10*(1), 29. PMID: 33374186

Liu, J., Liu, C., Zhang, J., Zhang, Y., Liu, K., Song, J. X., & Liu, G. (2020). A self-assembled α-synuclein nanoscavenger for Parkinson's disease. *ACS Nano*, *14*(2), 1533–1549. PMID: 32027482

Liu, Z. J., Li, Z. H., Liu, L., Tang, W. X., Wang, Y., Dong, M. R., & Xiao, C. (2016). Curcumin attenuates beta-amyloid-induced neuroinflammation via activation of peroxisome proliferator-activated receptor-gamma function in a rat model of Alzheimer's disease. *Frontiers in Pharmacology, 7*, 261. PMID: 27594837

Loma, I., & Heyman, R. (2011). Multiple sclerosis: Pathogenesis and treatment. *Current Neuropharmacology, 9*(3), 409–416. PMID: 22379455

Maciulaitis, R., D'apote, L., Buchanan, A., Pioppo, L., & Schneider, C. K. (2012). Clinical development of advanced therapy medicinal products in Europe: Evidence that regulators must be proactive. *Molecular Therapy : the Journal of the American Society of Gene Therapy, 20*(3), 479–482. PMID: 22378030

Mak, K. K., & Pichika, M. R. (2019). Artificial intelligence in drug development: Present status and future prospects. *Drug Discovery Today, 24*(3), 773–780. PMID: 30472429

Matsuura, H. N., & Fett-Neto, A. G. (2015). Plant alkaloids: main features, toxicity, and mechanisms of action. In *Plant toxins* (pp. 1–15). Springer.

Mendes, B. B., Zhang, Z., Conniot, J., Sousa, D. P., Ravasco, J. M., Onweller, L. A., & Conde, J. (2024). A large-scale machine learning analysis of inorganic nanoparticles in preclinical cancer research. *Nature Nanotechnology, 19*(6), 867–878. PMID: 38750164

Mijit, M., Caracciolo, V., Melillo, A., Amicarelli, F., & Giordano, A. (2020). Role of p53 in the regulation of cellular senescence. *Biomolecules, 10*(3), 420. PMID: 32182711

Muller, R. H., & Keck, C. M. (2004).. . *Journal of Nanoscience and Nanotechnology, 4*(5), 471–483. PMID: 15503432

Nataraj, J., Manivasagam, T., Justin Thenmozhi, A., & Essa, M. M. (2017). Neurotrophic effect of asiatic acid, a triterpene of Centella asiatica against chronic 1-methyl 4-phenyl 1, 2, 3, 6-tetrahydropyridine hydrochloride/probenecid mouse model of Parkinson's disease: The role of MAPK, PI3K-Akt-GSK3β and mTOR signalling pathways. *Neurochemical Research, 42*(5), 1354–1365. PMID: 28181071

Neelam, Khatkar, A., & Sharma, K. K. (2020). Phenylpropanoids and its derivatives: Biological activities and its role in food, pharmaceutical and cosmetic industries. *Critical Reviews in Food Science and Nutrition, 60*(16), 2655–2675. PMID: 31456411

Ohnishi, S., & Takano, K. (2004). Amyloid fibrils from the viewpoint of protein folding. *Cellular and Molecular Life Sciences : CMLS, 61*(5), 511–524. PMID: 15004691

Pal, A., & Das, S. (2019). Terpenoids in Treatment of Neurodegenerative Disease. In *Terpenoids Against Human Diseases* (pp. 95–117). CRC Press.

Pangalos, M. N., Schechter, L. E., & Hurko, O. (2004). Nat. Rev. Drug Discovery 2007, 6, 521–32. (6) Muller, R. H.; Keck, C. M. *Journal of Nanoscience and Nanotechnology, 4*(5), 471–483. PMID: 15503432

Pardridge, W. M. (2007). Blood–brain barrier delivery. *Drug Discovery Today*, *12*(1-2), 54–61. PMID: 17198973

Parkinson's Disease Mechanism Discovered," HHMI Research News June 22, 2006.

Patel, T. A., Kevadiya, B. D., Bajwa, N., Singh, P. A., Zheng, H., Kirabo, A., & Patel, K. P. (2023). Role of nanoparticle-conjugates and nanotheranostics in abrogating oxidative stress and ameliorating neuroinflammation. *Antioxidants*, *12*(10), 1877. PMID: 37891956

Patel, V., Chavda, V., & Shah, J. (2021). Nanotherapeutics in neuropathologies: Obstacles, challenges and recent advancements in CNS targeted drug delivery systems. *Current Neuropharmacology*, *19*(5), 693–710. PMID: 32851949

Petracca, M., Quarantelli, M., Moccia, M., Vacca, G., Satelliti, B., D'Ambrosio, G., & Morra, V. B. (2021). ProspeCtive study to evaluate efficacy, safety and tOlerability of dietary suppleme NT of Curcumin (BCM95) in subjects with Active relapsing MultIple Sclerosis treated with subcutaNeous Interferon beta 1a 44 mcg TIW (CONTAIN): A randomized, controlled trial. *Multiple Sclerosis and Related Disorders*, *56*, 103274. PMID: 34583214

Pupillo, E., Messina, P., Logroscino, G., & Beghi, E.SLALOM Group. (2014). Long-term survival in amyotrophic lateral sclerosis: A population-based study. *Annals of Neurology*, *75*(2), 287–297. PMID: 24382602

Puri, A., Loomis, K., Smith, B., Lee, J. H., Yavlovich, A., Heldman, E., & Blumenthal, R. (2009). Lipid-based nanoparticles as pharmaceutical drug carriers: from concepts to clinic. *Critical Reviews™ in Therapeutic Drug Carrier Systems, 26*(6).

Reinhardt, J., Flory, E., Büttel, I., Schröder, C., Fricke, S., Vucinic, V., & Niederwieser, D. (2012). MSCs: clinical applications and European regulatory aspects. In *Mesenchymal Stromal Cells: Biology and Clinical Applications* (pp. 355–364). Springer New York.

Ren, X., & Chen, J. F. (2020). Caffeine and Parkinson's disease: Multiple benefits and emerging mechanisms. *Frontiers in Neuroscience*, *14*, 602697. PMID: 33390888

Rizvi, S. A., Einstein, G. P., Tulp, O. L., Sainvil, F., & Branly, R. (2022). Introduction to traditional medicine and their role in prevention and treatment of emerging and re-emerging diseases. *Biomolecules*, *12*(10), 1442. PMID: 36291651

Saikia, S., Prajapati, J. B., Prajapati, B. G., Padma, V. V., & Pathak, Y. V. (2022). The role of artificial intelligence in therapeutic drug monitoring and clinical toxicity. In *Recent Advances in Therapeutic Drug Monitoring and Clinical Toxicology* (pp. 67–85). Springer International Publishing.

Salehi, B., Calina, D., Docea, A. O., Koirala, N., Aryal, S., Lombardo, D., & Sharifi-Rad, J. (2020). Curcumin's nanomedicine formulations for therapeutic application in neurological diseases. *Journal of Clinical Medicine*, *9*(2), 430. PMID: 32033365

Sanyal, A., Lajoie, B. R., Jain, G., & Dekker, J. (2012). The long-range interaction landscape of gene promoters. *Nature*, *489*(7414), 109–113. PMID: 22955621

Sathya, S., Shanmuganathan, B., Saranya, S., Vaidevi, S., Ruckmani, K., & Pandima Devi, K. (2018). Phytol-loaded PLGA nanoparticle as a modulator of Alzheimer's toxic Aβ peptide aggregation and fibrillation associated with impaired neuronal cell function. *Artificial Cells, Nanomedicine, and Biotechnology*, *46*(8), 1719–1730. PMID: 29069924

Saudou, F., & Humbert, S. (2016). The biology of huntingtin. *Neuron*, *89*(5), 910–926. PMID: 26938440

Sauer, A. V., Di Lorenzo, B., Carriglio, N., & Aiuti, A. (2014). Progress in gene therapy for primary immunodeficiencies using lentiviral vectors. *Current Opinion in Allergy and Clinical Immunology*, *14*(6), 527–534. PMID: 25207699

Shah, D., Mahurkar, N., Gadhave, D., Nikhate, R., & Kakad, K. (2023, September). Hepatoprotective activity of Balsamodendron mukul extract against Paracetamol-induced liver toxicity in rats: In vivo pharmacological and toxicological evaluation. [). Elsevier Masson.]. *Annales Pharmaceutiques Francaises*, *81*(5), 814–825. PMID: 37060939

Shameem, M. M., Sasikanth, S. M., Annamalai, R., & Raman, R. G. (2021). A brief review on polymer nanocomposites and its applications. *Materials Today: Proceedings*, *45*, 2536–2539.

Siegal, T., & Zylber-Katz, E. (2002). Strategies for increasing drug delivery to the brain: Focus on brain lymphoma. *Clinical Pharmacokinetics*, *41*(3), 171–186. PMID: 11929318

Singh, A. K., Mishra, S. K., Mishra, G., Maurya, A., Awasthi, R., Yadav, M. K., & Singh, S. K. (2020). Inorganic clay nanocomposite system for improved cholinesterase inhibition and brain pharmacokinetics of donepezil. *Drug Development and Industrial Pharmacy*, *46*(1), 8–19. PMID: 31809608

Skarlis, C., & Anagnostouli, M. (2020). The role of melatonin in Multiple Sclerosis. *Neurological Sciences*, *41*(4), 769–781. PMID: 31845043

Sohail, S., & Shah, F. A., uz Zaman, S., Almari, A. H., Malik, I., Khan, S. A., ... & ud Din, F. (2023). Melatonin delivered in solid lipid nanoparticles ameliorated its neuroprotective effects in cerebral ischemia. *Heliyon*, *9*(9). PMID: 37809765

Sriamornsak, P., & Waijanya, S. (2021). Applying information and computer technologies in various industrial sectors. *Science, Engineering and Health Studies*, 21010004-21010004.

Tiwari, S. K., Agarwal, S., Seth, B., Yadav, A., Nair, S., Bhatnagar, P., & Gupta, K. C. (2014). Curcumin-loaded nanoparticles potently induce adult neurogenesis and reverse cognitive deficits in Alzheimer's disease model via canonical Wnt/β-catenin pathway. *ACS Nano*, *8*(1), 76–103. PMID: 24467380

Wang, J., Zhang, R., Chen, X., Sun, X., Yan, Y., Shen, X., & Yuan, Q. (2020). Biosynthesis of aromatic polyketides in microorganisms using type II polyketide synthases. *Microbial Cell Factories*, *19*(1), 110. PMID: 32448179

Weinshenker, B. G. (1994). Natural history of multiple sclerosis. *Annals of Neurology: Official Journal of the American Neurological Association and the Child Neurology Society*, *36*(S1), S6–S11. PMID: 8017890

Wenk, G. L. (2003). Neuropathologic changes in Alzheimer's disease. *The Journal of Clinical Psychiatry*, *64*, 7–10. PMID: 12934968

Xiao, Y., Liu, F., Yang, J., Zhong, M., Zhang, E., Li, Y., & Yan, H. (2015). Over-activation of TLR5 signaling by high-dose flagellin induces liver injury in mice. *Cellular & Molecular Immunology*, *12*(6), 729–742. PMID: 25418468

Xie, H., Yang, X., Cao, Y., Long, X., Shang, H., & Jia, Z. (2022). Role of lipoic acid in multiple sclerosis. *CNS Neuroscience & Therapeutics*, *28*(3), 319–331. PMID: 34964271

Xie, J., Tang, Z., Chen, Q., Jia, X., Li, C., Jin, M., & Huang, S. (2023). Clearance of stress-induced premature senescent cells alleviates the formation of abdominal aortic aneurysms. *Aging and Disease*, *14*(5), 1778. PMID: 37196124

Yuan, Y., Wu, Y., Cheng, J., Yang, K., Xia, Y., Wu, H., & Pan, X. (2024). Applications of artificial intelligence to lipid nanoparticle delivery. *Particuology*, *90*, 88–97.

Zhao, L., Wu, C., Lin, K., & Chang, J. (2012). The effect of poly (lactic-co-glycolic acid)(PLGA) coating on the mechanical, biodegradable, bioactive properties and drug release of porous calcium silicate scaffolds. *Bio-Medical Materials and Engineering*, *22*(5), 289–300. PMID: 23023146

Zhao, X., Zhang, M., Li, C., Jiang, X., Su, Y., & Zhang, Y. (2019). Benefits of Vitamins in the Treatment of Parkinson's Disease. *Oxidative Medicine and Cellular Longevity*, *2019*(1), 9426867. PMID: 30915197

Chapter 4

Role of Plant-Derived Secondary Metabolites in Neuroprotection Against Neurodegenerative Disorders

Hajar Azraida
https://orcid.org/0000-0002-2208-8264
Department of Biology, Cadi Ayyad University, Morocco

Ouijdane El Hatimy
https://orcid.org/0009-0007-2446-6058
Department of Biology, Cadi Ayyad University, Morocco

Yasmina Jaouhari
https://orcid.org/0009-0007-4795-3051
Department of Biology, Cadi Ayyad University, Morocco

Fouad Bouazzaoui
Department of Biology, Cadi Ayyad University, Morocco

Majda Badaoui
Department of Biology, Cadi Ayyad University, Morocco

Hamid Kabdy
https://orcid.org/0009-0004-7438-3638
Department of Biology, Cadi Ayyad University, Morocco

Souad Moubtakir
Department of Biology, Cadi Ayyad University, Morocco

Loubna El Yazouli
Department of Biology, Cadi Ayyad University, Morocco

Rachida Aboufatima
Faculty of Sciences and Technology, Sultan Moulay Slimane University, Morocco

Mohamed Loukid
Department of Biology, Cadi Ayyad University, Morocco

Abderrahman Chait
Department of Biology, Cadi Ayyad University, Morocco

DOI: 10.4018/979-8-3373-1365-8.ch004

ABSTRACT

Neurodegenerative diseases, including Alzheimer's, Parkinson's, and amyotrophic lateral sclerosis, are increasing, highlighting the need for innovative treatments. Plant-derived phytochemicals, such as flavonoids, alkaloids, terpenoids, and polyphenols, offer neuroprotective properties by reducing oxidative stress, modulating neuroinflammation, preventing neuronal apoptosis, and enhancing neurotrophic signaling and mitochondrial function. However, their clinical application faces challenges like low bioavailability and rapid metabolism. This chapter explores these issues and potential solutions, including advanced drug delivery systems and structural modifications. It also highlights the need for interdisciplinary research to fully harness phytochemicals' therapeutic potential in neurodegenerative disease treatment.

1. INTRODUCTION

Neurodegenerative diseases such as Alzheimer's disease, Parkinson's disease, and amyotrophic lateral sclerosis (ALS) are becoming increasingly prevalent, driving the urgent need for innovative therapeutic strategies. Among the most promising avenues of exploration are secondary metabolites derived from plants, commonly known as phytochemicals. These natural compounds have gained significant attention for their multifaceted neuroprotective properties (Laadraoui et al., 2025a; Gopalakrishnan et al., 2022). This chapter aims to provide an in-depth examination of the role of these metabolites in neuroprotection, shedding light on their mechanisms of action, therapeutic potential, and the challenges associated with their clinical application.

The discussion will begin by highlighting the diverse classes of phytochemicals, including flavonoids, alkaloids, terpenoids, and polyphenols, which have shown remarkable neuroprotective effects in both preclinical and clinical studies (Vauzour, 2012). These compounds, abundant in various medicinal plants, have demonstrated the ability to modulate critical pathways implicated in the progression of neurodegenerative diseases. For example, flavonoids such as quercetin and epicatechin are renowned for their antioxidant and anti-inflammatory activities, while alkaloids like galantamine have already been approved for clinical use in cognitive enhancement therapies (Heinrich & Teoh, 2019). Similarly, terpenoids, including curcuminoids and ginsenosides, have shown potential in mitigating neuronal damage and promoting synaptic plasticity, offering a natural means of counteracting neurodegeneration.

A central focus of this chapter will be the exploration of the molecular mechanisms underlying the neuroprotective effects of these secondary metabolites. These mechanisms include their capacity to scavenge reactive oxygen species (ROS) and bolster endogenous antioxidant defenses, thereby protecting neurons from oxidative stress (Rehman et al., 2019). Additionally, their anti-inflammatory effects, mediated through the suppression of microglial activation and the downregulation of pro-inflammatory cytokines, will be analyzed in detail. Their role in inhibiting apoptotic pathways and preserving neuronal survival under pathological conditions will also be emphasized, alongside other mechanisms such as the enhancement of neurotrophic factor signaling and the maintenance of mitochondrial function. Collectively, these actions highlight the vast therapeutic potential of secondary metabolites in mitigating the multifactorial nature of neurodegenerative diseases (Laadraoui et al., 2025b; Brondino et al., 2018). This chapter will critically evaluate these limitations and discuss innovative strategies to overcome them, such as the development of advanced drug delivery systems like nanoparticles and liposomes, as well as structural modifications designed to enhance their pharmacokinetic and pharmacodynamic profiles (Youssef et al., 2023). By addressing these challenges, the chapter aims to provide actionable insights into how these natural compounds can be effectively harnessed for therapeutic purposes.

this chapter will synthesize the current state of knowledge regarding the neuroprotective properties of secondary metabolites, identifying key gaps in research that must be addressed. It will advocate for interdisciplinary approaches that combine pharmacology, biotechnology, and clinical research to fully unlock the potential of these compounds (Sahoo et al., 2022).

2. CLASSIFICATION OF NEUROPROTECTIVE SECONDARY METABOLITES

Secondary metabolites are bioactive compounds synthesized by plants, fungi, and other organisms, not for their primary metabolic processes, but to serve specialized roles such as defense, communication, and adaptation to environmental challenges (Erb & Kliebenstein, 2020). These phytochemicals have gained considerable attention due to their ability to modulate key mechanisms involved in neuroprotection (Howes et al., 2020; Venkatesan et al., 2015).

The classification of neuroprotective secondary metabolites is based on their chemical structure and biosynthetic origins. Major classes include flavonoids, terpenoids, alkaloids, phenolic compounds, tannins, and saponins, each demonstrating unique interactions with neuronal pathways.

2.1 Phenolic Compounds

Phenolic compounds are a diverse group of naturally occurring chemical substances characterized by the presence of one or more hydroxyl groups attached to an aromatic ring (Kumar & Goel, 2019). They are categorized into two main groups: simple phenols and polyphenols, based on the number of aromatic rings and the complexity of their structures (Nájera-Maldonado et al., 2024). These bioactive compounds have shown potential neuroprotective effects by scavenging free radicals, modulating signaling pathways, and inhibiting protein aggregation of proteins. Polyphenols, in turn, are further subdivided into several subclasses, including flavonoids, phenolic acids, tannins, lignans and stilbenes (Nájera-Maldonado et al., 2024).

2.2 Flavonoids

Flavonoids are a diverse group of polyphenolic compounds that represent a major class of plant secondary metabolites (Roy et al., 2022). They are found in the tissues and organs of various plant species and are widely recognized for their potent antioxidant, anti-inflammatory, and neuroprotective properties (Ullah et al., 2020). These bioactive compounds exhibit a wide array of chemical structures and biological functions (Minocha et al., 2022). Flavonoids are systematically classified into various categories, each demonstrating unique mechanisms of action and significant neuroprotective potential (Minocha et al., 2022). They are further divided into several subclasses (Table 1). Figure 1 depicts how flavonoids influence neuronal signaling pathways, reduce oxidative stress, and modulate neuroinflammation.

Table 1. Classification and neuroprotective effects of flavonoids

Flavonoid Class	Example	Effects on Neurodegeneration	References
Flavones	Luteolin	Reduces neuroinflammation, inhibits microglial activation, protects against oxidative stress, improves memory.	(Jang et al., 2010) (He et al., 2023)
	Apigenin	Enhances neurogenesis, exhibits anti-inflammatory effects, aids cognitive function.	(Olasehinde & Olaokun, 2024)
Flavonols (flavan-3-ols)	Quercetin	Protects neurons from oxidative stress, inhibits amyloid-beta aggregation, reduces inflammation.	(Safarzadeh et al., 2024)
	Kaempferol	Reduces neuroinflammation, promotes neuronal survival.	(Jin et al., 2023)
Flavanones	Naringenin	Protects against oxidative damage, reduces neuroinflammation, modulates dopamine levels	(Khan et al., 2012) (Nouri et al., 2019)
	Hesperidin	Enhances memory and learning, reduces oxidative stress, protects in cerebral ischemia models.	(Roohbakhsh et al., 2014)
Isoflavones	Genistein	Acts as a phytoestrogen, modulates estrogen receptors, protects neurons, enhances synaptic plasticity.	(Konstantinou et al., 2024) (Li et al., 2022)
Flavanols	Catechin	Enhances cognitive performance, protects against oxidative damage, improves cerebral blood flow.	(Drouin et al., 2011)
	Epicatechin	Supports brain function, reduces oxidative stress.	(Haskell-Ramsay et al., 2018)
Anthocyanidins	Cyanidin	Protects against oxidative stress, inhibits neuroinflammation.	(Cásedas et al., 2019)
	Delphinidin	Modulates signaling pathways, protects against amyloid-beta toxicity.	(Kim et al., 2009)
Chalcones	Phloridzin	Exhibits anti-inflammatory properties, supports glucose metabolism in the brain.	(Zhang et al., 2025)

Figure 1. Mechanisms of Flavonoid-mediated neuroprotection against neurodegenerative disorders

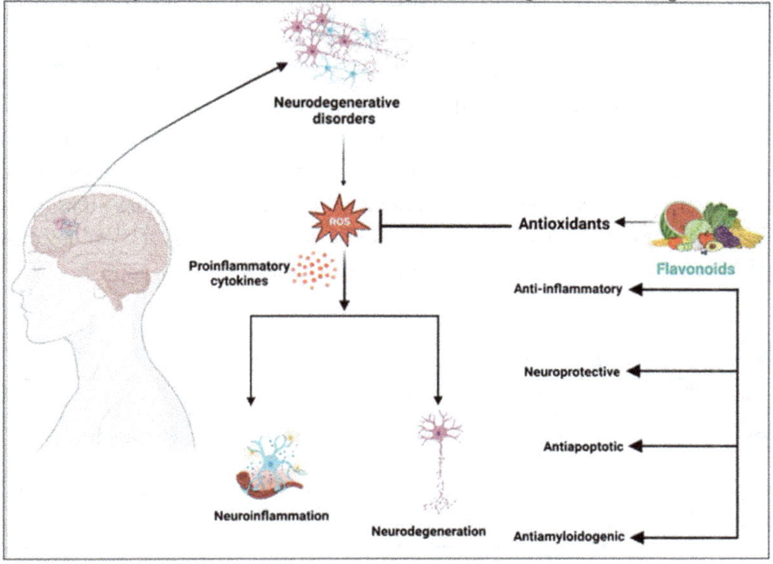

Source: Minocha et al. (2022)

2.3 Phenolic Acids, Stilbenes, Lignans and Tannins

The table details key compounds from each class, their sources, and evidence supporting their roles in reducing oxidative stress and promoting neuronal health (Table 2).

Table 2. Neuroprotective effects of phenolic acids, stilbenes, lignans and tannins

Class	Examples	Effects on Neurodegeneration	References
Phenolic acids	Ferulic acid	Protects dopaminergic neurons, reduces inflammatory markers (IL-1β, IL-6, TNF-α), and attenuates oxidative stress.	(Ojha et al., 2015)
	Caffeic acid	Exhibits neuroprotective effects by reducing oxidative stress, improving cognitive function, and inhibiting NF-κB and GSK3β signaling, along with suppressing acetylcholinesterase and butyrylcholinesterase activity.	(Castro et al., 2023; Pavlíková, 2022)
Stilbenes	Resveratol	Modulates inflammation via the TLR4/NF-κB pathway, promotes antioxidant responses, and reduces neuroinflammation.	(Moussa et al., 2017)
Tannins	Condensed tannins (Proanthocyanidins)	Enhance cognitive performance, reduce oxidative damage and modulate cholinergic pathways.	(Reshma et al., s. d.)
	Hydrolyzable tannins (Tannic Acid)	Demonstrates anti-amyloidogenic properties, protects neuroblastoma cells, and suppresses apoptosis and oxidative stress.	(Ono et al., 2004; Yulak & Ergul, 2024)
Lignans	Secoisolariciresinol	Antioxidant activity and potential regulation of neuroinflammatory pathways	(Kesavan et al., 2018)

2.4 Terpenoids

Terpenoids, also known as isoprenoids, are one of the largest and most diverse classes of secondary metabolites, exhibiting remarkable structural diversity (Bohlmann & Keeling, 2008). These bioactive compounds play a significant role in protecting plants from biotic and abiotic stresses (Bohlmann & Keeling, 2008). Furthermore, terpenoids have demonstrated a variety of health-promoting effects, including anti-inflammatory and neuroprotective activities, each class or example exhibiting distinct and specific effects (Kiyama, 2017). Table 3 provide examples of terpenoid compounds, their natural sources, and the mechanisms by which they contribute to neuroprotection in various neurodegenerative models (Table 3).

Table 3. Neuroprotective effects of terpenoids

Terpenoids Class	Examples	Effects on Neurodegeneration	References
Monoterpenoids	Limonene	Regulates neurotransmitter activity, enhances serotonin signaling pathway.	(Gu et al., 2019)
	Catalpol	Improves behavior and neurodevelopment, reduces hyperactivity and impulsivity, enhances spatial learning and memory, protects neurons, and promotes myelination.	(Yuan et al., 2019)
Sesquiterpenoids	β-Caryophyllene	Inhibits amyloid-beta aggregation, protects neurons from oxidative stress and reduces inflammation.	(Hu et al., 2017)
	Artemisinin	Reduces neurotoxicity, inhibits microglial migration, suppresses inflammation, and improves cognitive function.	(Zhao et al., 2022)
Diterpenoids	Forskolin	Protects against oxidative damage, modulates dopamine levels, improves cognitive performance, enhances mitochondrial enzyme activity, attenuates neuroinflammation.	(Dudi & Mehan, 2018)
Triterpenoids	Asiatic Acid	Reduces the risk of major cognitive impairment.	(Umka Welbat et al., 2016)
	Lupeol	Reduces neuroinflammation, inhibits inflammatory mediators, improves cognitive function, and modulates biochemical pathways.	(Choe et al., 2024)

Figure 2 summarize the cellular pathways through which terpenoids exert antioxidant, anti-inflammatory, and neuroprotective effects (Figure 2).

Figure 2. Neuroprotective mechanisms of terpenoids in neurodegenerative disorders

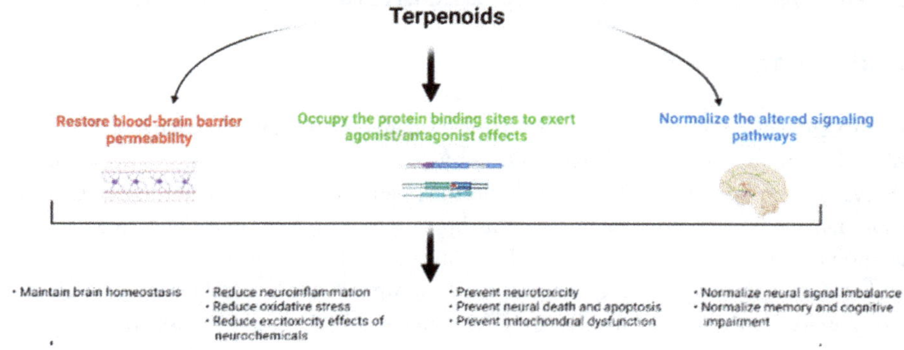

Source: Mony et al. (2022)

2.5 Alkaloids

Alkaloids, a significant class of plant-derived medicinal compounds, are biologically active substances that vary in structure and function (Pelletier, 1996). They can be classified based on their chemical ring structure (Pelletier, 1996). The concentration and diversity of alkaloids differ among plant species and are influenced by factors such as growth stage and season (Pundarikakshudu & Bhavsar, 1988). Research has identified various alkaloids from different plant families, highlighting their importance in in neuroprotective effects (Table 4) (Al-Mamoori et al., 2023) summarizes the major alkaloid compounds, their sources, and their demonstrated effects on neuronal survival, oxidative stress, and neuroinflammatory pathways.

Table 4. Neuroprotective properties of alkaloids class

Alkaloids Class	Examples	Effects on Neurodegeneration	references
Indole Alkaloids	Conophylline	Activates autophagy and suppresses protein aggregation, protecting the neural cells from cell death.	(Umezawa et al., 2018)
	Geissoschizoline	Prevents neurodegeneration and restores neurotransmission.	(Lima et al., 2020)
Tropane Alkaloids	Atropine	Reduces excessive cholinergic activity and prevents excitotoxic damage.	(Kumari et al., 2007)
Quinazoline alkaloid	Dehydroevodiamine	Protects the central nervous system and exhibits significant blood-brain barrier permeability.	(Fu et al., 2022)
	Vasicinone	Exhibits neuroprotection in paraquat-induced Parkinson's models and supports neural cell survival.	(Huang et al., 2020)
Protoalkaloids	Capsaicin	Exerts neuroprotective effects by inhibiting apoptosis in Parkinson's disease models and modulating amyloid precursor protein processing to prevent amyloid-beta formation.	(Liu et al., 2020)

3. NEUROPROTECTIVE EFFECT OF SEVERAL PHYTOCHEMICALS AND THE PREVENTION OF NEURODEGENERATIVE DISEASES

3.1 Antioxydant Activity

Phytochemicals are important for scavenging oxidative stress, a major contributor to neuronal injury in neurodegenerative diseases. Oxidative stress arises from an increased production of reactive oxygen species (ROS) and a diminished capacity of antioxidant defenses. Excess ROS can damage cellular components such as lipids, proteins, and DNA, leading to neuronal dysfunction and death. Phytochemicals such as flavonoids and polyphenols demonstrate strong antioxidant activity, acting on free radicals, reducing lipid peroxidation, and upregulating endogenous antioxidant enzymes like superoxide dismutase (SOD) and catalase (Rahman et al., 2019). These actions decrease oxidative stress, protect neurons, and support brain function. Additionally, certain phytochemicals activate the Nrf2 signaling pathway, upregulating genes involved in cellular antioxidant defense (Satoh & Lipton, 2013). The neuroprotective potential of compounds like curcumin and resveratrol has been documented in in vitro and in vivo studies (Khan et al., 2020). Figure 3 illustrates how plant-derived antioxidants reduce oxidative stress and inflammatory responses in neurodegenerative disorders (Figure 3).

Figure 3. Role of natural antioxidants on neuroprotection and neuroinflammation

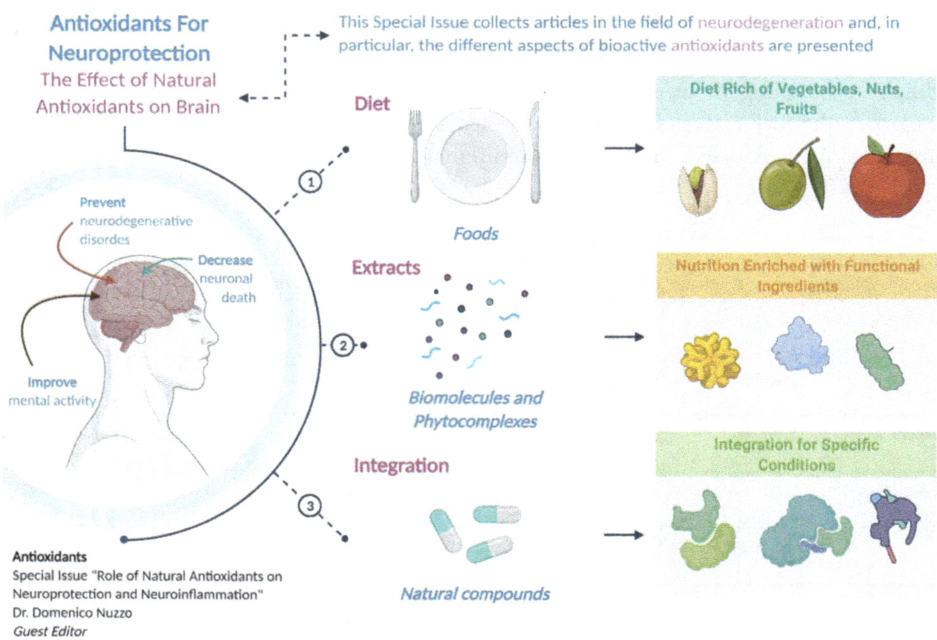

Source: Nuzzo et al. (2021)

3.2 Anti-Inflammatory Effects

Chronic neuroinflammation is a hallmark of numerous neurodegenerative diseases, including Alzheimer's disease and Parkinson's disease (Heneka et al., 2015). Activated microglia, the brain's primary immune cells, secrete pro-inflammatory cytokines, including tumor necrosis factor-alpha (TNF-α) and interleukin-1 beta (IL-1β) (Glass et al., 2010). These inflammatory molecules can trigger cascades that result in neuronal damage. Phytochemicals like curcumin and ginsenosides can inhibit inflammation by modulating the nuclear factor-kappa B (NF-κB) signaling pathway, reducing the production of inflammatory mediators (Zhang et al., 2013). Furthermore, phytochemicals help shift microglia from a neurotoxic profile to a neuroprotective one, mitigating chronic inflammation and neuronal injury (Yang et al., 2020).

3.3 Preventing Cell Death (Cytoprotection)

Neuronal apoptosis, or programmed cell death, is a significant factor in neurodegenerative diseases (Bredesen, 2009). Phytochemicals help inhibit apoptosis by reducing caspase activation and maintaining mitochondrial membrane stability to prevent the release of pro-apoptotic factors like cytochrome c (Chandra et al., 2015; Kelsey et al., 2017). These compounds can increase anti-apoptotic protein expression, such as Bcl-2, while downregulating pro-apoptotic proteins like Bax, ensuring neuronal survival under stress (Kumar et al., 2018). Through multiple pathways, phytochemicals present a promising strategy for preventing neurodegenerative diseases (Rahman et al., 2019).

3.4 Improving Brain Connections (Synaptic Plasticity and Neurotrophin Modulation)

Synaptic plasticity is essential for learning and memory, but it is often impaired in neurodegenerative diseases, leading to cognitive decline. Phytochemicals enhance synaptic plasticity by modulating neurotrophins such as brain-derived neurotrophic factor (BDNF), which promote neuron survival and synaptic connectivity (Mattson et al., 2008). Compounds such as ginsenosides and flavonoids stimulate BDNF expression, improving synaptic communication and neuronal adaptability (Wang et al., 2016).

3.5 Protecting Mitochondria

Mitochondria are essential for neuronal energy production and cell survival. Dysfunctional mitochondria contribute to neurodegenerative diseases due to disrupted energy metabolism and excess ROS production (Bhat et al., 2015). Phytochemicals like curcumin and resveratrol enhance mitochondrial function by promoting mitochondrial biogenesis and stabilizing mitochondrial membranes, preventing apoptotic signaling (Zheng et al., 2017).

Through these mechanisms, phytochemicals represent multi-target agents that provide holistic protection to the brain. They reduce oxidative stress and inflammatory chronicity in the brain, protect against neuronal apoptosis, promote synaptic plasticity, and prevent mitochondrial damage. Such a holistic approach renders them indispensable for tackling the multifaceted pathologies of neurodegenerative disorders and emphasizes their reparative potential. This involvment can lead to effective & potential interventions against neurodegenerative disorders as the current research is still revealed out their therapy in particular diseases (Table 5).

Table 5. Neuroprotective role of phytochemicals via management of oxidative stress

Compound	Methodology	Results	References
Apocynin	Mice and rats were administered varying doses of apocynin, ranging from 4 to 100 mg/kg, through the intraperitoneal (i.p.) route for pretreatment. Post-trauma interventions in mice involved an i.p. dose of 5 mg/kg or a range of 0.05–5 mg/kg administered subcutaneously (s.c.).	1. Inhibit NOX activity upregulated following TBI 2. Decrease of oxidative stress	Ferreira et al. (2013); Loane, Stoica, Byrnes, Jeong and Faden. (2013)
Caffeic acid phenethyl ester (cape)	Adult male rats received CAPE (10 μmol/kg) intraperitoneally (i.p.) immediately following a head injury	1. Reduction in lipid peroxidation. 2. Enhancement of essential antioxidants	Alam et al., 2022; Kerman et al. (2012)
Curcumin	Adult rats underwent a mild injury and were fed a diet containing curcumin (500 ppm) for a 4-week pretreatment period or received post-treatment for 2 weeks	The 4-week pretreatment resulted in a reduction of oxidative stress, an increase in BDNF levels, and the preservation of synaptic proteins	Alam et al., 2022; Kerman et al. (2012)
Formononetin (fp)	Following a closed head injury, FN treatment was administered intraperitoneally (i.p.) at 10 or 20 mg/kg doses, starting 5 days after the TBI.	Increased antioxidant enzymes (GPx and SOD) and decreased lipid peroxidation (LP)	Li et al. (2014); Singh et al. (2023)
Ginseng	Two weeks after a significant injury, the subjects orally consumed 100 or 200 mg/kg of ginseng	Eliminated oxidative stress and neuroinflammation in adult rats with injury using a modified Marmarou injury model	Kumar et al. (2014)
Hydroxysafflor yellow A (HSYA)	The animals received pretreatment with HYSA intravenously (i.v.) 30 min before and 6 h after the injury	HSYA treatment significantly improved mitochondrial function, boosting antioxidant activity and reducing lipid peroxidation (LP)	Bie et al. (2010)
Luteolin	A modified weight-drop closed head injury model was utilized; animals received treatment with luteolin at doses of 10, 30, or 50 mg/kg intraperitoneally (i.p.) 30 min after the trauma	Decrease in oxidative stress and an elevation in the antioxidant GPx levels	Kempuraj et al. (2021)
Pycnogenol (PYC)	Young adult rats experienced a moderate unilateral CCI injury. They were then promptly treated with PYC (100 mg/kg, intraperitoneal), followed by two additional doses administered 3 and 6 h after the injury	She has elevated several antioxidants (GSH, GPx, and catalase) while concurrently reducing oxidative stress markers	Scheff et al. (2013); Çolak et al. (2021)

4. THERAPEUTIC POTENTIAL OF PHYTOCHEMICALS

Phytochemicals or plant-derived compounds have gained considerable attention for their protective potential against neurodegenerative diseases, such as Parkinson's and Alzheimer's disease. Their therapeutic effects are mainly due to their antioxidant, anti-inflammatory, and anti-amyloidogenic effects.

4.1 Approved Phytochemical-Based Interventions

Some phytochemicals have already arrived in approved treatments. Galantamine: Galantamine (also known as Gentes) is an alkaloid obtained from a variety of plants (for example, Galanthus nivalis (snowdrop)) and is approved for clinical use as a treatment for memory impairment in Alzheimer's disease. It acts as a cholinesterase inhibitor, increasing acetylcholine levels in the brain, a neurotransmitter crucial for memory and learning (Mangialasche et al., 2010).

4.2 Phytochemicals as Potential Therapeutic Agents

Other phytochemicals are being studied for possible use as natural complements or substitutes to standard treatments. These compounds may also act on multiple targets, affecting various pathological features of neurodegenerative diseases simultaneously (Singh et al., 2018).

4.3 Neuroprotective Effects

Polyphenols have shown promising neuroprotective effects. These compounds are found in various plant-based foods and beverages, including fruits, vegetables, tea, and spices, and have been extensively studied for their ability to combat neurodegenerative diseases. Examples include (Table 6):

Table 6. Examples for some polyphenols and their neuroprotective effects

Curcumin	Found in turmeric, curcumin exhibits antioxidant, anti-inflammatory, and anti-amyloidogenic properties. It reduces oxidative stress, a key contributor to neuronal damage, and inhibits the aggregation of amyloid-beta peptides, a hallmark of Alzheimer's disease (Pawar et al., 2012; Velmurugan et al., 2018).
Epigallocatechin-3-gallate	The major catechin in green tea, EGCG possesses potent antioxidant properties and can inhibit Aβ aggregation and tau protein phosphorylation, both implicated in Alzheimer's pathology (Velmurugan et al., 2018).
Quercetin	A flavonoid found in apples, onions, and berries, quercetin has been reported to cross the blood-brain barrier, exerting antioxidant and anti-inflammatory effects within the central nervous system. Studies suggest that quercetin can protect against excitotoxicity and neuronal apoptosis, mechanisms involved in diseases such as Parkinson's and Huntington's disease (Costa et al., 2016)
Resveratrol	Present in grapes, red wine, and peanuts, resveratrol has been shown to activate sirtuins, a class of enzymes associated with longevity and neuroprotection. It can mitigate oxidative stress, modulate microglial activation, and reduce neuroinflammation, making it a promising candidate for combating neurodegenerative diseases like Alzheimer's and Parkinson's (Bastianetto et al., 2000).

These polyphenols modulate various neuroinflammatory pathways, reducing the inflammatory response that contributes to neuronal dysfunction in dementia (Libro et al., 2016; Calfío et al., 2020)

4.4 Isothiocyanates and Mitochondrial Function

Isothiocyanates, found in cruciferous vegetables like broccoli and cauliflower, also show promise in neuroprotection. Sulforaphane, a well-studied isothiocyanate, may improve mitochondrial function and reduce neurodegeneration, particularly in Parkinson's disease. Mitochondrial dysfunction plays a significant role in neuronal death in neurodegenerative diseases.

It's important to note that while many phytochemicals show promise in preclinical studies, further research, including large-scale clinical trials, is needed to confirm their efficacy and safety in humans. Additionally, the bioavailability of some phytochemicals can be limited, requiring strategies to enhance their absorption and utilization in the body.

5. CHALLENGES IN USING PHYTOCHEMICALS CLINICALLY

Despite the potential of phytochemicals to ameliorate neurodegenerative disorders because of their antioxidant, anti-inflammatory, and neuroprotective properties, several obstacles must be overcome for their clinical implementation. A major problem is low bioavailability, since many phytochemicals are poorly absorbed and rapidly metabolized, restricting their therapeutic potential. For instance, preclinical studies have shown that curcumin, derived from turmeric, possesses strong neuroprotective effects, but that it has low bioavailability resulting from poor water solubility and a high rate of metabolism in the liver and intestines (Anand et al., 2007). The next major challenge is that the blood-brain barrier (BBB) keeps most nutrients, including phytochemicals, out of the CNS. Although resveratrol and epigallocatechin gallate (EGCG) display promising neuroprotective properties, their limited brain penetration may make them less effective (Frolinger et al., 2020). Moreover, standardization issues have impeded its clinical applications since the various plant sources, growth conditions, and extraction methods result in different concentrations and contents of active ingredients, raising issues regarding reproducibility and therapeutic reliability (Li et al., 2021). Although they are deemed safe at typical concentrations in foods, many phytonutrients might be toxic and have adverse effects if taken in excess. Some Other alkaloids have been implicated in hepatotoxicity and neurotoxicity, indicating that they may be harmful at doses exceeding (Teschke et al., 2015). Moreover, the scarcity of clinical trials remains a critical obstacle to the acceptance of them in conventional medicine. However, limited large-scale clinical trials assessing their safety and efficacy in human neurodegenerative disease prevented their widespread adoption (Cragg & Newman, 2013).

To address these challenges, researchers have explored the use of nanotechnology-based delivery systems to enhance the bioavailability, brain penetration, and therapeutic efficacy of neuroprotective phytochemicals. Advances in nanomedicine have led to the development of a variety of nanocarriers, including liposomes, polymeric nanoparticles, and solid lipid nanoparticles, which have been investigated as potential vehicles for improving the solubility, stability, and targeted delivery of phytochemicals to the central nervous system (Paramanick et al., 2022), although scientists are investigating novel strategies to improve phytochemicals' clinical applicability including:

<u>Cutting-Edge Drug Delivery Methods</u>: To improve the bioavailability and targeted administration of phytochemicals, novel approaches are being investigated, including polymer-based carriers, liposomal formulations, and nanoparticle encapsulation. For example, curcumin delivered by nanoparticles has shown increased stability and brain penetration, increasing its therapeutic efficacy (Kakkar et al., 2018).

<u>Alterations to the Structure</u>: Phytochemicals are modified chemically to maximize their pharmacokinetic and pharmacodynamic characteristics. The production of curcumin analogs

with increased hydrophilicity is one noteworthy example, which has greatly increased solubility and bioavailability (Sharma et al., 2021).

Combination Treatments: Promising therapeutic results can be obtained by combining phytochemicals with other pharmacological drugs in a synergistic manner. For example, resveratrol co-administration. For instance, co-administration of resveratrol with anti-inflammatory drugs has been shown to enhance neuroprotective effects in preclinical animal models (Berman et al., 2017).

Standardization of Extracts: To guarantee the constant quality and repeatability of plant-derived products, sophisticated analytical methods like mass spectrometry and high-performance liquid chromatography (HPLC) must be used. Standardizing the composition and potency of phytochemical extracts is made possible in large part by these techniques (Li et al., 2021).

Studies on Safety and Controlled Dosing: Determining the best therapeutic dosage schedules while reducing possible toxicity requires thorough preclinical and clinical research. For substances like quercetin and EGCG, ongoing clinical trials are improving dosage guidelines to optimize therapeutic advantages while preserving safety (Singh et al., 2011).

Cooperation Attempts: Accelerating the transfer of phytochemicals from the lab to clinical practice is made possible by public-private partnerships between pharmaceutical corporations and academic research institutions. These collaborations facilitate resource sharing, funding opportunities, and the execution of large-scale clinical trials, thereby advancing the clinical applications of plant-derived compounds (Cragg & Newman, 2013).

6. CONCLUSION

In conclusion, secondary metabolites from medicinal plants represent a valuable resource in the management of neurodegenerative diseases. These natural compounds exhibit potent neuroprotective effects through a variety of molecular mechanisms, including antioxidant, anti-inflammatory, and anti-apoptotic actions. Despite the promising preclinical and clinical evidence, significant challenges remain in translating these findings into effective therapies. The development of advanced drug delivery systems and the optimization of the pharmacological properties of phytochemicals will be critical to their clinical success.

REFERENCES

Abadie, V., Sollid, L. M., Barreiro, L. B., & Jabri, B. (2011). Integration of genetic and immunological insights into a model of celiac disease pathogenesis. *Annual Review of Immunology, 29*, 493–525. DOI: 10.1146/annurev-immunol-040210-092915

Aboul-Enein, H. Y., & Abu El-Enein, M. A. (2016). *Herbal medicine: Clinical applications and evidence*. Nova Science Publishers.

Aeschlimann, D., & Paulsson, M. (1994). Transglutaminases: Protein cross-linking enzymes in tissues and body fluids. *Thrombosis and Haemostasis, 71*(4), 402–415. DOI: 10.1055/s-0038-1642545

Ahmed, S., Anuntiyo, J., Malemud, C. J., & Haqqi, T. M. (2005). Biological basis for the use of botanicals in osteoarthritis and rheumatoid arthritis: A review. *Evidence-Based Complementary and Alternative Medicine : eCAM, 2*(3), 301–308. DOI: 10.1093/ecam/neh117

Akhtar, N., & Haqqi, T. M. (2011). Current nutraceuticals in the management of osteoarthritis: A review. *Therapeutic Advances in Musculoskeletal Disease, 3*(1), 51–63. DOI: 10.1177/1759720X10363052

Al-Dabbagh, B., Elhatimy, O., & Kacem, M. (2020). Traditional medicinal plants used for the treatment of autoimmune diseases in Morocco: An ethnopharmacological survey. *Journal of Ethnopharmacology, 259*, 112955. DOI: 10.1016/j.jep.2020.112955

Al-Mamoori, F., Qasem, A. M. A., Al-Mamoori, F., & Qasem, A. M. A. (2023). Alkaloids as New Leads for Neurodegenerative Diseases. In Medicinal Plants—Chemical, Biochemical, and Pharmacological Approaches. IntechOpen. https://doi.org/DOI: 10.5772/intechopen.112584

Al-Snafi, A. E. (2016). Nutritional value and pharmacological importance of *Peganum harmala*—A review. *International Journal of Scientific Research, 5*(2), 202–211.

Alam, M., Ahmed, S., Elasbali, A. M., Adnan, M., Alam, S., Hassan, M. I., & Pasupuleti, V. R. (2022). Therapeutic implications of caffeic acid in cancer and neurological diseases. *Frontiers in Oncology, 12*, 860508.

Alonso-Llamazares, J., Gibson, L. E., & Rogers, R. S. (2007). Clinical, pathologic, and immunopathologic study of 52 North American patients with dermatitis herpetiformis. *International Journal of Dermatology, 46*(9), 927–933. DOI: 10.1111/j.1365-4632.2007.03347.x

Amadi, B. C., & Nwosu, J. N. (2019). Nutritional and health benefits of sesame (*Sesamum indicum* L.). *International Journal of Nutrition and Food Sciences, 8*(1), 14–20. DOI: 10.11648/j.ijnfs.20190801.13

Anand, P., Kunnumakkara, A. B., Newman, R. A., & Aggarwal, B. B. (2007). Bioavailability of curcumin: Problems and promises. *Molecular Pharmaceutics, 4*(6), 807–818. DOI: 10.1021/mp700113r

Anderson, R. P. (2008). Coeliac disease: Current approach and future prospects. *Internal Medicine Journal, 38*(10), 790–799. DOI: 10.1111/j.1445-5994.2008.01723.x

Anwar, F., Ali, M., Hussain, A. I., & Shahid, M. (2010). Antioxidant and antimicrobial activities of essential oil and extracts of fennel (*Foeniculum vulgare* Mill.) seeds from Pakistan. *Flavour and Fragrance Journal, 24*(4), 170–176. DOI: 10.1002/ffj.1928

Arnett, F. C., Edworthy, S. M., Bloch, D. A., McShane, D. J., Fries, J. F., Cooper, N. S., & Hunder, G. G. (1988). The American Rheumatism Association 1987 revised criteria for the classification of rheumatoid arthritis. *Arthritis and Rheumatism, 31*(3), 315–324. DOI: 10.1002/art.1780310302

Ashraf, M. U., Muhammad, G., Hussain, M. A., & Bukhari, S. N. A. (2017). Celiac disease: A comprehensive review. *Critical Reviews in Food Science and Nutrition, 52*(2), 153–168. DOI: 10.1080/10408398.2014.888112

Aslam, M., & Janbaz, K. H. (2020). Pharmacological basis for the medicinal use of *Peganum harmala* seeds in gastrointestinal and respiratory disorders. *Journal of Ethnopharmacology, 248*, 112334. DOI: 10.1016/j.jep.2019.112334

Atkinson, M. A., Eisenbarth, G. S., & Michels, A. W. (2014). Type 1 diabetes. *Lancet, 383*(9911), 69–82. DOI: 10.1016/S0140-6736(13)60591-7

Bastianetto, S., Zheng, W. H., & Quirion, R. (2000). Neuroprotective abilities of resveratrol and other red wine constituents against nitric oxide-related toxicity in cultured hippocampal neurons. *British Journal of Pharmacology, 131*(4), 711–720.

Berman, A. Y., Motechin, R. A., Wiesenfeld, M. Y., & Holz, M. K. (2017). The therapeutic potential of resveratrol: A review of clinical trials. npj Precision Oncology, 1(1), 35. https://doi .org/DOI: 10.1038/s41698-017-0038-6

Bhat, A. H., Iqbal, K., & Zubair, H. (2015). Mitochondrial dysfunction in neurodegeneration: Pathology and therapeutics. *Neuroscience and Biobehavioral Reviews, 57*, 128–146. DOI: 10.1016/j.neubiorev.2015.08.004

Bie, X. D., Han, J., & Dai, H. B. (2010). Effects of hydroxysafflor yellow A on the experimental traumatic brain injury in rats. *Journal of Asian Natural Products Research, 12*(3), 239–247.

Bohlmann, J., & Keeling, C. I. (2008). Terpenoid biomaterials. *The Plant Journal, 54*(4), 656–669. DOI: 10.1111/j.1365-313X.2008.03449.x

Bredesen, D. E. (2009). Apoptosis in neurodegenerative disorders: Cellular and molecular mechanisms. Nature Reviews Neuroscience, 10(4), 204 212. https://doi.org/DOI: 10.1038/nrn2616

Bunaciu, A. A., Aboul-Enein, H. Y., & Fleschin, S. (2011). Recent Applications of Fourier Transform Infrared Spectrophotometry in Herbal Medicine Analysis. *Applied Spectroscopy Reviews, 46*(4), 251–260. DOI: 10.1080/05704928.2011.565532

Calfío, C., García, G., & Castillo, R. (2020). Polyphenolic compounds and their impact on neurodegenerative diseases. *Frontiers in Neuroscience, 14*, 387. DOI: 10.3389/fnins.2020.00387

Cásedas, G., Les, F., González-Burgos, E., Gómez-Serranillos, M. P., Smith, C., & López, V. (2019). Cyanidin-3-O-glucoside inhibits different enzymes involved in central nervous system pathologies and type-2 diabetes. *South African Journal of Botany*, *120*, 241–246. DOI: 10.1016/j.sajb.2018.07.001

Castro, M. F. V., Assmann, C. E., Stefanello, N., Reichert, K. P., Palma, T. V., da Silva, A. D., Miron, V. V., Mostardeiro, V. B., Morsch, V. M. M., & Schetinger, M. R. C. (2023). Caffeic acid attenuates neuroinflammation and cognitive impairment in streptozotocin-induced diabetic rats : Pivotal role of the cholinergic and purinergic signaling pathways. *The Journal of Nutritional Biochemistry*, *115*, 109280. DOI: 10.1016/j.jnutbio.2023.109280

Catassi, C., Kryszak, D., Bhatti, B., Sturgeon, C., Helzlsouer, K., Clipp, S. L., & Fasano, A. (2010). Natural history of celiac disease autoimmunity in a USA cohort followed since 1974. *Annals of Medicine*, *42*(7), 530–538. DOI: 10.3109/07853890.2010.514285

Celik, G., Aydin, T., Oz, E. K., Yildirim, S., & Ozkan, M. T. (2019). Therapeutic potential of *Nigella sativa* seeds and oil in autoimmune disorders. *Phytotherapy Research*, *33*(10), 2530–2544. DOI: 10.1002/ptr.6438

Chandra, S., Ghosh, S., & Roy, M. (2015). Caspase inhibition by natural compounds: Therapeutic implications in neurodegenerative diseases. Journal of Neurochemistry, 134(6), 891 901. https://doi.org/DOI: 10.1111/jnc.13201

Chatterjee, S., Bhattacharjee, P., & Das, S. (2018). Phytochemical and pharmacological properties of *Sesamum indicum* Linn: A review. *Journal of Pharmacognosy and Phytochemistry*, *7*(2), 3191–3197.

Chen, H., & Xu, H. (2019). *Rheumatoid arthritis: Pathogenesis and clinical management.* Springer., DOI: 10.1007/978-981-13-7567-0

Choe, K., Park, J. S., Park, H. Y., Tahir, M., Park, T. J., & Kim, M. O. (2024). Lupeol protect against LPS-induced neuroinflammation and amyloid beta in adult mouse hippocampus. *Frontiers in Nutrition*, *11*. Advance online publication. DOI: 10.3389/fnut.2024.1414696

Ciccocioppo, R., Di Sabatino, A., & Corazza, G. R. (2005). The immune recognition of gluten in coeliac disease. *Clinical and Experimental Immunology*, *140*(3), 408–416. DOI: 10.1111/j.1365-2249.2005.02796.x

Ciccocioppo, R., Di Sabatino, A., & Corazza, G. R. (2005). The immune recognition of gluten in coeliac disease. *Clinical and Experimental Immunology*, *140*(3), 408–416. DOI: 10.1111/j.1365-2249.2005.02796.x

Cid, M. C., & Campo, E. (2012). Immunopathology of autoimmune diseases. In Buchanan, R. W. (Ed.), *Autoimmune diseases* (pp. 51–82). Springer., DOI: 10.1007/978-3-642-21061-2_3

Çolak, R., Celik, A., Diniz, G., Özdemir, S. A., Yilmaz, O., & Calkavur, S. (2023). Evaluation of the neuroprotective effect of pycnogenol in a hypoxic–ischemic brain injury model in newborn rats. *American Journal of Perinatology*, *40*(06), 612–618.

Collin, P., Kaukinen, K., Vogelsang, H., Korponay-Szabó, I., Sommer, R., Schreier, E., & Mäki, M. (2005). Antiendomysial and anti-transglutaminase antibodies in the diagnosis of coeliac disease: A prospective study. *Scandinavian Journal of Gastroenterology*, *40*(6), 564–571. DOI: 10.1080/00365520510012164

Combe, B., Landewe, R., Daien, C. I., Hua, C., Aletaha, D., Álvaro-Gracia, J. M., & Emery, P. (2017). 2016 update of the EULAR recommendations for the management of early arthritis. *Annals of the Rheumatic Diseases*, *76*(6), 948–959. DOI: 10.1136/annrheumdis-2016-210602

Cooke, W. T., & Holmes, G. K. T. (1984). *Coeliac disease*. Churchill Livingstone.

Cooper, G. S., Bynum, M. L., & Somers, E. C. (2009). Recent insights in the epidemiology of autoimmune diseases: Improved prevalence estimates and understanding of clustering of diseases. *Journal of Autoimmunity*, *33*(3–4), 197–207. DOI: 10.1016/j.jaut.2009.09.008

Costa, L. G., Garrick, J. M., Roquè, P. J., & Pellacani, C. (2016). Mechanisms of neuroprotection by quercetin: Counteracting oxidative stress and more. *Oxidative Medicine and Cellular Longevity*, *2016*(1), 2986796.

Cragg, G. M., & Newman, D. J. (2013). Natural products: A continuing source of novel drug leads. *Biochimica et Biophysica Acta*, *1830*(6), 3670–3695. DOI: 10.1016/j.bbagen.2013.02.008

Cragnolini, A. B., & Lampit, A. (2018). Lifestyle factors and autoimmune diseases. In Esser, A., & Müller, M. (Eds.), *Lifestyle medicine* (pp. 219–236). Springer., DOI: 10.1007/978-3-319-70227-5_15

Czaja, A. J. (2014). Diagnosis and management of autoimmune hepatitis: Current status and future directions. *Gut and Liver*, *8*(5), 573–589. DOI: 10.5009/gnl14058

De Almeida, A. B. A., da Silva, C. B., de Oliveira, R. G., Silva, C. O., de Freitas, R. M., & Barbosa-Filho, J. M. (2017). Medicinal plants and natural compounds with anti-inflammatory activities on rheumatoid arthritis. *Current Pharmaceutical Design*, *23*(24), 3652–3666. DOI: 10 .2174/1381612823666170309104303

De Silva, P. S. A., Luben, R., Shrestha, S. S., Khaw, K. T., & Hart, A. R. (2010). Dietary intake of antioxidants and risk of celiac disease: A prospective cohort study. *European Journal of Clinical Nutrition*, *64*(7), 770–775. DOI: 10.1038/ejcn.2010.87

Drouin, A., Bolduc, V., Thorin-Trescases, N., Bélanger, É., Fernandes, P., Baraghis, E., Lesage, F., Gillis, M.-A., Villeneuve, L., Hamel, E., Ferland, G., & Thorin, E. (2011). Catechin treatment improves cerebrovascular flow-mediated dilation and learning abilities in atherosclerotic mice. *American Journal of Physiology. Heart and Circulatory Physiology*, *300*(3), H1032–H1043. DOI: 10.1152/ajpheart.00410.2010

Dudi, R., & Mehan, S. (2018). Neuroprotection of brain permeable Forskolin ameliorates behavioral, biochemical and histopathological alterations in rat model of intracerebral hemorrhage. 10(2).

El-Ghazaly, M. A., Fadel, N. A., Abdel-Naby, D. H., & Al-Kassas, M. (2018). Protective effect of sesame oil against adjuvant-induced arthritis in rats: Histological and biochemical evidence. *International Immunopharmacology*, *56*, 205–211. DOI: 10.1016/j.intimp.2018.01.009

El-Sayed, S. M. (2014). Potential health benefits and scientific aspects of sesame: A review. *International Journal of Food Science, Nutrition and Dietetics*, *3*(3), 131–146. DOI: 10.19070/2326-3350-1400026

Emami, S. A., Sahebkar, A., & Javadi, B. (2010). Antioxidant and free radical scavenging activities of *Peganum harmala* seeds extracts. *Acta Medica Iranica*, *48*(1), 19–24.

Erb, M., & Kliebenstein, D. J. (2020). Plant Secondary Metabolites as Defenses, Regulators, and Primary Metabolites : The Blurred Functional Trichotomy1 [OPEN]. *Plant Physiology*, *184*(1), 39–52. DOI: 10.1104/pp.20.00433

Fasano, A. (2012). Clinical presentation of celiac disease in the pediatric population. *Gastroenterology*, *143*(3), 620–628. DOI: 10.1053/j.gastro.2012.06.008

Ferreira, R., & Cunha, C. (2013). Immune responses and therapeutic approaches in celiac disease. *Clinical Reviews in Allergy & Immunology*, *45*(2), 199–210. DOI: 10.1007/s12016-012-8346-6

Ferretti, G., Bacchetti, T., Masciangelo, S., & Bicchiega, V. (2012). Lipid peroxidation in rheumatoid arthritis: New insights into the pathogenesis of chronic inflammation. *Therapeutic Advances in Musculoskeletal Disease*, *4*(5), 315–318. DOI: 10.1177/1759720X12461005

Figueiredo, A. C., Barroso, J. G., & Pedro, L. G. (2017). Sesquiterpenes: Natural products with therapeutic potential in rheumatoid arthritis. *Current Pharmaceutical Design*, *23*(30), 4481–4493. DOI: 10.2174/1381612823666170711152222

Firestein, G. S., & McInnes, I. B. (2017). Immunopathogenesis of rheumatoid arthritis. *Immunity*, *46*(2), 183–196. DOI: 10.1016/j.immuni.2017.02.006

Frolinger, T., Herman, F., Sharma, S., & Ghiso, J. (2020). Blood-brain barrier transport of bioactive phytochemicals. *Frontiers in Aging Neuroscience*, *12*, 500785. DOI: 10.3389/fnagi.2020.500785

Frustaci, A., Cuoco, L., Chimenti, C., Pieroni, M., Fioravanti, G., Gentiloni, N., & Maseri, A. (2002). Celiac disease associated with autoimmune myocarditis. *Circulation*, *105*(22), 2611–2618. DOI: 10.1161/01.CIR.0000017880.86119.C0

Fu, S., Liao, L., Yang, Y., Bai, Y., Zeng, Y., Wang, H., & Wen, J. (2022). The pharmacokinetics profiles, pharmacological properties, and toxicological risks of dehydroevodiamine : A review. *Frontiers in Pharmacology*, *13*. Advance online publication. DOI: 10.3389/fphar.2022.1040154

Furuzawa-Carballeda, J., Vargas-Rojas, M. I., & Cabral, A. R. (2007). Autoimmune inflammation from the gut to the joints: The role of T cells in spondyloarthritis. *Arthritis Research & Therapy*, *9*(6), 1–13. DOI: 10.1186/ar2347

Glass, C. K., Saijo, K., Winner, B., Marchetto, M. C., & Gage, F. H. (2010). Mechanisms underlying inflammation in neurodegeneration. *Neuron*, *64*(1), 7–20. DOI: 10.1016/j.neuron.2009.09.021

Gu, S. M., Kim, S. Y., Lamichhane, S., Hong, J. T., & Yun, J. (2019). Limonene Inhibits Methamphetamine-Induced Sensitizations via the Regulation of Dopamine Receptor Supersensitivity. *Biomolecules & Therapeutics*, *27*(4), 357–362. DOI: 10.4062/biomolther.2018.213

Guandalini, S., & Assiri, A. (2014). Celiac disease: A review. *JAMA Pediatrics, 168*(3), 272–278. DOI: 10.1001/jamapediatrics.2013.3858

Hamer, R. J. (2005). Coeliac disease: Background and biochemical aspects. *Biotechnology Advances, 23*(6), 401–408. DOI: 10.1016/j.biotechadv.2005.05.005

Harbige, L. S. (1998). Dietary n-6 and n-3 fatty acids in immunity and autoimmune disease. *The Proceedings of the Nutrition Society, 57*(4), 555–562. DOI: 10.1079/PNS19980080

Haskell-Ramsay, C. F., Schmitt, J., & Actis-Goretta, L. (2018). The Impact of Epicatechin on Human Cognition : The Role of Cerebral Blood Flow. *Nutrients, 10*(8), 8. Advance online publication. DOI: 10.3390/nu10080986

Heneka, M. T., Carson, M. J., El Khoury, J., & Landreth, G. E. (2015). Neuroinflammation in Alzheimer's disease. *Lancet Neurology, 14*(4), 388–400. DOI: 10.1016/S1474-4422(15)00012-4

Howes, M.-J. R., Perry, N. S. L., Vásquez-Londoño, C., & Perry, E. K. (2020). Role of phytochemicals as nutraceuticals for cognitive functions affected in ageing. *British Journal of Pharmacology, 177*(6), 1294–1315. DOI: 10.1111/bph.14898

Huang, C.-Y., Sivalingam, K., Shibu, M. A., Liao, P.-H., Ho, T.-J., Kuo, W.-W., Chen, R.-J., Day, C.-H., Viswanadha, V. P., & Ju, D.-T. (2020). Induction of Autophagy by Vasicinone Protects Neural Cells from Mitochondrial Dysfunction and Attenuates Paraquat-Mediated Parkinson's Disease Associated α-Synuclein Levels. *Nutrients, 12*(6), 6. Advance online publication. DOI: 10.3390/nu12061707

Hviid, A., Svanström, H., & Frisch, M. (2008). Antibiotic use and inflammatory bowel diseases in childhood. *Gut, 60*(1), 49–54. DOI: 10.1136/gut.2009.180307

Iqbal, M., & Athar, M. (2005). Attenuation of immunological and biochemical alterations in rats treated with *Peganum harmala* seeds during experimental arthritis. *Inflammopharmacology, 13*(4), 411–419. DOI: 10.1163/156856005774423003

Jang, S., Dilger, R. N., & Johnson, R. W. (2010). Luteolin Inhibits Microglia and Alters Hippocampal-Dependent Spatial Working Memory in Aged Mice123. *The Journal of Nutrition, 140*(10), 1892–1898. DOI: 10.3945/jn.110.123273

Jiang, H., & Chess, L. (2006). Regulation of immune responses by T cells. *The New England Journal of Medicine, 354*(11), 1166–1176. DOI: 10.1056/NEJMra055446

Jin, S., Zhang, L., & Wang, L. (2023). Kaempferol, a potential neuroprotective agent in neurodegenerative diseases : From chemistry to medicine. *Biomedicine and Pharmacotherapy, 165*, 115215. DOI: 10.1016/j.biopha.2023.115215

Kakkar, V., Singh, S., Singla, D., & Kaur, I. P. (2018). Exploring solid lipid nanoparticles to enhance the oral bioavailability of curcumin. *Molecular Nutrition & Food Research, 62*(6), 1701000. DOI: 10.1002/mnfr.201701000

Kang, J. Y., Kang, A. H. Y., Green, A., Gwee, K. A., & Ho, K. Y. (2013). Systematic review: Worldwide variation in the frequency of coeliac disease and changes over time. *Alimentary Pharmacology & Therapeutics*, *38*(3), 226–245. DOI: 10.1111/apt.12373

Kaviarasan, S., Naik, G. H., Gangabhagirathi, R., Anuradha, C. V., & Priyadarsini, K. I. (2007). In vitro studies on antiradical and antioxidant activities of fenugreek (*Trigonella foenum graecum*) seeds. *Food Chemistry*, *103*(1), 31–37. DOI: 10.1016/j.foodchem.2006.05.064

Kelsey, D. C., Hsu, H. C., & Chang, J. Y. (2017). Mitochondrial membrane integrity and neurodegenerative diseases. *Molecular Neurodegeneration*, *12*(1), 10. DOI: 10.1186/s13024-017

Kempuraj, D., Thangavel, R., Kempuraj, D. D., Ahmed, M. E., Selvakumar, G. P., Raikwar, S. P., & Zaheer, A. (2021). Neuroprotective effects of flavone luteolin in neuroinflammation and neurotrauma. *BioFactors (Oxford, England)*, *47*(2), 190–197.

Kerman, M., Kanter, M., Coşkun, K. K., Erboga, M., & Gurel, A. (2012). Neuroprotective effects of caffeic acid phenethyl ester on experimental traumatic brain injury in rats. *Journal of Molecular Histology*, *43*(1), 49–57.

Kesavan, P., Banerjee, A., Banerjee, A., Murugesan, R., Marotta, F., & Pathak, S. (2018). An Overview of Dietary Polyphenols and Their Therapeutic Effects. In Watson, R. R., Preedy, V. R., & Zibadi, S. (Eds.), *Polyphenols : Mechanisms of Action in Human Health and Disease* (2nd ed., pp. 221–235). Academic Press., DOI: 10.1016/B978-0-12-813006-3.00017-9

Khan, F., Ahmad, A., & Alam, S. (2020). The neuroprotective potential of curcumin, resveratrol, and EGCG in neurodegenerative diseases. *Brain Research Bulletin*, *163*, 88–97. DOI: 10.1016/j.brainresbull.2020.04.013

Khan, M. B., Khan, M. M., Khan, A., Ahmed, M. E., Ishrat, T., Tabassum, R., Vaibhav, K., Ahmad, A., & Islam, F. (2012). Naringenin ameliorates Alzheimer's disease (AD)-type neurodegeneration with cognitive impairment (AD-TNDCI) caused by the intracerebroventricular-streptozotocin in rat model. *Neurochemistry International*, *61*(7), 1081–1093. DOI: 10.1016/j.neuint.2012.07.025

Kim, H. R., Kim, H. S., Jung, K. W., Lee, E. J., & Kang, K. H. (2013). Protective effects of sesame oil against arthritis induced by collagen in rats. *Journal of Medicinal Food*, *16*(6), 549–556. DOI: 10.1089/jmf.2012.2667

Kingwell, K. (2017). Autoimmune diseases: Coordinated T-cell response drives autoimmunity. *Nature Reviews. Drug Discovery*, *16*(5), 324–325. DOI: 10.1038/nrd.2017.91

Kiyama, R. (2017). Estrogenic terpenes and terpenoids : Pathways, functions and applications. *European Journal of Pharmacology*, *815*, 405–415. DOI: 10.1016/j.ejphar.2017.09.049

Konstantinou, E. K., Gioxari, A., Dimitriou, M., Panoutsopoulos, G. I., & Panagiotopoulos, A. A. (2024). Molecular Pathways of Genistein Activity in Breast Cancer Cells. *International Journal of Molecular Sciences*, *25*(10), 10. Advance online publication. DOI: 10.3390/ijms25105556

Kumar, A., Rinwa, P., & Dhar, H. (2014). Microglial inhibitory effect of ginseng ameliorates cognitive deficits and neuroinflammation following traumatic head injury in rats. *Inflammopharmacology*, *22*(3), 155–167.

Kumari, A., Sreetama, S., & Mohanakumar, K. P. (2007). Atropine, a muscarinic cholinergic receptor antagonist increases serotonin, but not dopamine levels in discrete brain regions of mice. *Neuroscience Letters*, *423*(2), 100–103. DOI: 10.1016/j.neulet.2007.06.027

Laadraoui, J., Kabdy, H., Oufquir, S., Aitlarradia, M., Agouram, F., Ait Baba, A., & El-Mansoury, B. (2025a). The Therapeutic Role of Secondary Metabolites in Neurodegenerative and Inflammatory Diseases, Mechanisms, and Potential Applications. In *Secondary Metabolites and Their Applications in Various Diseases* (pp. 129–192). IGI Global Scientific Publishing.

Laadraoui, J., Laaradia, M. A., El Gabbas, Z., El Mansoury, B., El Rasafi, T., Kabdy, H., & Chait, A. (2025b). Effect of Nigella sativa Consumption During Pregnancy on Fertility, Morphological and Behaviors of Mice Offspring. *Birth Defects Research*, *117*(9), e2530.

Lebwohl, B., Sanders, D. S., & Green, P. H. (2018). Coeliac disease. *Lancet*, *391*(10115), 70–81. DOI: 10.1016/S0140-6736(17)31796-8

Li, S., Shahzad, N., & Wang, D. (2021). Extraction techniques for bioactive compounds from medicinal and aromatic plants: A review. *Critical Reviews in Analytical Chemistry*, *51*(4), 279–294. DOI: 10.1080/10408347.2020.1717053

Li, Y., Sun, R., & Zheng, X. (2014). T-cell immunity and its regulation in autoimmune diseases. *Journal of Immunology Research*, *2014*, 1–12. DOI: 10.1155/2014/931351

Li, Y., Zhang, J.-J., Chen, R.-J., Chen, L., Chen, S., Yang, X.-F., & Min, J.-W. (2022). Genistein mitigates oxidative stress and inflammation by regulating Nrf2/HO-1 and NF-κB signaling pathways in hypoxic-ischemic brain damage in neonatal mice. *Annals of Translational Medicine*, *10*(2), 32. DOI: 10.21037/atm-21-4958

Libro, R., Moretti, R., & Guglielmi, V. (2016). The role of polyphenols in neuroprotection: A review of clinical evidence. *European Journal of Neurology*, *23*(7), 1079–1086. DOI: 10.1111/ene.13094

Lima, J. A., & Costa, R., T. W., da Fonseca, A. C. C., do Amaral, R. F., Nascimento, M. do D. S. B., Santos-Filho, O. A., de Miranda, A. L. P., Ferreira Neto, D. C., Lima, F. R. S., Hamerski, L., & Tinoco, L. W. (2020). Geissoschizoline, a promising alkaloid for Alzheimer's disease : Inhibition of human cholinesterases, anti-inflammatory effects and molecular docking. *Bioorganic Chemistry*, *104*, 104215. DOI: 10.1016/j.bioorg.2020.104215

Liu, J., Liu, H., Zhao, Z., Wang, J., Guo, D., & Liu, Y. (2020). Regulation of Actg1 and Gsta2 is possible mechanism by which capsaicin alleviates apoptosis in cell model of 6-OHDA-induced Parkinson's disease. *Bioscience Reports*, *40*(6), BSR20191796. Advance online publication. DOI: 10.1042/BSR20191796

Loane, D. J., Stoica, B. A., Byrnes, K. R., Jeong, W., & Faden, A. I. (2013). Activation of mGluR5 and inhibition of NADPH oxidase improves functional recovery after traumatic brain injury. *Journal of Neurotrauma*, *30*(5), 403–412.

Ludvigsson, J. F., Leffler, D. A., Bai, J. C., Biagi, F., Fasano, A., Green, P. H., & Murray, J. A. (2013). The Oslo definitions for coeliac disease and related terms. *Gut*, *62*(1), 43–52. DOI: 10.1136/gutjnl-2011-301346

Makhlof, A., Tozuka, Y., & Takeuchi, H. (2009). Design and evaluation of novel pH-sensitive chitosan nanoparticles for oral insulin delivery. *European Journal of Pharmaceutical Sciences*, *37*(3–4), 177–185. DOI: 10.1016/j.ejps.2009.02.008

Malekzadeh, R., Sachdev, A., & Fahid, M. (2005). Coeliac disease in developing countries: Middle East, India and North Africa. *Best Practice & Research. Clinical Gastroenterology*, *19*(3), 351–358. DOI: 10.1016/j.bpg.2005.01.003

Mangialasche, F., Solomon, A., Winblad, B., Mecocci, P., & Kivipelto, M. (2010). Alzheimer's disease: Clinical trials and drug development. *Lancet Neurology*, *9*(7), 702–716.

Mattson, M. P., & Cheng, A. (2008). Mitochondrial contributions to synaptic plasticity and cognitive function. *Journal of Neuroscience Research*, *85*(8), 1810–1819. DOI: 10.1002/jnr.21593

McInnes, I. B., & Schett, G. (2011). The pathogenesis of rheumatoid arthritis. *The New England Journal of Medicine*, *365*(23), 2205–2219. DOI: 10.1056/NEJMra1004965

Meresse, B., Ripoche, J., Heyman, M., & Cerf-Bensussan, N. (2009). Celiac disease: From oral tolerance to intestinal inflammation, autoimmunity and lymphomagenesis. *Mucosal Immunology*, *2*(1), 8–23. DOI: 10.1038/mi.2008.75

Minocha, T., Birla, H., Obaid, A. A., Rai, V., Sushma, P., Shivamallu, C., Moustafa, M., Al-Shehri, M., Al-Emam, A., Tikhonova, M. A., Yadav, S. K., Poeggeler, B., Singh, D., & Singh, S. K. (2022). Flavonoids as Promising Neuroprotectants and Their Therapeutic Potential against Alzheimer's Disease. *Oxidative Medicine and Cellular Longevity*, *2022*(1), 6038996. DOI: 10.1155/2022/6038996

Mony, T. J., Elahi, F., Choi, J. W., & Park, S. J. (2022). Neuropharmacological Effects of Terpenoids on Preclinical Animal Models of Psychiatric Disorders : A Review. *Antioxidants*, *11*(9), 1834. DOI: 10.3390/antiox11091834

Moussa, C., Hebron, M., Huang, X., Ahn, J., Rissman, R. A., Aisen, P. S., & Turner, R. S. (2017). Resveratrol regulates neuro-inflammation and induces adaptive immunity in Alzheimer's disease. *Journal of Neuroinflammation*, *14*(1), 1. DOI: 10.1186/s12974-016-0779-0

Murray, J. A., & Rubio-Tapia, A. (2012). Diarrhea due to small bowel diseases. *Gastroenterology Clinics of North America*, *41*(3), 629–642. DOI: 10.1016/j.gtc.2012.06.009

Nájera-Maldonado, J. M., Salazar, R., Alvarez-Fitz, P., Acevedo-Quiroz, M., Flores-Alfaro, E., Hernández-Sotelo, D., Espinoza-Rojo, M., & Ramírez, M. (2024). Phenolic Compounds of Therapeutic Interest in Neuroprotection. *Journal of Xenobiotics*, *14*(1), 1. Advance online publication. DOI: 10.3390/jox14010014

Niewinski, M. M. (2008). Advances in celiac disease and gluten-free diet. *Journal of the American Dietetic Association*, *108*(4), 661–672. DOI: 10.1016/j.jada.2008.01.011

Nouri, Z., Fakhri, S., El-Senduny, F. F., Sanadgol, N., Abd-ElGhani, G. E., Farzaei, M. H., & Chen, J.-T. (2019). On the Neuroprotective Effects of Naringenin : Pharmacological Targets, Signaling Pathways, Molecular Mechanisms, and Clinical Perspective. *Biomolecules*, *9*(11), 690. DOI: 10.3390/biom9110690

Nuzzo, D. (2021). Role of natural antioxidants on neuroprotection and neuroinflammation. *Antioxidants, 10*(4), 608. DOI: 10.3390/antiox10040608

Ojha, S., Javed, H., Azimullah, S., Abul Khair, S. B., & Haque, M. E. (2015). Neuroprotective potential of ferulic acid in the rotenone model of Parkinson's disease. *Drug Design, Development and Therapy, 9*, 5499–5510. DOI: 10.2147/DDDT.S90616

Olasehinde, T. A., & Olaokun, O. O. (2024). The Beneficial Role of Apigenin against Cognitive and Neurobehavioural Dysfunction : A Systematic Review of Preclinical Investigations. *Biomedicines, 12*(1), 178. DOI: 10.3390/biomedicines12010178

Ono, K., Hasegawa, K., Naiki, H., & Yamada, M. (2004). Anti-amyloidogenic activity of tannic acid and its activity to destabilize Alzheimer's β-amyloid fibrils in vitro. Biochimica et Biophysica Acta (BBA) -. *Molecular Basis of Disease, 1690*(3), 193–202. DOI: 10.1016/j.bbadis.2004.06.008

Ortiz-Sánchez, E., Martínez-González, J. J., & Sánchez-Nieto, J. M. (2019). Antioxidant effects of fenugreek seed extract on experimental arthritis in rats. *Phytotherapy Research, 33*(12), 3231–3240. DOI: 10.1002/ptr.6517

Paramanick, D., Singh, V. D., & Singh, V. K. (2022). Neuroprotective effect of phytoconstituents via nanotechnology for treatment of Alzheimer diseases. *Journal of Controlled Release, 351*, 638–655.

Pavlíková, N. (2022). Caffeic Acid and Diseases—Mechanisms of Action. *International Journal of Molecular Sciences, 24*(1), 588. DOI: 10.3390/ijms24010588

Pawar, R., Kulkarni, M., & Shinde, S. (2012). Curcumin: A potent polyphenolic compound for the treatment of Alzheimer's disease. *Neurochemical Research, 37*(4), 1092–1099. DOI: 10.1007/s11064-012-0716-2

Pelletier, S. W. (Ed.). (1996). Alkaloids : Chemical and biological perspectives. Volume 11 (Online-Ausg). Pergamon.

Pundarikakshudu, K., & Bhavsar, G. C. (1988). Variations in Pharrnacognostic Characters and Alkaloid Content in Two Types of Adhatoda vasica Nees. *International Journal of Crude Drug Research, 26*(2), 88–96. DOI: 10.3109/13880208809053898

Rahman, S., Patel, D., & Islam, M. (2019). Phytochemicals in neuroprotection and brain health. *Frontiers in Pharmacology, 10*, 232. DOI: 10.3389/fphar.2019.00232

Reshma, A., Subramanian, A., Kumarasamy, V., Tamilanban, T., Sekar, M., Gan, S. H., Subramaniyan, V., Wong, L. S., Rani, N. N. I. M., & Wu, Y. S. (s. d.). Neurocognitive effects of proanthocyanidin in Alzheimer's disease : A systematic review of preclinical evidence. Brazilian Journal of Medical and Biological Research, 57, e13587. https://doi.org/DOI: 10.1590/1414-431X2024e13587

Roohbakhsh, A., Parhiz, H., Soltani, F., Rezaee, R., & Iranshahi, M. (2014). Neuropharmacological properties and pharmacokinetics of the citrus flavonoids hesperidin and hesperetin—A mini-review. *Life Sciences, 113*(1-2), 1–6.

Roy, A., Khan, A., Ahmad, I., Alghamdi, S., Rajab, B. S., Babalghith, A. O., Alshahrani, M. Y., Islam, S., & Islam, Md. (2022). R. (2022). Flavonoids a Bioactive Compound from Medicinal Plants and Its Therapeutic Applications. *BioMed Research International, 5445291.* Advance online publication. DOI: 10.1155/2022/5445291

Rubio-Tapia, A., Hill, I. D., Kelly, C. P., Calderwood, A. H., & Murray, J. A. (2013). ACG clinical guidelines: Diagnosis and management of celiac disease. *The American Journal of Gastroenterology, 108*(5), 656–676. DOI: 10.1038/ajg.2013.79

Safarzadeh, E., Ataei, S., Akbari, M., Abolhasani, R., Baziar, M., Asghariazar, V., & Dadkhah, M. (2024). Quercetin ameliorates cognitive deficit, expression of amyloid precursor gene, and pro-inflammatory cytokines in an experimental models of Alzheimer's disease in Wistar rats. *Experimental Gerontology, 193*, 112466. DOI: 10.1016/j.exger.2024.112466

Sapone, A., Bai, J. C., Ciacci, C., Dolinsek, J., Green, P. H., Hadjivassiliou, M., & Fasano, A. (2012). Spectrum of gluten-related disorders: Consensus on new nomenclature and classification. *BMC Medicine, 10*, 13. DOI: 10.1186/1741-7015-10-13

Satoh, T., & Lipton, S. A. (2013). Redox regulation of neuronal survival mediated by Nrf2. *Nature Reviews. Neuroscience, 14*(11), 759–771. DOI: 10.1038/nrn3645

Scheff, S. W., Ansari, M. A., & Roberts, K. N. (2013). Neuroprotective effect of Pycnogenol® following traumatic brain injury. *Experimental Neurology, 239*, 183–191.

Schuppan, D., & Zimmer, K. P. (2013). The diagnosis and treatment of celiac disease. *Deutsches Ärzteblatt International, 110*(49), 835–846. DOI: 10.3238/arztebl.2013.0835

Sek¬kout, Z., El Hamsas El Youbi, A., Boudaia, O., Janani, S., Radallah, D., & El Amrani, N. (2024). Traditional medicinal plants used for rheumatoid arthritis and immune system disorders treatment in the Casablanca-Settat region, Morocco: An ethnopharmacological study. European Journal of Medicinal Chemistry Reports, 11, 100146.

Sfikakis, P. P., Souliotis, V. L., & Vlachoyiannopoulos, P. G. (2005). Genetic factors in autoimmune diseases. *Autoimmunity Reviews, 4*(2), 109–114. DOI: 10.1016/j.autrev.2004.09.003

Sharifi-Rad, J., et al. (2021). Peganum spp.: A comprehensive review on bioactivities and therapeutic potential. (Review, open access).

Sharma, R. A., Steward, W. P., & Gescher, A. J. (2021). Pharmacokinetics and pharmacodynamics of curcumin. *Advances in Experimental Medicine and Biology, 595*, 453–470. DOI: 10.1007/978-1-4419-6518-9_20

Singh, B. N., Shankar, S., & Srivastava, R. K. (2011). Green tea catechin, epigallocatechin-3-gallate (EGCG): Mechanisms, perspectives and clinical applications. *Biochemical Pharmacology, 82*(12), 1807–1821. DOI: 10.1016/j.bcp.2011.07.093

Singh, P., Arora, A., Strand, T. A., Leffler, D. A., Catassi, C., Green, P. H., & Makharia, G. K. (2018). Global prevalence of celiac disease: Systematic review and meta-analysis. *Clinical Gastroenterology and Hepatology, 16*(6), 823–836.e2. DOI: 10.1016/j.cgh.2017.06.037

Sollid, L. M., & Jabri, B. (2013). Triggers and drivers of autoimmunity: Lessons from coeliac disease. *Nature Reviews. Immunology, 13*(4), 294–302. DOI: 10.1038/nri3407

Teschke, R., Eickhoff, A., & Wolff, A. (2015). Herbal hepatotoxicity in traditional and modern medicine: Actual key issues and new encouraging steps. *Frontiers in Pharmacology, 6*, 72. DOI: 10.3389/fphar.2015.00072

Tovoli, F., Negrini, G., Farì, R., Guidetti, E., Faggiano, C., & Bolondi, L. (2013). Autoimmune liver diseases and celiac disease: A review of the literature. *World Journal of Gastroenterology, 19*(42), 7116–7124. DOI: 10.3748/wjg.v19.i42.7116

Turner, M. D., Nedjai, B., Hurst, T., & Pennington, D. J. (2014). Cytokines and chemokines: At the crossroads of cell signalling and inflammatory disease. *Biochimica et Biophysica Acta (BBA) –. Molecular Cell Research, 1843*(11), 2563–2582. DOI: 10.1016/j.bbamcr.2014.05.014

Ullah, A., Munir, S., Badshah, S. L., Khan, N., Ghani, L., Poulson, B. G., Emwas, A.-H., & Jaremko, M. (2020). Important Flavonoids and Their Role as a Therapeutic Agent. *Molecules (Basel, Switzerland), 25*(22), 5243. DOI: 10.3390/molecules25225243

Umezawa, K., Kojima, I., Simizu, S., Lin, Y., Fukatsu, H., Koide, N., Nakade, Y., & Yoneda, M. (2018). Therapeutic activity of plant-derived alkaloid conophylline on metabolic syndrome and neurodegenerative disease models. *Human Cell, 31*(2), 95–101. DOI: 10.1007/s13577-017-0196-4

Umka Welbat, J., Sirichoat, A., Chaijaroonkhanarak, W., Prachaney, P., Pannangrong, W., Pakdeechote, P., Sripanidkulchai, B., & Wigmore, P. (2016). Asiatic Acid Prevents the Deleterious Effects of Valproic Acid on Cognition and Hippocampal Cell Proliferation and Survival. *Nutrients, 8*(5), 5. Advance online publication. DOI: 10.3390/nu8050303

Velmurugan, B., Guillemin, G. J., & Braidy, N. (2018). Epigallocatechin-3-gallate as a neuroprotective agent: A comprehensive review. *Journal of Alzheimer's Disease, 62*(2), 485–497. DOI: 10.3233/JAD-170973

Venkatesan, R., Ji, E., & Kim, S. Y. (2015). Phytochemicals That Regulate Neurodegenerative Disease by Targeting Neurotrophins : A Comprehensive Review. *BioMed Research International, 814068*. Advance online publication. DOI: 10.1155/2015/814068

Ventura, A., Neri, E., Ughi, C., Leopaldi, A., Citta, A., & Not, T. (2001). Gluten-dependent diabetes-related and thyroid-related autoantibodies in patients with celiac disease. *Pediatrics, 107*(1), E12. DOI: 10.1542/peds.107.1.e12

Volta, U., & Villanacci, V. (2011). Celiac disease: Diagnostic criteria in progress. *Cellular & Molecular Immunology, 8*(2), 96–102. DOI: 10.1038/cmi.2010.64

Wang, M., Yang, Y., & Li, Z. (2016). Flavonoids and ginsenosides in brain-derived neurotrophic factor (BDNF) modulation: Implications for neuroprotection. *Biochemical Pharmacology, 109*, 1–8. DOI: 10.1016/j.bcp.2016.07.014

Wei, P., Zhao, S., et al. (2022). Sesame (Sesamum indicum L.): A comprehensive review of nutritional value, phytochemical composition, health benefits, and food/industrial uses. (Open access review).

Yamamoto-Furusho, J. K. (2020). Role of cytokines in inflammatory bowel disease. *World Journal of Gastroenterology*, *26*(34), 4731–4749. DOI: 10.3748/wjg.v26.i34.4731

Yang, Y., Zhang, Z., & Xie, Z. (2020). Microglial activation and its role in neuroinflammation. *Neuropharmacology*, *172*, 107848. DOI: 10.1016/j.neuropharm.2020.107848

Yuan, H., Ni, X., Zheng, M., Han, X., Song, Y., & Yu, M. (2019). Effect of catalpol on behavior and neurodevelopment in an ADHD rat model. *Biomedicine and Pharmacotherapy*, *118*, 109033. DOI: 10.1016/j.biopha.2019.109033

Zhang, L., Chen, D., & Yang, C. (2013). Curcumin inhibits neuroinflammation via NF-kB signaling. *Biochimica et Biophysica Acta*, *1830*(10), 4930–4938. DOI: 10.1016/j.bbagen.2013.07.010

Zhao, X., Huang, X., Yang, C., Jiang, Y., Zhou, W., & Zheng, W. (2022). Artemisinin Attenuates Amyloid-Induced Brain Inflammation and Memory Impairments by Modulating TLR4/NF-κB Signaling. International Journal of Molecular Sciences, Zheng, J., Liu, W., & Chen, X. (2017). Resveratrol and mitochondrial protection in neurodegeneration. Journal of Molecular Neuroscience, 63(2), 293-302. https://doi.org/, Article 11. https://doi.org/DOI: 10.1007/s12031-017-0953-123(11)

Chapter 5
Secondary Metabolites: Therapeutic Tools for Tackling Neurodegenerative Diseases

Adil El Housseini
https://orcid.org/0009-0007-9294-2685
AQUAMAR Laboratory, Faculty of Sciences, Ibn Zohr University, Agadir, Morocco

Khawla Rachmoune
https://orcid.org/0009-0001-6483-1060
Higher Normal School, Cadi Ayyad University, Morocco

Abdelfatah Ait Baba
https://orcid.org/0000-0002-5650-3868
Laboratory of Ecotoxicology, Bioresources, and Coastal Geomorphology, Polydisciplinary Faculty of Safi, Cadi Ayyad University, Morocco

Youssef Ait Hamdan
Higher Normal School, Cadi Ayyad University, Morocco

Imade Armadi
https://orcid.org/0009-0003-3063-5058
Cadi Ayyad University, Morocco

Chaymaa Harkat
Higher School of Technology Essaouira, Cadi Ayyad University, Morocco

El-Mustapha Laghzaoui
https://orcid.org/0000-0002-7776-5218
Higher Institute of Nursing Professions and Health Techniques of Marrakech, Morocco

Kholoud Kahime
LIREMET Laboratory, Higher School of Technology, Cadi Ayyad University, Morocco

Moulay Abdelmonaim El Hidan
Ibn Zohr University, Agadir, Morocco

Mehdi Ait Larradia
https://orcid.org/0000-0003-2390-2659
Higher Institute of Nursing Professions and Health Techniques, Beni Mellal, Morocco

El Hassan El Mouden
Faculty of Sciences, Cadi Ayyad University, Morocco

Mustapha Agnaou
AQUAMAR Laboratory, Faculty of Sciences, Ibn Zohr University, Agadir, Morocco

ABSTRACT

Alzheimer's, Parkinson's, and various neurodegenerative diseases are generally characterized by neuronal loss, which directly induces cognitive and motor dysfunctions. Secondary metabolites, as natural products of aromatic and medicinal plants and other origins, could present a promising

DOI: 10.4018/979-8-3373-1365-8.ch005

alternative source to combat neurodegenerative diseases. With their important phytochemical composition, especially in alkaloids, terpenes, flavonoids, and polyphenols, these secondary metabolites pave the way for the development of new therapies. Their neuroprotective properties and their antioxidant and anti-inflammatory activities constitute a solid basis for their use in the fight against neurodegenerative diseases. This chapter aims to explore the therapeutic effects of secondary metabolites in the fight against neurodegenerative diseases while discussing the responsible mechanisms, challenges, perspectives, as well as the limits of this application.

INTRODUCTION

Science progresses through a succession of trials and errors, with experimentation generally being at the origin of the production of scientific knowledge. For centuries, humanity has exploited nature and its products, particularly plants, to meet its food and medicinal needs. Food and medicine have thus been central concerns, generation after generation. During these transitions, knowledge about medicinal plants, as well as their appropriate uses, has been transmitted and enriched over time, forming a precious heritage (Jamshidi-Kia et al., 2018).

Humans have also utilized other natural resources to heal themselves, including animal products, minerals, and in particularly medicinal plants (Ait Hamdan et al., 2024; Chaachouay and Zidane, 2024). The use of plants for human health, known as pharmacognosy, is among the oldest practiced sciences, present in nearly all major civilizations, including those of China, Greece, Egypt, India, and the Muslim world (Hamilton, 2004).

Medicinal plants owe their effectiveness to a multitude of chemical compounds, particularly secondary metabolites. These molecules, which mainly serve as defense mechanisms against pathogens (bacteria, fungi, etc.) or as a deterrent against predators (via toxicity), are not essential for the survival of plant cells, unlike primary metabolites (Anjali et al., 2023). However, secondary metabolites play a crucial role in the interactions between plants and their environment, particularly in response to biotic or abiotic stresses (Saurabh Pagare et al., 2015).

Moreover, more than 20,000 secondary metabolites have been identified (Oladipo et al., 2022). Among them, alkaloids, terpenes, flavonoids, phenolic compounds and other key molecules have various biological activities, such as antibacterial (Daghbouche Selma et al., 2017), antioxidant, immunostimulant, aphrodisiac and antifungal (Fox and Howlett, 2008).

In addition to these properties, secondary metabolites have been shown to play a role in the treatment of neurodegenerative diseases, a group of pathologies whose incidence has increased considerably in recent decades. Among the most common diseases are Parkinson's disease (PD), Alzheimer's disease (AD), Huntington's disease (HD) and prion diseases. Some natural metabolites, such as galantamine and hupezine, from aromatic medicinal plants, have already proven their efficacy. These compounds act as Acetylcholinesterase (AChE) inhibitors, thereby increasing the availability of acetylcholine (ACh) in the central nervous system (CNS) (Gao et al., 2025).

This chapter aims to highlight the impact of secondary metabolites, from various sources, on neurodegenerative diseases. It proposes to explore their potential as promising and innovative therapeutic avenues in the medical field.

I. NEURODEGENERATIVE DISEASES AND CURRENT THERAPEUTIC APPROACHES

1. Overview of Neurodegenerative Diseases

Neurologic and neurodegenerative diseases (NDD) constitute a major public health issue. Neurodegeneration refers to the progressive loss of both structure and function of the CNS leading to cognitive dysfunction and motor impairment among others (Castellani et al., 2010). The histopathological hallmark of NDD, such as AD, multiple sclerosis (MS), amyotrophic lateral sclerosis (ALS) and PD, is neurodegeneration or neuronal cell death. Neurodegeneration has also been reported in stroke, neurotrophic infections, prion diseases, traumatic brain and spinal cord injuries as well as neuropsychiatric and genetic disorders (Castellani et al., 2010; Ciurea et al., 2023). An overwhelming body of research suggests that reactive oxygen species (ROS), which cause oxidative stress, are a significant contributor in the development of neurological disorders. While most cases of NDD are multifactorial and involve the interaction between external environmental factors and internal genetic ones. A strong body of evidence sustains the paramount role of oxidative stress in the development of neurologic disorders.

Neurodegeneration is a slow, progressive loss of neurons in particular brain regions and it is the main pathologic feature of several brain disorders such as AD and PD. It is also seen in other conditions, including MS and ALS, as well as neuropsychiatric and genetic disorders (Amor et al., 2014). It is believed that chronic activation of innate immune responses, including those mediated by microglia cells in the CNS, is a prevalent feature of these disorders. This activation may open up neurotoxic pathways that lead to the gradual degeneration of neurons (Amor et al., 2014).

Additionally, toxic protein aggregates are highly established to underlie the development of NDD such as AD (Khanam et al., 2016). β-amyloidepeptide-42 aggregates could trigger oxidative stress, which is in turn boosts the activity of the β-amyloid cleavage enzyme (BACE-1) and increasing the generation of β-amyloid, leading to plaque amyloid formation (Chami and Checler, 2012).

a. Alzheimer's Disease (AD)

AD is the most prevalent neurodegenerative disorder in humans, accounting for approximately 70% of major cognitive impairments (Castellani et al., 2010). In 1901, neurologist Alois Alzheimer documented the unusual symptoms exhibited by a 51-year-old patient, including sleep disturbances, progressive memory loss, and increasing disorientation. Post-mortem examination revealed neurofibrillary tangles and neuritic plaques, prompting Alzheimer to associate these structural abnormalities with the disease. Neurofibrillary tangles are composed of hyperphosphorylated tau proteins, while neuritic plaques result from the accumulation of beta-amyloid peptides (Ciurea et al., 2023).

A study conducted by Gatz and colleagues demonstrated that environmental factors can influence the risk of developing AD, even among individuals with a genetic predisposition (Gatz et al., 2006). For decades, the amyloid cascade hypothesis, which centers on the formation of beta-amyloids, served as the primary framework for understanding the disease. However, recent research suggests that this model oversimplifies the complex mechanisms underlying Alzheimer's pathology. Emerging studies highlight additional factors, such as the role of the glymphatic system and the involvement of specific proteins, including the low-density lipoprotein receptor-related protein 1 (LRP-1) and the receptor for advanced glycation end products (RAGE) (An et al., 2022).

b. Parkinson's Disease (PD)

PD is a progressive neurodegenerative condition predominantly affecting older individuals. It is estimated to impact approximately 0.3% of the general population and between 1% and 3% of those aged over 65. Currently, the global number of cases is approximately 8.7 million, with projections suggesting an increase to 9.3 million by 2030 (Ciurea et al., 2023). First documented in 1817 by James Parkinson, this disease is primarily characterized by motor disturbances, including muscle rigidity, bradykinesia (slowness of movement), postural instability, gait difficulties, and tremors. As a progressive disorder, PD significantly impairs daily functioning, particularly in elderly populations. PD is marked by the aggregation of α-synuclein proteins, forming structures known as Lewy bodies and Lewy neurites, which accumulate in various regions of the central nervous system. These areas include the basal ganglia, the dorsal motor nucleus of the vagus (DMV), the olfactory bulb (OB), the locus coeruleus (LC), the intermediolateral nucleus of the spinal cord (IML), the celiac ganglia, and the enteric nervous system (ENS) (Beitz, 2014). Recent studies suggest a strong association between PD, environmental factors, and aging. Exposure to toxins or age-related stress can induce chronic neuroinflammation referred to as "inflammaging"—which contributes to neuronal degeneration (Ciurea et al., 2023). Pathologically, PD is characterized by significant damage in the substantia nigra pars compacta (SNpc) and the pontine locus coeruleus. These regions exhibit depigmentation, severe neuronal loss, and gliosis. At the onset of clinical symptoms, approximately 60% to 70% of neurons in these areas have already been destroyed (McColgan and Tabrizi, 2018). Lewy bodies and Lewy neurites in surviving dopamine (DA) producing neurons consist the pathological hallmark of PD in addition to dopaminergic neurons loss in the SNpc of the midbrain (Hu and Wang, 2016). Neuron's death in PD could be resulted from mitochondria dysfunction (Migliore and Coppedè, 2009), metabolic dysfunctions related to alpha-synuclein aggregations and ubiquitin-proteasome system dysfunction or oxidative stress (Hu and Wang, 2016). Moreover, oxidative stress brought on by ROS may also be a significant factor in neurodegenerative diseases like PD. Furthermore, progressive neuronal cell death in neocortex and striatum is associated with motor and cognitive impairments as well as psychiatric abnormalities are the major features of HD (Kumar et al., 2015). HD is a NDD that is considered relatively a dopaminergic predominantly related disease (SPOKES, 1980). Even though there are no changes in the DA producing neurons of the striatal DA level, and the mechanism leading to neuronal cell death precisely in the striatum remains unknown (Kumar et al., 2015).

c. Huntington's Disease (HD)

HD, named after the American physiologist George Huntington (1850–1916), is a neurodegenerative disorder. Huntington was the first to describe its hallmark features, including involuntary movements resembling a dance referred to as chorea alongside personality changes and cognitive impairments. The disease progresses through two distinct stages: an initial phase characterized by a hyperkinetic syndrome, resulting from the degeneration of medium spiny neurons (MSNs) in the indirect pathway, followed by an advanced phase marked by a hypokinetic syndrome due to the loss of MSNs in the direct pathway (Wexler et al., 2016). Huntington also recognized the hereditary and progressive nature of the condition, famously stating, "Once it begins, it continues to its conclusion" (Basavarajappa and Subbanna, 2021).

d. Amyotrophic Lateral Sclerosis (ALS)

ALS is a neurodegenerative disorder that initially presents as fatigue accompanied by localized muscle weakness, gradually progressing to involve both upper and lower motor neurons. In some cases, complications such as dysarthria, dysphonia, or swallowing difficulties (dysphagia) may also occur (Masrori and Van Damme, 2020). The disease typically advances from distal muscles toward those closer to the trunk, leading to a progressive decline in the patient's overall health. Life expectancy following diagnosis is estimated at 2 to 5 years, with respiratory failure often being the primary cause of death.

ALS can be classified into two categories: familial ALS, accounting for approximately 10% of cases, and sporadic ALS, which represents the remaining 90%. Despite their similar clinical manifestations, this distinction underscores the critical role of epigenetic research in developing novel therapeutic approaches for neurodegenerative disorders. The interaction between epigenetic factors and environmental elements is particularly significant in the onset of ALS and other neurodegenerative conditions (Renton et al., 2011). Studies suggest that ALS slightly affects women more often than men. Geographically, notable disparities are observed, with a higher prevalence in Europe, while Asia and the Middle East report significantly fewer cases (Balendra and Isaacs, 2018; Woollacott and Mead, 2014). Additionally, exposure to various environmental factors, such as pathogens or heavy metals, can induce genetic alterations that promote the development of neurodegenerative diseases. The key genes implicated in ALS include C9orf72, SOD1, TARDBP, FUS, and TBK1 (Masrori and Van Damme, 2020; Renton et al., 2011).

2. Current Therapeutic Approaches: Successes and Challenges

The majority of neurodegenerative diseases are characterized by cognitive decline, significant motor impairments, deterioration of social interactions, and major challenges in performing daily activities. To date, these conditions remain incurable. Current treatments primarily aim to slow disease progression, enhance patients' quality of life, and manage associated comorbidities. Over recent decades, the number of diagnosed cases of neurodegenerative disorders has risen significantly. This increase is largely attributed to longer life expectancies, which heighten the likelihood of developing such diseases, and advancements in early diagnostic techniques, allowing for more frequent detection by healthcare systems (Checkoway et al., 2011). AD, while 5% of those over 65 suffer from PD (Shusharina et al., 2023). The growing number of patients, coupled with the need for comprehensive management approaches, necessitates high standards for medical data management systems and the artificial intelligence algorithms integrated into them (Termine et al., 2021).

The treatment of neurodegenerative diseases, given the irreversible nature of neuronal tissue degeneration, primarily focuses on slowing disease progression while enhancing patient well-being. Pharmacological management plays a critical role by targeting the metabolism of neurons and glial cells to mitigate damage and partially compensate for the functions of lost neurons. Complementary to drug therapies, non-pharmacological strategies aim to improve patients' autonomy, facilitate their adaptation to their environment, and enhance their overall quality of life. These approaches often involve comprehensive rehabilitation programs that include aerobic physical exercises, cognitive training, and therapeutic activities. Additionally, brain stimulation techniques may be employed to augment these measures (Clare et al., 2019).

From a pharmacological perspective, commonly used treatments include cholinesterase inhibitors, such as donepezil and rivastigmine, as well as NMDA receptor antagonists like memantine, which are often administered alone or in combination (Dhiman et al., 2022). To alleviate motor symptoms associated with parkinsonism, therapies may involve dopaminergic receptor agonists (e.g., apomorphine), dopamine precursors (e.g., levodopa), and monoamine oxidase type B (MAO-B) inhibitors (Shusharina et al., 2023). Although ongoing research aims to develop neuroprotective drugs, their clinical application remains a subject of debate, with no definitive consensus achieved to date.

II. INVOLVEMENT OF SECONDARY METABOLITES IN NEUROPROTECTION

Neurodegenerative diseases, including Alzheimer's, Parkinson's, and Huntington's, are chronic conditions marked by progressive neuronal loss. While the specific triggers may vary, shared pathological features include oxidative stress, neuroinflammation, protein aggregation, and neurotransmitter imbalances (Rauf et al., 2022; Rekatsina et al., 2020; Teleanu et al., 2022). In addition, several natural products are recognized for their importance in neurodegenerative diseases (Figure 1). Secondary metabolite compounds such as alkaloids, flavonoids, terpenes, and polyphenols are emerging as powerful tools to address these challenges. These metabolites not only exhibit multiple protective effects on neuronal integrity but also offer a basis for the development of novel therapeutic strategies (Hilal et al., 2024).

Figure 1. Neuroprotective mechanisms of secondary metabolites

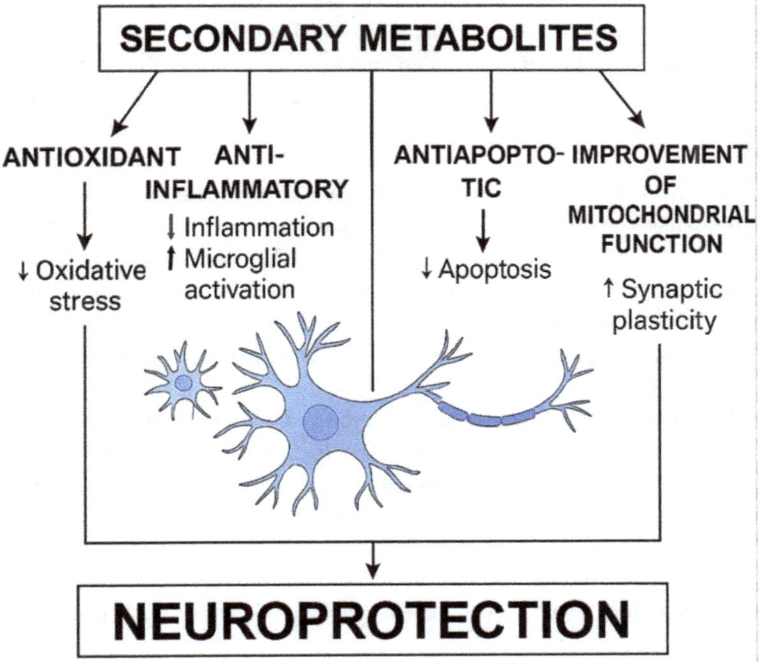

1. Antioxidant Properties

Oxidative stress arises when ROS and reactive nitrogen species (RNS) overwhelm the brain's antioxidant defenses. ROS can damage DNA, proteins, and lipids, leading to mitochondrial dysfunction and apoptosis. Oxidative stress is one of the most powerful factors responsible for the loss of neuronal structure and function (Teleanu et al., 2019). Thus, the antioxidant activity of secondary metabolites could contribute to neuroprotection. In fact, Secondary metabolites mitigate this damage through different mechanisms (Figure 1) (Ramalingam and Kim, 2012). For example, flavonoids (e.g., quercetin, catechins, and rutin) neutralize ROS by donating electrons, forming stable intermediates, and terminating radical chain reactions (Heim et al., 2002). Polyphenols like resveratrol exhibit dual mechanisms: scavenging ROS directly and boosting endogenous antioxidant systems via Nrf2 activation (Sandoval-Acuña et al., 2014). Nrf2 regulates the transcription of antioxidant enzymes such as glutathione peroxidase and heme oxygenase-1 (HO^{-1}) (Alonso-Piñeiro et al., 2021).

Studies on experimental models of AD reveal that flavonoids reduce lipid peroxidation, decrease amyloid beta (Aβ)-induced oxidative stress, and improve cognitive performance (Ma et al., 2018). Similarly, curcumin from turmeric exhibits metal-chelating properties, reducing iron- and copper-mediated ROS generation (Zhong et al., 2024).

Quercetin, a flavonoid found in onions and apples, has been extensively studied for its neuroprotective effects. In rodent models of ischemia-reperfusion injury, quercetin administration restored mitochondrial function and prevented neuronal apoptosis.

Epigallocatechin gallate (EGCG) from green tea demonstrates ROS scavenging and inhibits ROS-producing enzymes like NADPH oxidase (Naoi et al., 2019). In Parkinson's models, EGCG protected dopaminergic neurons from oxidative damage and improved motor deficits (Xu et al., 2017).

Flavonoids are noted to activate Nrf2 (Nuclear factor erythroid 2-like 2), that increase producing antioxidant enzymes and enhance cellular resilience to oxidative stress, as a result reducing neurodegenerative inflammation (Bellavite, 2023). Antioxidants including vitamin C and E are marked by their effectiveness to inhibit cellular damaging by neutralizing reactive oxygen species (ROS). Furthermore, these antioxidants can attenuate brain inflammation by decreasing the release of pro-inflammatory cytokines via activated glial cells (Bhandari et al., 2024 and Wang et al., 2006).

Antioxidants as well, affect microglial activation and convert them to a neuroprotective M2 phenotype in order to reduce the inflammation (Fang et al., 2024). By reducing the production of inflammatory mediators, cytokines also reduce the secondary neuronal damage associated with neurodegenerative diseases (Wang et al., 2006). According to studies using neonatal mouse models, activating PPAR-γ can significantly reduce oxidative stress and improve neurobehavioral outcomes (Fang et al., 2024). Antioxidants may shield neuronal cells from oxidative damage and inflammation, according to some experimental evidence (Gelain et al., 2012).

2. Anti-inflammatory Activities

Chronic neuroinflammation is perpetuated by the overactivation of microglia and astrocytes, leading to the sustained release of cytokines such as TNF-α, IL-1β, and IL-6 (Hassamal, 2023). These pro-inflammatory signals amplify oxidative stress, disrupt neuronal homeostasis, and exacerbate disease progression (Simpson & Oliver, 2020). Secondary metabolites counteract neuroinflammation through several pathways. They inhibit the Nuclear factor kappa (NF-κB) pathway, a master regulator of inflammatory responses, and activate anti-inflammatory pathways like PPARγ signaling (Sen Singh et al., 2020).

Terpenoids, such as β-caryophyllene and ginkgolides, show exceptional promise in mitigating neuroinflammation.

- **β-Caryophyllene:** This sesquiterpene, found in cloves and black pepper, acts as a CB2 receptor agonist. By selectively targeting CB2 receptors on glial cells, β-caryophyllene reduces the production of pro-inflammatory cytokines while preserving neuronal function. In models of AD, β-caryophyllene ameliorated memory deficits and decreased microglial activation (Cheng et al., 2014).
- **Ginkgolides:** Extracted from Ginkgo biloba, these terpenoids inhibit platelet-activating factor (PAF), a molecule implicated in inflammatory cascades. Clinical studies suggest improved cognitive function and reduced neuroinflammation in patients with early-stage AD treated with Ginkgo biloba extracts (Li et al., 2020).

- **Berberine:** an isoquinoline alkaloid, has shown similar anti-inflammatory effects in Parkinson's models, reducing microglial activation and protecting dopaminergic neurons from degeneration (Ahmed et al., 2015).

The impact of non-steroidal anti-inflammatory drugs (NSAIDs) on minimizing neuroinflammation is a research contradictory subject for now. Suggesting that the NSAIDs may be able to target neuroinflammation in some neurodegenerative diseases like Alzheimer's, Parkinson's, and ALS. Nevertheless, their effectiveness in clinical trials have been inconsistent and conflicting results, thus, necessitating additional research, despite animal studies which have been suggesting protective effects (Hanna et al., 2022 and Klegeris & McGeer, 2005).

The activation of GLP-1 receptors in the brain, which have been shown to be anti-inflammatory, could be a novel therapy option for neurodegenerative disorders. The activation of GLP-1 receptor is a promising treatment target for diseases such as Alzheimer's and Parkinson's because of its anti-inflammatory power particularly for these NDDs. It could affect the activity of microglia and astrocytes, decreasing as well the production of pro-inflammatory cytokines, and reestablishes the homeostasis of brain cells (Diz-Chaves et al., 2022).

Microglia inflammation can be prevented by Quercetin, Ranolazine, and Naringenin, potentially reducing neurodegeneration conditions. In exception of the long-term side effects of corticosteroids, it is possible that these compounds can augment the anti-inflammatory functions of microglia, consequently, minimizing inflammation and neurodegeneration (Piano et al., 2023). Inflammatory responses contributing to neurodegenerative disorders can be decreased by tocotrienols (T3). The neuroprotective properties of these drugs are well known, but their low oral bioavailability necessitates looking into advanced drug delivery systems, especially nanotechnology, to get the most out of their therapeutic potential (Mathew et al., 2023).

3. Regulation of Protein Aggregation and Misfolding

Protein misfolding and aggregation disrupt neuronal signaling and induce cytotoxicity (Zaman et al., 2019). In AD, amyloid-beta peptides aggregate into plaques, while tau proteins form neurofibrillary tangles. Secondary metabolites can inhibit these aggregation processes and promote protein clearance (Roda et al., 2022).

Studies have demonstrated that curcumin has significant neuroprotective properties. It binds to Aβ peptides, preventing their oligomerization and fibrillation, two processes involved in the formation of the amyloid plaques characteristic of AD. In addition, curcumin stimulates autophagic clearance of misfolded proteins by activating AMP-activated protein kinase (AMPK) pathways. In studies carried out on transgenic mouse models, curcumin administration reduced the burden of Aβ plaques and improved cognitive function (Azzini et al., 2024). Furthermore, research has revealed that resveratrol exerts a neuroprotective effect by activating SIRT1, an enzyme involved in the degradation of hyperphosphorylated tau proteins, characteristic of neurodegenerative diseases such as AD. Its anti-inflammatory and antioxidant properties reinforce its overall action in protecting neurons and preventing degenerative processes (Rahman et al., 2020).

In a clinical trial involving patients with mild cognitive impairment, curcumin supplementation improved memory and reduced Aβ accumulation, as confirmed by PET imaging (Voulgaropoulou et al., 2019). Similarly, resveratrol's ability to cross the blood-brain barrier has positioned it as a promising candidate for therapeutic interventions in proteinopathies (Katila et al., 2022).

4. Modulation of Neurotransmitter Systems

Neurotransmitter imbalances are central to the pathology of neurodegenerative diseases. Dopamine depletion in PD leads to motor dysfunction (Ramesh & Arachchige, 2023), while cholinergic deficits in AD impair memory and learning. Secondary metabolites offer a promising therapeutic approach by acting directly or indirectly on these imbalances (Dunnett and Fibiger, 1993).

Nicotine-derived compounds, for example, enhance dopaminergic signaling by stimulating nicotinic acetylcholine receptors (nAChRs) present on dopaminergic neurons. Research has shown that nicotine improves motor performance while protecting dopaminergic neurons from oxidative damage in experimental models of PD (Ross and Petrovitch, 2001).

On the other hand, galantamine, an alkaloid derived from plants of the Galanthus genus, acts by inhibiting AChE, thereby increasing the availability of ACh in the brain. In addition, this compound also modulates nicotinic receptors, providing neuroprotection against toxicity induced by amyloid (Aβ) protein aggregates, a key factor in the progression of AD (D'Onofrio et al., 2017).

In preclinical studies, nicotine and related alkaloids reduced alpha-synuclein aggregation, a hallmark of PD. Moreover, nicotine-treated animals exhibited improved motor function and decreased dopaminergic neuronal loss, highlighting its potential as a neurotherapeutic agent (Khodadadi and Javadi, 2025).

III. PROMINENT SOURCES OF NEUROPROTECTIVE SECONDARY METABOLITES

Studies have demonstrated that natural antioxidant compounds such as flavonoids, coumarins, alkaloids and essential fatty acids have promising neuroprotective potential in the treatment of neurodegenerative diseases such as ALS. By modulating oxidative stress, neurogenic inflammation and apoptotic pathways, these agents help protect neuronal cells and stabilize cell membranes (Silva et al., 2020).

1. Compounds of Plant Origin

Plants have long been a valuable source of medicines, and secondary plant metabolites are at the heart of traditional treatments for various ailments. These bioactive substances play a crucial role in protecting against oxidative and inflammatory stress, two key factors contributing to neurodegenerative diseases. Polyphenols, one of the most abundant classes of SBMs in plant foods, are becoming a promising approach to the development of effective agents for the prevention of neurodegenerative diseases (Câmara et al., 2020). This family of chemicals includes flavonoids (e.g. epigallocatechin, epigallocatechin gallate, catechin, quercetin, luteolin), non-flavonoids (e.g. caffeic acid, gallic acid, ferulic acid, syringic acid) and anthocyanins (e.g. malvidin), which are pharmacologically important SBMs (Baghel et al., 2023; Parsafar et al., 2020).

Research has highlighted the neuroprotective potential of secondary compounds extracted from *Melicope lunu-ankenda*, a member of the Rutaceae family. These compounds include phenolic acid derivatives, flavonoids, coumarins and alkaloids, all of which have shown beneficial effects in the treatment of neurodegenerative diseases. These agents exert their neuroprotective effects primarily by modulating oxidative stress, reducing inflammation and inhibiting apoptotic pathways. For instance, flavonoids have been reported to reduce neuronal damage by neutralizing free radicals, while coumarins have anti-inflammatory effects, helping to reduce inflammation associated with neurodegenerative diseases. Alkaloids and phenolic derivatives also showed potential for protecting neurons against oxidative damage and inflammation. Such results point to the importance of *M. lunu-ankenda* as a potential source of new treatments for neurological diseases (Abdulwanis Mohamed et al., 2019).

Some studies have also investigated the neuroprotective properties of secondary sulfur-containing metabolites of plant origin, with particular emphasis on glucosinolates, garlic (*Allium sativum L.*) and other sulfur compounds. The results of in vitro and in vivo research indicate that these metabolites can modulate several systems involved in the pathogenesis of neurological diseases, such as oxidative stress, inflammation and apoptosis. The study reports on the compounds' activity as direct and indirect antioxidants, their ability to modulate mediators of apoptosis and their potential to inhibit amyloid protein formation, with a particular focus on the underlying mechanisms (Venditti and Bianco, 2020).

Ginkgo biloba is recognized for its benefits in cognitive impairment and AD prevention, thanks to its standardized extracts (GBE) rich in flavonoids and terpenoids with antioxidant properties. Such compounds reduce oxidative stress, promote cerebral microcirculation and modulate inflammation, thus slowing neuronal degeneration (S. K. Singh et al., 2019). EGb 761 extract (20-30 mg/kg) showed promising effects in mice with AD, improving memory, reducing amyloid plaques and stimulating neurogenesis in the serrated gyrus (Ge et al., 2021). These findings confirm its neuroprotective potential and its role in cognitive support.

A recent study has shown that curcumin, a polyphenol extracted from the rhizome of *Curcuma longa*, has strong neuroprotective potential in the treatment of neurodegenerative diseases such as PD, AD, Huntington's (HD), MS and ALS. Traditionally used in Asia for its anti-inflammatory, antioxidant and neuroprotective properties, it remains limited by its low solubility, reduced bioavailability and difficulty in crossing the Blood-brain barrier BBB. Nanotransporters such as liposomes and nanoparticles have been developed to improve its delivery to the brain, thereby increasing its efficacy while reducing side effects (Genchi et al., 2024).

In addition, studies have revealed that adaptogens such as *Schisandra chinensis*, *Eleutherococcus senticosus*, *Rhodiola rosea* and *Withania somnifera* possess anti-neuroinflammatory potential, modulating cytokine expression and NF-κB pathways to reduce chronic inflammation implicated in neurodegenerative diseases, depression and anxiety (Wróbel-Biedrawa and Podolak, 2024). *Withania somnifera* (*ashwagandha*) contains withanolides and alkaloids that reduce oxidative stress and neuronal inflammation, potentially delaying diseases such as AD and PD. It also acts as an adaptogen, improving stress resilience and cognitive function, offering promising therapeutic hope in neuroscience (Mikulska et al., 2023).

2. Compounds Derived from Marine Organisms

Marine organisms, such as algae, sponges and corals, produce secondary metabolites that are chemically distinct from those of land plants (Karthikeyan et al., 2022). These compounds have unique structures and innovative mechanisms of action, making them particularly attractive for neuroprotective research.

Sea algae, which include a wide variety of types, have shown significant potential as natural antioxidants to combat the free radicals responsible for various chronic human diseases, such as cancer, inflammation, as well as neurological disorders such as Alzheimer's, Parkinson's, schizophrenia and depression. Brown, red and green algae are the types most studied for their antioxidant properties. These include fucoidane from *Sargassum fusiforme* and *Sargassum muticum* (brown algae), phycoerythrin from *Porphyra sp.* and *Gracilaria gracilis* (red algae), and loliolide from *Codium tomentosum* (green algae). As antioxidants, these compounds have been shown to work by a variety of mechanisms, helping to reduce the harmful effects of free radicals without damaging the body's organs (Khairinisa et al., 2023). In some cases, compounds such as flavonoids and phenols present in various species of seaweed have been identified for their ability to neutralize free radicals (Pratama & Busman, 2020). Combining the neurotherapeutic effects of compounds extracted from seaweed (Figure 2) (Khairinisa et al., 2023).

Studies have shown that *spirulina* exerts a significant neuroprotective effect, mainly thanks to its antioxidant and anti-inflammatory properties. Rich in bioactive peptides, it protects the brain from oxidative stress, improves cognitive and emotional functions, and reduces mental fatigue. It also contributes to cerebral vascular protection by preserving endothelial walls and regulating internal pressure. In the case of malnourished children, its intake appears to promote motor, linguistic and cognitive development. In addition, spirulina may also play a central role in regulating appetite, indicating a direct action on the nervous system (Sorrenti et al., 2021).

In neuroprotection, compounds extracted from marine sponges have shown remarkable potential. For instance, dysideamine, extracted from Dysidea sp., stimulates neurite extension in Neuro 2A cells, with 40% efficiency at 3 μM, although it is slightly cytotoxic at 10 μM (Suna et al., 2009). Pyridoacridine alkaloids from Biemna fortis induce neuritogenesis at 0.01 μM, with a fourfold increase in AChE activity at 0.03 μM, but become cytotoxic above 0.3 μM (Grosso et al., 2014). Furthermore, lembehyne A, extracted from Haliclona sp., enhances neuritogenesis at 2 μg/mL in PC12 cells and 0.1 μg/mL in Neuro 2A cells, involving actin polymerization and de novo protein synthesis (Grosso et al., 2014).

Figure 2. Neurotherapeutic potential of seaweed-derived compounds

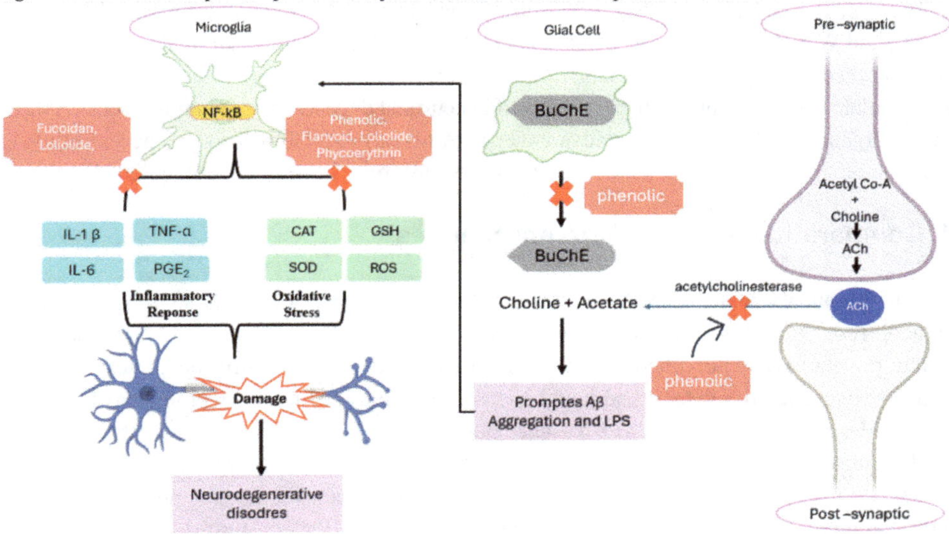

Source: Khairinisa et al. (2023)

3. Compounds Derived From Micro-Organisms

Microorganisms, including fungi and bacteria, are another rich and diverse source of secondary metabolites with properties that are neuroprotective. Such organisms produce metabolites that can modulate neuroinflammation, reduce oxidative stress and protect neurons against degeneration (Parker et al., 2020).

In addition, the gut microbiota plays a key role in intestinal homeostasis and brain health. Bacteria such as *Lactobacillus rhamnosus* strengthen the intestinal barrier (Laval et al., 2015), while the proliferation of Clostridium is linked to inflammatory diseases (Nash et al., 2018). It also influences the BBB, whose permeability increases in the absence of microbiota, but recovers after inoculation (Parker et al., 2020).

Studies have highlighted that in neurodegenerative diseases such as Alzheimer's and Parkinson's, alterations in the microbiota, such as a decrease in Prevotella and an increase in Enterobacteriaceae, are associated with neuroinflammation (Parker et al., 2020). Such links between microbiota and brain open up prospects for therapies targeting this interaction. Moreover, the studies focused on alkaloids from ergot (*Claviceps purpurea*), a fungus that infects cereals and grasses. These alkaloids, whose ergoline structure is similar to that of neurotransmitters such as noradrenaline, dopamine and serotonin, are used in the treatment of neurological disorders such as migraines and PD. In addition, these compounds have applications in obstetrics, gynecology, prolactinomas, type II diabetes and cancer. They are also among the "smart drugs" used to improve cognitive function and treat age-related brain disorders (Sharma et al., 2016).

Additionally, medicinal champignon species such as *Ganoderma lucidum* (Curtis) P. Karst, *Lentinula edodes* (Berk.) Pegler, *Pleurotus* spp, *Coriolus versicolor*, *Grifola frondosa* (Dicks.)

Gray, *Hericium erinaceus* (Bull.) Pers, *Antrodia cinnamomea* T.T. Chang and W.N. Chou, *Agaricus bisporus*, and *Agaricus blazei* Murrill have also demonstrated neuroprotective effects. Chang and W.N. Chou, Agaricus bisporus, and Agaricus blazei Murrill have also demonstrated neuroprotective effects. Research has revealed that these fungi attenuate neuronal inflammation and regulate cell signalling pathways involved in neuronal degeneration (Venturella et al., 2021). Micro-organisms offer significant advantages in terms of production costs and accessibility, as they can be grown on a large scale in controlled environments (Venturella et al., 2021).

4. Comparative Study of Different Sources

The secondary metabolites derived from plant; marine and microbial sources represent a valuable reserve of neuroprotective compounds. Plants such as *Ginkgo biloba* and *Curcuma longa*, rich in flavonoids and polyphenols, act on the inflammatory and oxidative processes associated with neurodegenerative diseases, although their efficacy may be limited by problems of bioavailability. Marine organisms, such as algae and sponges, produce original molecular structures that hold promise for limiting neuronal damage. In addition, certain micro-organisms, notably in the intestinal microbiota, regulate neuroinflammation and offer new therapeutic prospects. Despite their potential, further research is needed to overcome extraction constraints and promote their clinical application. Table 1 summarizes these sources, highlighting their strengths and limitations.

Table 1. Advantages and limitations of neuroprotective secondary metabolite sources

Source	Metabolites	Avantages	Limites	Typical species	References
Compounds of Plant Origin	Flavonoids, coumarins, alkaloids, polyphenols, anthocyanins, fatty acids	A readily available natural source, rich in antioxidants, anti-inflammatories and neuroprotective agents.	Bioavailability limited, variations in efficacy depending on extraction and plant.	*Melicope lunu-ankenda* (Rutaceae), *Allium sativum* (Garlic), *Ginkgo biloba, Curcuma longa* (Curcuma), *Schisandra chinensis*	(Câmara et al., 2020; Genchi et al., 2024; Silva et al., 2020; Wróbel-Biedrawa and Podolak, 2024)
Compounds Derived from Marine Organisms	Fucoidane, phycocerythrine, loliolide, bioactive peptides, alkaloids	Powerful antioxidant effects, innovative mechanisms for neuroprotection.	Difficult to extract and supply, some compounds are cytotoxic in high doses.	*Sargassum fusiforme, Porphyra sp., Codium tomentosum, Dysidea sp., Haliclona sp.*	(Grosso et al., 2014; Khairinisa et al., 2023; Suna et al., 2009)
Compounds derived from micro-organisms	Alkaloids, phenolic compounds, products of microbiota-brain interaction	Potential modulation of microbiota, neuroprotective effects through modulation of inflammation and oxidative stress.	Variability of effects depending on the microbiotic strain, risk of microbial side effects.	*Lactobacillus rhamnosus, Clostridium, Prevotella, Enterobacteriaceae, Claviceps purpurea* (ergot)	(Laval et al., 2015; Nash et al., 2018; Parker et al., 2020)

IV. ADVANCES IN RESEARCH ON SECONDARY METABOLITES IN NEURODEGENERATIVE THERAPY

1. Bioavailability and Pharmacokinetics

Technologies such as permeabilizers, osmotic pumps, slow polymer release systems, and cell transplantation, were being employed to cross successfully blood-brain barrier enhancing delivery of bioactive molecules into the central nervous system, in order to reestablish normal function. These advancements were essential and have been foundational in evolving treatment for neurodegenerative diseases (Tan and Aebischer, 2007). Recently, Nanosystems had a significant impact in biomedicine and maintain an essential strategy in the pathogenic drug delivery to treat neurodegenerative disorders (NDDs) and cancer and minimize side effects. This is especially true for therapeutics that target the BBB, a significant barrier in the body, selectively preventing the majority of drugs from entering the brain. However, the roles of efflux transporters and nanosystem toxicity are still difficult to overcome. However, long-term effects, safety in the brain, cost-effectiveness, and the scalability of these nanosystems need to be assessed to promote their clinical application (Rafati et al., 2024). Moreover, Gajbhiye et al (2015) highlighted the beneficial treatment model that can be achieved by using nanocarriers for delivering drugs like paclitaxel and doxorubicin across the BBB and help overcoming its problems. Functionalizing nanocarriers with target-specific ligands to attain enhanced targeting and control release of drugs has been discovered to improve their ability to cross the BBB and deliver therapeutic agents to target cells in the diseased brain. demonstrating the application of liposomes and dendrimers as potential drug carriers. Nonetheless, the safety, as well as the efficacy, of nanocarrier systems, is yet to be established in order to validate their clinical use (Gajbhiye, 2015).

Figure 3. Schematic representation of the different mechanisms of nanoparticles for crossing the blood–brain barrier (BBB). AMT, adsorptive-mediated transcytosis; TMT, transporter-mediated transcytosis; RMT, receptor-mediated transcytosis

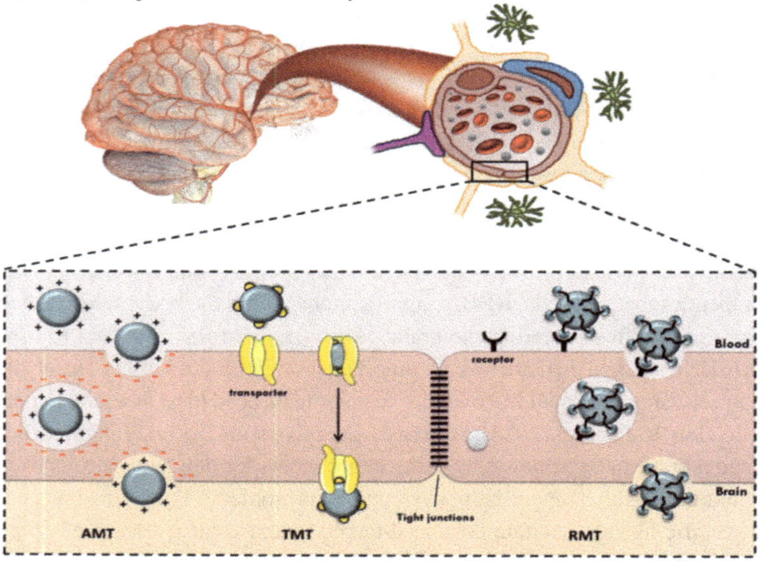

Source: Pinheiro et al. (2021)

In the same context, the treatments potential of nanoparticle drug delivery systems on Alzheimer and Parkinson, could enhancing the symptomatic treatment and improving drug delivery to the brain through ameliorating pharmacokinetics, bioavailability, and brain penetration (Boyton et al., 2024). Other studies are presently concentrating on secondary bioactive products (SBMs) within foods to relieve the effects of neurodegenerative diseases. Due to their varied biological effects such as antioxidant, antidiabetic, antihypertensive, anti-Alzheimer's, antiproliferative, and antimicrobial, SBMs like terpenoids, polyphenols, glucosinolates, phytosterols, and capsaicinoids could be a preventive strategy for oxidative stress-related diseases or chronic diseases such as those with Alzheimer and Parkinson (Gomes et al., 2024). The safety and beneficial effects of rosemary and some other medicinal herbs poses their advantages in controlling some of the symptoms in NDDs. Similarly, (Alvi et al., 2019) reported medicinal properties of Rosmarinus officinalis that is full of bio-active secondary metabolites, including carnosol, carnosic acid, and rosmarinic acid, which are polyphenolic diterpenes, and provides therapeutic benefits against various NDDs such as epilepsy, dementia, Alzheimer, Parkinson and migrain. In this setting, Alzheimer's and Parkinson's is specifically characterized by the aggregation of amyloid-beta, neuronal cell death, and neuroinflammation; on the other hand, bio-active compounds such as carnosol, carnosic acid, and rosmarinic acid can target those fundamental mechanisms characteristic of neurological diseases, coming to light as potential therapeutic options (Alvi et al., 2019). In the same context, flavonoids are highly transforming by gut microbiome which can active their bioavailability and therapeutic effects, it has been shown that specific flavonoid metabolites, like protocatechuic acid and urolithins agents can cross the blood-brain barrier and maximizing neuroprotective properties (Melrose, 2023).

2. Synergistic Effects of Metabolites

Monotherapy may not be effective for treating neurodegenerative diseases due its complicated pathology, highlighting the requirement for combination treatment strategy instead of individual drug. Last papers likewise provided valuable insights to guide future clinical toward new therapeutic strategies for neurodegenerative disorders. Eventually, Xie et al. (2024) recommended that synergistic impact of curcumin and mitoquinol mesylate was exhibited to essentially work on mental capacities in a triple-transgenic Alzheimer model. Where the findings suggested that this drug synergy effectively diminish Aβ aggregation, tau phosphorylation, and mitigation of synaptic damage which are key pathological features of the AD. Underscoring that curcumin and mitoquinol synergy could be a promising for an effective Alzheimer's drug. In the same context, the authors thought that cortexin, citicoline, and antioxidants interactions have the potential advantages when combining these agents in treating cerebrovascular diseases due to their synergistic spectra. Highlighting the the necessity of comprehension of molecules' mechanisms in each drug to identifying the most effective combinations which can lead to enhance treatments (Putilina and Teplova, 2022). Similarly, Kundu and Dubey (2021) suggested that purines and pyrimidines are distinguished as essential possibility for developing future drugs that can impact neurodegenerative cycles especially concern relation to neuromodulation and fetal neurodevelopment. Biochemical mechanisms for these metabolites focusing on protein misfolding and amyloidosis which largely unexplored. Calling for the need for a deeper comprehension of its mechanism synergy for both compounds to fill the gaps (Kundu and Dubey, 2021). As contrasted, Pierozan et al. (2018) tested the synergistic toxicity of the neurometabolites quinolinic acid and homocysteine in cortical neurons and astrocytes and their implications in AD. They thought that combination of quinolinic acid and homocysteine can leads to increase toxicity in both neurons and astrocytes, suggesting a synergistic effect that probably aggravate cell damaging. As the mechanisms of toxicity depend to the concentration of these metabolites that could provoke neurons primarily undergo necrosis and astrocytes apoptosis. The oxidative stress and mitochondrial dysfunction induced by quinolinic acid and homocysteine increased reactive oxygen species production and lipid peroxidation leading to neuronal death (Pierozan et al., 2018).

3. High-Throughput Screening for Novel Compounds

Molecular docking plays a crucial rule in understanding how bioactive secondary metabolites interact with the target enzymes, and guide for future drug development efforts for neurodegenerative diseases generally. The interaction between six compounds and the target enzymes was achieved using molecular docking, which was used to understand their potential therapeutic effects. The key findings emphasized that the ultimate candidate was Harmalol owing to its good binding affinity on AChE and Butyrylcholinesterase (BChE), the two key enzymes intertwined with Alzheimer disease. Moreover, it fitted well with various drug property parameters, and might be regarded as a prospective lead compound for the further exploration lead compound for the treatment of neurodegenerative disorders, especially AD (Jyothi and Yellamma, 2016). Protocatechuic acid and Urolithins have been shown exhibiting neuroprotective properties and combatting mood, anxiety, and cognitive deficits common in neurological disorders. Also, therapeutic probiotics and nutraceuticals that promote the levels of beneficial flavonoid metabolites may be therapeutic candidates for improving overall health and well-being, in particular for patients with cognitive decline (Melrose, 2023). Ceramide and Sphingosine-1-phosphate (S1P) levels increased has been shown that induced cell death and consequently the loss of neuroprotection. This elevation aggravates the symptomatic and the progression of neurodegenerative situation particularly with regard to Alzheimer Parkinson and -ALS. Monitoring the metabolism of these lipid levels can give a perspective solution on therapeutic strategies of neurodegenerative diseases by changing their levels and restoring balance and protecting neurons. Furthermore, innovative therapeutic approach targeting the metabolic involvement of these illnesses depends basically on the development of new bioinformatics and molecular docking strategies (Tringali and Giussani, 2022).

Figure 4. Ceramide and sphingosine-1-phospate metabolism and signaling in CNS. Ceramide (Cer) is generated from serine and palmitoyl Coenzyme A (de novo pathway) and through the acylation of sphingosine (Sph) (salvage pathway) at the endoplasmic reticulum. Cer is a substrate for sphingomyelin (SM) and glycosphingolipids synthesis at the Golgi apparatus. Sphingosine 1-phosphate (S1P) is synthesized via phosphorylation of sphingosine (Sph). S1P can be dephosphorylated back to Sph and then reacylated to Cer, or it can be degraded to hexadecenal and ethanolamine-1-phosphate (ethanolamine-1P). Many of the biological effects of S1P are mediated via specific receptors, designated S1PR

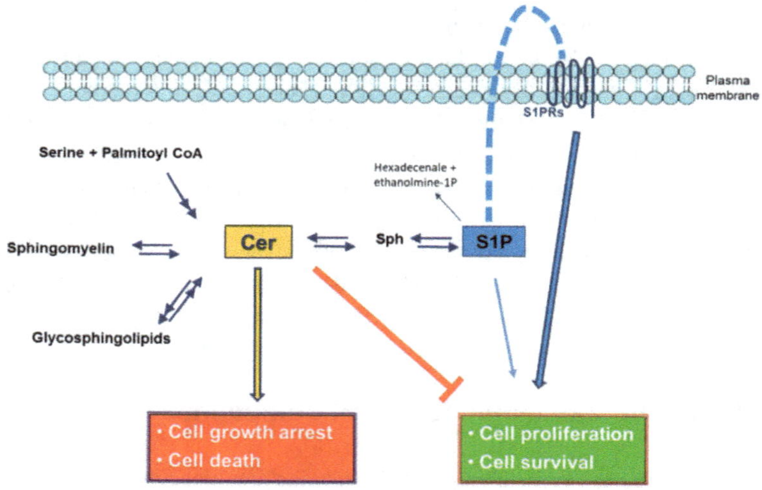

Source: Tringali and Giussani (2022)

Eventually, using computational tools in drug discovery is underscoring their prospective rule to identifying leads from medicinal plants. Computational methods such as homology modeling, molecular docking, and molecular dynamics simulations have been showing their advantage in assessing the binding affinities of multiply secondary metabolites. In their study, Subramaniyan and (Vijayakumar and Rajalakshmi, 2021) applied suitable computational methods to screen natural molecules that could block DJ1 receptor. Results indicated that Punigluconin was the most promising candidate, as it binded more than other molecules with the DJ1 receptor. This study highlighted potential of Punigluconin as a promising therapeutic agent for neurodegenerative disorders such as Parkinson (Vijayakumar & Rajalakshmi, 2021). In the same vein, Oyinloye et al. (2021) tested the Polypharmacology of Gongronema latifolium leaf secondary metabolites against protein kinases implicated in PD and AD, through computational screening. Nine molecules were selected based on molecular docking results. Main finding suggested that the secondary metabolites from Gongronema latifolium could be promising candidates for future therapies targeting neurodegenerative diseases like PD and AD due to their potential inhibitory activities against protein kinases (PKs) involved in neurodegenerative diseases (Oyinloye et al., 2021).

Flavonoids such as Catechin, Gallocatechin, Butein, and Isorhamnetin demonstrate favorable drug-likeness attributes, indicating their potential for further development as drug candidates including low risk of drug-drug interactions and high gastrointestinal absorption, indicating their potential effectiveness as therapeutic agents. Emphasizing the importance of exploring natural compounds for drug discovery in this field. Computational screening using QSAR models and 2D QSAR developed using a Kernel-based PLS regression which predicted both pIC50 values for these compounds, underscoring that these secondary metabolites could play an important role in therapeutic strategies for these diseases (Oyinloye et al., 2021).

Iqbal et al. (2022) successfully screened 44 natural compounds that can induce brain-derived neurotrophic factor (BDNF) to find potential treatments for AD. Among these, physicochemical properties, including low toxicity and good absorption potential, making Soyasapogenol B (Soy-B) emerged as the most promising candidate for crossing the blood-brain barrier. In molecular docking studies, Soy-B had better binding energy scores compared to other compounds against key proteins involved in AD, such as AChE, BuChE, GSK3β, MAO-A, MAO-B, and NMDA. Showing a stability in interaction with these proteins by strong hydrogen bonds and hydrophobic interactions which can form strong complexes with these target proteins. Further experimental validation studies are necessaire to confirm multi-target therapeutic ability for Soy-B against AD (Iqbal et al., 2022).

V. CHALLENGES AND PERSPECTIVES

A. Challenges

Although secondary metabolites offer considerable therapeutic potential in the treatment of neurodegenerative diseases, their integration as a viable therapeutic option faces several challenges. These challenges reflect biological limitations, constraints related to their bioavailability, and environmental issues concerning their production. This chapter section presents the major challenges impacting the use of secondary metabolites and then suggests alternative perspectives for their future integration into therapeutic strategies against neurodegenerative diseases.

1. Scientific and Biological Challenges

One of the main obstacles, widely encountered in the literature, is the identification of metabolites required to establish a link between metabolites and neuroprotection. Indeed, in non-targeted metabolomics studies, researchers often identify only a subset of metabolites due to the limitations of detection methods and the complexity of biological samples. This leads to incomplete results, in turn contributing to ignorance of the relationship between metabolites and their biological effects (Johnson et al., 2016).

In this very sense, it should be noted that exploring how metabolites interact within metabolic pathways remains a very important step in this research. According to Donatti et al. (2020) metabolites such as ATP and acetyl-CoA are known to regulate various cellular processes, in particular signal transduction and epigenetic modifications. However, in-depth, multi-level analyses are required for a complete understanding.

2. Challenges Related to the Properties of Secondary Metabolites

Crossing the BBB

Recent studies have clearly shown that crossing the BBB remains a major challenge. Due to its neuroprotective functions, drug delivery, for example, may not reach target tissues or reach sufficient concentration in the brain (Achar et al., 2021; Stamatovic et al., 2016).

Low Bioavailability

Studies such as those by (Sharifi-Rad et al., 2020) and (Gomes et al., 2024) highlight the obstacle of the bioavailability of secondary metabolites. The latter are often criticized for their solubility and chemical stability. As explained in the previous sections, compounds such as flavonoids and polyphenols can be rapidly metabolized or excreted, which significantly minimizes their therapeutic efficacy. In addition, intestinal absorption of these compounds is often insufficient to reach the levels required to exert significant neuroprotective effects.

Rapid Degradation

Secondary metabolites may be rapidly degraded in the body, limiting their duration of action. For example, studies have shown that certain plant extracts containing bioactive metabolites can be broken down by intestinal enzymes before reaching the circulatory system. This breakdown is influenced by various factors, including intestinal microbiota and diet (Gomes et al., 2024; Missiego-Beltrán & Beltrán-Velasco, 2024).

Inefficient Metabolism

The metabolism of secondary metabolites can also lead to their being converted into inactive forms. The complex metabolic pathways involved in the processing of these compounds vary considerably between individuals, making it difficult to predict their therapeutic efficacy. Consequently, it is essential to study the metabolite interactions with liver enzymes to optimize their clinical use (Sharifi-Rad et al., 2020).

3. Limits and Constraints of Secondary Metabolites: Models, Sources and Production

This section focuses on the most practical challenges. First of all, there are the constraints associated with the preclinical models (in vitro and in vivo) used in the studies. These models do not reliably reflect human physiology or the complex interactions within biological systems.

This makes it difficult to reconcile laboratory and clinical results. This obstacle makes it difficult to understand the therapeutic potential of secondary metabolites and their mechanisms of action (de Oliveira et al., 2023; Han and Miao, 2024). Turning from this limitation to one related to the scarcity of natural sources, it is important to note that many specific plant species represent the raw material for many metabolites. However, their scarcity contributes to a concrete challenge, limiting their availability for therapeutic purposes. As highlighted by recent studies, there is an urgent need to think about alternative production methods, such as plant cell and tissue cultures, which can provide a more sustainable and controlled environment for the synthesis of secondary metabolites without depleting natural resources (Cardoso et al., 2019; Hussain et al., 2012). Finally, the problem not to be overlooked is that of cost. Extracting and synthesizing secondary metabolites is very costly, especially on a large scale. Nevertheless, advances in biotechnological approaches and metabolic engineering offer innovative solutions for improving production efficiency and reducing costs (Mipeshwaree Devi et al., 2023). However, these technologies require research funding, which unfortunately is not always available for all potential applications (Méndez-Hernández et al., 2023).

Figure 5. Challenges in utilizing secondary metabolites for treating neurodegenerative diseases

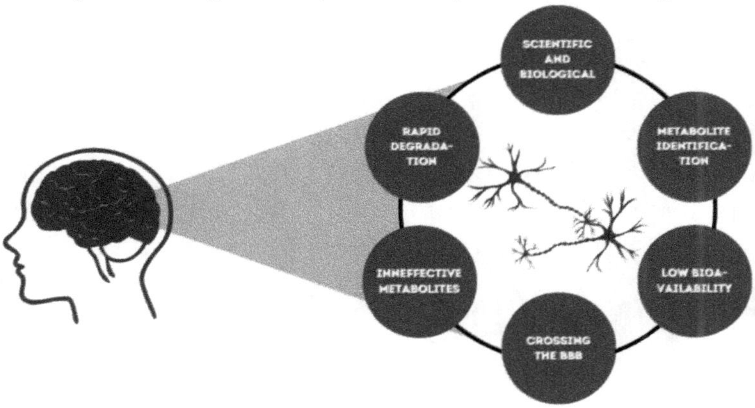

B. Perspectives

As perspectives of this study, it is important to highlight that, despite the therapeutic potential of secondary metabolites, the challenges previously cited remain major obstacles to overcome. However, liposomal nano-encapsulation remains a novel technology among others that can improve their absorption and passage across the blood-brain barrier. Researchers need to focus their studies on identifying optimal dosages, as well as fully understanding absorption, chemical variations and, not least, interactions with other components. Longitudinal studies are also needed to assess the efficacy of secondary metabolites in humans, as well as their interactions with existing drugs (Donatti et al., 2020; Johnson et al., 2016; Sharifi-Rad et al., 2020).

CONCLUSION

In summary, and despite their different origins, secondary metabolites represent a particularly promising avenue for the development of new treatments for neurodegenerative diseases such as AD and PD... Their neuroprotective, antioxidant and anti-inflammatory properties, combined with their richness in phytochemicals, such as alkaloids, terpenes, flavonoids and polyphenols, and especially the combination of the properties of these different compounds could constitute a solid basis for innovative preventive and therapeutic approaches. However, despite their undeniable potential, the clinical application of these compounds still raises a number of challenges, particularly regarding their bioavailability, long-term efficacy and safety. Thus, in-depth research and well-designed preclinical and clinical studies are needed to overcome these obstacles and allow an optimal exploitation of secondary metabolites in the fight against neurodegenerative diseases. These initiatives could not only help improve the quality of life of patients, but also open the way to new perspectives in personalized medicine and advanced herbal medicine.

REFERENCES

Abdulwanis Mohamed, Z., Mohamed Eliaser, E., Mazzon, E., Rollin, P., Cheng Lian Ee, G., & Abdull Razis, A. F. (2019). Neuroprotective Potential of Secondary Metabolites from Melicope lunu-ankenda (Rutaceae). *Molecules (Basel, Switzerland)*, *24*(17), 3109. DOI: 10.3390/molecules24173109

Achar, A., Myers, R., & Ghosh, C. (2021). Drug Delivery Challenges in Brain Disorders across the Blood–Brain Barrier: Novel Methods and Future Considerations for Improved Therapy. *Biomedicines*, *9*(12), 1834. DOI: 10.3390/biomedicines9121834

Ahmed, T., Gilani, A. U. H., Abdollahi, M., Daglia, M., Nabavi, S. F., & Nabavi, S. M. (2015). Berberine and neurodegeneration: A review of literature. *Pharmacological Reports*, *67*(5), 970–979. DOI: 10.1016/J.PHAREP.2015.03.002

Ait Hamdan, Y., El-Mansoury, B., Elouali, S., Rachmoune, K., Belbachir, A., Oudadesse, H., & Rhazi, M. (2024). A review of chitosan polysaccharides: Neuropharmacological implications and tissue regeneration. *International Journal of Biological Macromolecules*, *279*, 135356. DOI: 10.1016/J.IJBIOMAC.2024.135356

Alonso-Piñeiro, J. A., Gonzalez-Rovira, A., Sánchez-Gomar, I., Moreno, J. A., & Durán-Ruiz, M. C. (2021). Nrf2 and Heme Oxygenase-1 Involvement in Atherosclerosis Related Oxidative Stress. *Antioxidants*, *10*(9), 1463. DOI: 10.3390/antiox10091463

Alvi, S. S., Ahmad, P., Ishrat, M., Iqbal, D., & Khan, M. S. (2019). Secondary Metabolites from Rosemary (Rosmarinus officinalis L.): Structure, Biochemistry and Therapeutic Implications Against Neurodegenerative Diseases. In *Natural Bio-active Compounds* (pp. 1–24). Springer Singapore. https://doi.org/DOI: 10.1007/978-981-13-7205-6_1

Amor, S., Peferoen, L. A. N., Vogel, D. Y. S., Breur, M., van der Valk, P., Baker, D., & van Noort, J. M. (2014). Inflammation in neurodegenerative diseases – an update. *Immunology*, *142*(2), 151–166. DOI: 10.1111/imm.12233

An, F., Zhao, R., Xuan, X., Xuan, T., Zhang, G., & Wei, C. (2022). Calycosin ameliorates advanced glycation end product-induced neurodegenerative changes in cellular and rat models of diabetes-related Alzheimer's disease. *Chemico-Biological Interactions*, *368*, 110206. DOI: 10.1016/J.CBI.2022.110206

Anjali, Kumar, S., Korra, T., Thakur, R., Arutselvan, R., Kashyap, A. S., Nehela, Y., Chaplygin, V., Minkina, T., & Keswani, C. (2023). Role of plant secondary metabolites in defence and transcriptional regulation in response to biotic stress. *Plant Stress*, *8*, 100154. DOI: 10.1016/J.STRESS.2023.100154

Azzini, E., Peña-Corona, S. I., Hernández-Parra, H., Chandran, D., Saleena, L. A. K., Sawikr, Y., Peluso, I., Dhumal, S., Kumar, M., Leyva-Gómez, G., Martorell, M., Sharifi-Rad, J., & Calina, D. (2024). Neuroprotective and anti-inflammatory effects of curcumin in Alzheimer's disease: Targeting neuroinflammation strategies. *Phytotherapy Research*, *38*(6), 3169–3189. DOI: 10.1002/ptr.8200

Baghel, R. S., Choudhary, B., Pandey, S., Pathak, P. K., Patel, M. K., & Mishra, A. (2023). Rehashing Our Insight of Seaweeds as a Potential Source of Foods, Nutraceuticals, and Pharmaceuticals. *Foods, 12*(19), 3642. DOI: 10.3390/foods12193642

Balendra, R., & Isaacs, A. M. (2018). C9orf72-mediated ALS and FTD: Multiple pathways to disease. *Nature Reviews. Neurology, 14*(9), 544–558. DOI: 10.1038/s41582-018-0047-2

Basavarajappa, B. S., & Subbanna, S. (2021). Histone Methylation Regulation in Neurodegenerative Disorders. *International Journal of Molecular Sciences, 22*(9), 4654. DOI: 10.3390/ijms22094654

Beitz, J. M. (2014). Parkinson s disease a review. *Frontiers in Bioscience, S6*(1), S415. DOI: 10.2741/S415

Bellavite, P. (2023). Neuroprotective Potentials of Flavonoids: Experimental Studies and Mechanisms of Action. *Antioxidants, 12*(2), 280. DOI: 10.3390/antiox12020280

Bhandari, U. R., Danish, S. H., Ahmad, S., Ikram, M., Nadaf, A., Hasan, N., Kesharwani, P., & Ahmad, F. J. (2024). New opportunities for antioxidants in amelioration of neurodegenerative diseases. *Mechanisms of Ageing and Development, 111961.* Advance online publication. DOI: 10.1016/j.mad.2024.111961

Boyton, I., Rennie, C., Collins-Praino, L. E., & Care, A. (2024). Nanoparticle Formulations for the Improvement of Symptomatic Treatments of Neurodegenerative Disorders. *ChemNanoMat : Chemistry of Nanomaterials for Energy, Biology and More, 10*(12). Advance online publication. DOI: 10.1002/cnma.202400147

Câmara, J. S., Albuquerque, B. R., Aguiar, J., Corrêa, R. C. G., Gonçalves, J. L., Granato, D., Pereira, J. A. M., Barros, L., & Ferreira, I. C. F. R. (2020). Food Bioactive Compounds and Emerging Techniques for Their Extraction: Polyphenols as a Case Study. *Foods, 10*(1), 37. DOI: 10.3390/foods10010037

Cardoso, J. C., de Oliveira, M. E. B., & Cardoso, F. de C. (2019). Advances and challenges on the in vitro production of secondary metabolites from medicinal plants. *Horticultura Brasileira, 37*(2), 124–132. DOI: 10.1590/s0102-053620190201

Castellani, R. J., Rolston, R. K., & Smith, M. A. (2010). Alzheimer Disease. *Disease-a-Month, 56*(9), 484–546. DOI: 10.1016/j.disamonth.2010.06.001

Chaachouay, N., & Zidane, L. (2024). Plant-Derived Natural Products: A Source for Drug Discovery and Development. *Drugs and Drug Candidates, 3*(1), 184–207. DOI: 10.3390/ddc3010011

Chami, L., & Checler, F. (2012). BACE1 is at the crossroad of a toxic vicious cycle involving cellular stress and β-amyloid production in Alzheimer's disease. *Molecular Neurodegeneration, 7*(1), 52. DOI: 10.1186/1750-1326-7-52

Checkoway, H., Lundin, J. I., & Kelada, S. N. (2011). Neurodegenerative diseases. *IARC Scientific Publications, 163*, 407–419.

Cheng, Y., Dong, Z., & Liu, S. (2014). β-Caryophyllene Ameliorates the Alzheimer-Like Phenotype in APP/PS1 Mice through CB2 Receptor Activation and the PPARγ Pathway. *Pharmacology, 94*(1–2), 1–12. DOI: 10.1159/000362689

Ciurea, A. V., Mohan, A. G., Covache-Busuioc, R.-A., Costin, H.-P., Glavan, L.-A., Corlatescu, A.-D., & Saceleanu, V. M. (2023). Unraveling Molecular and Genetic Insights into Neurodegenerative Diseases: Advances in Understanding Alzheimer's, Parkinson's, and Huntington's Diseases and Amyotrophic Lateral Sclerosis. *International Journal of Molecular Sciences, 24*(13), 10809. DOI: 10.3390/ijms241310809

Ciurea, V. A., Covache-Busuioc, R.-A., Mohan, A. G., Costin, H. P., & Voicu, V. (2023). Alzheimer's disease: 120 years of research and progress. *Journal of Medicine and Life, 16*(2), 173–177. DOI: 10.25122/jml-2022-0111

Clare, L., Teale, J. C., Toms, G., Kudlicka, A., Evans, I., Abrahams, S., Goldstein, L. H., Hindle, J. V., Ho, A. K., Jahanshahi, M., Langdon, D., Morris, R., Snowden, J. S., Davies, R., Markova, I., Busse, M., & Thompson-Coon, J. (2019). Cognitive rehabilitation, self-management, psychotherapeutic and caregiver support interventions in progressive neurodegenerative conditions: A scoping review. *NeuroRehabilitation, 43*(4), 443–471. DOI: 10.3233/NRE-172353

D'Onofrio, G., Sancarlo, D., Ruan, Q., Yu, Z., Panza, F., Daniele, A., Greco, A., & Seripa, D. (2017). Phytochemicals in the Treatment of Alzheimer's Disease: A Systematic Review. *Current Drug Targets, 18*(13). Advance online publication. DOI: 10.2174/1389450117666161102121553

de Oliveira, I., Albuquerque, B., Ueda, J. M., Alves, M. J., Ferreira, I. C. F. R., Barros, L., & Heleno, S. A. (2023). New Challenges and Opportunities from Secondary Metabolites. In *Natural Secondary Metabolites* (pp. 925–965). Springer International Publishing., DOI: 10.1007/978-3-031-18587-8_29

Dhiman, A., Handa, M., Ruwali, M., Singh, D. P., Kesharwani, P., & Shukla, R. (2022). Recent trends of natural based therapeutics for mitochondria targeting in Alzheimer's disease. *Mitochondrion, 64*, 112–124. DOI: 10.1016/J.MITO.2022.03.006

Diz-Chaves, Y., Mastoor, Z., Spuch, C., González-Matías, L. C., & Mallo, F. (2022). Anti-Inflammatory Effects of GLP-1 Receptor Activation in the Brain in Neurodegenerative Diseases. *International Journal of Molecular Sciences, 23*(17), 9583. DOI: 10.3390/ijms23179583

Donatti, A., Canto, A. M., Godoi, A. B., da Rosa, D. C., & Lopes-Cendes, I. (2020). Circulating Metabolites as Potential Biomarkers for Neurological Disorders—Metabolites in Neurological Disorders. *Metabolites, 10*(10), 389. DOI: 10.3390/metabo10100389

Dunnett, S. B., & Fibiger, H. C. (1993). Chapter 49: Role of forebrain cholinergic systems in learning and memory: relevance to the cognitive deficits of aging and Alzheimer's dementia. *Progress in Brain Research, 98*(C), 413–420. DOI: 10.1016/S0079-6123(08)62425-5

Fang, M., Yu, Q., Ou, J. T., Lou, J., Zhu, J., & Lin, Z. (2024). The Neuroprotective Mechanisms of PPAR-γ: Inhibition of Microglia-Mediated Neuroinflammation and Oxidative Stress in a Neonatal Mouse Model of Hypoxic-Ischemic White Matter Injury. *CNS Neuroscience & Therapeutics, 30*(11). Advance online publication. DOI: 10.1111/cns.70081

Fox, E. M., & Howlett, B. J. (2008). Secondary metabolism: Regulation and role in fungal biology. *Current Opinion in Microbiology*, *11*(6), 481–487. DOI: 10.1016/J.MIB.2008.10.007

Gajbhiye, V. G. K. (2015). Targeted Brain Delivery of Bioactive Molecules Using Nanocarriers. *Journal of Bioequivalence & Bioavailability*, *07*(03). Advance online publication. DOI: 10.4172/jbb.1000224

Gao, W., Jing, S., He, C., Saberi, H., Sharma, H. S., Han, F., & Chen, L. (2025). Advancements in neurodegenerative diseases: Pathogenesis and novel neurorestorative interventions. *Journal of Neurorestoratology*, *13*(2), 100176. DOI: 10.1016/J.JNRT.2024.100176

Gatz, M., Reynolds, C. A., Fratiglioni, L., Johansson, B., Mortimer, J. A., Berg, S., Fiske, A., & Pedersen, N. L. (2006). Role of Genes and Environments for Explaining Alzheimer Disease. *Archives of General Psychiatry*, *63*(2), 168. DOI: 10.1001/archpsyc.63.2.168

Ge, W., Ren, C., Xing, L., Guan, L., Zhang, C., Sun, X., Wang, G., Niu, H., & Qun, S. (2021). Ginkgo biloba extract improves cognitive function and increases neurogenesis by reducing Aβ pathology in 5×FAD mice. *American Journal of Translational Research*, *13*(3), 1471–1482.

Gelain, D. P., Behr, G. A., de Oliveira, R. B., & Trujillo, M. (2012). Antioxidant Therapies for Neurodegenerative Diseases: Mechanisms, Current Trends, and Perspectives. *Oxidative Medicine and Cellular Longevity*, *895153*. Advance online publication. DOI: 10.1155/2012/895153

Genchi, G., Lauria, G., Catalano, A., Carocci, A., & Sinicropi, M. S. (2024). Neuroprotective Effects of Curcumin in Neurodegenerative Diseases. *Foods*, *13*(11), 1774. DOI: 10.3390/foods13111774

Gomes, T. M., Sousa, P., Campos, C., Perestrelo, R., & Câmara, J. S. (2024). Secondary Bioactive Metabolites from Foods of Plant Origin as Theravention Agents against Neurodegenerative Disorders. *Foods*, *13*(14), 2289. DOI: 10.3390/foods13142289

Grosso, C., Valentão, P., Ferreres, F., & Andrade, P. (2014). Bioactive Marine Drugs and Marine Biomaterials for Brain Diseases. *Marine Drugs*, *12*(5), 2539–2589. DOI: 10.3390/md12052539

Hamilton, A. C. (2004). Medicinal plants, conservation and livelihoods. *Biodiversity and Conservation*, *13*(8), 1477–1517. DOI: 10.1023/B:BIOC.0000021333.23413.42

Han, T., & Miao, G. (2024). Strategies, Achievements, and Potential Challenges of Plant and Microbial Chassis in the Biosynthesis of Plant Secondary Metabolites. *Molecules (Basel, Switzerland)*, *29*(9), 2106. DOI: 10.3390/molecules29092106

Hanna, L., Poluyi, E. O., Ikwuegbuenyi, C. A., Morgan, E., & Imaguezegie, G. (2022). Peripheral inflammation and neurodegeneration; a potential for therapeutic intervention in Alzheimer's disease (AD), Parkinson's disease (PD) and amyotrophic lateral sclerosis (ALS). *Egyptian Journal of Neurosurgery*, *37*(1), 1–9. DOI: 10.1186/s41984-022-00150-4

Hassamal, S. (2023). Chronic stress, neuroinflammation, and depression: An overview of pathophysiological mechanisms and emerging anti-inflammatories. *Frontiers in Psychiatry*, *14*. Advance online publication. DOI: 10.3389/fpsyt.2023.1130989

Heim, K. E., Tagliaferro, A. R., & Bobilya, D. J. (2002). Flavonoid antioxidants: Chemistry, metabolism and structure-activity relationships. *The Journal of Nutritional Biochemistry, 13*(10), 572–584. DOI: 10.1016/S0955-2863(02)00208-5

Hilal, B., Khan, M. M., & Fariduddin, Q. (2024). Recent advancements in deciphering the therapeutic properties of plant secondary metabolites: Phenolics, terpenes, and alkaloids. *Plant Physiology and Biochemistry, 211*, 108674. DOI: 10.1016/J.PLAPHY.2024.108674

Hu, Q., & Wang, G. (2016). Mitochondrial dysfunction in Parkinson's disease. *Translational Neurodegeneration, 5*(1), 14. DOI: 10.1186/s40035-016-0060-6

Hussain, M. S., Fareed, S., Ansari, S., Rahman, M. A., Ahmad, I. Z., & Saeed, M. (2012). Current approaches toward production of secondary plant metabolites. *Journal of Pharmacy & Bioallied Sciences, 4*(1), 10–20. DOI: 10.4103/0975-7406.92725

Iqbal, D., Rizvi, S. M. D., Rehman, M. T., Khan, M. S., Bin Dukhyil, A., AlAjmi, M. F., Alshehri, B. M., Banawas, S., Zia, Q., Alsaweed, M., Madkhali, Y., Alsagaby, S. A., & Alturaiki, W. (2022). Soyasapogenol-B as a Potential Multitarget Therapeutic Agent for Neurodegenerative Disorders: Molecular Docking and Dynamics Study. *Entropy (Basel, Switzerland), 24*(5), 593. DOI: 10.3390/e24050593

Jamshidi-Kia, F., Lorigooini, Z., & Amini-Khoei, H. (2018). Medicinal plants: Past history and future perspective. *Journal of Herbmed Pharmacology, 7*(1), 1–7. DOI: 10.15171/jhp.2018.01

Johnson, C. H., Ivanisevic, J., & Siuzdak, G. (2016). Metabolomics: Beyond biomarkers and towards mechanisms. *Nature Reviews. Molecular Cell Biology, 17*(7), 451–459. DOI: 10.1038/nrm.2016.25

Jyothi, P., & Yellamma, K. (2016). MOLECULAR DOCKING STUDIES ON THE THERAPEUTIC TARGETS OF ALZHEIMER'S DISEASE (AChE AND BChE) USING NATURAL BIOACTIVE ALKALOIDS. *International Journal of Pharmacy and Pharmaceutical Sciences, 8*(12), 108. DOI: 10.22159/ijpps.2016v8i12.14833

Karthikeyan, A., Joseph, A., & Nair, B. G. (2022). Promising bioactive compounds from the marine environment and their potential effects on various diseases. *Journal of Genetic Engineering and Biotechnology, 20*(1), 14. DOI: 10.1186/S43141-021-00290-4

Katila, N., Duwa, R., Bhurtel, S., Khanal, S., Maharjan, S., Jeong, J. H., Lee, S., Choi, D. Y., & Yook, S. (2022). Enhancement of blood–brain barrier penetration and the neuroprotective effect of resveratrol. *Journal of Controlled Release, 346*, 1–19. DOI: 10.1016/J.JCONREL.2022.04.003

Khairinisa, M. A., Latarissa, I. R., Athaya, N. S., Charlie, V., Musyaffa, H. A., Prasedya, E. S., & Puspitasari, I. M. (2023). Potential Application of Marine Algae and Their Bioactive Metabolites in Brain Disease Treatment: Pharmacognosy and Pharmacology Insights for Therapeutic Advances. *Brain Sciences, 13*(12), 1686. DOI: 10.3390/brainsci13121686

Khanam, H., Ali, A., & Asif, M., & Shamsuzzaman. (2016). Neurodegenerative diseases linked to misfolded proteins and their therapeutic approaches: A review. *European Journal of Medicinal Chemistry, 124*, 1121–1141. DOI: 10.1016/J.EJMECH.2016.08.006

Khodadadi, M., & Javadi, B. (2025). A Review of the Protective Effects of Alkaloids against Alpha-synuclein Toxicity in Parkinson's Disease. *Mini-Reviews in Medicinal Chemistry*, *25*(2), 112–127. DOI: 10.2174/0113895575306884240604065754

Klegeris, A., & McGeer, P. L. (2005). Non-Steroidal Anti-Inflammatory Drugs (NSAIDs) and Other Anti- Inflammatory Agents in the Treatment of Neurodegenerative Disease. *Current Alzheimer Research*, *2*(3), 355–365. DOI: 10.2174/1567205054367883

Kumar, A., Kumar Singh, S., Kumar, V., Kumar, D., Agarwal, S., & Rana, M. K. (2015). Huntington's disease: An update of therapeutic strategies. *Gene*, *556*(2), 91–97. DOI: 10.1016/J.GENE.2014.11.022

Kundu, D., & Dubey, V. K. (2021). Purines and Pyrimidines: Metabolism, Function and Potential as Therapeutic Options in Neurodegenerative Diseases. *Current Protein & Peptide Science*, *22*(2), 170–189. DOI: 10.2174/1389203721999201208200605

Laval, L., Martin, R., Natividad, J., Chain, F., Miquel, S., de Maredsous, C. D., Capronnier, S., Sokol, H., Verdu, E., van Hylckama Vlieg, J., Bermúdez-Humarán, L., Smokvina, T., & Langella, P. (2015). *Lactobacillus rhamnosus* CNCM I-3690 and the commensal bacterium *Faecalibacterium prausnitzii* A2-165 exhibit similar protective effects to induced barrier hyper-permeability in mice. *Gut Microbes*, *6*(1), 1–9. DOI: 10.4161/19490976.2014.990784

Li, C., Liu, K., Liu, S., Aerqin, Q., & Wu, X. (2020). Role of Ginkgolides in the Inflammatory Immune Response of Neurological Diseases: A Review of Current Literatures. *Frontiers in Systems Neuroscience*, *14*. Advance online publication. DOI: 10.3389/fnsys.2020.00045

Ma, Y., Ma, B., Shang, Y., Yin, Q., Hong, Y., Xu, S., Shen, C., Hou, X., & Liu, X. (2018). Flavonoid-rich ethanol extract from the leaves of Diospyros kaki attenuates cognitive deficits, amyloid-beta production, oxidative stress, and neuroinflammation in APP/PS1 transgenic mice. *Brain Research*, *1678*, 85–93. DOI: 10.1016/J.BRAINRES.2017.10.001

Mathew, A., Bhuvanendran, S., Nair, R. S., & Radhakrishnan, A. K. (2023). Exploring the anti-inflammatory activities, mechanism of action and prospective drug delivery systems of tocotrienol to target neurodegenerative diseases. *F1000 Research*, *12*, 338. DOI: 10.12688/f1000research.131863.1

McColgan, P., & Tabrizi, S. J. (2018). Huntington's disease: A clinical review. *European Journal of Neurology*, *25*(1), 24–34. DOI: 10.1111/ene.13413

Melrose, J. (2023). The Potential of Flavonoids and Flavonoid Metabolites in the Treatment of Neurodegenerative Pathology in Disorders of Cognitive Decline. *Antioxidants*, *12*(3), 663. DOI: 10.3390/antiox12030663

Méndez-Hernández, J. E., Rodríguez-Durán, L. V., Páez-Lerma, J. B., & Soto-Cruz, N. O. (2023). Strategies for Supplying Precursors to Enhance the Production of Secondary Metabolites in Solid-State Fermentation. *Fermentation (Basel, Switzerland)*, *9*(9), 804. DOI: 10.3390/fermentation9090804

Migliore, L., & Coppedè, F. (2009). Genetics, environmental factors and the emerging role of epigenetics in neurodegenerative diseases. *Mutation Research. Fundamental and Molecular Mechanisms of Mutagenesis, 667*(1–2), 82–97. DOI: 10.1016/J.MRFMMM.2008.10.011

Mikulska, P., Malinowska, M., Ignacyk, M., Szustowski, P., Nowak, J., Pesta, K., Szeląg, M., Szklanny, D., Judasz, E., Kaczmarek, G., Ejiohuo, O. P., Paczkowska-Walendowska, M., Gościniak, A., & Cielecka-Piontek, J. (2023). Ashwagandha (Withania somnifera)—Current Research on the Health-Promoting Activities: A Narrative Review. *Pharmaceutics, 15*(4), 1057. DOI: 10.3390/pharmaceutics15041057

Mipeshwaree Devi, A., Khedashwori Devi, K., Premi Devi, P., Lakshmipriyari Devi, M., & Das, S. (2023). Metabolic engineering of plant secondary metabolites: Prospects and its technological challenges. *Frontiers in Plant Science, 14*. Advance online publication. DOI: 10.3389/fpls.2023.1171154

Missiego-Beltrán, J., & Beltrán-Velasco, A. I. (2024). The Role of Microbial Metabolites in the Progression of Neurodegenerative Diseases—Therapeutic Approaches: A Comprehensive Review. *International Journal of Molecular Sciences, 25*(18), 10041. DOI: 10.3390/ijms251810041

Naoi, M., Wu, Y., Shamoto-Nagai, M., & Maruyama, W. (2019). Mitochondria in Neuroprotection by Phytochemicals: Bioactive Polyphenols Modulate Mitochondrial Apoptosis System, Function and Structure. *International Journal of Molecular Sciences, 20*(10), 2451. DOI: 10.3390/ijms20102451

Nash, V., Ranadheera, C. S., Georgousopoulou, E. N., Mellor, D. D., Panagiotakos, D. B., McKune, A. J., Kellett, J., & Naumovski, N. (2018). The effects of grape and red wine polyphenols on gut microbiota – A systematic review. *Food Research International, 113*, 277–287. DOI: 10.1016/J.FOODRES.2018.07.019

Oladipo, A., Enwemiwe, V., Ejeromedoghene, O., Adebayo, A., Ogunyemi, O., & Fu, F. (2022). Production and Functionalities of Specialized Metabolites from Different Organic Sources. *Metabolites, 12*(6), 534. DOI: 10.3390/metabo12060534

Oyinloye, B. E., Iwaloye, O., & Ajiboye, B. O. (2021). Polypharmacology of Gongronema latifolium leaf secondary metabolites against protein kinases implicated in Parkinson's disease and Alzheimer's disease. *Scientific African, 12*, e00826. DOI: 10.1016/J.SCIAF.2021.E00826

Pagare, S., Bhatia, M., Tripathi, N., & Bansal, Y. K. (2015). Secondary Metabolites of Plants and their Role: Overview. *Current Trends in Biotechnology and Pharmacy, 9*(3), 293–304.

Parker, A., Fonseca, S., & Carding, S. R. (2020). Gut microbes and metabolites as modulators of blood-brain barrier integrity and brain health. *Gut Microbes, 11*(2), 135–157. DOI: 10.1080/19490976.2019.1638722

Parsafar, S., Nayeri, Z., Aliakbari, F., Shahi, F., Mohammadi, M., & Morshedi, D. (2020). Multiple neuroprotective features of Scutellaria pinnatifida–derived small molecule. *Heliyon, 6*(8), e04737. DOI: 10.1016/j.heliyon.2020.e04737

Piano, I., Votta, A., Colucci, P., Corsi, F., Vitolo, S., Cerri, C., Puppi, D., Lai, M., Maya-Vetencourt, J. F., Leigheb, M., Gabellini, C., & Ferraro, E. (2023). Anti-inflammatory reprogramming of microglia cells by metabolic modulators to counteract neurodegeneration; a new role for Ranolazine. *Dental Science Reports, 13*. Advance online publication. DOI: 10.1038/s41598-023-47540-8

Pierozan, P., Biasibetti-Brendler, H., Schmitz, F., Ferreira, F., Netto, C. A., & Wyse, A. T. S. (2018). Synergistic Toxicity of the Neurometabolites Quinolinic Acid and Homocysteine in Cortical Neurons and Astrocytes: Implications in Alzheimer's Disease. *Neurotoxicity Research, 34*(1), 147–163. DOI: 10.1007/s12640-017-9834-6

Pinheiro, R. G. R., Coutinho, A. J., Pinheiro, M., & Neves, A. R. (2021). Nanoparticles for Targeted Brain Drug Delivery: What Do We Know? *International Journal of Molecular Sciences, 22*(21), 11654. DOI: 10.3390/ijms222111654

Pratama, A. N., & Busman, H. (2020). Potensi Antioksidan Kedelai (Glycine Max L) Terhadap Penangkapan Radikal Bebas. *Jurnal Ilmiah Kesehatan Sandi Husada, 11*(1), 497–504. DOI: 10.35816/jiskh.v11i1.333

Putilina, M. V., & Teplova, N. V. (2022). Drug Synergism as the Basis of Rational Neuroprotection. *Neuroscience and Behavioral Physiology, 52*(8), 1207–1211. DOI: 10.1007/s11055-023-01349-0

Rafati, N., Zarepour, A., Bigham, A., Khosravi, A., Naderi-Manesh, H., Iravani, S., & Zarrabi, A. (2024). Nanosystems for targeted drug Delivery: Innovations and challenges in overcoming the Blood-Brain barrier for neurodegenerative disease and cancer therapy. *International Journal of Pharmaceutics, 666*, 124800. DOI: 10.1016/J.IJPHARM.2024.124800

Rahman, M. H., Akter, R., Bhattacharya, T., Abdel-Daim, M. M., Alkahtani, S., Arafah, M. W., Al-Johani, N. S., Alhoshani, N. M., Alkeraishan, N., Alhenaky, A., Abd-Elkader, O. H., El-Seedi, H. R., Kaushik, D., & Mittal, V. (2020). Resveratrol and Neuroprotection: Impact and Its Therapeutic Potential in Alzheimer's Disease. *Frontiers in Pharmacology, 11*. Advance online publication. DOI: 10.3389/fphar.2020.619024

Ramalingam, M., & Kim, S.-J. (2012). Reactive oxygen/nitrogen species and their functional correlations in neurodegenerative diseases. *Journal of Neural Transmission (Vienna, Austria), 119*(8), 891–910. DOI: 10.1007/s00702-011-0758-7

Ramesh, S., & Arachchige, A. S. P. M. (2023). Depletion of dopamine in Parkinson's disease and relevant therapeutic options: A review of the literature. *AIMS Neuroscience, 10*(3), 200–231. DOI: 10.3934/Neuroscience.2023017

Rauf, A., Badoni, H., Abu-Izneid, T., Olatunde, A., Rahman, M. M., Painuli, S., Semwal, P., Wilairatana, P., & Mubarak, M. S. (2022). Neuroinflammatory Markers: Key Indicators in the Pathology of Neurodegenerative Diseases. *Molecules (Basel, Switzerland), 27*(10), 3194. DOI: 10.3390/molecules27103194

Rekatsina, M., Paladini, A., Piroli, A., Zis, P., Pergolizzi, J. V., & Varrassi, G. (2020). Pathophysiology and Therapeutic Perspectives of Oxidative Stress and Neurodegenerative Diseases: A Narrative Review. *Advances in Therapy, 37*(1), 113–139. DOI: 10.1007/s12325-019-01148-5

Renton, A. E., Majounie, E., Waite, A., Simón-Sánchez, J., Rollinson, S., Gibbs, J. R., Schymick, J. C., Laaksovirta, H., van Swieten, J. C., Myllykangas, L., Kalimo, H., Paetau, A., Abramzon, Y., Remes, A. M., Kaganovich, A., Scholz, S. W., Duckworth, J., Ding, J., Harmer, D. W., & Traynor, B. J. (2011). A Hexanucleotide Repeat Expansion in C9ORF72 Is the Cause of Chromosome 9p21-Linked ALS-FTD. *Neuron*, *72*(2), 257–268. DOI: 10.1016/j.neuron.2011.09.010

Roda, A., Serra-Mir, G., Montoliu-Gaya, L., Tiessler, L., & Villegas, S. (2022). Amyloid-beta peptide and tau protein crosstalk in Alzheimer's disease. *Neural Regeneration Research*, *17*(8), 1666. DOI: 10.4103/1673-5374.332127

Ross, G. W., & Petrovitch, H. (2001). Current Evidence for Neuroprotective Effects of Nicotine and Caffeine Against Parkinson??s Disease. *Drugs & Aging*, *18*(11), 797–806. DOI: 10.2165/00002512-200118110-00001

Sandoval-Acuña, C., Ferreira, J., & Speisky, H. (2014). Polyphenols and mitochondria: An update on their increasingly emerging ROS-scavenging independent actions. *Archives of Biochemistry and Biophysics*, *559*, 75–90. DOI: 10.1016/J.ABB.2014.05.017

Selma, D., Abd El Salem, D., Abd El Malek, B., Ahmed, S. S., & Zahr-eddine, D. (2017). Variation Phénologique Du Contenu Phytochimique Et De L'activité Antibacteriènne De Cytisus Triflorus L'her. *AGROBIOLOGIA*, *7*(2), 548–561.

Sen Singh, S., Rai, S. N., Birla, H., Zahra, W., Rathore, A. S., & Singh, S. P. (2020). NF-κB-Mediated Neuroinflammation in Parkinson's Disease and Potential Therapeutic Effect of Polyphenols. *Neurotoxicity Research*, *37*(3), 491–507. DOI: 10.1007/s12640-019-00147-2

Sharifi-Rad, M., Lankatillake, C., Dias, D. A., Docea, A. O., Mahomoodally, M. F., Lobine, D., Chazot, P. L., Kurt, B., Boyunegmez Tumer, T., Catarina Moreira, A., Sharopov, F., Martorell, M., Martins, N., Cho, W. C., Calina, D., & Sharifi-Rad, J. (2020). Impact of Natural Compounds on Neurodegenerative Disorders: From Preclinical to Pharmacotherapeutics. *Journal of Clinical Medicine*, *9*(4), 1061. DOI: 10.3390/jcm9041061

Sharma, N., Sharma, V., Manikyam, H., & Krishna, A. (2016). Ergot Alkaloids: A Review on Therapeutic Applications. *European Journal of Medicinal Plants*, *14*(3), 1–17. DOI: 10.9734/EJMP/2016/25975

Shusharina, N., Yukhnenko, D., Botman, S., Sapunov, V., Savinov, V., Kamyshov, G., Sayapin, D., & Voznyuk, I. (2023). Modern Methods of Diagnostics and Treatment of Neurodegenerative Diseases and Depression. *Diagnostics (Basel)*, *13*(3), 573. DOI: 10.3390/diagnostics13030573

Silva, S. S., Gomes, J. M., Rodrigues, L. C., & Reis, R. L. (2020). Marine-Derived Polymers in Ionic Liquids: Architectures Development and Biomedical Applications. *Marine Drugs*, *18*(7), 346. DOI: 10.3390/md18070346

Simpson, D. S. A., & Oliver, P. L. (2020). ROS Generation in Microglia: Understanding Oxidative Stress and Inflammation in Neurodegenerative Disease. *Antioxidants*, *9*(8), 743. DOI: 10.3390/antiox9080743

Singh, S. K., Srivastav, S., Castellani, R. J., Plascencia-Villa, G., & Perry, G. (2019). Neuroprotective and Antioxidant Effect of Ginkgo biloba Extract Against AD and Other Neurological Disorders. *Neurotherapeutics; the Journal of the American Society for Experimental NeuroTherapeutics, 16*(3), 666–674. DOI: 10.1007/S13311-019-00767-8

Sorrenti, V., Castagna, D. A., Fortinguerra, S., Buriani, A., Scapagnini, G., & Willcox, D. C. (2021). Spirulina Microalgae and Brain Health: A Scoping Review of Experimental and Clinical Evidence. *Marine Drugs, 19*(6), 293. DOI: 10.3390/md19060293

SPOKES, E. G. S. (1980). NEUROCHEMICAL ALTERATIONS IN HUNTINGTON'S CHOREA. *Brain, 103*(1), 179–210. DOI: 10.1093/brain/103.1.179

Stamatovic, S. M., Johnson, A. M., Keep, R. F., & Andjelkovic, A. V. (2016). Junctional proteins of the blood-brain barrier: New insights into function and dysfunction. *Tissue Barriers, 4*(1), e1154641. DOI: 10.1080/21688370.2016.1154641

Suna, H., Arai, M., Tsubotani, Y., Hayashi, A., Setiawan, A., & Kobayashi, M. (2009). Dysideamine, a new sesquiterpene aminoquinone, protects hippocampal neuronal cells against iodoacetic acid-induced cell death. *Bioorganic & Medicinal Chemistry, 17*(11), 3968–3972. DOI: 10.1016/J.BMC.2009.04.025

Tan, S. A., & Aebischer, P. (2007). *The Problems of Delivering Neuroactive Molecules to the CNS.*, DOI: 10.1002/9780470514863.ch14

Teleanu, R. I., Chircov, C., Grumezescu, A. M., Volceanov, A., & Teleanu, D. M. (2019). Antioxidant therapies for neuroprotection—A review. *Journal of Clinical Medicine, 8*(10), 1659.

Teleanu, R. I., Niculescu, A.-G., Roza, E., Vladâcenco, O., Grumezescu, A. M., & Teleanu, D. M. (2022). Neurotransmitters—Key Factors in Neurological and Neurodegenerative Disorders of the Central Nervous System. *International Journal of Molecular Sciences, 23*(11), 5954. DOI: 10.3390/ijms23115954

Termine, A., Fabrizio, C., Strafella, C., Caputo, V., Petrosini, L., Caltagirone, C., Giardina, E., & Cascella, R. (2021). Multi-Layer Picture of Neurodegenerative Diseases: Lessons from the Use of Big Data through Artificial Intelligence. *Journal of Personalized Medicine, 11*(4), 280. DOI: 10.3390/jpm11040280

Tringali, C., & Giussani, P. (2022). Ceramide and Sphingosine-1-Phosphate in Neurodegenerative Disorders and Their Potential Involvement in Therapy. *International Journal of Molecular Sciences, 23*(14), 7806. DOI: 10.3390/ijms23147806

Venditti, A., & Bianco, A. (2020). Sulfur-containing Secondary Metabolites as Neuroprotective Agents. *Current Medicinal Chemistry, 27*(26), 4421–4436. DOI: 10.2174/0929867325666180912105036

Venturella, G., Ferraro, V., Cirlincione, F., & Gargano, M. L. (2021). Medicinal Mushrooms: Bioactive Compounds, Use, and Clinical Trials. *International Journal of Molecular Sciences, 22*(2), 634. DOI: 10.3390/ijms22020634

Vijayakumar, S., & Rajalakshmi, S. (2021). Exploring novel natural compound inhibitors for Parkinsonian receptor (DJ1) by homology modeling, molecular docking and MD simulations. *Vegetos*, *34*(4), 959–970. DOI: 10.1007/s42535-021-00263-5

Voulgaropoulou, S. D., van Amelsvoort, T. A. M. J., Prickaerts, J., & Vingerhoets, C. (2019). The effect of curcumin on cognition in Alzheimer's disease and healthy aging: A systematic review of pre-clinical and clinical studies. *Brain Research*, *1725*, 146476. DOI: 10.1016/J.BRAINRES.2019.146476

Wang, J. Y., Wen, L. L., Huang, Y. N., Chen, Y. T., & Ku, M. C. (2006). Dual effects of antioxidants in neurodegeneration: Direct neuroprotection against oxidative stress and indirect protection via suppression of glia-mediated inflammation. *Current Pharmaceutical Design*, *12*(27), 3521–3533. DOI: 10.2174/138161206778343109

Wexler, A., Wild, E. J., & Tabrizi, S. J. (2016). George Huntington: A legacy of inquiry, empathy and hope. *Brain*, *139*(8), 2326–2333. DOI: 10.1093/brain/aww165

Woollacott, I. O. C., & Mead, S. (2014). The C9ORF72 expansion mutation: Gene structure, phenotypic and diagnostic issues. *Acta Neuropathologica*, *127*(3), 319–332. DOI: 10.1007/s00401-014-1253-7

Wróbel-Biedrawa, D., & Podolak, I. (2024). Anti-Neuroinflammatory Effects of Adaptogens: A Mini-Review. *Molecules (Basel, Switzerland)*, *29*(4), 866. DOI: 10.3390/molecules29040866

Xie, Y., Li, X., Shi, Q., Le, L., Wang, C., Xu, H., Wu, G., Du, X., & Chen, Z. (2024). The synergistic effect of curcumin and mitoquinol mesylate on cognitive impairment and the neuropathology of Alzheimer's disease. *Brain Research*, *1837*, 148959. DOI: 10.1016/J.BRAINRES.2024.148959

Xu, Q., Langley, M., Kanthasamy, A. G., & Reddy, M. B. (2017). Epigallocatechin Gallate Has a Neurorescue Effect in a Mouse Model of Parkinson Disease. *The Journal of Nutrition*, *147*(10), 1926–1931. DOI: 10.3945/JN.117.255034

Zaman, M., & Khan, A. N., Wahiduzzaman, Zakariya, S. M., & Khan, R. H. (2019). Protein misfolding, aggregation and mechanism of amyloid cytotoxicity: An overview and therapeutic strategies to inhibit aggregation. *International Journal of Biological Macromolecules*, *134*, 1022–1037. DOI: 10.1016/J.IJBIOMAC.2019.05.109

Zhong, G., Wang, X., Li, J., Xie, Z., Wu, Q., Chen, J., Wang, Y., Chen, Z., Cao, X., Li, T., Liu, J., & Wang, Q. (2024). Insights Into the Role of Copper in Neurodegenerative Diseases and the Therapeutic Potential of Natural Compounds. *Current Neuropharmacology*, *22*(10), 1650–1671. DOI: 10.2174/1570159X22666231103085859

ABBREVIATIONS

ACh: Acetylcholine
AChE: Acetylcholinesterase
AD: Alzheimer's disease

ALS: amyotrophic lateral sclerosis
AMT: Adsorptive-mediated transcytosis
APP: Amyloid precursor proteins
BACE-1: β-amyloid cleavage enzyme
BBB: Blood-brain barrier
BChE: Butyrylcholinesterase
CNS: Central nervous system
Hsp-70: Heat shock protein 70
MS: multiple sclerosis
NDD: Neurologic and neurodegenerative diseases
NF-Kb: Nuclear factor kappa
PD: Parkinson's disease
PKs: Protein kinases
RMT: Receptor-mediated transcytosis
ROS: Reactive Oxygen Species
SCI: Spinal cord injury
SNpc: substantia nigra pars compacta
TMT: Transporter-mediated transcytosis

Chapter 6
The Neuroprotective Role of Resveratrol:
Therapeutic Implications in Parkinson's and Alzheimer's Diseases

Hanane El Fatimi
Mohamed VI University, Morocco

Martin Ndayambaje
Faculty of Science Aïn Chock, Morocco

Ayyoub Skaou
https://orcid.org/0009-0003-7981-2473
Faculty of Science, Semlalia, Morocco

Bilal El-Mansoury
https://orcid.org/0000-0002-2583-7515
Royal College of Surgeons in Ireland, University of Medicine and Health Sciences, Dublin, Ireland & Faculty of Sciences, Chouaib Doukkali University, Morocco

ABSTRACT

Parkinson's disease (PD) and Alzheimer's disease (AD) are progressive neurodegenerative disorders and major global health concerns due to their complex pathogenesis and lack of effective treatments. AD is marked by amyloid-beta (Aβ) accumulation and neurofibrillary tangles, while PD involves dopaminergic neuronal loss in the substantia nigra and Lewy body formation. Despite distinct hallmarks, both share mechanisms such as oxidative stress, neuroinflammation, and impaired protein clearance. Resveratrol (RSV), a natural polyphenol from grapes, berries, and peanuts, shows strong neuroprotective properties through antioxidant, anti-inflammatory, and mitochondrial-stabilizing effects. It also regulates pathways like SIRT1, AMPK, and autophagy, essential for neuronal survival. Preclinical findings revealed that RSV enhances cognitive and

DOI: 10.4018/979-8-3373-1365-8.ch006

motor functions, reduces amyloid burden, protects dopaminergic neurons, and improves synaptic plasticity. This chapter outlines shared AD–PD mechanisms and highlights RSV's therapeutic potential as an adjunct strategy against neurodegeneration.

1. INTRODUCTION

Alzheimer's disease (AD) and Parkinson's disease (PD) are the most common neurodegenerative disorders, affecting millions of people worldwide (El-Mansoury et al., 2023; El Khiat et al., 2023; Masters et al., 2015; Oukhrib et al., 2023; Scheltens et al., 2021). The pathogenesis of PD and AD is believed to result from a complex interplay between genetic predisposition and environmental factors (Aimrane et al., 2023; Grant et al., 2002; Migliore & Coppedè, 2022; Moore et al., 2005; Saad et al., 2023; Vázquez-Vélez & Zoghbi, 2021). Although several therapeutic approaches are currently available, they primarily offer symptomatic relief rather than halting or reversing disease progression (El-Mansoury, El Hiba, & Jayakumar, 2023; El-Mansoury et al., 2023; Ben Maloui et al., 2023; Stoker & Barker, 2020; Yiannopoulou & Papageorgiou, 2020). Nevertheless, their therapeutic options remain limited. Resveratrol (RSV), a hydrophobic stilbene flavonoid found in various sources such as red wine, peanuts, and grapes, has demonstrated potential benefits in multiple diseases, including neurological disorders. Owing to its capacity to cross the blood-brain barrier (BBB), RSV gains access to the brain parenchyma, where it exerts neuroprotective effects by modulating neuroinflammation, oxidative stress, and neuronal survival. However, its therapeutic application is hindered by extensive metabolism, leading to the rapid formation of glucuronide and sulfate conjugates, preventing effective concentrations in the brain.

Neurodegenerative diseases are among the most challenging and heterogeneous disorders, characterized by progressive degeneration of nervous system function and structure, leading to neuronal cell death (Kalia & Lang, 2016) which impose a severe global health and socioeconomic burden, with conditions such as AD, PD diseases sharing common pathological features. These disorders involve abnormal protein aggregates that induce selective neuronal damage, with contributing factors including neuroinflammation, oxidative stress, and disrupted neuronal signaling. Current treatment strategies primarily focus on symptomatic relief, often targeting neurotransmitter replacement, but they do not halt the disease progression. Therapeutic interventions typically include antioxidants, anti-inflammatory agents, and anti-amyloid drugs (Maiti & Dunbar, 2018).

RSV has been widely studied for its protective effects against inflammation and oxidative stress (Das & Das, 2007). It has demonstrated potential benefits in neurodegenerative disorders (Vingtdeux et al., 2008). Its neuroprotective properties make it a promising candidate for AD, PD diseases. RSV readily enters the bloodstream as glucuronide conjugates and can cross the BBB, enabling its therapeutic effects (Baur & Sinclair, 2006). Studies have reported that RSV at concentrations of 10–100 μM exerts neuroprotective effects (Richard et al., 2011).

This chapter will discuss the potential neuroprotective role of RSV in AD and PD, highlighting its several roles in neuroprotection. It will also examine emerging nanotechnology-based approaches to enhance the delivery of RSV to the brain, which could improve its therapeutic efficacy. By addressing pharmacokinetic challenges, these innovative strategies may contribute to the development of more effective treatments for neurodegenerative disorders.

2. PATHOPHYSIOLOGY OF PD AND AD

PD and AD, while clinically distinct, they share multiple overlapping pathophysiological features. Both conditions are characterized by progressive neuronal degeneration, pathological protein aggregation, neurotransmitter system dysfunction, and widespread network disruption. In addition, common molecular processes such as oxidative stress and neuroinflammation play central roles in these diseases' progression. These converging mechanisms underscore the existence of shared biological pathways and highlight potential common targets for therapeutic intervention in neurodegeneration.

Dysfunction in the basal ganglia (BG), a network of subcortical nuclei that includes the substantia nigra, putamen, caudate nucleus, globus pallidus, and subthalamic nucleus, is the main etiology of PD (Brown & Marsden, 1998). The globus pallidus internus (GPi) and substantia nigra pars reticulata (SNr) produce excessive amounts of BG when dopamine (DA) is depleted. This suppresses the thalamocortical and brainstem motor systems and results in motor dysfunction (Bergman & Deuschl, 2002). In addition to Lewy bodies aggregates of α-synuclein and neurofilaments (Olanow et al., 2009 ; Rodriguez-Oroz et al., 2009 ; Stoessl et al., 2014 ; Obeso et al., 2014). Neuronal degeneration in the substantia nigra pars compacta (SNc), where 50–80% of dopaminergic neurons are lost (Cheng et al., 2010), causes decreased DA release in the striatum, and the onset of motor and non-motor symptoms (NMS)(El-Mansoury et al., 2023; El-Mansoury et al., 2025; El Ghachi et al., 2023; El Khiat et al., 2023, 2025). However, PD pathophysiology is complex and involves multiple altered mechanisms and pathways, including mitochondrial dysfunction, oxidative stress, impaired protein degradation, neuroinflammation, abnormal α-synuclein aggregation, disrupted autophagy, and dysregulated neurotransmission within the basal ganglia circuitry (El-Mansoury et al., 2023; El-Mansoury et al., 2023).

Similarly, AD is characterized by neuronal loss, especially in the entorhinal cortex, hippocampus, amygdala, and cortical association areas. Additionally, subcortical nuclei like the cholinergic basal nucleus, noradrenergic locus coeruleus (LC), and serotonergic dorsal raphe are affected (Anand et al., 2014). As tau tangles spread from the trans-entorhinal cortex to the hippocampus and cortical areas, they are associated with cognitive atrophy and deterioration. Due to imbalances in production and clearance, amyloid β (Aβ) builds up and forms toxic plaques, which cause neuroinflammation and activate monocytes, macrophages, and microglia. Both disorders cause disruptions in neurotransmitter systems: AD is characterized by a large loss of cholinergic neurons and decreased acetylcholine-mediated neurotransmission (Chase et al., 2017), which contributes to cognitive and behavioral symptoms (Sultzer & Marder, 2017) (Figure 1), while PD is characterized by early serotonergic degeneration (Draoui et al., 2020), with mood disorders and REM sleep behavior disorder (Alexander, 2004 ; Halliday et al., 1990 ; Paulus & Jellinger, 1991). Noradrenergic depletion also happens in PD, where loss of LC neurons lowers norepinephrine (NE) content by up to 80%. This affects projections to the cortex, STN, SNc, and VTA (Gaspar et al., 1991 ; Delaville et al., 2012) and it may have an impact on motor symptoms such as rigidity and akinesia (Paulus & Jellinger, 1991). Additionally, AD affects serotonergic and noradrenergic nuclei, which contribute to non-cognitive symptoms (Imbimbo et al., 2005). Both disorders also include mitochondrial malfunction and oxidative stress: neurons are susceptible to ROS and RNS, which can result in lipid peroxidation, DNA damage, and apoptosis, particularly in the brain with high oxygen consumption (Hardy, 2009 ; Imbimbo et al., 2005). According to Prakash et al. (2017), oxidative damage in AD is further aggravated by trace metal dyshomeostasis (copper, iron, manganese, aluminum, and zinc). One recognized genetic risk factor for late-onset AD is the APOE genotype, which also affects how well cholinesterase inhibitors work (Imbimbo et al., 2005). Beyond motor symptoms, PD pathology affects cognition, emotion, and motivation through disturbances in mesolimbic and mesocortical circuits involving dopaminergic projections from the VTA and SNc to the nucleus accumbens, hippocampus, amygdala, and prefrontal cortex (Foffani & Obeso, 2018 ; Surmeier et al., 2017; Tritsch & Sabatini, 2012 ; Chaudhuri & Schapira, 2009). PD patients also experience postural instability, cognitive impairment, and sleep disturbances due to degeneration in the cholinergic pedunculopontine nucleus (PPN), which projects to the thalamus, striatum, cerebellum, and brainstem (Yarnall et al., 2011 ; Month & Monti, 2012 ; Nandi et al., 2002).

Figure 1. This figure illustrates the key pathological mechanisms involved in AD, including amyloid-β plaque accumulation, tau hyperphosphorylation, oxidative stress, inflammation, cholinergic dysfunction, metal dyshomesostasis and neuronal loss. The diagram was created using BioRender.

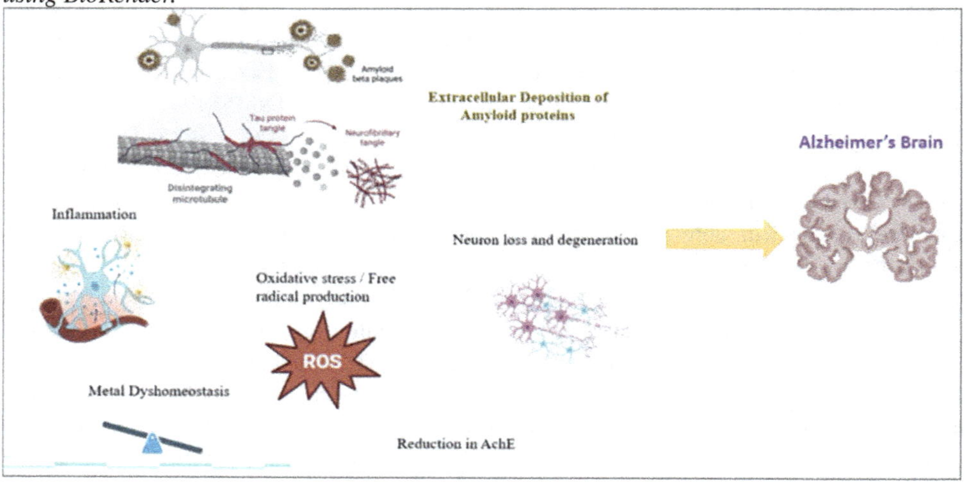

3. NEUROPROTECTIVE MECHANISMS OF RSV ON AD AND PD

RSV, also known as 3,5,4-trihydroxy-trans-stilbene, is a naturally occurring polyphenolic compound that is a member of the phytoalexin superfamily. In 1940, Dr. Michio Takaoka identified this compound after it was initially isolated from the roots of white hellebore (Veratrum grandiflorum O. LOES) (Rege et al., 2014). According to Chen et al. (2002) and Soleas et al. (1997), RSV can be found in the skin and seeds of over 70 different plant species, such as peanuts, berries, cereals, tea, and grapes. It controls a variety of disease pathology-related pathways because of its many target strategies. Non-flavonoid RSV is well known for its pharmacological properties, which include anti-inflammatory, neuroprotective, antiviral, anticancer, and antiphotoaging activities

(Chinraj & Raman, 2022). It has shown neuroprotective effects in AD, the most prevalent neurodegenerative disease in older adults, where tau and β-amyloid peptide misfolding is the primary cause of the disease progression (Hardy, 2006), through multiple mechanisms. It reduces oxidative stress by inhibiting ROS production, decreasing glutathione levels, and enhancing antioxidant enzymes' activities such as glutathione and superoxide dismutase. This leads to a decline in lipid peroxidation and oxidative damage (Kim et al., 2006). RSV also mitigates neuroinflammation by suppressing inflammatory cytokines (IL-6, IL-1β, TNF-α) and inhibiting microglial activation via NF-kB, reducing COX-2 and nitric oxide synthase expression (Chinraj & Raman, 2022). Moreover, RSV exerts anti-apoptotic effects by increasing Bcl-2 and decreasing pro-apoptotic Bax protein, thereby regulating cell survival. Additionally, it activates SIRT1, which protects against oxidative stress and apoptosis, contributing to its anti-aging properties. Its cognitive benefits are linked to its role in reducing amyloid-β (Aβ) accumulation and aggregation while preventing tau protein misfolding. Studies indicate RSV facilitates Aβ clearance by stimulating dephosphorylation through inhibition of CaMKII and GSK-3β and activation of PP2A (Jhang et al., 2017). A phase-2 clinical trial reported that RSV reduces amyloid-β levels in plasma and cerebrospinal fluid (Chinraj & Raman, 2022). RSV crosses the BBB, enhancing antioxidant enzymes' activity (Palle & Neerati, 2018), preventing neuronal cell death, and improving cognition in AD models. It has been shown to decrease amyloid precursor protein expression and improve spatial working memory in Tg6799 mice (Chen et al., 2019). Indeed, RSV directly binds to monomeric and fibrillary β-amyloids, remodeling toxic oligomers into non-toxic forms. Other studies reveal that RSV incubation reduces Aβ1-40 and Aβ1-42 fibril length and quantity (Ahmed et al., 2017 ; Ge et al., 2012 ; Pessentheiner et al., 2013), supporting its potential to slow AD progression (Ahmed et al., 2017 ; J. P. Guo et al., 2010). Hence, by counteracting oxidative stress, reducing neuroinflammation, and preventing Aβ accumulation, RSV emerges as a promising therapeutic agent for AD (Sönmez et al., 2007 ; Venturini et al., 2010).

PD is the second most common neurological movement disorder, affecting approximately 1% of the global population over 65 years (de Rijk et al., 1997). It is characterized by dopaminergic neuron loss and the presence of Lewy bodies (El-Mansoury et al., 2023). RSV exerts neuroprotective effects by activating AMP-activated protein kinase (AMPK) and SIRT1, which promote mitochondrial clearance and reduce oxidative stress. These pathways also activate PGC-1α, leading to mitochondrial biogenesis and oxidative stress regulation (Ferretta et al., 2014 ; Wu et al., 2011).RSV supports neuronal survival by modulating apoptotic proteins, increasing Bcl-2 while decreasing Bax, and enhancing the release of neurotrophic factors, including brain-derived, glial-cell-derived, and astroglia-derived neurotrophic factors (Bournival et al., 2009 ; F. Zhang et al., 2012). It also restores antioxidant defenses, normalizing DA levels in the striatal region and protecting against oxidative stress (Khan et al., 2010). Additionally, RSV reduces neuroinflammation by decreasing TNF-α, IL-6, IL-1β, and COX-2 expression, which are key contributors to PD progression (Jin et al., 2008 ; Lofrumento et al., 2014). One of RSV's primary protective mechanisms against PD is reducing oxidative stress. It directly scavenges ROS, though its low bioavailability limits this effect (Fukui et al., 2010). Beyond ROS neutralization, RSV inhibits ROS production by modulating gene expression and protein activity (Lin et al., 2014), upregulating antioxidant enzymes (e.g., SOD1, glutathione peroxidase 1) while suppressing ROS-generating enzymes such as nitric oxide synthase (Figure 2) (Y. Li et al., 2006 ; Park et al., 2009). Lin et al. demonstrated that rotenone, a neurotoxin inducing PD-like symptoms, disrupts autophagic flux and promotes apoptosis. RSV counteracts rotenone-induced neuronal death by facilitating autophagic induction and restoring autophagic flux (Lin et al., 2014 ; Heinz et al., 2017). However, when an autophagosome-lysosome fusion inhibitor (bafilomycin A1) was introduced, RSV's protective effects were blocked. Through its antioxidant, anti-inflammatory, mitochondrial protective, and autophagy-enhancing effects, RSV shows potential as a therapeutic agent for PD and other neurodegenerative diseases (Chinraj & Raman, 2022) .

Figure 2. Mechanism of action of resveratrol in both Alzheimer's Disease (AD) and Parkinson's Disease (PD).

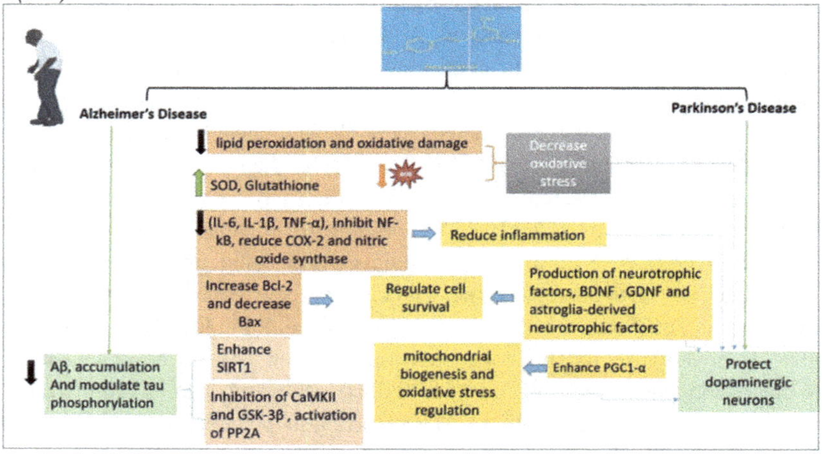

4. ANIMAL MODEL OF PARKINSON'S DISEASES

Animal models have been crucial in advancing our understanding of PD and developing treatments. These models include neurotoxin-based approaches using 6-OHDA, MPTP, rotenone, and paraquat, which induce dopaminergic cell death and motor deficits (Blesa et al., 2012). Genetic models based on PD-linked genes like SNCA, LRRK2, and PINK1 have also been developed to study disease mechanisms and potential therapies (Konnova EA & Swanberg M, 2018). While these models have been instrumental in discovering treatments for motor symptoms and investigating disease pathology, they have limitations in fully replicating the progressive nature and complexity of human PD (Duty & Jenner, 2011). The study conducted histopathological analysis demonstrated that RSV can reduce neuronal damage, improve motor function, and protect against mitochondrial dysfunction, which are all hallmarks of PD (Pirmoradi et al., 2024). The effectiveness of animal models in predicting successful translation to clinical treatments remains a key consideration. Combining different models may help address PD's heterogeneity and elucidate the interplay between genetic and environmental factors (Konnova EA & Swanberg M, 2018).

RSV, a natural polyphenol found in various plants, has been extensively studied for its potential neuroprotective effects in PD. This section provides a comprehensive overview of the role of RSV in animal models of PD, focusing on its mechanisms of action, therapeutic effects, and potential as a treatment for this neurodegenerative disorder.

5. MECHANISMS OF RSV IN PD

5.1. Anti-Inflammatory Effects

RSV has been shown to reduce neuroinflammation, a key contributor to PD progression. In MPTP-induced PD models, RSV decreases the levels of pro-inflammatory cytokines such as IL-1β, IL-6, and TNF-α, while upregulating the suppressor of cytokine signaling-1 (SOCS-1), which further mitigates inflammatory responses (Lofrumento et al., 2014 ; L. Liu et al., 2023). Additionally, RSV inhibits the TLR4/MyD88/NF-κB signaling pathway, which is a major pathway involved in neuroinflammation (Gui et al., 2024). LR4, MyD88, and NF-κB are essential elements of a signaling pathway particularly important in the immune system's response to infectious agents and injury. This pathway, TLR4/MyD88/NF-κB, is responsible for stimulating the innate immune response and the inflammatory reaction (Tian et al., 2021).

5.2. Antioxidant Properties

Oxidative stress is a hallmark of PD, and RSV's antioxidant properties play a significant role in mitigating this stress. RSV scavenges free radicals, such as hydroxyl radicals, and enhances the activity of antioxidant enzymes like glutathione peroxidase (GPx) and superoxide dismutase (SOD) (Lu et al., 2008 ; Anandhan et al., 2010). These effects help protect dopaminergic neurons from oxidative damage in the substantia nigra.

5.3. Anti-Apoptotic Effects

RSV modulates apoptotic pathways to prevent the death of dopaminergic neurons. It downregulates pro-apoptotic genes like Bax and upregulates anti-apoptotic genes like Bcl-2, thereby reducing apoptosis induced by neurotoxins such as MPP+ (Bournival et al., 2009). Additionally, RSV inhibits the release of cytochrome c and the nuclear translocation of apoptosis-inducing factor (AIF), further protecting neurons from apoptosis (Bournival et al., 2009).

5.4. Modulation of α-Synuclein

α-Synuclein aggregation is a key feature of PD pathology. RSV reduces the levels of total α-synuclein and its toxic oligomers, thereby alleviating the cytotoxicity associated with α-synuclein aggregates (Y. J. Guo et al., 2016 ; L. F. Zhang et al., 2018). This effect is partly mediated by the activation of SIRT1, which promotes the autophagic degradation of α-synuclein (Y. J. Guo et al., 2016).

5.5. Gut-Brain Axis Regulation

RSV has been shown to repair the intestinal barrier by inhibiting the TLR4/MyD88/NF-κB signaling pathway, which reduces gut inflammation and improves motor function in PD models (Gui et al., 2024). The gut-brain axis has become central to PD research, with evidence suggesting that pathological α-synuclein may spread from the gastrointestinal tract to the brain in some PD cases (Menozzi et al., 2025). A RSV-cyclodextrin inclusion complex (RHSD) shows improved efficacy in treating PD through the gut-brain axis, increasing beneficial gut bacteria like Lactobacillus species and reversing abnormal metabolite levels associated with motor function and intestinal health in transgenic mice (Sun et al., 2024). This suggests that RSV's effects may also involve the gut-brain axis, a recently recognized factor in PD pathogenesis.

6. THERAPEUTIC EFFECTS IN ANIMAL MODELS

6.1. Motor Function Improvement

RSV has been shown to significantly improve motor function in various PD animal models. In MPTP-induced PD models, RSV enhances motor coordination and reduces rotational behavior (Gui et al., 2024; Lofrumento et al., 2014). Similarly, in rotenone-induced models, RSV improves motor performance in tasks such as the rotarod test (Palle & Neerati, 2018).

6.2. Neuroprotection of Dopaminergic Neurons

RSV protects dopaminergic neurons in the substantia nigra from neurotoxin-induced damage. It increases the number of tyrosine hydroxylase (TH)-positive neurons and restores dopamine levels in the striatum (Gui et al., 2024 ; Lofrumento et al., 2014 ; Dong et al., 2022). This neuroprotection is associated with reduced neuroinflammation and oxidative stress.

6.3. Cognitive Deficit Alleviation

In addition to motor deficits, RSV alleviates cognitive impairments in PD models. For instance, in the A53T α-synuclein mouse model, RSV improves cognitive function by reducing neuroinflammation and oxidative stress (L. F. Zhang et al., 2018).

6.4. Enhanced Bioavailability with Nanoparticles

The bioavailability of RSV can be enhanced using nanoparticles, which have been shown to improve its therapeutic effects in rotenone-induced PD models. RSV nanoparticles (NRSV) are more effective than free RSV in reducing oxidative stress and mitochondrial dysfunction (Palle & Neerati, 2018).

Table 1. Comparative analysis of resveratrol's effects in different models

Model	Key Effects	Mechanism	Reference
MPTP-induced PD	Improves motor function, reduces neuroinflammation, protects dopaminergic neurons	Inhibition of TLR4/MyD88/NF-κB and SOCS-1 upregulation	(Lofrumento et al., 2014)(Gui et al., 2024)
Rotenone-induced PD	Alleviates motor deficits, reduces oxidative stress, improves mitochondrial function	Antioxidant effects, inhibition of α-synuclein aggregation	(Palle & Neerati, 2018)
6-OHDA-induced PD	Reduces rotational behavior, protects nigral dopaminergic neurons	Modulation of pro-inflammatory cytokines and antioxidant activity	(Faghihi A., Ph.D., Joghataie M.T., Ph.D.*, Darabi S., M.Sc., Mehdizadeh M., Ph.D., Roghani M., Ph.D., Bakhtiari M., 2007)
A53T α-synuclein PD	Improves motor and cognitive deficits, reduces α-synuclein cytotoxicity	Inhibition of α-synuclein aggregation and neuroinflammation	(L. F. Zhang et al., 2018)

7. ANIMAL MODEL OF ALZHEIMER'S DISEASE

Animal models are crucial for understanding AD pathogenesis and developing potential therapies. Various models have been developed, including transgenic mice, rats, rabbits, and non-human primates, each with unique strengths and limitations (Woodruff-pak, 2008). Transgenic mice overexpressing human genes associated with familial AD are the most commonly used models, focusing primarily on amyloid plaque formation (Drummond & Wisniewski, 2017). Other models include chemical-induced and spontaneous animal models (Xu et al., 2023). While these models have provided valuable insights, they often fail to fully replicate the complex interplay of amyloid plaques and neurofibrillary tangles characteristic of human AD (Drummond & Wisniewski, 2017). This limitation may contribute to the high failure rate of AD clinical trials (Drummond & Wisniewski, 2017). To enhance the success of therapy translation, researchers should consider using multiple complementary models and recognize the constraints of each approach (Woodruff-pak, 2008).

7.1. RSV in Animal Models of Alzheimer's Disease: Mechanisms of Action

This review synthesizes findings from animal models to explore RSV's mechanisms of action, optimal dosage, comparative efficacy, and behavioral outcomes in the context of AD. RSV exhibits a multifaceted approach in combating AD, targeting various pathological features of the disease.

7.1.1. Antioxidant and Anti-Inflammatory Effects

RSV's antioxidant properties help mitigate oxidative stress, a key contributor to AD progression. It enhances the activity of antioxidant enzymes such as superoxide dismutase (SOD), glutathione peroxidase (GSH-Px), and catalase (CAT), while reducing malondialdehyde (MDA) levels, a marker of oxidative stress (Kong et al., 2019 ; Kumar et al., 2007). Additionally, RSV suppresses inflammatory pathways, including NF-κB, IL-6, and IL-1β, which are often upregulated in AD (Zeini etal., 2024 ; Qi et al., 2019).

7.1.2. Modulation of the AMPK/PGC-1α/SIRT1 Pathway

For cognitive function, the AMPK/PGC-1α/SIRT1 pathway is essential for preserving mitochondrial integrity and preventing oxidative stress. This pathway's activation, frequently through exercise or calorie restriction, can improve energy metabolism, reduce oxidative damage, and promote mitochondrial biogenesis, all of which can improve cognitive function (Yang et al., 2020). RSV activates the AMPK/PGC-1α/SIRT1 pathway, which plays a critical role in energy metabolism and mitochondrial function. This activation enhances hippocampal function and reverses the decrease in AMPK expression caused by AD induction (Rashet et al., 2024 ; Corpas et al., 2019).

7.1.3. Proteostasis Enhancement

RSV improves proteostasis by increasing the levels of neprilysin, an Aβ-degrading enzyme, and reducing BACE1, an amyloidogenic secretase. It also enhances proteasome activity, aiding in the clearance of aberrant proteins (Corpas et al., 2019 ; Sarroca et al., 2021).

7.1.4. Tau Pathology Modulation

RSV inhibits tau aggregation and reduces phosphorylated tau levels, addressing one of the hallmark features of AD. It also blocks the uptake of extracellular tau oligomers, further mitigating tau-induced neurotoxicity (Sun et al., 2019).

7.1.5. Mitochondrial Protection

By activating SIRT1 and PGC-1α, RSV improves mitochondrial biogenesis and function, countering the mitochondrial dysfunction observed in AD (Corpas et al., 2019 ; Qi et al., 2019).

7.1.6. Neurogenesis and Synaptic Plasticity

RSV has shown promising effects on neuroplasticity and cognitive function. Studies have demonstrated that RSV enhances hippocampal neurogenesis by increasing cell survival, promoting neuronal differentiation, and reducing apoptosis (Torres-Pérez et al., 2015). It also improves dendritic spine formation and maturation in the dentate gyrus, leading to better memory performance (Torres-Pérez et al., 2015). RSV's neuroprotective effects extend to chronic cerebral hypoperfusion models, where it restores synaptic plasticity and activates the PKA-CREB pathway (H. Li et al., 2016). RSV enhances synaptic plasticity, which are crucial for cognitive function. It achieves this by activating the Wnt signaling pathway and increasing the expression of synaptic proteins like synaptophysin and PSD95 (Chen et al., 2019). This activation is believed to be mediated by SIRT1, a protein associated with stress response and cell longevity. Activation of the Wnt pathway is known to increase the expression of synaptic proteins such as synaptophysin and PSD95, which are essential for the formation and functioning of synapses (Chen et al., 2019).

7.1.7. Blood-Brain Barrier (BBB) Protection

RSV strengthens the BBB, preventing the leakage of Aβ and other neurotoxic substances into the brain, thereby reducing neuroinflammation and oxidative stress (Azargoonjahromi & Abutalebian, 2024).

8. DOSAGE AND ADMINISTRATION

8.1. Dosage Range

The efficacy of RSV in animal models varies with dosage and administration route. Oral Administration, The most common oral doses range from ≤15 mg/kg/day or higher doses of 100–200 mg/kg/day (X. Liu et al., 2024) . Intracerebroventricular (ICV) Injection: Direct brain delivery has been used at lower doses, such as 10 mg/kg, to target Aβ-induced neurodegeneration (Qi et al., 2019).

8.2. Administration Route

Oral Gavage: This is the most frequently used method, with RSV administered daily or every other day (Corpas et al., 2019 ; Chen et al., 2019).Dietary Supplementation: Some studies incorporate RSV into the diet, typically at 0.1% concentration, to mimic long-term exposure (Sarroca et al., 2021). Despite its efficacy, RSV's bioavailability is limited. Novel formulations, such as JOTROL, have been developed to enhance brain exposure, showing promising results in 3xTg-AD mice (Dennison et al., 2022).

9. COMPARATIVE EFFICACY

9.1. Comparison with Donepezil

RSV's efficacy has been compared to other treatments. RSV has shown comparable or superior cognitive benefits to donepezil, a standard AD medication, in animal models. For example, RSV improved memory in Aβ-induced AD models, while donepezil had limited effects (Qi et al., 2019 ; Foudah et al., 2023).

9.2. Synergy with Vitamin E

Combining RSV with vitamin E enhances its nootropic effects, as seen in scopolamine-induced AD models, where the combination improved recognition index and inflection ratio in behavioral tests (Foudah et al., 2023).

9.3. Comparison with Aerobic Training

RSV and aerobic training synergistically enhance AMPK/PGC-1α/SIRT1 expression, suggesting that combining lifestyle interventions with RSV may maximize therapeutic benefits (Rashet et al., 2024).

10. BEHAVIORAL OUTCOMES

10.1. Cognitive Improvement

RSV consistently improves cognitive and behavioral deficits in AD models. Morris Water Maze: RSV-treated animals showed improved spatial memory and reduced escape latency (Chen et al., 2019 ; Sun et al., 2019). Y-Maze and Object Recognition Tests: RSV enhanced working memory and recognition memory in AD models (Zeini et al., 2024 ; Sun et al., 2019).

10.2. Anxiety and Exploratory Behavior

RSV reduced anxiety-like behaviors and improved exploratory behavior in both AD and healthy mice, indicating its anxiolytic effects (Corpas et al., 2019).

10.3. Neuroprotection

Histopathological studies reveal that RSV protects against hippocampal and cortical damage, reducing Aβ plaques and neuroinflammation (Corpas et al., 2019 ; Chen et al., 2019).

Table 2. Summary of key studies

Model	Dose	Effects	References
Scopolamine-induced AD	50-75 mg/kg, oral	Improved cognitive function, reduced oxidative stress, and inflammation	(Foudah et al., 2023)
3xTg-AD mice	100 mg/kg, dietary supplementation	Reduced Aβ and tau pathology, enhanced proteostasis	(Corpas et al., 2019)
Tg6799 mice	60 mg/kg, oral	Reduced amyloid plaques and improved spatial memory	(Chen et al., 2019)
LPS-induced neuroinflammation	30 mg/kg, oral	Reduced glial density and neuronal loss, improved spatial and working memory	(Zeini et al., 2024)
Colchicine-induced AD	10-20 mg/kg, oral	Improved cognitive function and reduced oxidative stress	(Kumar et al., 2007)

CONCLUSION

RSV holds significant promise as a neuroprotective agent for treating AD and PD. In pre-clinical research, RSV has shown antioxidant, anti-inflammatory, anti-apoptotic, and anti-protein aggregation properties, addressing critical mechanisms like oxidative stress, mitochondrial dysfunction, and neuroinflammation. In AD models, RSV decreases amyloid-β buildup, tau phosphorylation, and cognitive deterioration, while in PD models, it mitigates α-synuclein toxicity and supports neuronal survival. Although its bioavailability is limited, innovative formulations such as JOTROL and nanotechnology-based delivery systems, present promising ways to enhance its therapeutic effects. Combination therapies might also lead to better outcomes. While results from various studies are promising, additional clinical trials are necessary to verify RSV's effectiveness in humans. Overall, RSV emerges as a promising candidate for future treatments targeting neurodegenerative diseases.

REFERENCES

Ahmed, T., Javed, S., Javed, S., Tariq, A., Šamec, D., Tejada, S., Nabavi, S. F., Braidy, N., & Nabavi, S. M. (2017). Resveratrol and Alzheimer's Disease: Mechanistic Insights. *Molecular Neurobiology, 54*(4), 2622–2635. DOI: 10.1007/s12035-016-9839-9 PMID: 26993301

Aimrane, A., Sabir, E.-S., Laadraoui, J., El-Mansoury, B., Smimih, K., Khanouchi, M., El Amine, S., Lafhal, K., Oublal, K., & El Baz, S. (2023). An Overview of the Genetics of Parkinson's Disease. *Experimental and Clinical Evidence of the Neuropathology of Parkinson's Disease*, 164–181.

Alexander, G. E. (2004). Biology of Parkinson's disease: Pathogenesis and pathophysiology of a multisystem neurodegenerative disorder. *Dialogues in Clinical Neuroscience, 6*(3), 259–280. DOI: 10.31887/dcns.2004.6.3/galexander PMID: 22033559

Anand, R., Gill, K. D., & Mahdi, A. A. (2014). Therapeutics of Alzheimer's disease: Past, present and future. *Neuropharmacology, 76*(PART A), 27–50. https://doi.org/DOI: 10.1016/j.neuropharm.2013.07.004

Anandhan, A., Tamilselvam, K., Vijayraja, D., Ashokkumar, N., Rajasankar, S., & Manivasagam, T. (2010). Resveratrol attenuates oxidative stress and improves behaviour in 1 -methyl-4-phenyl-1,2,3,6-tetrahydropyridine (MPTP) challenged mice. *Annals of Neurosciences, 17*(3), 113. PMID: 25205886

Azargoonjahromi, A., & Abutalebian, F. (2024). Unraveling the therapeutic efficacy of resveratrol in Alzheimer's disease: An umbrella review of systematic evidence. *Nutrition & Metabolism, 21*(1), 1–15. DOI: 10.1186/s12986-024-00792-1 PMID: 38504306

Baur, J. A., & Sinclair, D. A. (2006). Therapeutic potential of resveratrol: The in vivo evidence. *Nature Reviews. Drug Discovery, 5*(6), 493–506. DOI: 10.1038/nrd2060 PMID: 16732220

Bergman, H., & Deuschl, G. (2002). Pathophysiology of parkinsion's disease: From clinical neurology to basic neuroscience and back. *Movement Disorders, 17*(SUPPL. 3), 28–40. DOI: 10.1002/mds.10140 PMID: 11948753

Blesa, J., Phani, S., Jackson-Lewis, V., & Przedborski, S. (2012). Classic and new animal models of Parkinson's disease. *Journal of Biomedicine & Biotechnology, 2012*. Advance online publication. DOI: 10.1155/2012/845618 PMID: 22536024

Bournival, J., Quessy, P., & Martinoli, M. G. (2009). Protective effects of resveratrol and quercetin against MPP+ -induced oxidative stress act by modulating markers of apoptotic death in dopaminergic neurons. *Cellular and Molecular Neurobiology, 29*(8), 1169–1180. DOI: 10.1007/s10571-009-9411-5 PMID: 19466539

Brown, P., & Marsden, C. D. (1998). What do the basal ganglia do? *Lancet, 351*(9118), 1801–1804. DOI: 10.1016/S0140-6736(97)11225-9 PMID: 9635969

Chase, T. N., Farlow, M. R., & Clarence-Smith, K. (2017). Donepezil Plus Solifenacin (CPC-201) Treatment for Alzheimer's Disease. *Neurotherapeutics; the Journal of the American Society for Experimental NeuroTherapeutics, 14*(2), 405–416. DOI: 10.1007/s13311-016-0511-x PMID: 28138837

Chaudhuri, K. R., & Schapira, A. H. (2009). Non-motor symptoms of Parkinson's disease: Dopaminergic pathophysiology and treatment. *Lancet Neurology, 8*(5), 464–474. DOI: 10.1016/S1474-4422(09)70068-7 PMID: 19375664

Chen, Y., Shi, G. W., Liang, Z. M., Sheng, S. Y., Shi, Y. S., Peng, L., Wang, Y. P., Wang, F., & Zhang, X. M. (2019). Resveratrol improves cognition and decreases amyloid plaque formation in Tg6799 mice. *Molecular Medicine Reports, 49*(5), 3783–3790. DOI: 10.3892/mmr.2019.10010 PMID: 30864708

Cheng, H. C., Ulane, C. M., & Burke, R. E. (2010). Clinical progression in Parkinson disease and the neurobiology of axons. *Annals of Neurology, 67*(6), 715–725. DOI: 10.1002/ana.21995 PMID: 20517933

Chinraj, V., & Raman, S. (2022). Neuroprotection by resveratrol: A review on brain delivery strategies for Alzheimer's and Parkinson's disease. *Journal of Applied Pharmaceutical Science, 12*(7), 1–17. DOI: 10.7324/JAPS.2022.120701

Corpas, R., Griñán-Ferré, C., Rodríguez-Farré, E., Pallàs, M., & Sanfeliu, C. (2019). Resveratrol Induces Brain Resilience Against Alzheimer Neurodegeneration Through Proteostasis Enhancement. *Molecular Neurobiology, 56*(2), 1502–1516. DOI: 10.1007/s12035-018-1157-y PMID: 29948950

Das, S., & Das, D. K. (2007). Anti-inflammatory responses of resveratrol. *Inflammation & Allergy Drug Targets, 6*(3), 168–173. DOI: 10.2174/187152807781696464 PMID: 17897053

de Rijk, M. C., Tzourio, C., Breteler, M. M. B., Dartigues, J. F., Amaducci, L., Lopez-Pousa, S., Manubens-Bertran, J. M., Alperovitch, A., & Rocca, W. A. (1997). Prevalence of parkinsonism and Parkinson's. *Journal of Neurology, Neurosurgery, and Psychiatry, 62*, 10–15. PMID: 9010393

Delaville, C., Navailles, S., & Benazzouz, A. (2012). Effects of noradrenaline and serotonin depletions on the neuronal activity of globus pallidus and substantia nigra pars reticulata in experimental parkinsonism. *Neuroscience, 202*, 424–433. DOI: 10.1016/j.neuroscience.2011.11.024 PMID: 22138505

Dennison, J. L., Volmar, C. H., Modarresi, F., Ke, D., Wang, J., Gravel, E., Hammond-Vignini, S., Li, Z., Timmons, J. A., Lohse, I., Hayward, M. A., Brothers, S. P., & Wahlestedt, C. (2022). Erratum: JOTROL, a Novel Formulation of Resveratrol, Shows Beneficial Effects in the 3xTg-AD Mouse Model (Journal of Alzheimer's Disease (2022) 86:1 (173-190) DOI:). *Journal of Alzheimer's Disease, 88*(1), 385. https://doi.org/DOI: 10.3233/JAD-215370

Dong, Y., Wang, Z., Yan, Z., Wu, C., Yang, T., & Qiu, H. (2022). Neuroprotective Effects of Resveratrol on a Mouse Model of Parkinson's Disease via the Wnt/Beta-Catenin Signaling Pathway. *Indian Journal of Pharmaceutical Sciences, 84*(4), 959–968. DOI: 10.36468/pharmaceutical-sciences.990

Draoui, A., El Hiba, O., Aimrane, A., El Khiat, A., & Gamrani, H. (2020). Parkinson's disease: From bench to bedside. *Revue Neurologique*, *176*(7–8), 543–559. PMID: 31928746

Drummond, E., & Wisniewski, T. (2017). Alzheimer's disease: Experimental models and reality. *Acta Neuropathologica*, *133*(2), 155–175. DOI: 10.1007/s00401-016-1662-x PMID: 28025715

Duty, S., & Jenner, P. (2011). Animal models of Parkinson's disease: A source of novel treatments and clues to the cause of the disease. *British Journal of Pharmacology*, *164*(4), 1357–1391. DOI: 10.1111/j.1476-5381.2011.01426.x PMID: 21486284

El Ghachi, H., Oukhrib, M., Tamegart, L., Ben Maloui, A., El-Mansoury, B., & Gamrani, H. (2023). Diagnosing Methods of Parkinson's Disease: Revolutionary Techniques. In *Experimental and Clinical Evidence of the Neuropathology of Parkinson's Disease* (pp. 46–63). IGI Global.

El Khiat, A., Ait Ali, D., El-Mansoury, B., Ait Hamdan, Y., Hakem, A., Tamegart, L., El Koutbi, M., El Houate, B., Kissani, N., & Gamrani, H. (2025). Psychological assessment of patients with Parkinson's disease. In *Essential Guide to Neurodegenerative Disorders* (pp. 257–268). Elsevier.

El Khiat, A., El-Mansoury, B., Ali, D. A., El Koutbi, M., Kaoutar, K., Hamdan, Y. A., El Houate, B., Tamegart, L., Smimih, K., & Hakam, A. (2023). Symptomatology and Clinical Features of Parkinson's Disease. In *Experimental and Clinical Evidence of the Neuropathology of Parkinson's Disease* (pp. 14–24). IGI Global.

El-Mansoury, B., Ben Maloui, A., Draoui, A., Hamdan, Y. A., Oukhrib, M., Smimih, K., Elmourid, A., El Khiat, A., Boulbaroud, S., & Meftah, I. (2023). Non-Motor Symptoms in Parkinson's Disease: The Other Side of the Disease. In *Experimental and Clinical Evidence of the Neuropathology of Parkinson's Disease* (pp. 25–45). IGI Global.

El-Mansoury, B., El Hiba, O., Agnaou, M., Jayakumar, A. R., El Khiat, A., Kahim, K., Boulbaroud, S., Zouhairi, N., Tibssirte, O., & Bouighajd, H. (2023). Neuropathology of Parkinson's Disease: An Update of the Recent Findings. In *Experimental and Clinical Evidence of the Neuropathology of Parkinson's Disease* (pp. 82–101). IGI Global.

El-Mansoury, B., El Hiba, O., & Jayakumar, A. R. (2023). *Physiology and Function of Glial Cells in Health and Disease*. IGI Global.

El-Mansoury, B., El Khiat, A., Skaou, A., El Hiba, O., Jayakumar, A. R., Ait Hamdan, Y., Bitar, A., & Ferssiwi, A. (2025). Non-motor symptoms of Parkinson's disease. In *Essential Guide to Neurodegenerative Disorders* (pp. 269–280). Elsevier.

El-Mansoury, B., Smimih, K., El Khiat, A., El Fatimi, H., Aimrane, A., El Baz, S., Elmourid, A., Saad, F. E., El Ghachi, H., & Khanouchi, M. (2023). Potential Role of Probiotics and Fecal Microbiota Transplantation in the Treatment of Parkinson's Disease. In *Experimental and Clinical Evidence of the Neuropathology of Parkinson's Disease* (pp. 296–317). IGI Global.

Faghihi, A., Ph.D., Joghataie M.T., Ph.D.*, Darabi S., M.Sc., Mehdizadeh M., Ph.D., Roghani M., Ph.D., Bakhtiari M., P. (2007). Evaluation of Behavioral Effects of Trans-Resveratrol in the Hemi- Parkinsonian Rat Model Faghihi. *Journal of Iranian Anatomical Sciences*, *5*(19), 107–114.

Ferretta, A., Gaballo, A., Tanzarella, P., Piccoli, C., Capitanio, N., Nico, B., Annese, T., Di Paola, M., Dell'Aquila, C., De Mari, M., Ferranini, E., Bonifati, V., Pacelli, C., & Cocco, T. (2014). Effect of resveratrol on mitochondrial function: Implications in parkin-associated familiar Parkinson's disease. *Biochimica et Biophysica Acta. Molecular Basis of Disease, 1842*(7), 902–915. DOI: 10.1016/j.bbadis.2014.02.010 PMID: 24582596

Foffani, G., & Obeso, J. A. (2018). A Cortical Pathogenic Theory of Parkinson's Disease. *Neuron, 99*(6), 1116–1128. DOI: 10.1016/j.neuron.2018.07.028 PMID: 30236282

Foudah, A. I., Devi, S., Alam, A., Salkini, M. A., & Ross, S. A. (2023). Anticholinergic effect of resveratrol with vitamin E on scopolamine-induced Alzheimer's disease in rats: Mechanistic approach to prevent inflammation. *Frontiers in Pharmacology, 14*(February), 1–11. DOI: 10.3389/fphar.2023.1115721 PMID: 36817151

Fukui, M., Choi, H. J., & Zhu Bao Ting, B. T. (2010). Mechanism for the protective effect of resveratrol against oxidative stress-induced neuronal death. *Free Radical Biology & Medicine, 49*(5), 800–813. DOI: 10.1016/j.freeradbiomed.2010.06.002 PMID: 20542495

Gaspar, P., Duyckaerts, C., Alvarez, C., Javoy-Agid, F., & Berger, B. (1991). Alterations of dopaminergic and noradrenergic innervations in motor cortex in parkinson's disease. *Annals of Neurology, 30*(3), 365–374. DOI: 10.1002/ana.410300308 PMID: 1683212

Ge, J. F., Qiao, J. P., Qi, C. C., Wang, C. W., & Zhou, J. N. (2012). The binding of resveratrol to monomer and fibril amyloid beta. *Neurochemistry International, 61*(7), 1192–1201. DOI: 10.1016/j.neuint.2012.08.012 PMID: 22981725

Grant, W. B., Campbell, A., Itzhaki, R. F., & Savory, J. (2002). The significance of environmental factors in the etiology of Alzheimer's disease. *Journal of Alzheimer's Disease, 4*(3), 179–189. PMID: 12226537

Gui, J., Sun, X., Wen, S., Liu, X., Qin, B., & Sang, M. (2024). Resveratrol protects dopaminergic neurons in a mouse model of Parkinson's disease by regulating the gut-brain axis via inhibiting the TLR4 signaling pathway. *Nan Fang Yi Ke da Xue Xue Bao = Journal of Southern Medical University, 44*(2), 270–279. DOI: 10.12122/j.issn.1673-4254.2024.02.09 PMID: 38501412

Guo, J. P., Yu, S., & McGeer, P. L. (2010). Simple in vitro assays to identify amyloid-β aggregation blockers for Alzheimer's disease therapy. *Journal of Alzheimer's Disease, 19*(4), 1359–1370. DOI: 10.3233/JAD-2010-1331 PMID: 20061605

Guo, Y. J., Dong, S. Y., Cui, X. X., Feng, Y., Liu, T., Yin, M., Kuo, S. H., Tan, E. K., Zhao, W. J., & Wu, Y. C. (2016). Resveratrol alleviates MPTP-induced motor impairments and pathological changes by autophagic degradation of α-synuclein via SIRT1-deacetylated LC3. *Molecular Nutrition & Food Research, 60*(10), 2161–2175. DOI: 10.1002/mnfr.201600111 PMID: 27296520

Halliday, G. M., Li, Y. W., Blumbergs, P. C., Joh, T. H., Cotton, R. G. H., Howe, P. R. C., Blessing, W. W., & Geffen, L. B. (1990). Neuropathology of immunohistochemically identified brainstem neurons in Parkinson's disease. *Annals of Neurology, 27*(4), 373–385. DOI: 10.1002/ana.410270405 PMID: 1972319

Hardy, J. (2006). Alzheimer's disease: The amyloid cascade hypothesis - An update and reappraisal. *Journal of Alzheimer's Disease, 9*(SUPPL. 3), 151–153. DOI: 10.3233/jad-2006-9s317 PMID: 16914853

Hardy, J. (2009). The amyloid hypothesis for Alzheimer's disease: A critical reappraisal. *Journal ofNeurochemistry, 110*(4), 1129–1134. DOI: 10.1111/j.1471-4159.2009.06181.x PMID: 19457065

Heinz, S., Freyberger, A., Lawrenz, B., Schladt, L., Schmuck, G., & Ellinger-Ziegelbauer, H. (2017). Mechanistic Investigations of the Mitochondrial Complex i Inhibitor Rotenone in the Context of Pharmacological and Safety Evaluation. *Scientific Reports, 7*(February), 1–13. DOI: 10.1038/srep45465 PMID: 28374803

Imbimbo, B. P., Lombard, J., & Pomara, N. (2005). Pathophysiology of Alzheimer's disease. *Neuroimaging Clinics of North America, 15*(4), 727–753. DOI: 10.1016/j.nic.2005.09.009 PMID: 16443487

Jhang, K. A., Park, J. S., Kim, H. S., & Chong, Y. H. (2017). Resveratrol Ameliorates Tau Hyperphosphorylation at Ser396 Site and Oxidative Damage in Rat Hippocampal Slices Exposed to Vanadate: Implication of ERK1/2 and GSK-3β Signaling Cascades. *Journal of Agricultural and Food Chemistry, 65*(44), 9626–9634. DOI: 10.1021/acs.jafc.7b03252 PMID: 29022339

Jin, F., Wu, Q., Lu, Y. F., Gong, Q. H., & Shi, J. S. (2008). Neuroprotective effect of resveratrol on 6-OHDA-induced Parkinson's disease in rats. *European Journal of Pharmacology, 600*(1–3), 78–82. DOI: 10.1016/j.ejphar.2008.10.005 PMID: 18940189

Kalia, L. V., & Lang, A. E. (2016). Parkinson disease in 2015: Evolving basic, pathological and clinical concepts in PD. *Nature Reviews. Neurology, 12*(2), 2–3. DOI: 10.1038/nrneurol.2015.249 PMID: 26782330

Khan, M. M., Ahmad, A., Ishrat, T., Khan, M. B., Hoda, M. N., Khuwaja, G., Raza, S. S., Khan, A., Javed, H., Vaibhav, K., & Islam, F. (2010). Resveratrol attenuates 6-hydroxydopamine-induced oxidative damage and dopamine depletion in rat model of Parkinson's disease. *Brain Research, 1328*, 139–151. DOI: 10.1016/j.brainres.2010.02.031 PMID: 20167206

Kim, Y. A., Lim, S. Y., Rhee, S. H., Park, K. Y., Kim, C. H., Choi, B. T., Lee, S. J., Park, Y. M., & Choi, Y. H. (2006). Resveratrol inhibits inducible nitric oxide synthase and cyclooxygenase-2 expression in β-amyloid-treated C6 glioma cells. *International Journal of Molecular Medicine, 17*(6), 1069–1075. DOI: 10.3892/ijmm.17.6.1069 PMID: 16685418

Kong, D., Yan, Y., He, X. Y., Yang, H., Liang, B. Y., Wang, J., He, Y., Ding, Y., & Yu, H. (2019). Effects of Resveratrol on the Mechanisms of Antioxidants and Estrogen in Alzheimer's Disease. *BioMed Research International, 2019*. Advance online publication. DOI: 10.1155/2019/8983752 PMID: 31016201

Konnova, E. A., & Swanberg, M. (2018). Parkinson's Disease: Pathogenesis and Clinical Aspects, chapter 5:Animal Models of Parkinson's Disease. *Parkinson's Disease: Pathogenesis and Clinical Aspects*, 83–106.

Kumar, A., Naidu, P. S., Seghal, N., & Padi, S. S. V. (2007). Neuroprotective effects of resveratrol against intracerebroventricular colchicine-induced cognitive impairment and oxidative stress in rats. *Pharmacology*, *79*(1), 17–26. DOI: 10.1159/000097511 PMID: 17135773

Li, H., Wang, J., Wang, P., Rao, Y., & Chen, L. (2016). Resveratrol Reverses the Synaptic Plasticity Deficits in a Chronic Cerebral Hypoperfusion Rat Model. *Journal of Stroke and Cerebrovascular Diseases*, *25*(1), 122–128. DOI: 10.1016/j.jstrokecerebrovasdis.2015.09.004 PMID: 26456198

Li, Y., Cao, Z., & Zhu, H. (2006). Upregulation of endogenous antioxidants and phase 2 enzymes by the red wine polyphenol, resveratrol in cultured aortic smooth muscle cells leads to cytoprotection against oxidative and electrophilic stress. *Pharmacological Research*, *53*(1), 6–15. DOI: 10.1016/j.phrs.2005.08.002 PMID: 16169743

Lin, T. K., Der Chen, S., Chuang, Y. C., Lin, H. Y., Huang, C. R., Chuang, J. H., Wang, P. W., Huang, S. T., Tiao, M. M., Chen, J. B., & Liou, C. W. (2014). Resveratrol partially prevents rotenone-induced neurotoxicity in dopaminergic SH-SY5Y cells through induction of heme oxygenase-1 dependent autophagy. *International Journal of Molecular Sciences*, *15*(1), 1625–1646. DOI: 10.3390/ijms15011625 PMID: 24451142

Liu, L., Rajkumar, M., Pandi, B., Jayaraman, A., & Sonamuthu, J. (2023). Combined neuroprotective potential of Levodopa and Resveratrol to ameliorates cognitive impairments in rotenone-induced Parkinson's disease via targeting TLR mediated neuroinflammation. *Authorea Preprints*, 1–15.

Liu, X., Gu, X., Zhang, J., Li, X., Wei, X., Jiang, S., & Li, W. (2024). Resveratrol delays the progression of diabetic nephropathy through multiple pathways: A dose–response meta-analysis based on animal models. *Journal of Diabetes*, *16*(9). Advance online publication. DOI: 10.1111/1753-0407.13608 PMID: 39264004

Lofrumento, D. D., Nicolardi, G., Cianciulli, A., De Nuccio, F., La Pesa, V., Carofiglio, V., Dragone, T., Calvello, R., & Panaro, M. A. (2014). Neuroprotective effects of resveratrol in an MPTP mouse model of Parkinson's-like disease: Possible role of SOCS-1 in reducing pro-inflammatory responses. *Innate Immunity*, *20*(3), 249–260. DOI: 10.1177/1753425913488429 PMID: 23764428

Lu, K. T., Ko, M. C., Chen, B. Y., Huang, J. C., Hsieh, C. W., Lee, M. C., Chiou, R. Y. Y., Wung, B. S., Peng, C. H., & Yang, Y. L. (2008). Neuroprotective effects of resveratrol on MPTP-induced neuron loss mediated by free radical scavenging. *Journal of Agricultural and Food Chemistry*, *56*(16), 6910–6913. DOI: 10.1021/jf8007212 PMID: 18616261

Maiti, P., & Dunbar, G. L. (2018). Use of curcumin, a natural polyphenol for targeting molecular pathways in treating age-related neurodegenerative diseases. In *International Journal of Molecular Sciences* (Vol. 19, Issue 6). https://doi.org/DOI: 10.3390/ijms19061637

Masters, C. L., Bateman, R., Blennow, K., Rowe, C. C., Sperling, R. A., & Cummings, J. L. (2015). Alzheimer's disease. *Nature Reviews. Disease Primers*, *1*(1), 1–18. PMID: 27188934

Menozzi, E., Schapira, A. H. V., & Borghammer, P. (2025). The Gut-Brain Axis in Parkinson disease: Emerging Concepts and Therapeutic Implications. *Movement Disorders Clinical Practice*, *12*(March), 904–916. DOI: 10.1002/mdc3.70029 PMID: 40079755

Migliore, L., & Coppedè, F. (2022). Gene–environment interactions in Alzheimer disease: The emerging role of epigenetics. *Nature Reviews. Neurology, 18*(11), 643–660. PMID: 36180553

Month, J. M., & Monti, D. (2012). Role of Dorsal Raphe Nucleus Serotonin 5Htia Receptor in the Regulation of Rem Sleep. *Elsevier, 66*(21), 1999–2012.

Moore, D. J., West, A. B., Dawson, V. L., & Dawson, T. M. (2005). Molecular pathophysiology of Parkinson's disease. *Annual Review of Neuroscience, 28*, 57–87. DOI: 10.1146/annurev.neuro.28.061604.135718 PMID: 16022590

Nandi, D., Liu, X., Winter, J. L., Aziz, T. Z., & Stein, J. F. (2002). Deep brain stimulation of the pedunculopontine region in the normal non-human primate. *Journal of Clinical Neuroscience, 9*(2), 170–174. DOI: 10.1054/jocn.2001.0943 PMID: 11922707

Obeso, J. A., Rodriguez-Oroz, M. C., Stamelou, M., Bhatia, K. P., & Burn, D. J. (2014). The expanding universe of disorders of the basal ganglia. *Lancet, 384*(9942), 523–531. DOI: 10.1016/S0140-6736(13)62418-6 PMID: 24954674

Olanow, C. W., Stern, M. B., & Sethi, K. (2009). The scientific and clinical basis for the treatment of Parkinson disease (2009). *Neurology, 72*(21, SUPPL. 4). Advance online publication. DOI: 10.1212/WNL.0b013e3181a1d44c PMID: 19470958

Oukhrib, M., Tamegart, L., El Ghachi, H., Maloui, A. Ben, El-Mansoury, B., Chraa, M., & Gamrani, H. (2023). Epidemiology of Parkinson's Disease: An Updated Overview. *Experimental and Clinical Evidence of the Neuropathology of Parkinson's Disease*, 1–12.

Palle, S., & Neerati, P. (2018). Improved neuroprotective effect of resveratrol nanoparticles as evinced by abrogation of rotenone-induced behavioral deficits and oxidative and mitochondrial dysfunctions in rat model of Parkinson's disease. *Naunyn-Schmiedeberg's Archives of Pharmacology, 391*(4), 445–453. DOI: 10.1007/s00210-018-1474-8 PMID: 29411055

Park, D. W., Baek, K., Kim, J. R., Lee, J. J., Ryu, S. H., Chin, B. R., & Baek, S. H. (2009). Resveratrol inhibits foam cell formation via NADPH oxidase 1-mediated reactive oxygen species and monocyte chemotactic protein-1. *Experimental & Molecular Medicine, 41*(3), 171–179. DOI: 10.3858/emm.2009.41.3.020 PMID: 19293636

Paulus, W., & Jellinger, K. (1991). Correspondance anatomique _groupe rigide et gr tremblement (1991)_.pdf. In *Neuropathology and Experimental Neurology* (Vol. 50, Issue 6, pp. 743–755).

Pessentheiner, A. R., Pelzmann, H. J., Walenta, E., Schweiger, M., Groschner, L. N., Graier, W. F., Kolb, D., Uno, K., Miyazaki, T., Nitta, A., Rieder, D., Prokesch, A., & Bogner-Strauss, J. G. (2013). NAT8L (N-acetyltransferase 8-like) accelerates lipid turnover and increases energy expenditure in brown adipocytes. *The Journal of Biological Chemistry, 288*(50), 36040–36051. DOI: 10.1074/jbc.M113.491324 PMID: 24155240

Pirmoradi, Z., Nakhaie, M., Ranjbar, H., Kalantar-Neyestanaki, D., Kohlmeier, K. A., Asadi-Shekaari, M., Hassanshahi, A., & Shabani, M. (2024). Resveratrol and 1,25-dihydroxyvitamin D decrease Lingo-1 levels, and improve behavior in harmaline-induced Essential tremor, suggesting potential therapeutic benefits. *Scientific Reports, 14*(1), 1–17. DOI: 10.1038/s41598-024-60518-4 PMID: 38684734

Qi, Y., Shang, L., Liao, Z., Su, H., Jing, H., Wu, B., Bi, K., & Jia, Y. (2019). Intracerebroventricular injection of resveratrol ameliorated Aβ-induced learning and cognitive decline in mice. *Metabolic Brain Disease*, *34*(1), 257–266. DOI: 10.1007/s11011-018-0348-6 PMID: 30460524

Rashet, A., Abdi, A., & Barari, A. (2024). Synergistic Role of Aerobic Training and Resveratrol on AMPK/PGC1-α/SIRT1 Pathway in the Hippocampus of Rats with Alzheimer's Disease. *Journal of Archives in Military Medicine*, *12*(1), 1–8. DOI: 10.5812/jamm-144281

Rege, S. D., Geetha, T., Griffin, G. D., Broderick, T. L., & Babu, J. R. (2014). Neuroprotective effects of resveratrol in Alzheimer disease pathology. *Frontiers in Aging Neuroscience*, *6*(AUG), 1–27. DOI: 10.3389/fnagi.2014.00218 PMID: 25309423

Richard, T., Pawlus, A. D., Iglésias, M. L., Pedrot, E., Waffo-Teguo, P., Mérillon, J. M., & Monti, J. P. (2011). Neuroprotective properties of resveratrol and derivatives. *Annals of the New York Academy of Sciences*, *1215*(1), 103–108. DOI: 10.1111/j.1749-6632.2010.05865.x PMID: 21261647

Rodriguez-Oroz, M. C., Jahanshahi, M., Krack, P., Litvan, I., Macias, R., Bezard, E., & Obeso, J. A. (2009). Initial clinical manifestations of Parkinson's disease: Features and pathophysiological mechanisms. *Lancet Neurology*, *8*(12), 1128–1139. DOI: 10.1016/S1474-4422(09)70293-5 PMID: 19909911

Saad, F. E.-Z., Lelo, A. P., Smimih, K., El-Mansoury, B., El Amine, S., & Bitar, A. (2023). Pesticide Exposure as a Risk Factor for Parkinson's Disease. In *Experimental and Clinical Evidence of the Neuropathology of Parkinson's Disease* (pp. 182–195). IGI Global.

Sarroca, S., Gatius, A., Rodríguez-Farré, E., Vilchez, D., Pallàs, M., Griñán-Ferré, C., Sanfeliu, C., & Corpas, R. (2021). Resveratrol confers neuroprotection against high-fat diet in a mouse model of Alzheimer's disease via modulation of proteolytic mechanisms. *The Journal of Nutritional Biochemistry*, *89*, 108569. DOI: 10.1016/j.jnutbio.2020.108569 PMID: 33321185

Scheltens, P., De Strooper, B., Kivipelto, M., Holstege, H., Chételat, G., Teunissen, C. E., Cummings, J., & van der Flier, W. M. (2021). Alzheimer's disease. *Lancet*, *397*(10284), 1577–1590. PMID: 33667416

Sönmez, Ü., Sönmez, A., Erbil, G., Tekmen, I., & Baykara, B. (2007). Neuroprotective effects of resveratrol against traumatic brain injury in immature rats. *Neuroscience Letters*, *420*(2), 133–137. DOI: 10.1016/j.neulet.2007.04.070 PMID: 17531385

Stoessl, A. J., Lehericy, S., & Strafella, A. P. (2014). Imaging insights into basal ganglia function, Parkinson's disease, and dystonia. *Lancet*, *384*(9942), 532–544. DOI: 10.1016/S0140-6736(14)60041-6 PMID: 24954673

Stoker, T. B., & Barker, R. A. (2020). Recent developments in the treatment of Parkinson's Disease. [-Faculty.]. *F1000 Research*, *9*, F1000. PMID: 32789002

Sultzer, D. L., & Marder, S. R. (2017). Older Brains are Different: Brain–Behavior Studies and Their Clinical Utility. *The American Journal of Geriatric Psychiatry*, *25*(1), 11–12. DOI: 10.1016/j.jagp.2016.10.002 PMID: 28231870

Sun, X., Feng, S., Qin, B., Ye, J., Xie, L., Gui, J., & Sang, M. (2024). Integrated multi-omics profiling highlights the benefits of resveratrol hydroxypropyl-β-cyclodextrin inclusion complex for A53T transgenic mice through the microbiota-gut-brain axis. *Food & Function, 15*(3), 1489–1512. DOI: 10.1039/d3fo03667g PMID: 38227477

Sun, X. Y., Dong, Q. X., Zhu, J., Sun, X., Zhang, L. F., Qiu, M., & Liu, R. T. (2019). Resveratrol rescues tau-induced cognitive deficits and neuropathology in a mouse model of tauopathy. *Current Alzheimer Research, 16*(8), 710–722. DOI: 10.3390/ijms241411275 PMID: 31368873

Surmeier, D. J., Obeso, J. A., & Halliday, G. M. (2017). Selective neuronal vulnerability in Parkinson disease. *Nature Reviews. Neuroscience, 18*(2), 101–113. DOI: 10.1038/nrn.2016.178 PMID: 28104909

Tian, J., Zhao, Y., Wang, L., & Li, L. (2021). Role of TLR4/MyD88/NF-κB signaling in heart and liver-related complications in a rat model of type 2 diabetes mellitus. *The Journal of International Medical Research, 49*(3). Advance online publication. DOI: 10.1177/0300060521997590 PMID: 33787393

Torres-Pérez, M., Tellez-Ballesteros, R. I., Ortiz-López, L., Ichwan, M., Vega-Rivera, N. M., Castro-García, M., Gómez-Sánchez, A., Kempermann, G., & Ramirez-Rodriguez, G. B. (2015). Resveratrol enhances neuroplastic changes, including hippocampal neurogenesis, and memory in Balb/C mice at six months of age. *PLoS One, 10*(12), 1–21. DOI: 10.1371/journal.pone.0145687 PMID: 26695764

Tritsch, N. X., & Sabatini, B. L. (2012). Dopaminergic Modulation of Synaptic Transmission in Cortex and Striatum. *Neuron, 76*(1), 33–50. DOI: 10.1016/j.neuron.2012.09.023 PMID: 23040805

Vázquez-Vélez, G. E., & Zoghbi, H. Y. (2021). Parkinson's disease genetics and pathophysiology. *Annual Review of Neuroscience, 44*(1), 87–108. PMID: 34236893

Venturini, C. D., Merlo, S., Souto, A. A., Fernandes, M. D. C., Gomez, R., & Rhoden, C. R. (2010). Resveratrol and red wine function as antioxidants in the central nervous system without cellular proliferative effects during experimental diabetes. *Oxidative Medicine and Cellular Longevity, 3*(6), 434–441. DOI: 10.4161/oxim.3.6.14741 PMID: 21307644

Vingtdeux, V., Dreses-Werringloer, U., Zhao, H., Davies, P., & Marambaud, P. (2008). Therapeutic potential of resveratrol in Alzheimer's disease. *BMC Neuroscience, 9*(SUPPL. 2), 1–5. DOI: 10.1186/1471-2202-9-S2-S6 PMID: 19090994

Woodruff-pak, D. S. (2008). Animal Models of Alzheimer ' s Disease. *Therapeutic Implications., 15*, 507–521.

Wu, Y., Li, X., Zhu, J. X., Xie, W., Le, W., Fan, Z., Jankovic, J., & Pan, T. (2011). Resveratrol-activated AMPK/SIRT1/autophagy in cellular models of Parkinson's disease. *Neuro-Signals, 19*(3), 163–174. DOI: 10.1159/000328516 PMID: 21778691

Xu, Q. Q., Yang, W., Zhong, M., Lin, Z. X., Gray, N. E., & Xian, Y. F. (2023). Animal models of alzheimer's disease: Preclinical insights and challenges. *Acta Materia Medica, 2*(2), 192–215. DOI: 10.15212/AMM-2023-0001

Yang, X. Y., Li, Q. J., Zhang, W. C., Zheng, S. Q., Qu, Z. J., Xi, Y., & Wang, G. (2020). AMPK-SIRT1-PGC1α Signal Pathway Influences the Cognitive Function of Aged Rats in Sevoflurane-Induced Anesthesia. *Journal of Molecular Neuroscience, 70*(12), 2058–2067. DOI: 10.1007/s12031-020-01612-w PMID: 32514740

Yarnall, A., Rochester, L., & Burn, D. J. (2011). The interplay of cholinergic function, attention, and falls in Parkinson's disease. *Movement Disorders, 26*(14), 2496–2503. DOI: 10.1002/mds.23932 PMID: 21898597

Yiannopoulou, K. G., & Papageorgiou, S. G. (2020). Current and future treatments in Alzheimer disease: An update. *Journal of Central Nervous System Disease, 12*, 1179573520907397. PMID: 32165850

Zeini, S., Davoodian, N., Kazemi, H., Brojeni, M. S., Ghani, E., Firouzjaei, M. A., & Atashabparvar, A. (2024). Resveratrol prevents cognitive impairment and hippocampal inflammatory response induced by lipopolysaccharide in a mouse model of chronic neuroinflammation. *Physiology & Behavior, 278*(1), 114508. DOI: 10.1038/s41392-024-01911-3 PMID: 38460779

Zhang, F., Wang, Y. Y., Liu, H., Lu, Y. F., Wu, Q., Liu, J., & Shi, J. S. (2012). Resveratrol produces neurotrophic effects on cultured dopaminergic neurons through prompting astroglial BDNF and GDNF release. *Evidence-Based Complementary and Alternative Medicine : eCAM, 2012*. Advance online publication. DOI: 10.1155/2012/937605 PMID: 23304227

Zhang, L. F., Yu, X. L., Ji, M., Liu, S. Y., Wu, X. L., Wang, Y. J., & Liu, R. T. (2018). Resveratrol alleviates motor and cognitive deficits and neuropathology in the A53T α-synuclein mouse model of Parkinson's disease. *Food & Function, 9*(12), 6414–6426. DOI: 10.1039/C8FO00964C PMID: 30462117

Chapter 7
Effects of Saffron's (Crocus sativus) Bioactive Compounds on Parkinson's Disease:
Inhibition of Monoamine Oxidase

Hanane Moummou
https://orcid.org/0009-0002-9252-4292
Private University of Marrakesh, Morocco

Hanane Ait Hammou
Higher Institute of Nursing Professions and Health, Ministry of Health, Morocco & Chouaib Doukkali University, Morocco

Omar El Hiba
Chouaib Doukkali University, Morocco

Ilyass Britel
Private University of Marrakesh, Morocco

Oumnia Akhallaayoune
Cadi Ayyad University of Marrakech, Morocco

Mounia Akhallaa Youne
Cadi Ayyad University of Marrakech, Morocco

Mounia Achoch
Private University of Marrakesh, Morocco

Es-said Sabir
https://orcid.org/0000-0003-2246-5645
Private University of Marrakesh, Morocco

Iman Meftah
Sultan Moulay Sliman University, Beni Mellal, Morocco

Jamal El Karroumi
https://orcid.org/0009-0006-8574-2426
Private University of Marrakesh, Morocco

Hicham Chatoui
https://orcid.org/0000-0002-4707-1645
Higher Institute of Nursing Professions and Health, Ministry of Health, Morocco

Abdelilah El Abbassi
Cadi Ayyad University of Marrakech, Morocco

ABSTRACT

Parkinson's disease (PD) is often accompanied by neuropsychiatric symptoms, including mild depression, which significantly affect patients' quality of life. Saffron (Crocus sativus L.) has emerged

DOI: 10.4018/979-8-3373-1365-8.ch007

as a natural therapeutic option due to its bioactive compounds crocin, crocetin, and safranal which exhibit antidepressant effects by inhibiting monoamine oxidase, modulating neurotransmitter reuptake, and enhancing brain-derived neurotrophic factor (BDNF) signaling. Extracts containing these compounds have shown efficacy comparable to conventional antidepressants in alleviating depression-related symptoms in PD. This chapter discusses saffron's neuroprotective properties, its impact on neurotransmitter systems, and interactions with monoamine oxidase enzymes, highlighting its potential for depression management. The aim of this chapter is to provide a comprehensive overview of saffron's phytochemical composition, mechanisms of action, and its therapeutic potential in mitigating depression and related neuropsychiatric symptoms in PD.

1. INTRODUCTION

Depression is a universal mental disorder with a global impact, afflicting millions worldwide. Its effects extend beyond the individual, influencing those in their immediate circle. The disorder's etiology is multifaceted, encompassing various factors such as heredity (Shadrina et al., 2018), biology (Szymkowicz et al., 2023), and everyday life events (Lynall & McIntosh, 2023). Additionally, depression is a common symptom of neurological conditions, especially Parkinson's disease (PD) (Poewe, 2008; Chaudhuri & Schapira, 2009; El-Mansoury et al., 2023, 2025).

Currently, the incidence of depression associated with PD is increasing (Chaudhuri & Schapira, 2009 ; Cai et al., 2023), with treatment-resistant cases frequently observed despite conventional drug therapies (McIntyre et al., 2023). Moreover, standard antidepressant treatments are often associated with side effects and limited efficacy. Hence, there is a growing need to identify natural bioactive substances that may offer alternative or complementary solutions to this problem.

Emerging epidemiological evidence underscores depression as a prevalent and clinically significant neuropsychiatric comorbidity in PD, with prevalence estimates ranging from 35% to 85%, depending on demographic variables and diagnostic frameworks employed (Prange et al., 2022; *Neuropsychiatric Disease and Treatment*, 2025). Notably, over 20% of individuals with PD meet diagnostic criteria for major depressive disorder (MDD), representing a fourfold increase relative to the general population (Prange et al., 2022).

Regional disparities further accentuate the burden of affective disturbances in PD. For instance, in Saudi Arabia, recent data reveal an alarming depression prevalence of 84.8%, with 37% of cases classified as severe attributable primarily to psychosocial stressors such as social isolation and functional decline (*Neuropsychiatric Disease and Treatment*, 2025). Importantly, depressive symptomatology frequently emerges in the prodromal and early stages of PD. Within the first-year post-diagnosis, approximately 38% of patients exhibit clinically relevant depressive symptoms, a prevalence more than double that observed in matched non-PD cohorts (Prange et al., 2022). These findings suggest that mood disorders may constitute an integral component of PD pathophysiology rather than a mere reactive consequence.

Natural remedies have attracted attention for their potential in managing depression associated with PD, with saffron emerging as a promising candidate (Jackson et al., 2021). Saffron, derived from the *Crocus sativus* plant, has long been used in traditional medicine and culinary practices. It is rich in bioactive compounds, including crocin, crocetin, and safranal, which have been widely studied for their antidepressant and anxiolytic properties (Jackson et al., 2021; Samarghandian

& Farkhondeh, 2020). Understanding the pharmacological mechanisms of these compounds is essential for elucidating their therapeutic potential in the treatment of depression.

One of the first types of antidepressants developed were monoamine oxidase (MAO) inhibitors. MAO inhibitors are a class of compounds with significant therapeutic potential for various neuropsychiatric conditions, including depression, anxiety, and neurodegenerative disorders (Cesura, 2007). Recent studies have demonstrated the inhibitory effects of saffron bioactive compounds on MAO activity, suggesting a potential mechanism underlying its psychotropic effects (Siddiqui et al., 2022). Understanding the precise mechanism of MAO inhibition by saffron is essential for effectively harnessing its therapeutic benefits.

This chapter aims to provide an overview of the phytochemical composition and bioactive compounds of saffron. In addition, it explores the antidepressant effects of saffron compounds in the context of PD, their mechanisms of action, and their interactions with monoamine oxidase enzymes. By focusing on elucidating the mechanism by which saffron exerts its inhibitory effects on MAO, a crucial enzyme involved in the degradation of neurotransmitters such as serotonin, dopamine, and norepinephrine this chapter highlights the underlying biochemical processes and outlines avenues for future research and clinical applications. Through an in-depth analysis of recent studies, it examines the bioactive constituents of saffron responsible for MAO inhibition, explores their interactions with the enzyme, and discusses implications for potential therapeutic applications in neuropsychiatric disorders, especially in PD.

2. *CROCUS SATIVUS* PLANT

2.1. Phenetic Taxonomy

Before delving into the pharmacological relevance of saffron, it is important to contextualize the plant within the broader botanical framework. *Crocus sativus*, commonly known as saffron, belongs to the plant kingdom (Kingdom Plantae) and falls under the Magnoliophyta division (Angiosperms), which encompasses flowering plants bearing enclosed seeds (Saxena, 2010). Within the Angiosperms, *C. sativus* is classified under the class Liliopsida (monocots), characterized by parallel leaf venation and floral organs typically arranged in multiples of three. It further belongs to the order Asparagales, a diverse group of flowering plants, and the family Iridaceae (the iris family), known for showy flowers composed of six tepals (Saxena, 2010). Finally, *C. sativus* represents both the genus and species designation within the Iridaceae family. It is a perennial flowering plant of significant economic and cultural value, primarily as the natural source of the saffron spice, which is derived from its dried stigmas (Fujii et al., 2022).

2.2. *Crocus sativus* Flower Parts

In the botanical anatomy of the *Crocus sativus* flower (Figure 1), the reproductive structures include the stigma, perianth, stamen, and surrounding foliage. The stigma, located at the apex of the pistil, serves as the receptive surface for pollen grains during fertilization. Surrounding it, the perianth forms the floral envelope typically displaying a striking purple hue in *Crocus sativus* which performs both protective and attractive functions during pollination (Saxena, 2010). The stamen, composed of the filament and anther, constitutes the male reproductive organ responsible for producing and dispersing pollen grains. Finally, the green leaves, situated below the flower, play a crucial role in photosynthesis and provide structural support to the reproductive organs.

Figure 1. The flower of crocus sativus

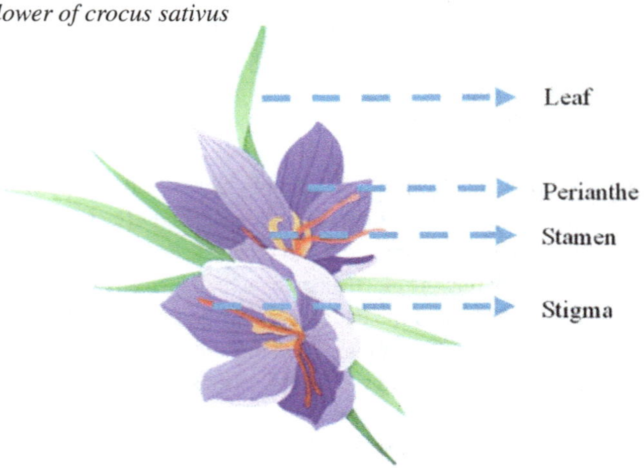

3. PHYTOCHEMICAL COMPOSITION OF SAFFRON

The main phytochemical groups found in saffron include:

Alkaloids. Saffron contains alkaloids such as safranal and picrocrocin, which contribute to its distinct flavor and aroma (Xing et al., 2021). Alkaloids often possess significant pharmacological activities and can influence various physiological processes.

Polyphenols. These include flavonoids such as quercetin, kaempferol, and catechins, as well as phenolic acids such as ferulic acid and gallic acid (Fujii et al., 2022). Polyphenols are known for their antioxidant properties and potential health benefits, including anti-inflammatory and neuroprotective effects (Spagnuolo et al., 2016).

Triterpenoids. Saffron contains several triterpenoid compounds, including β-sitosterol and lupeol (Xing et al., 2021). Triterpenoids have been studied for their diverse biological activities, including anti-inflammatory, anticancer, and antimicrobial properties.

Glycosides. Saffron includes glycosidic compounds such as crocetin glycosides and crocin glycosides (Xing et al., 2021). These compounds contribute to saffron's vibrant color and possess strong antioxidant properties.

Volatile compounds. In addition to safranal, saffron contains other volatile compounds, including terpenes such as pinene, cineole, and limonene (Cozzolino et al., 2022). These compounds are responsible for saffron's characteristic aroma and exhibit a range of biological activities.

Carbohydrates. Saffron contains simple carbohydrates such as glucose and fructose, which provide energy and contribute to its nutritive value.

Amino acids and proteins. The amino acids and proteins in saffron play essential roles in physiological processes and contribute to its nutritional properties.

Minerals and vitamins. Saffron contains minerals such as potassium, calcium, and magnesium (D'Archivio et al., 2014), as well as vitamins including vitamin C and vitamin A, which enhance its nutritional profile and potential health benefits.

These phytochemicals act synergistically to produce the various biological and health-promoting effects associated with saffron consumption (Xing et al., 2021). As a medicinal plant, saffron represents a promising natural alternative for the treatment of depression.

The efficacy of saffron has been demonstrated across multiple populations and conditions. For instance, in young adults (Moghadam et al., 2021) and individuals with diabetes, a dose of 300 mg of saffron per day for one week significantly reduced anxiety compared to placebo (Shahdadi et al., 2018). Another study showed that women with mild to moderate postpartum depression who received 20 mg of saffron daily for six weeks experienced a 50% reduction in depression symptoms, comparable to those treated with fluoxetine (40 mg/day) (Kashani et al., 2016). These health-promoting properties of saffron are attributed to its bioactive constituents. The red stigmas of *Crocus sativus* flowers are rich in bioactive compounds that act as natural antidepressants, such as crocin (Siddiqui et al., 2022), a crocetin derivative from the carotenoid group responsible for saffron's color picrocrocin (Pandey et al., 2020), a monoterpene aldehyde contributing to its bitter taste, and safranal (Samarghandian et al., 2017), the volatile compound responsible for its aroma.

4. BIOACTIVE COMPOUNDS

Crocin. Water-soluble carotenoid compounds responsible for the characteristic color of saffron. Crocin possesses antioxidant properties and has been associated with anti-inflammatory and neuroprotective effects (Lu et al., 2020).

Picrocrocin. This compound contributes to saffron's bitter taste. During the drying process or cooking, picrocrocin converts into safranal, which enhances saffron's aroma (Chrysanthou et al., 2016). Picrocrocin has demonstrated antioxidant and anticancer effects (Lambrianidou et al., 2020).

Safranal. A volatile compound that gives saffron its distinctive aroma and flavor. Formed from picrocrocin degradation during drying, safranal exhibits antioxidant, antidepressant, and neuroprotective activities (Yang et al., 2024).

Crocin-1. A specific type of crocin known for its antioxidant properties and potential protective role against oxidative stress and inflammation (Bastani et al., 2022).

Terpenes. Saffron contains terpenes such as pinene, cineole, and limonene, which contribute to its aroma and possess antimicrobial, anti-inflammatory, and antitumor properties (Środa-Pomianek et al., 2020).

Carotenoids. In addition to crocin, saffron contains other carotenoids, such as zeaxanthin and lycopene, which contribute to its color and antioxidant capacity. These compounds have been associated with protective effects against age-related macular degeneration and the promotion of eye health (Broadhead et al., 2024).

Among these, crocin, crocetin, and safranal have been extensively studied for their roles in reducing stress and depression particularly in PD. These compounds exhibit antidepressant effects by inhibiting MAO, modulating neurotransmitter reuptake, and enhancing brain-derived neurotrophic factor (BDNF) signaling (Cerdá-Bernad et al., 2022).

Studies have shown that saffron extracts containing these bioactive compounds improve depression-related behaviors, such as increased consumption of sweet solutions and enhanced escape responses (Siddiqui et al., 2022). Moreover, saffron and its constituents have shown efficacy comparable to that of conventional antidepressants (Kashani et al., 2016), emphasizing their potential as natural treatments for mild depression. Among these molecules, crocin stands out as a particularly promising candidate for alleviating depressive symptoms associated with PD (Orio et al., 2020).

5. ANTIDEPRESSANT EFFECT OF THE STIGMA COMPONENTS OF *C. SATIVUS*

5.1. Effect of Crocin on Depression

Crocin, one of the primary active compounds in saffron, has demonstrated antidepressant effects comparable to medications such as fluoxetine and imipramine in studies of mild depression associated with PD (Tao et al., 2023). This suggests saffron's potential as a natural therapeutic option. Although the precise mechanisms remain under investigation, they likely extend beyond monoamine oxidase inhibition (Chen et al., 2023). Further clinical trials are needed to confirm saffron's efficacy and safety in managing Parkinson's-related depression.

Stress inhibits pituitary adenylate cyclase-activating polypeptide (PACAP), a member of the vasoactive intestinal polypeptide family (Lutfy & Shankar, 2019). PACAP is distributed throughout several brainstem regions involved in sensory and motor control, as well as stress regulation (Vaudry et al., 2009). Reduced PACAP activity disrupts the phosphorylation of extracellular signal-regulated kinases (ERK) and cAMP response element-binding protein (CREB), leading to diminished synaptic plasticity and the onset of depressive behaviors.

As illustrated in Figure 2, based on a corticosterone-induced neuronal stress model (the rodent analog of human cortisol), crocin counteracts these effects by enhancing endogenous PACAP production and activating ERK and CREB pathways. This activation promotes neuronal survival and restores synaptic plasticity impaired by stress (Lu et al., 2020).

Figure 2. The effect of crocin on depressive mouse neuronal cells. Crocin enhances the expression of pituitary adenylate cyclase-activating polypeptide (PACAP), leading to activation of extracellular signal-regulated kinases (ERK) and cAMP response CREB, thereby promoting synaptic plasticity and neuronal survival under stress conditions. PACAP, pituitary adenylate cyclase-activating polypeptide; ERK, extracellular signal-regulated kinases; CREB, cAMP response element-binding protein; cAMP, cyclic adenosine monophosphate; ATP, adenosine triphosphate; AC, adenylyl cyclase

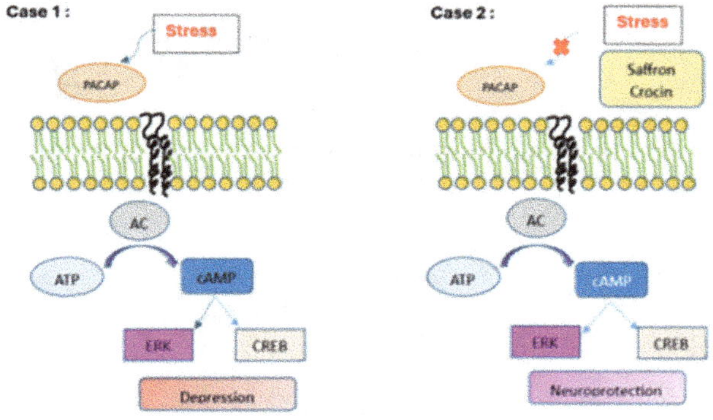

Stress inhibits the regulatory function of pituitary adenylate cyclase-activating polypeptide (PACAP), a member of the intestinal polypeptide family involved in stress modulation. The suppression of PACAP disrupts the phosphorylation of ERKs and CREB, thereby reducing synaptic plasticity and contributing to the development of depression. This mechanism is illustrated in the first part of Figure 2, based on a corticosterone-induced mouse neuronal cell model analogous to human cortisol activity. Conversely, the second part of the figure demonstrates the effect of crocin, which enhances endogenous PACAP production and activates ERK and CREB signaling pathways. This activation improves neuronal survival and reverses stress-induced impairments in synaptic plasticity.

5.2. Effect of Safranal on Synapses

Safranal, a monoterpene aldehyde with a molecular weight of 150.21 g/mol and a melting point below 25°C, exhibits significant neuroprotective potential by counteracting oxidative stress in neuronal tissues. In a study conducted by Cerdá-Bernad et al. (2022), stress was shown to markedly decrease the levels of glutathione (GSH) and the activities of endogenous antioxidant enzymes, including superoxide dismutase (SOD) and catalase (CAT), in the brains of mice (Figure 3). Concurrently, the concentration of malondialdehyde (MDA) a byproduct of lipid peroxidation, was significantly elevated, reflecting heightened oxidative damage.

Following treatment with safranal, MDA levels decreased substantially, while the activities of SOD, CAT, and GSH were restored toward normal levels. These results indicate that safranal mitigates oxidative stress in brain tissue by enhancing antioxidant defenses and preventing lipid peroxidation. Consequently, safranal protects neuronal cells from oxidative damage, supporting its neuroprotective and antidepressant effects (Samarghandian et al., 2017).

Figure 3. Activity of safranal in inhibiting antioxidant stress. MDA, Malondialdehyde; GSH, Glutathione; SOD, superoxide dismutase

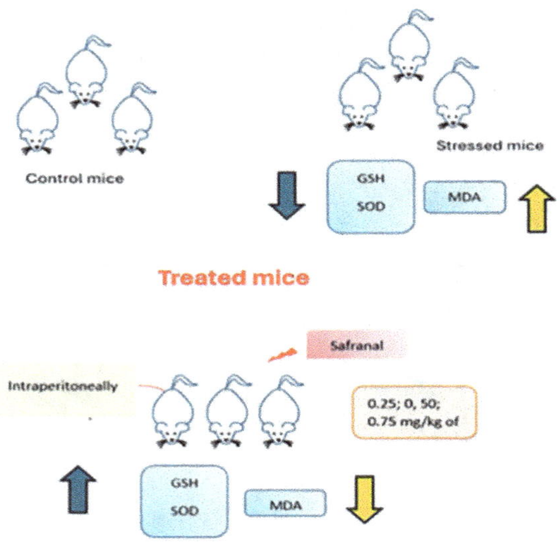

5.3. Crocetin and Its Antidepressant Effects

Crocetin, one of the major carotenoid constituents of saffron, possesses a wide spectrum of therapeutic properties, including neuroprotective and antidepressant effects, particularly relevant in the context of PD (Lin et al., 2021). The complex interplay between nutrition and mental health highlights the pivotal role of gut–brain interactions, with the gastrointestinal system often described as the *"second brain."* Communication between the gut and the central nervous system occurs through four primary pathways: neuronal, immune, endocrine, and circulatory routes.

Alterations in gut microbiota composition have been strongly correlated with depressive symptoms, suggesting that microbial imbalance can influence neurochemical signaling and emotional regulation. Lin et al. (2021) investigated the effects of depression on synaptic neurotransmitter secretion and assessed the antidepressant potential of crocetin in murine models.

Under normal physiological conditions, neurotransmitters mediate the transfer of information between neurons across synapses. However, in depressive states, this communication becomes dysregulated, leading to impaired synaptic transmission and functional neuronal alterations (Figure 4). Experimental administration of crocetin at graded doses (20, 40, and 80 mg/kg) significantly alleviated depressive-like behaviors in mice. This effect was accompanied by increased neurotransmitter levels associated with mood regulation and improved neuronal integrity in the hippocampus, a brain region critical for cognitive and emotional processing.

These findings suggest that crocetin exerts its antidepressant effects through modulation of neurotransmitter systems and preservation of hippocampal neuronal function, potentially mediated via gut–brain axis regulation.

Figure 4. Effect of crocetin in the transmission of nerve impulse

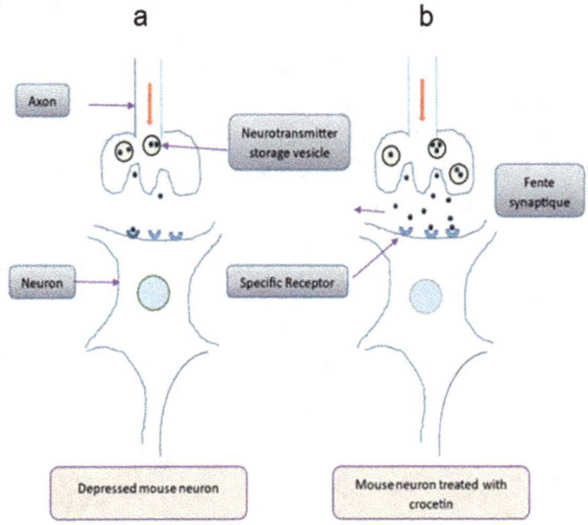

In mice experiencing depression, crocetin has been shown to significantly enhance nerve impulse transmission within the nervous system. In treated animals, crocetin increases neurotransmitter release at the synaptic cleft, the junction between neurons (Lin et al., 2021). It is thought to act on specific presynaptic receptors, facilitating the release of neurotransmitters from storage vesicles into the synaptic cleft. This process promotes efficient nerve impulse propagation along the neuronal axon, leading to improved neuronal communication. Additionally, crocetin may modulate synaptic plasticity by influencing signaling pathways involved in neuronal survival and synaptic function, thereby contributing to its neuroprotective effects.

5.4. Picrocrocin

Picrocrocin, another bioactive compound in saffron, contributes to its characteristic bitter taste (Chrysanthou et al., 2016). Emerging research suggests that it may also possess neuroprotective properties, although the precise mechanisms underlying its action remain unclear. While its role in taste perception is well established, studies have proposed additional functions in neuronal health. Further investigations are needed to elucidate picrocrocin's molecular mechanisms and its potential implications for neurological function and protection (Hamedani et al., 2024).

6. NEUROPROTECTIVE MECHANISM AND MAO INTERACTIONS

As described previously, saffron bioactive compounds particularly crocin, crocetin, and safranal have received substantial attention for their therapeutic potential in depression and anxiety

(Tao et al., 2023). Extensive research demonstrates their efficacy in alleviating mild depression associated with PD, often serving as an adjuvant therapy alongside conventional treatments.

These compounds, derived from *Crocus sativus*, not only exhibit antidepressant effects but also confer neuroprotective benefits (Figure 5). Their mechanisms involve modulation of synaptic signaling, reduction of oxidative stress, and interaction with MAO enzymes, which are crucial for the degradation of neurotransmitters such as serotonin, dopamine, and norepinephrine. By inhibiting MAO activity, these compounds help maintain neurotransmitter balance, supporting neuronal function and resilience in depressive states.

Figure 5. Neuroprotective effects of stigma bioactive compounds

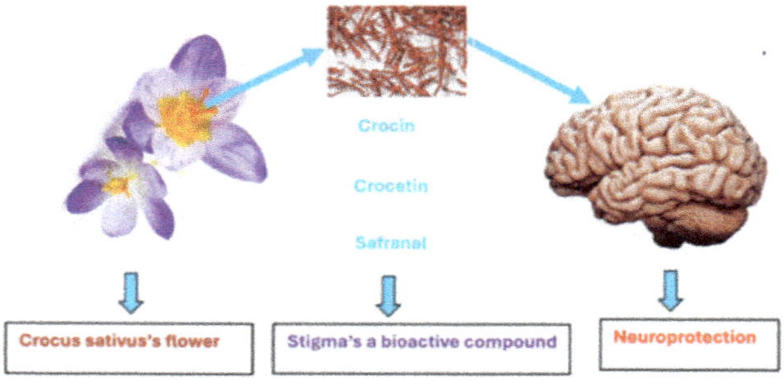

7. INTERACTIONS BETWEEN THE BIOACTIVE COMPOUNDS OF SAFFRON AND MAO

The bioactive compounds of saffron particularly crocin, crocetin, and safranal, interact with MAO, providing a multifaceted mechanism for their antidepressant efficacy. By inhibiting monoamine reuptake and modulating neurotransmitter levels, especially serotonin, dopamine, and norepinephrine, these compounds exert significant antidepressant effects. Their anti-inflammatory and antioxidant properties further contribute to neuroprotection, enhancing BDNF signaling and promoting neuronal growth and plasticity essential for emotional well-being.

Saffron compounds inhibit MAO activity through multiple mechanisms. Molecular docking studies indicate that interactions with MAO's active site involve hydrogen bonding, π–π stacking, and hydrophobic interactions, which determine binding affinity (Carradori et al., 2018). In vitro experiments confirm that saffron extracts directly modulate MAO enzyme kinetics (Krishnaswamy et al., 2020). These effects are particularly relevant for both MAO-A and MAO-B isoforms, which regulate the degradation of neurotransmitters associated with mood and emotional regulation.

MAO Inhibition and Neurotransmitter Modulation

MAO inhibition by saffron bioactive compounds results in elevated neurotransmitter levels in the synaptic cleft, contributing to improved mood and alleviation of depressive symptoms:

Serotonin: Increased synaptic serotonin enhances mood stabilization (Naoi et al., 2018).

Norepinephrine: Elevated norepinephrine levels improve alertness, attention, and energy (Noori et al., 2022).

Dopamine: Higher dopamine availability contributes to pleasure, motivation, and overall well-being (Monchaux De Oliveira et al., 2021).

While classical MAO inhibitors (MAOIs) have been used as antidepressants, their use is limited due to potential dietary and drug interactions (Shafiee et al., 2018). Saffron offers a natural alternative that may partially inhibit MAO activity, increase neurotransmitter availability and promoting mood regulation while potentially minimizing adverse effects. Nevertheless, clinical validation and professional guidance are essential before its use for depression management.

The likely mechanism involves reversible binding of saffron's bioactive components to MAO's catalytic site, preventing the oxidation of monoaminergic neurotransmitters. This neurochemical modulation underlies the observed antidepressant and anxiolytic effects in preclinical and clinical studies (Jackson et al., 2021; Xing et al., 2021; Kashani et al., 2016).

8. PATENT SURVEY

The recent increase in patents related to saffron highlights its therapeutic potential for mental health conditions, particularly depression and mood disorders associated with PD. Numerous patents focus on saffron extracts and bioactive compounds in formulations designed to prevent and treat depression (Table 1). These range from herbal compositions to high-dose crocin preparations, demonstrating the versatility of saffron in mental health therapeutics.

Table 1. The main patents about Crocus sativa use to treat or prevent different brain health disorders

Patent ID	Title	Publication date
JP-H1180009-A	Agent for improving brain function and agent for preventing lowering of brain function	**1999-03-23**
US-2004180104-A1	Herbal composition for preventing and/or treating conditions in associated with depression	**2004-09-16**
DE-102007040798-A1	Herb mixture, useful e.g. to slow the aging process, comprises e.g. *Crocus sativus, Elettaria cardamomum, Cinnamomum zeylanicum, Hedychium spicatum, Terminalia chebula, Trichosanthes dioica, Coleus forskolin* and *Berberis aristata*	**2009-04-02**
CA-2667397-C	Oleanane triterpene saponin compounds which are effective on treatment of dementia and mild cognitive impairment (mci), and improvement of cognitive function	**2013-01-22**
FR-3004110-A1	Composition for combating stress	**2014-10-10**
CN-104623670-A	Compositions containing enriched natural crocin and/or crocetin, and their therapeutic or nutraceutical uses	**2015-05-20**

continued on following page

Table 1. Continued

Patent ID	Title	Publication date
WO-2015124318-A1	Edible composition including safranal, crocin, picrocrocin and a vitamin b complex for treating the initial phase of depression	**2015-08-27**
US-9211298-B2	Compositions containing enriched natural crocin and/or crocetin, and their therapeutic or nutraceutical Uses	**2015-12-15**
CN-105194277-A	Composition for treating sympathetic nervous cervical spondylosis and preparation method of composition	**2015-12-30**
WO-2019185439-A1	Natural base composition for use in the treatment of chronical stress characterized by mood and/or anxiety disorders	**2019-10-03**
WO-2020089831-A1	Phyto-therapeutic compositions useful to improve the quality of sleep and in the treatment of insomnia, anxiety and depression	**2020-05-07**
CN-111109483-A	Beverage for relieving fatigue, promoting blood circulation and removing blood stasis	**2020-05-08**
CN-112869162-A	Composition with functions of resisting fatigue and improving sleep and preparation method thereof	**2021-06-01**
GR-1010023-B	Stable orally administrated composition containing probiotics, crocus sativus and magnesium to treat mild anxiety and depression	**2021-06-01**
CN-113349381-A	Food composition capable of relieving depression and preparation method thereof	**2021-09-07**
CN-113509438-A	Saffron medicinal composition emulsion for treating insomnia and preparation and application thereof	**2021-10-19**
CN-114146042-A	Saffron crocus fermentation product and preparation method and application thereof	**2022-03-08**
EP-3967320-A1	Food supplement for improving the mood of an individual	**2022-03-16**
BE-1028612-B1	**Composition comprising a saffron extract, a casein peptide and another plant compound for the treatment of sleep disorders, mood disorders or anxiety**	**2022-04-06**

In addition to mood disorders, patents have explored saffron's potential in cognitive health, including interventions for dementia and mild cognitive impairment (MCI). Certain saffron-derived oleanane triterpene saponins have shown neuroprotective efficacy, suggesting a dual role in addressing depression and cognitive decline, conditions that frequently coexist in neurodegenerative disorders.

Beyond depression and cognitive disorders, saffron-based formulations target fatigue, sleep disturbances, and general mental well-being, reflecting its broader applicability in mental health. Innovative combinations with probiotics and other bioactive compounds highlight potential synergistic effects on gut-brain axis modulation, a rapidly emerging area in neuropsychiatric therapeutics.

This expanding patent landscape underscores the multifaceted contributions of saffron to mental well-being, neuroprotection, and the holistic management of neurological and psychiatric conditions.

The growing body of patents related to saffron highlights its therapeutic promise, particularly its neuroprotective effects in the context of depression and mood disorders. The diverse formulations and applications reflected in these patents underscore saffron's potential to address multifaceted challenges in mental health and neurodegenerative disorders. Continued research is essential to elucidate the mechanisms of action and clinical efficacy of saffron, which may ultimately unlock its full potential in promoting brain health and well-being.

Although classical MAOIs have demonstrated efficacy in managing depression, their use is limited by potential side effects and interactions with certain foods and medications. Consequently, newer antidepressants such as selective serotonin reuptake inhibitors (SSRIs) and serotonin-norepinephrine reuptake inhibitors (SNRIs) are often preferred due to their improved safety profiles and ease of use. In this context, natural compounds, including those found in saffron, have garnered attention for their moderate MAO-inhibitory activity. Saffron is enriched with bioactive constituents such as crocin, safranal, crocetin, and picrocrocin, which can reversibly bind to the catalytic site of MAO enzymes, reducing the oxidation of monoaminergic neurotransmitters and increasing synaptic levels of serotonin, dopamine, and norepinephrine. This neurochemical modulation underlies the antidepressant and anxiolytic effects observed in preclinical and clinical studies, providing a compelling rationale for saffron's integration into depression management strategies. However, it is important to note that saffron's interaction with MAO may not be as complete or potent as that of classical MAOIs, and other molecular pathways may contribute to its effects on mood regulation.

Current Limitations and Clinical Perspectives

Despite promising findings, several methodological limitations restrict the clinical translation of saffron in managing depression among individuals with PD. A key limitation is the lack of large-scale, rigorously designed randomized controlled trials (RCTs), which limits the generalizability of existing data (Bej et al., 2024; Bahrami et al., 2023). Additionally, variability in extract formulations and processing techniques, including the concentration, bioavailability, and stability of key bioactive compounds such as crocin, picrocrocin, and safranal, poses challenges for establishing standardized and reproducible dosing protocols (Bej et al., 2024).

Potential pharmacokinetic and pharmacodynamic interactions with antiparkinsonian medications also require careful evaluation, given the polypharmacy common in PD management. Saffron-based supplements could influence therapeutic efficacy or exacerbate adverse effects.

Despite these challenges, the available evidence suggests that saffron possesses neuroprotective and antidepressant properties. If future clinical trials confirm its safety and efficacy, and if standardization of saffron extracts is achieved alongside thorough drug interaction profiling, saffron could emerge as a valuable adjunctive component in personalized therapeutic strategies for individuals with PD.

9. CONCLUSION

Saffron, a natural source of bioactive compounds such as crocin, crocetin, and safranal, shows promise as an antidepressant and anxiolytic agent, particularly in PD. Its therapeutic effects involve MAO inhibition, modulation of neurotransmitters, and enhancement of neurotrophic signaling, contributing to improved mood and neuronal health. Saffron extracts have demonstrated efficacy comparable to conventional antidepressants for mild depression. Further research, including mechanistic studies and clinical trials, is needed to confirm its safety and effectiveness. Overall, saffron represents a promising natural approach for the development of novel treatments for depression and related neuropsychiatric conditions.

REFERENCES

Bahrami, F., Jafari Pour, F., Hassanpour, M., Saki, M., Ebrahimzadeh, F., & Jafaripour, L. (2023). The effect of saffron and corrective exercises on depression and quality of life in women with multiple sclerosis: A randomized controlled clinical trial. *Multiple Sclerosis and Related Disorders*, *79*, 105008.

Bastani, S., Vahedian, V., Rashidi, M., Nazemi, M., Kesharvarzi, S., Kashani, K., & Zarenezhad, E. (2022). An evaluation on potential anti-oxidant and anti-inflammatory effects of Crocin. *Biomedicine and Pharmacotherapy*, *153*, 113297. DOI: 10.1016/j.biopha.2022.113297

Bej, E., Volpe, A. R., Cesare, P., Cimini, A., d'Angelo, M., & Castelli, V. (2024). Therapeutic potential of saffron in brain disorders: From bench to bedside. *Phytotherapy Research*, *38*(4), e8169.

Broadhead, G. K., Grigg, J., McCluskey, P. J., Chang, A. A., Nong, J., Boon, C. J. F., & Kearns, L. S. (2024). Saffron therapy for the ongoing treatment of age-related macular degeneration. *BMJ Open Ophthalmology*, *9*(1), e001399. DOI: 10.1136/bmjophth-2023-001399

Cai, H., Jin, Y., Liu, R., Zhang, Q., Su, Z., Ungvari, G. S., Ng, C. H., Zhang, Y., Li, X. H., & Xiang, Y. T. (2023). Global prevalence of depression in older adults: A systematic review and meta-analysis of epidemiological surveys. *Asian Journal of Psychiatry*, *80*, 103417. DOI: 10.1016/j.ajp.2022.103417

Carradori, S., Secci, D., & Petzer, J. P. (2018). MAO inhibitors and their wider applications: A patent review. *Expert Opinion on Therapeutic Patents*, *28*(3), 211–226. DOI: 10.1080/13543776.2018.1427735

Cerdá-Bernad, D., Valero-Cases, E., Pastor, J. J., & Frutos, M. J. (2022). Saffron bioactives crocin, crocetin and safranal: Effect on oxidative stress and mechanisms of action. *Critical Reviews in Food Science and Nutrition*, *62*(12), 3232–3249. DOI: 10.1080/10408398.2020.1864279

Cesura, A. M. (2007). Monoamine oxidases. In *xPharm: The Comprehensive Pharmacology Reference* (pp. 1–5). Elsevier.

Chaudhuri, K. R., & Schapira, A. H. (2009). Non-motor symptoms of Parkinson's disease: Dopaminergic pathophysiology and treatment. *Lancet Neurology*, *8*(5), 464–474.

Chen, Z., Gu, J., Lin, S., Wang, S., Chen, S., Chen, W., & Kang, D. (2023). Saffron essential oil ameliorates CUMS-induced depression-like behavior in mice via the MAPK-CREB1-BDNF signaling pathway. *Journal of Ethnopharmacology*, *300*, 115719. DOI: 10.1016/j.jep.2022.115719

Chrysanthou, A., Pouliou, E., Kyriakoudi, A., & Tsimidou, M. Z. (2016). Sensory threshold studies of picrocrocin, the major bitter compound of saffron. *Journal of Food Science*, *81*(2), C240–C249. DOI: 10.1111/1750-3841.13152

Cozzolino, R., Stocchero, M., Perestrelo, R., & Câmara, J. S. (2022). Comprehensive evaluation of the volatomic fingerprint of saffron from Campania towards its authenticity and quality. *Foods*, *11*(3), 366. DOI: 10.3390/foods11030366

D'Archivio, A. A., Giannitto, A., Incani, A., & Nisi, S. (2014). Analysis of the mineral composition of Italian saffron by ICP-MS and classification of geographical origin. *Food Chemistry, 157*, 485–489. DOI: 10.1016/j.foodchem.2014.02.068

El-Mansoury, B., El Khiat, A., Skaou, A., El Hiba, O., Jayakumar, A. R., Ait Hamdan, Y., & Ferssiwi, A. (2025). Non-motor symptoms of Parkinson's disease. In *Essential Guide to Neurodegenerative Disorders* (pp. 269–280). Academic Press.

El-Mansoury, B., Maloui, A. B., Draoui, A., Ait Hamdan, Y., Oukhrib, M., Smimih, K., & Jayakumar, A. R. (2023). Non-motor symptoms in Parkinson's disease: The other side of the disease. In *Experimental and clinical evidence of the neuropathology of Parkinson's disease* (pp. 25–45). IGI Global.

Fujii, S., Morita, Y., Ohta, T., Yamazaki, T., Saburi, H., Yonekura, M., & Takamura, H. (2022). Saffron (*Crocus sativus* L.) as a valuable spice and food product: A narrative review. *Longhua Chinese Medicine*, *5*, 18. DOI: 10.21037/lcm-22-1

Hamedani, S. G., Pourmasoumi, M., Zarifi, S. H., Najafgholizadeh, A., Ghavami, A., Clark, C. C. T., & Aryaeian, N. (2024). Therapeutic effects of saffron and its components on neurodegenerative diseases. *Heliyon*, *10*(2), e24334. DOI: 10.1016/j.heliyon.2024.e24334

Jackson, P. A., Forster, J., Khan, J., Pouchieu, C., Dubreuil, S., Gaudout, D., Moras, B., Pourtau, L., Joffre, F., Vaysse, C., Bertrand, K., Abrous, H., Vauzour, D., Brossaud, J., Corcuff, J. B., Capuron, L., & Kennedy, D. O. (2021). Effects of saffron extract supplementation on mood, well-being, and response to a psychosocial stressor in healthy adults: A randomized, double-blind, parallel group, clinical trial. *Frontiers in Nutrition*, *7*, 606124. DOI: 10.3389/fnut.2020.606124

Kashani, L., Eslatmanesh, S., Saedi, N., Niroomand, N., Ebrahimi, M., Hosseinian, M., Foroughifar, T., & Salimi, S. (2016). Comparison of saffron versus fluoxetine in treatment of mild to moderate postpartum depression: A double-blind, randomized clinical trial. *Pharmacopsychiatry*, *50*(2), 64–68. DOI: 10.1055/s-0042-115306

Krishnaswamy, V. K. D., Alugoju, P., & Periyasamy, L. (2020). Multifaceted targeting of neurodegeneration with bioactive molecules of saffron (*Crocus sativus*): An in silico evidence-based hypothesis. *Medical Hypotheses*, *143*, 109872. DOI: 10.1016/j.mehy.2020.109872

Lambrianidou, A., Koutsougianni, F., Papapostolou, I., & Dimas, K. (2020). Recent advances on the anticancer properties of saffron (*Crocus sativus* L.) and its major constituents. *Molecules (Basel, Switzerland)*, *26*(1), 86. DOI: 10.3390/molecules26010086

Lin, S., Li, Q., Jiang, S., Xu, W., Zhang, N., Liu, Y., Wang, K., Wang, C., Jia, X., & Kang, D. (2021). Crocetin ameliorates chronic restraint stress-induced depression-like behaviors in mice by regulating MEK/ERK pathways and gut microbiota. *Journal of Ethnopharmacology*, *268*, 113608. DOI: 10.1016/j.jep.2020.113608

Lu, L., Wu, D., Wang, K., Jiang, S., Tan, X., Kang, D., & Liu, Y. (2020). Beneficial effects of crocin against depression via pituitary adenylate cyclase-activating polypeptide. *BioMed Research International*, *2020*, 1–10. DOI: 10.1155/2020/3903125

Lutfy, K., & Shankar, G. (2019). Emerging evidence for the role of pituitary adenylate cyclase-activating peptide in neuropsychiatric disorders. In *Current Topics in Neurotoxicity* (pp. 143–157). Springer.

Lynall, M. E., & McIntosh, A. M. (2023). The heterogeneity of depression. *The American Journal of Psychiatry, 180*(10), 703–704. DOI: 10.1176/appi.ajp.20230574

McIntyre, R. S., Alsuwaidan, M., Baune, B. T., Berk, M., Demyttenaere, K., Goldberg, J. F., Gorwood, P., Ho, R., Kasper, S., Kennedy, S. H., Ly-Uson, J., Mansur, R. B., McAllister-Williams, R. H., Murrough, J. W., Nemeroff, C. B., Nierenberg, A. A., Rosenblat, J. D., Sanacora, G., Schatzberg, A. F., & Maj, M. (2023). Treatment-resistant depression: Definition, prevalence, detection, management, and investigational interventions. *World Psychiatry; Official Journal of the World Psychiatric Association (WPA), 22*(3), 394–412. DOI: 10.1002/wps.21120

Moghadam, B. H., Bagheri, R., Roozbeh, B., Ashtary-Larky, D., Gaeini, A. A., Dutheil, F., & Wong, A. (2021). Impact of saffron (*Crocus sativus* Linn) supplementation and resistance training on markers implicated in depression and happiness levels in untrained young males. *Physiology & Behavior, 233*, 113352. DOI: 10.1016/j.physbeh.2021.113352

Monchaux De Oliveira, C., Pourtau, L., Vancassel, S., Kecidi, S., Gaudout, D., Gallet, S., Pouchieu, C., Capuron, L., & Castanon, N. (2021). Saffron extract-induced improvement of depressive-like behavior in mice is associated with modulation of monoaminergic neurotransmission. *Nutrients, 13*(3), 904. DOI: 10.3390/nu13030904

Naoi, M., Maruyama, W., & Shamoto-Nagai, M. (2018). Type A monoamine oxidase and serotonin are coordinately involved in depressive disorders: From neurotransmitter imbalance to impaired neurogenesis. *Journal of Neural Transmission (Vienna, Austria), 125*(1), 53–66. DOI: 10.1007/s00702-017-1709-8

Noori, T., Sureda, A., Sobarzo-Sánchez, E., & Shirooie, S. (2022). The role of natural products in treatment of depressive disorder. *Current Neuropharmacology, 20*(5), 929–949. DOI: 10.2174/1570159X20666220103140834

Orio, L., Alen, F., Ballesta, A., Tabernero, A., & Correa, M. (2020). Antianhedonic and antidepressant effects of Affron®, a standardized saffron (*Crocus sativus* L.) extract. *Molecules (Basel, Switzerland), 25*(14), 3207. DOI: 10.3390/molecules25143207

Pandey, D. K., Nandy, S., Mukherjee, A., & Dey, A. (2020). Advances in bioactive compounds from *Crocus sativus* (saffron): Structure, bioactivity and biotechnology. In *Bioactive Natural Products* (pp. 273–304). Springer.

Prange, S., Thobois, S., & Broussolle, E. (2022). Depression in patients with Parkinson's disease: Current understanding of its neurobiology and implications for treatment. *Drugs & Aging, 39*(6), 417–439. DOI: 10.1007/s40266-022-00942-1

Samarghandian, S., & Farkhondeh, T. (2020). Saffron and neurological disorders. In *Saffron: The Age-Old Panacea in a New Light* (pp. 103–116). Academic Press., DOI: 10.1016/B978-0-12-818462-2.00009-7

Samarghandian, S., Samini, F., Azimi-Nezhad, M., & Farkhondeh, T. (2017). Anti-oxidative effects of safranal on immobilization-induced oxidative damage in rat brain. *Neuroscience Letters*, *659*, 26–32. DOI: 10.1016/j.neulet.2017.08.065

Saxena, R. (2010). Botany, taxonomy and cytology of *Crocus sativus* series. [An International Quarterly Journal of Research in Ayurveda]. *Ayu*, *31*(3), 374. DOI: 10.4103/0974-8520.77153

Shadrina, M., Bondarenko, E. A., & Slominsky, P. A. (2018). Genetic factors in major depression disease. *Frontiers in Psychiatry*, *9*, 334. DOI: 10.3389/fpsyt.2018.00334

Shafiee, M., Arekhi, S., Omranzadeh, A., & Sahebkar, A. (2018). Saffron in the treatment of depression, anxiety and other mental disorders: Current evidence and potential mechanisms of action. *Journal of Affective Disorders*, *227*, 330–337. DOI: 10.1016/j.jad.2017.11.020

Shahdadi, H., Balouchi, A., & Dehghanmehr, S. (2018). Effect of saffron oral capsule on anxiety and quality of sleep of diabetic patients in a tertiary healthcare facility in southeastern Iran: A quasi-experimental study. *Tropical Journal of Pharmaceutical Research*, *16*(11), 2749. DOI: 10.4314/tjpr.v16i11.24

Siddiqui, S. A., Ali Redha, A., Snoeck, E. R., Singh, S., Simal-Gandara, J., Ibrahim, S. A., & Jambrak, A. R. (2022). Anti-depressant properties of crocin molecules in saffron. *Molecules (Basel, Switzerland)*, *27*(7), 2076. DOI: 10.3390/molecules27072076

Spagnuolo, C., Napolitano, M., Tedesco, I., Moccia, S., Milito, A., & Russo, G. L. (2016). Neuroprotective role of natural polyphenols. *Current Topics in Medicinal Chemistry*, *16*(17), 1943–1950. DOI: 10.2174/1568026616666160204122449

Chapter 8
Cannabinoids and Multiple Sclerosis:
Insights Into Therapeutics, Mechanisms, and Patient Outcomes

Zineb El Gabbas
https://orcid.org/0000-0002-3024-604X
Higher Institute of Nursing and Technical Health Occupations, Casablanca, Morocco

Sara Oufquir
https://orcid.org/0000-0001-5014-7025
Faculty of Sciences, University Cadi Ayyad, Morocco

Fatimzahra Agouram
https://orcid.org/0000-0002-8073-2126
Faculty of Sciences, University Cadi Ayyad, Morocco

Mehdi Ait Laaradia
https://orcid.org/0000-0003-2390-2659
Higher Institute of Nursing and Technical Health Occupations Béni Mellal, Morocco

Jawad Laadraoui
https://orcid.org/0000-0002-9784-1265
Faculty of Sciences Aïn Chock, University of Hassan II, Morocco

ABSTRACT

Multiple sclerosis (MS) is a chronic neurodegenerative disorder characterized by inflammation, demyelination, and progressive neurological impairment. This chapter explores the therapeutic potential of cannabinoid-based drugs in MS, focusing on their mechanisms of action, including anti-inflammatory, analgesic, and neuromodulatory effects. It provides an overview of current MS treatments informed by recent clinical studies, while emphasizing key challenges such as optimal dosing, toxicity, and limited bioavailability, which restrict effective symptom management. In parallel, the chapter examines advances in nanomedicine, particularly nano-cannabinoids, as innovative strategies to enhance drug delivery, improve safety profiles, and increase therapeutic

DOI: 10.4018/979-8-3373-1365-8.ch008

efficacy. By integrating evidence from pharmacology, clinical trials, and nanotechnology, this chapter highlights both the opportunities and challenges of cannabinoids and nanocannabis approaches, offering a comprehensive perspective on their potential in MS management.

I. INTRODUCTION

Multiple sclerosis (MS) is neurodegenerative disease and an autoimmune disorder that affects the central nervous system (CNS) and leads to a progressive functional decline. The disease manifests itself through a wide range of neurological symptoms, vary widely from patient to patient and include visual disturbances, movement disorder (Ataxia), numbness and tingling, neuropathic pain, urinary and intestinal dysfunction, and cognitive impairment like: depression, anxiety. Multiple sclerosis progresses in different ways: relapsing-remitting, primary progressive and secondary progressive (El-Mansoury et al., 2023; Albelo-Martínez & Rizvi, 2025).

Although various therapeutic strategies and pharmacological interventions exist to modulate the immune response and alleviate symptoms in multiple sclerosis (MS), they remain insufficient in promoting effective remyelination. The progressive nature of the disease, characterized by continuous neuropathological damage and recurrent relapses, leads to worsening disability over time. This ongoing deterioration not only compromises neurological function but also contributes to significant patient frustration and dissatisfaction due to the limited efficacy of current treatments in halting disease progression. (Longoria *et al.*,2022)

The use of cannabinoids as a symptomatic treatment is increasingly recognized, with numerous clinical trials and patient reports suggesting their potential in managing and alleviating the symptoms of multiple sclerosis (Jones *et al.*,2020). However, despite substantial evidence of their therapeutic benefits, uncertainties remain as to their precise effects, largely due to variations in patient response. There is growing interest in improving the efficacy of cannabinoid-based drugs in the treatment of multiple sclerosis is directing current scientific research towards nanomedicine and the development of nano-cannabinoids as a better solution for the management of this disease.

In this chapter, we will explore and discuss: Multiple Sclerosis (MS) and its impact on patients, the role of the Endocannabinoid System and its interaction with MS pathology.

On the other hand, the chapter will also evaluate clinical studies on the use of cannabinoids in MS treatment, as well as the challenges posed by side effects and variability in patient response.

Additionally, the chapter will examine advancements in cannabinoid therapy, notably the emerging use of nanoparticles derived from cannabis, and finally, the legal status of medical cannabis across different regions of the world.

II. THE ENDOCANNABINOID SYSTEM AND MS

Multiple sclerosis is a chronic degenerative disease in which inflammation causes demyelination of axons in the central nervous system (CNS). Current treatments for MS focus on preventing relapses at an early stage, mainly by stopping immune attacks using drugs that inhibit the release of inflammatory mediators or prevent the recruitment of leukocytes into the CNS. However, there are still virtually no drugs capable of preventing or slowing the progression of neurodegenerative symptoms. Given that preclinical data show that cannabinoids possess anti-inflammatory properties and neuroprotective effects, these substances could limit progressive neurodegeneration in MS (Garcia-Gutiérrez *et al* 2025).

1- Structure of the Endocannabinoid System

The endocannabinoid system (ECS) comprises three primary components: endocannabinoids, receptors, and enzymes.

a- ***Endocannabinoids***: These endocannabinoids are, in essence, lipid-based neurotransmitters, synthesized and released by cells in accordance with specific physiological necessities. The most widely recognized endocannabinoids are anandamide (AEA) and 2-arachidonoylglycerol (2-AG), as demonstrated, these are synthesized from lipid precursors. The responsibility of these molecules lies in activating cannabinoid receptors, with the result that various biological effects are triggered (Pete and Narouze, 2021).

b- ***Cannabinoid Receptor***: Two primary types of cannabinoid receptors within the endocannabinoid system (ECS) have been identified: CB1 and CB2. CB1 receptors, the focus of this discussion, are predominantly located within the central nervous system (CNS), encompassing the brain and spinal cord. These CB1 receptors play a pivotal role in various physiological processes, including pain modulation, appetite regulation, and mood (O'Brien et al., 2021). CB2 receptors, on the other hand, are predominantly located in peripheral tissues, particularly those involved in the immune system. These CB2 receptors serve the crucial function of modulating immune cell activity and inflammation within these contexts (Cabral *et al.*, 2008). It is important to note that both CB1 and CB2 receptors are classified as G-protein coupled receptors (GPCRs).

c- ***Enzymes***: Enzymes have been demonstrated to play a crucial role in the synthesis and degradation of endocannabinoids (Rezende *et al.*, 2023). The two main enzymes involved in this process are Fatty acid amide hydrolase (FAAH), responsible for the breakdown of anandamide, one of the primary endocannabinoids, and monoacylglycerol lipase (MAGL), responsible for the degradation of 2-AG, the other major endocannabinoid (Matheson *et al.*, 2021). The precise regulation of these enzymes ensures that the duration and intensity of endocannabinoid signaling is meticulously calibrated and occurs exclusively when required (Ciuffreda *et al.*, 2024).

2- Function of the Endocannabinoid System

The ECS plays a pivotal role in numerous physiological processes, which can be categorised into the following key areas:

Homeostasis: The primary function of the endocannabinoid system (ECS) is to maintain homeostasis, defined as the balance of physiological processes within the body (Hill, 2024). The ECS plays a regulatory role in various vital functions including body temperature, blood pressure, and glucose metabolism (Lowe *et al.*, 2021). Neuroprotection and Cognitive Function: The endocannabinoid system (ECS) has been demonstrated to play a pivotal role in brain function and neuroprotection. The presence of cannabinoid receptor 1 (CB1) in various regions of the brain suggests a significant influence on cognitive processes, including memory, learning, and mood regulation. Furthermore, the ECS has been shown to possess neuroprotective properties, contributing to the preservation of neurons from damage induced by stress, injury, or degenerative diseases (Vasincu *et al.*, 2022).

Immune System Modulation: CB2 receptors are predominantly located within immune cells, where they facilitate the regulation of immune responses (Rahaman and Ganguly, (2021). The endocannabinoid system (ECS) has the capacity to influence inflammation, immune cell proliferation, and cytokine release, thereby serving as a significant modulator of immune function (Braile *et al.*, 2021).

Metabolic Regulation: The endocannabinoid system (ECS) has been demonstrated to play a role in a number of physiological processes, including appetite control, energy storage and fat metabolism (Rakotoarivelo *et al.*, 2021). The ECS exerts its influence over these processes by modulating the hypothalamus and other brain regions, thus regulating food intake and energy balance (Berland *et al.*, 2022). The implications of this for obesity and metabolic disorders are significant (Miralpeix *et al.*, 2021).

Pain and Inflammation Modulation: The endocannabinoid system (ECS) plays an essential role in the perception of pain and the process of inflammation. By binding to the CB1 and CB2 receptors, endocannabinoids have the capacity to reduce pain sensitivity and limit inflammation (Quintero *et al.*, 2024). This makes the ECS a target for therapeutic interventions in the management of pain and inflammatory diseases (Rakotoarivelo *et al.*, 2024).

3- Role of CB1 and CB2 Receptors in Immune Modulation and Neuroprotection

a- CB1 and CB2 Receptors in Immune Modulation

The endocannabinoid system (ECS) comprises cannabinoid receptors (CB1 and CB2), endogenous cannabinoids (endocannabinoids), and enzymes responsible for their synthesis and degradation (Balezina *et al.*, 2024). The CB1 and CB2 receptors fulfill pivotal functions in the modulation of immune responses, yet their actions exhibit marked disparities in terms of their location, function, and consequences (Aghazadeh Tabrizi *et al.*, 2016).

The prevalence of CB1 receptors within the central nervous system (CNS) is well documented, with their presence also noted in various peripheral tissues. These immune cells include macrophages, T cells and dendritic cells (Kaplan, 2013). Activation of CB1 receptors in these immune cells has the capacity to influence the balance between pro-inflammatory and anti-inflammatory responses, as evidenced by the findings of several research studies (Kopustinskiene *et al.*, 2022). The pro-inflammatory effects of CB1 activation in certain immune cells, including T cells, have been well documented. This process is of particular significance in the context of autoimmune diseases and inflammatory disorders, where excessive inflammation can lead to tissue damage (Argenziano *et al.*, 2019). The regulation of cytokine release: The activation of the CB1 receptor has been demonstrated to exert influence over the production of cytokines, which are signalling molecules that regulate immune responses. In certain instances, CB1 activation has been observed to result in a decrease in the release of pro-inflammatory cytokines, such as TNF-α and IL-6. Conversely, in other contexts, CB1 activation may enhance the production of specific cytokines (Jean-Gilles *et al.*, 2015; Henshaw *et al.*, 2021).

The expression of CB2 receptors is primarily observed in peripheral tissues, with a high concentration noted in immune cells, including B cells, macrophages, neutrophils, and microglia. In comparison with CB1 receptors, CB2 receptors exhibit a more pronounced role in immune modulation, predominantly characterised by anti-inflammatory effects. The regulation of immune cell migration is an area of significant research. CB2 receptor activation has been shown to influence the migration of immune cells to sites of infection or injury (Barker and Ferraro, 2024). It has been demonstrated that CB2 receptor activation can influence chemotaxis, the movement of immune cells in response to chemical signals (Han *et al.*, 2009; Martínez-Aguilar *et al.*, 2023). This suggests that CB2 receptor activation may play a role in controlling immune responses at local sites of inflammation (Grabon *et al.*, 2023).

b- CB1 and CB2 Receptors in Neuroprotection

It is evident that both CB1 and CB2 receptors play a substantial neuroprotective role within the central nervous system (CNS). The effects of these receptors are associated with a variety of neuroprotective mechanisms, including the reduction of oxidative stress, the modulation of neurotransmitter release, the promotion of neurogenesis, and the regulation of apoptosis (Bietar *et al.*, 2023). Recent studies have demonstrated that the activation of CB1 receptors exerts a significant influence on the process of neurogenesis, with a particular emphasis on the hippocampus a brain region involved in memory and learning functions (Oddi *et al.*, 2023). It is hypothesized that this regulatory mechanism performs a crucial function in promoting brain plasticity following injury and in age-related cognitive decline. Furthermore, CB1 receptors have been demonstrated to play a pivotal protective role in a number of debilitating diseases, including Huntington's disease, Alzheimer's disease, and multiple sclerosis (Sharon *et al.*, 2024). CB1 receptor activation has been shown to have a marked effect on neuroinflammation as evidenced by its ability to reduce both oxidative stress and the level of neuronal cell death that occurs in response to periods of stress (Martínez-Torres and Morán, 2024). In pathologies such as Alzheimer's and Parkinson's, in which neuroinflammation plays an instrumental role in disease progression, CB2 receptor activation has been demonstrated to reduce the production of reactive oxygen species (ROS) and inhibit the inflammatory pathways that result in neuronal damage (Fernández-Ruiz *et al.*, 2015; Vallée *et al.*, 2021). CB2 activation also contributes to the maintenance of synaptic plasticity and neuronal integrity in such conditions (Vuic *et al.*, 2022).

4- Interaction of Cannabinoids With MS Pathology

The cannabinoid compounds present in cannabis have attracted significant scientific interest due to their demonstrated therapeutic potential in the treatment of a variety of neurological disorders, including multiple sclerosis (MS). The mechanism by which cannabinoids exert their therapeutic effects is through the action of two principal receptors, designated as CB1 and CB2, which are components of the endocannabinoid system (ECS). The ECS is a pivotal regulating mechanism that governs a broad spectrum of physiological processes, including immune response, pain perception and inflammation (Lowe *et al.*, 2021).

The interaction of cannabinoids and multiple sclerosis pathology has been a topic of investigation in several studies (Nouh *et al.*, 2023). The potential neuroprotective, anti-inflammatory and analgesic properties of cannabinoids have been a primary focus in these investigations. It is hypothesized that a key mechanism through which cannabinoids act is through the modulation of the immune system (Khan *et al.*, 2022). Cannabinoids have been observed to influence the activity of microglia and astrocytes, which play a crucial role in the inflammatory response associated with MS (Abd-Nikfarjam *et al.*, 2023). The activation of CB2 receptors by cannabinoids has been shown to reduce the activation of pro-inflammatory T cells and the release of cytokines, which may contribute to the attenuation of the autoimmune attack on myelin (Nouh *et al.*, 2024. It has been hypothesized that the neuroprotective properties of cannabinoids are mediated via the CB1 receptor, a neurotransmitter receptor which is found in high concentrations in the brain's motor control and cognitive regions (Abd-Nikfarjam *et al.*, 2023).

III. MECHANISMS OF CANNABINOID ACTION

1- Pharmacological Properties of Cannabinoids

Cannabinoids compounds have the capacity to produce a variety of biological effects that influence multiple physiological processes. Among the most studied cannabinoids are delta9-tetrahydrocannabinol (THC) and cannabidiol (CBD), although there are over 100 different cannabinoids that have been identified to date (Śmiarowska *et al.*, 2022).

Furthermore, the pharmacological properties of cannabinoids are multifaceted and are subject to variation based on the specific compound, administered dosage, and the route of administration. Cannabinoids are known to exert their effects through interaction with a pair of primary cannabinoid receptors, termed CB1 and CB2. These receptors are integral components of the endocannabinoid system (ECS), a complex network that plays a critical regulatory role in numerous physiological processes, including mood, appetite, pain sensation, immune response, and memory (Caruso et al., 2025) (Figure 1). CB1 receptors are found predominantly in the brain and the central nervous system, and they are involved in the regulation of neurotransmitter release (Yarar, 2021). The ingestion of cannabis, which contains the compound THC, results in the psychoactive effects of the drug, including euphoria, altered perception and memory impairment, due to the binding of THC to CB1 receptors. The cannabidiol receptor, CB2, is predominantly expressed in peripheral tissues and immune cells. CB2 receptor activation has been linked to anti-inflammatory effects, immune response modulation, and the potential for therapeutic benefits in conditions such as autoimmune diseases (Lefebvre *et al.*, 2024).

Figure 1. Signalling pathways and effects of cannabinoids. Despite the type of cannabinoid ligand (phytocannabinoid, endocannabinoid or synthetic), the primary function of these compounds is to interact with GPCR proteins, such as the CB1 and CB2 receptors and GPR55, or with TRPs, such as TRPV1, to elicit a cellular response. The pathways stimulated vary depending on the receptor activation and have physiological effects including pain, appetite, and mood, as well as many other effects in the body (Legare et al., 2022).

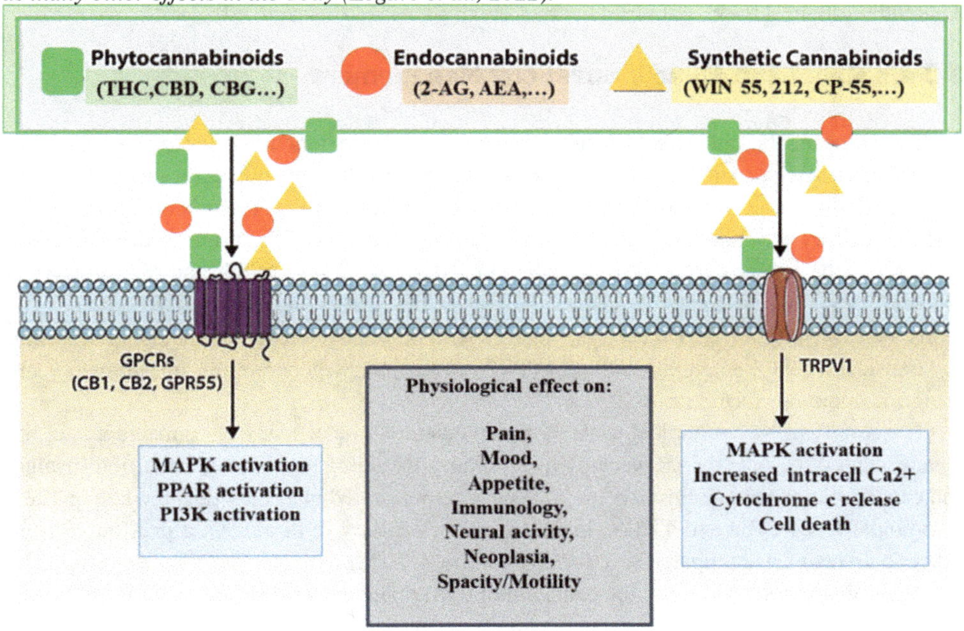

2- Anti-Inflammatory Effects

Cannabinoids, the active compounds present in cannabis, have been the subject of considerable scientific investigation due to evidence suggesting their capacity to exhibit anti-inflammatory properties. A number of recent studies have elucidated various cannabinoid-mediated mechanisms that regulate inflammatory responses, thereby offering insights into the potential therapeutic applications of cannabinoids in conditions characterized by chronic inflammation (Eddin *et al.*, 2024, Kuzumi *et al.*, 2024).

Receptor-mediated pathways: The interaction of cannabinoids, such as cannabidiol (CBD), with CB2 receptors is a significant phenomenon in the field of pharmacology. These receptors, which are predominantly located in peripheral tissues and immune cells, play a crucial role in various physiological processes (Eddin *et al.*, 2024).

Enzyme inhibition: It has been established that cannabinoids have the capacity to inhibit enzymes such as cyclooxygenases (COXs), including both COX-1 and COX-2 (Calapai *et al.*, 2022). These enzymes play a pivotal role in the synthesis of pro-inflammatory prostaglandins. The inhibitive effect of cannabinoids upon the enzymes in question has been demonstrated to result in reduced production of inflammatory mediators, thus leading to a state of reduced inflammation (Calapai *et al.*, 2022).

Modulation of Immune Cell Function: Cannabinoids have been demonstrated to exert an influence on the function of macrophages by modulating the release of cytokines and phagocytic activity (Anil *et al.*, 2021). For instance, it has been demonstrated that cannabis extracts have the capacity to reduce the release of skin inflammation mediators in keratinocytes, which suggests a potential role in skin-related inflammatory conditions (Martins *et al.*, 2022).

3- Pain Relief and Muscle Spasticity Management

The endocannabinoid system performs a pivotal modulatory function in the regulation of pain perception, whereby cannabinoids establish an interaction with cannabinoid receptors (CB1 and CB2) to regulate neurotransmitter release and reduce pain signaling. It has been demonstrated that the primary psychoactive compound in cannabis, known as delta-9-tetrahydrocannabinol (THC), has the capacity to bind to cannabinoid 1 (CB1) receptors within the central nervous system (CNS), thereby inducing analgesic effects. Conversely, cannabidiol (CBD), which exhibits a low affinity for CB1 receptors, engages with alternative receptors, including serotonin and transient receptor potential vanilloid 1 (TRPV1) receptor, to generate anti-inflammatory and analgesic outcomes (Narouze, 2021).

The results of recent clinical trials have demonstrated the efficacy of cannabinoids in the management of chronic pain, especially in conditions including neuropathic pain, fibromyalgia and cancer-related pain (Johnson *et al.* 2025). As demonstrated in a study documented in 2022, a combination of CBD and THC exhibited superior efficacy in diminishing pain intensity in cancer patients when compared to a placebo (O'Brien, 2022).

According to recent studies, the cannabidiol (CBD) and tetrahydrocannabinol (THC) compounds have the potential to assist in the management of muscle spasms. The mechanism by which THC exerts its effect involves the activation of CB1 receptors located within the central nervous system (CNS), thereby modulating the release of excitatory neurotransmitters that are implicated in the development of muscle spasms (Iannotti, 2023). In addition, cannabidiol (CBD) has demonstrated the capacity to enhance motor function and diminish muscular tension by interacting with non-cannabinoid receptors, such as serotonin and adenosine receptor subtypes, thus influencing the activity of the gamma-aminobutyric acid (GABAergic) system, which modulates muscle relaxation (Koppo, 2025). An important investigation, examined the clinical effectiveness of a THC/CBD-based spray (Sativex) in diminishing spasticity in individuals diagnosed with MS. It was determined by the study that Sativex significantly improved spasticity scores and the quality of life experienced by patients (Sacco *et al.*, 2024).

IV. CLINICAL STUDIES AND THERAPEUTIC EFFICACY

1- Review of Clinical Trials on Cannabinoid-Based Treatments for MS

Cannabinoids, particularly Δ9-tetrahydrocannabinol (THC) and cannabidiol (CBD), have been extensively investigated for their therapeutic potential in multiple sclerosis (MS), mainly due to their analgesic and muscle-relaxant properties. Several clinical trials have evaluated their efficacy, either individually or in combination, in various pharmaceutical forms, including nabiximols (Sativex®), an oromucosal spray and other medical cannabis extracts. Sativex has been shown to be well tolerated by MS patients, with a low incidence of side effects (Nouh *et al.*, 2024). Furthermore, multiple trials have assessed its role as an adjunctive treatment for controlling moderate to severe spasticity in MS patients (Markovà *et al.*, 2019; Manera & Bertini, 2021).

The literature provides substantial evidence supporting the use of cannabinoid-based treatments for MS. Clinical studies have confirmed their approved applications in managing chronic pain, cannabis use disorder (CUD), and weight loss in individuals with AIDS (Squintani *et al.*, 2016; Markovà *et al.*, 2019). A systematic review by Hakami & Alshehri (2025) analyzed 47 clinical trials investigating the therapeutic potential of cannabinoids, including 13 specifically focusing on THC and CBD. These phase 2 trials examined their effects on MS, fibromyalgia, and Parkinson's disease.

Several large-scale studies have provided valuable insights into the efficacy of cannabinoids in MS treatment. The CAMS (Cannabinoids in Multiple Sclerosis) trial, involving 630 patients, reported subjective benefits for spasticity, muscle spasms, and sleep; however, it found no significant changes on the Ashworth scale. A fter 12 months, slight benefits were observed with dronabinol but not with Cannador or placebo (Zajicek *et al.*, 2005). Similarly, the MUSEC (MUltiple Sclerosis and Extract of Cannabis) trial demonstrated a significant reduction in muscle stiffness in 29.4% of patients treated with Cannador, compared to 15.7% in the placebo group, with additional improvements in sleep quality (Zajicek *et al.*, 2012). Studies by Rice and Cameron (2018) and Collin *et al.* (2010) further confirmed these benefits with THC doses ranging from 20 to 40 mg per day, suggesting that an early response to treatment could predict long-term efficacy. Additionally, a multicenter, double-blind study on Sativex® (189 participants) reported reduced spasticity and an improved quality of life after six weeks.

According to the American Society of Pain and Neuroscience, the use of medical cannabis for the treatment of neuropathic pain is based on prospective studies, but the lack of high-quality evidence limits its recommendation to grade C. Similarly, although clinical trials have explored the effectiveness of cannabis in alleviating cancer-related pain, the results obtained only justify a grade D recommendation. Therefore, while some patients report improvements in spasticity, pain relief, and sleep quality (Strand *et al.*, 2023), a clinical trial conducted in Argentina further confirmed positive responses to cannabinoid-based treatments for spasticity, pain, and insomnia (Contentti *et al.*, 2021).

2- Effectiveness in Symptom Management and Disease Progression

Cannabinoids play a crucial role in managing symptoms of multiple sclerosis (MS) due to their interaction with the endocannabinoid system, which is involved in modulating pain, inflammation, and neuroprotective mechanisms (Khan *et al.*, 2022). Clinical studies have highlighted their effectiveness in reducing muscle spasticity, a common symptom in MS patients, by promoting smoother movements and alleviating associated pain. Notably, Shkodina *et al.* (2024) focus on the effects of cannabinoids in relieving neuropathic pain, while de Fátima dos Santos Sampaio *et al.* (2024) analyze their utility in various neurological disorders, including MS, mentioning the use of nabiximols (a combination of THC and CBD) to reduce spasticity.

The impact of cannabinoids on disease progression remains uncertain (Hakami and Alshehri, 2025). Sharon *et al.* (2024) report no direct data demonstrating an influence of cannabinoids on the clinical course of MS. Additionally, Kaszyńska (2024) highlights the potential effects of cannabidiol (CBD) combined with vitamin B12 in reducing certain pathological mechanisms of the disease, emphasizing neuroprotective and anti-inflammatory properties, particularly through microglial modulation and oxidative stress reduction. Nevertheless, while these interventions may contribute to the improvement of certain neurological symptoms, no clinical data confirm their effect on disease progression.

Neuropathic pain, which is particularly resistant to conventional treatments, appears to respond well to cannabinoids, leading to a reduction in its intensity and offering better tolerance compared to opioids or anticonvulsants. Some trials also indicate positive effects on sleep disturbances and mood disorders, thereby improving patients' quality of life (Fine and Rosenfeld, 2013). However, research on their influence on disease progression remains limited. To date, no clinical study has demonstrated a significant effect of cannabinoids on reducing demyelinating lesions, inhibiting inflammatory markers, or preserving neuronal function. Thus, while they are effective in symptom relief, their impact on disease progression appears to be limited.

Despite encouraging results obtained in animal models of MS, particularly experimental autoimmune encephalomyelitis (EAE), their translation to patients remains uncertain due to biological differences between species. Furthermore, available clinical studies mainly focus on symptomatic relief without incorporating comprehensive evaluations of disease progression using biomarkers such as MRI. The variability in CBD bioavailability depending on the route of administration and challenges related to its therapeutic efficacy further complicate its clinical assessment. Consequently, despite promising preclinical results, rigorous and long-term trials are necessary to determine whether cannabinoids can truly influence the progression of MS in humans (Nouh *et al.*, 2024).

3- Patient-Reported Outcomes and Quality of Life Improvements

The results of surveys conducted among patients treated with cannabinoids indicate a significant improvement in quality of life, particularly regarding pain reduction, improved functional mobility, and overall well-being. A significant proportion of patients report more effective symptom management, especially for spasticity and neuropathic pain, compared to standard treatments (Belgers *et al.*, 2023). Furthermore, one of the most frequently studied aspects is the impact of cannabinoids on daily activities and functional autonomy. Many patients report significant improvements in their ability to perform daily tasks, sleep better, and manage their emotional state (Bernerth and Walker, 2020).

Furthermore, the study by Tait *et al.* (2023) analyzed the impact of medicinal cannabis on the quality of life of 2327 patients, primarily suffering from chronic pain (68.7%), insomnia (22.9%), anxiety (21.5%), and depression (11%). After three months of treatment, the results showed a significant improvement in overall quality of life, measured by the EQ-5D-5L and QLQ-C30. Improvements included a marked reduction in pain (effect size 0.65), a decrease in fatigue (effect size 0.54), as well as improvements in anxiety and depression. These results contributed to a better quality of life in daily living, with a reduction in physical limitations and an increase in overall well-being, although no significant effect on sleep disorders was observed.

The overall improvement in the quality of life of patients treated with cannabinoids is primarily attributed to the effects of the main active compounds of cannabis, particularly CBD and THC, as evidenced by studies conducted by Tait *et al.* (2023). These results highlight the importance of cannabinoids as a therapeutic alternative in the treatment of MS and other chronic disorders.

V. CHALLENGES AND CONSIDERATIONS

1- Variability in Patient Response to Cannabinoids

The therapeutic response to cannabinoid-based treatments exhibits significant heterogeneity, necessitating an individualized approach. Several interconnected factors contribute to this variability, including genetic, metabolic, psychological, and environmental elements, each playing a key role in treatment efficacy and tolerance (Kitdumrongthum and Trachootham, 2023).

Genetic variations in cannabinoid receptors (CB1 and CB2) can influence the affinity of cannabinoids for their receptors, thereby modifying their therapeutic efficacy. Studies have highlighted the impact of these polymorphisms on treatment response, emphasizing the importance of a personalized approach based on the patient's genetic characteristics (Nouh *et al.*, 2023; Babayeva and Loewy, 2023).

The metabolism of cannabinoids also plays a crucial role in the variability of therapeutic response. Differences in the activity of cytochrome P450 enzymes, responsible for the degradation of THC and CBD, alter bioavailability and the duration of effects. These metabolic differences can influence not only treatment efficacy but also the profile of side effects, making the therapeutic response more or less favorable depending on the patient (Zhang *et al.*, 2024).

Tolerance development with prolonged cannabinoid use is another important factor. In some patients, therapeutic effects diminish over time, requiring dose escalation to maintain treatment benefits (Gottschling *et al.*, 2020). However, this also increases the risk of adverse effects, contributing to interpatient variability in response.

Psychological and environmental factors, such as anxiety, stress, and patient expectations, play a crucial role in the perception of cannabinoid effects. These elements can bias the evaluation of treatment efficacy and influence the overall therapeutic response. Additionally, the presence of comorbidities and differences in formulations used (such as nabiximols or oral/inhaled extracts) may further explain the variability observed in clinical outcomes (Filippini *et al.*, 2022; Bell *et al.*, 2024).

Consequently, this variability highlights the need for a personalized approach to maximize therapeutic benefits while minimizing the risk of side effects. Current recommendations suggest a gradual adjustment of dosage and product selection based on the specific characteristics of each patient, considering genetic and metabolic differences, prior experiences, and environmental factors (Nouh *et al.*, 2023; Bell *et al.*, 2024).

2- Issues Related to Dosing and Side Effects

Determining the optimal dosage of cannabinoids for multiple sclerosis remains a significant challenge due to the lack of standardized guidelines. Clinical trials have investigated various concentrations and THC/CBD ratios, complicating direct comparisons of findings. Insufficient dosing may result in limited therapeutic efficacy, whereas excessive intake increases the risk of adverse effects. Notably, THC is associated with psychotropic effects such as euphoria, confusion, and anxiety, potentially affecting patient adherence. Conversely, while CBD is generally well tolerated, it can interact with other medications, including anticoagulants and immunosuppressants (Murphy *et al.*, 2024).

Regarding dosage protocols, Gustavsen *et al.* (2021) proposed an initial THC dose of 2.5 mg once daily, typically administered at bedtime. Incremental increases of 1.25 to 2.5 mg every 3–4 days were permitted, provided no adverse effects emerged, up to a maximum of 22.5 mg/day. Similarly, CBD dosing started at 5 mg/day, with 5–10 mg increments every 3–4 days, reaching a maximum of 50 mg/day. Patients had flexibility in adjusting their doses according to symptomatic relief and tolerance. Reported side effects primarily included neurological, psychiatric, gastrointestinal, and cardiovascular symptoms (Gustavsen *et al.*, 2021).

Further studies have documented adverse effects associated with medical cannabis. Filippini *et al.* (2022) highlighted nervous system disorders, psychiatric symptoms, and potential physical dependence. Longoria *et al.* (2022) noted that CBD/THC combinations provided symptomatic relief comparable to conventional treatments, with a more favorable side effect profile. Reported adverse events included nausea (3.8%), dry mouth (3.2%), dizziness (2.2%), and drowsiness (2.2%) (Ergisi *et al.*, 2022). However, Longoria *et al.* (2022) emphasized several challenges in medical cannabis use, including a narrow therapeutic window, cytochrome P450 inhibition, variable tolerance thresholds, and heterogeneous treatment responses.

Haddad *et al.* (2022) explored the efficacy of cannabis for MS symptoms, identifying dizziness (14–59%), gastrointestinal disturbances (13–37%), dry mouth (4–26%), and urinary tract infections (15.4–34%) as prevalent adverse effects. Additional symptoms included fatigue, headaches, attention deficits, and disorientation. Among nabiximols users, 46.9% experienced side effects, primarily dizziness and fatigue, while 16.6% of 325 patients reported adverse events such as fatigue, drowsiness, nausea, and dry mouth, despite overall good treatment tolerance. For oral cannabinoids (CBD/THC), adverse effects were generally mild to moderate, yet 30 out of 144 patients discontinued treatment due to dizziness, dry mouth, fatigue, urinary infections, asthenia, and headaches (Haddad *et al.*, 2022).

Mustafa *et al.* (2021) investigated cannabis effects on clinical profiles and cytokine levels in MS patients, reporting side effects such as drowsiness, increased appetite, pain, euphoria, talkativeness, balance loss, concentration difficulties, weakness, blurred vision, paranoia, hallucinations, and nausea. These findings underscore the necessity for rigorous dosage monitoring and adverse effect management to optimize cannabinoid-based therapy for MS patients. While cannabinoids offer potential therapeutic benefits for MS symptom management, their use demands careful dose titration and monitoring to mitigate adverse effects. Standardized dosing protocols and individualized treatment plans are essential to enhance efficacy while minimizing risks.

3- Safety Concerns and Long-Term Implications

The prolonged use of cannabinoids in a therapeutic setting, particularly for multiple sclerosis, raises questions regarding their safety and long-term effects.

Among the major concerns, the neurocognitive effects of THC are particularly studied. Prolonged exposure to this compound could impair certain brain functions, especially in vulnerable patients, with deficits observed in working memory and information processing speed (Adam *et al.*, 2020). Furthermore, although the risk of dependence is moderate compared to other substances, it remains a significant issue.

The neurocognitive and psychomotor implications of cannabinoids also raise concerns, particularly regarding motor skills and safety in daily activities. Prolonged THC use has been associated with deficits in decision-making, processing speed, and executive functions, potentially compromising the ability to drive or operate machinery (Nouh *et al.*, 2023). Recommendations suggest an abstinence period of at least eight hours after THC consumption before engaging in such activities, although the intensity of effects may vary depending on the dose, route of administration, and individual metabolism. Studies on recreational use have also shown that cognitive deficits can persist beyond twelve hours after cessation, raising questions about residual effects in a therapeutic context (Gottschling *et al.*, 2020).

The assessment of cannabinoid safety is further complicated by the diversity of formulations and dosage protocols. Unlike recreational use, which involves variable doses and products often mixed with other substances, medical use relies on precise and controlled dosing regimens (Wang *et al.*, 2008; Millar *et al.*, 2019). However, even in a therapeutic setting, the lack of standardization in dosages and administration routes complicates risk analysis. This variability in practices also makes it difficult to extrapolate preclinical study results to humans, given differences in cannabinoid systems across species (Longoria *et al.*, 2022). Consequently, while existing preclinical and clinical data provide valuable insights, they remain insufficient to establish a clear consensus on the long-term safety of cannabinoids (Bonn-Miller *et al.*, 2019; Zeraatkar *et al.*, 2022).

In the context of multiple sclerosis, cannabinoids, particularly nabiximols (Sativex), have demonstrated efficacy in managing spasticity and other symptoms. However, long-term studies on their cognitive effects remain limited. Some research indicates a potential impact on memory and anxiety, although the relationship between prolonged use and neurodegeneration progression is not yet clearly defined (Alessandria *et al.*, 2020). Moreover, the effectiveness of nabiximols appears to be limited to certain patients, and controlled trials have encountered methodological challenges, including bias risk due to prior exposure of patients to cannabinoids before randomization. As a result, while cannabinoids may be considered a therapeutic alternative for some multiple sclerosis patients, their prescription should remain cautious, serving as a complement to conventional treatments rather than a first-line option (Friedman *et al.*, 2019).

Despite the reported benefits in pain management and relief of neurological symptoms, concerns about long-term tolerance persist. Pharmacovigilance studies and real-world follow-ups are essential to complement data from clinical trials and refine therapeutic recommendations. Furthermore, the effects of cannabinoids on reproduction and neurological development remain poorly understood, particularly in young adults and vulnerable populations (Gottschling *et al.*, 2020). In this context, further research is necessary to identify the most at-risk patient profiles and establish safety thresholds tailored to each medical indication.

VI. NANOPARTICLES AND CANNABINOIDS: INNOVATIONS IN DRUG DELIVERY AND THE LEGAL LANDSCAPE OF MEDICAL CANNABIS FOR MS

1- Nanoparticles and Cannabinoids: A Promising Approach for Multiple Sclerosis Treatment

Notable progress has been made in research on cannabinoids-based MS treatments. Recently, nanomedicine has emerged as a promising approache to overcoming the limitations of traditional drugs. The use of nanotechnology for therapeutic purposes allows for precise dose delivery, optimizing treatment effectiveness while minimizing toxicity. With their ability to cross the blood-brain barrier (BBB), nanomaterials are considered a promising strategy for multiple sclerosis (MS) treatment, offering both neuroprotection and enhanced efficacy (Zeng *et al.*,2020).

In order to address the factors influencing drug pharmacokinetics, nanomedicine, through nanoparticles, offers a therapeutic alternative by enhancing the bioavailability, stability and solubility of active drug substances, leading to improved safety and therapeutic efficacy (Gunasekaran *et al.*, 2014). Several studies have demonstrated that the galenic reformulation of drugs with water-solubility, into nanosuspension form is attracting growing interest due to its potential to reduce toxicity and improve treatment efficacy, by eliminating the need for co-solvents in the formulation (Rabinow *et al.*, 2004; Patravale *et al.*, 2004).

The nanoparticles have been investigated as a potential means of optimizing drug delivery to brain tissue. An alternative approach relies on exploiting a precise interaction between a receptor and its ligand to enable targeted and controlled delivery of nanoparticle-encapsulated therapeutic agents across an intact BBB (Tylawsky *et al.*, 2023). In the case of cannabinoids, results suggest that the use of nanoparticles targeting specific proteins like P-glycoprotein activity promote the transcytosis and delivery of nano-cannabinoids across the BBB, thus opening up a promising avenue for the treatment of central nervous system diseases (Tylawsky *et al.*, 2023; Calapai *et al.*, 2020).

Considering that cannabinoid-based drugs currently on the pharmaceutical market are active compounds with a lipophilic nature, this may influence their mode of administration, absorption and overall distribution within the body (Lucas *et al.*, 2018). Moreover, exposure to physical factors such as light and temperature fluctuation can affect and limit their systemic bioavailability through auto-oxidation and degradation (Grotenhermen *et al.*, 2003), These challenges contribute to the limited efficacy of cannabinoid-based galenic pharmaceutical formulations in the treatment of patients with multiple sclerosis (MS), and remains inconsistent across different stages of the disease, further limiting their widespread clinical application (Gajofatto *et al.*,2015 ; Ojha *et al.*, 2018).

To address this issue, and in order to minimize the required dosage, toxicity and improve the stability of cannabinoids, studies have focused on developing and integrating of nanoparticles with cannabinoids to facilitate their passage across the blood-brain barrier (BBB), with the aim of targeting the different lesions of multiple sclerosis. According to Rebelatto *et al* the use of nanoparticles, such as: lipids, micelles, silica nanoparticles and carbon nanotubes (Figure 2).

Figure 2. Categories of nano-enhanced cannabinoid formulations

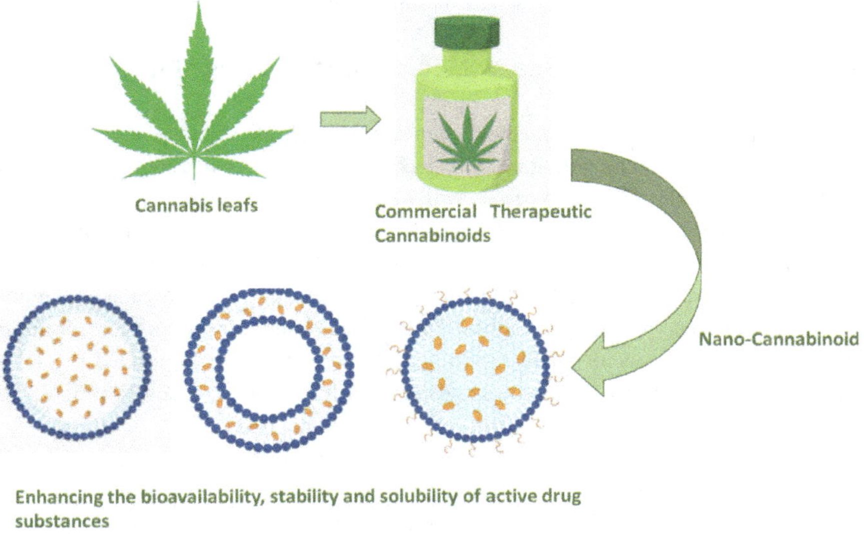

- Lipid-Based Nanocarriers for cannabidiol (CBD): lipid-based nanoparticles, such as Pro-Nanolipospheres, Nanoliposomes, Transferosomes, Nanoemulsions, micelles, silica nanoparticles and carbon nanotubes, have been reported to be effective for the specific delivery of CBD, showing promising therapeutic effects against various diseases. enhancing its therapeutic efficacy across various diseases and disorders (Kumar *et al.*,2023; Broadley *et al.*,2014; Aparicio-Blanco *et al.*,2019).
- Polymeric for CBD: A number of polymers and biopolymers have been studied for encapsulating CBD. At the same time, numerous scientific studies have led to the development of nanomaterials based on natural and synthetic polyesters as promising carriers to improve the delivery, stability and bioavailability of active compounds, thereby increasing the efficacy of our products (Fraguas-Sánchez *et al.*,2020).

Previous study has incorporated cannabidiol (CBD), a major cannabis compound into solid lipid nanocapsules, resulting in a slower, more prolonged release of CBD, reduced toxicity required for treatment and effective targeting of glioma cells (Aparicio-Blanco *et al.*,2019).

On the other hand, another recent study revealed that nanoparticle formulation of THC (Δ9-tetrahydrocannabinol) enabled potentiation and control of anticancer effects with prolonged cannabinoid release at tumor cell level (Durán-Lobato *et al.*,2022).

The future of cannabinoid-based nanomedicines is bright, with the promise of more effective, personalized treatments that can minimize side effects and improve patient outcomes, paving the way for better management of patients with multiple sclerosis.

Despite promising results from preclinical and early clinical studies, further research is needed to refine these formulations and ensure their safety, efficacy and optimal dosing. The future of cannabinoid-based nanomedicines is bright, with the promise of more effective, personalized treatments that can minimize side effects and improve patient outcomes, paving the way for better management of neurological diseases such as multiple sclerosis.

2- Legal Authorization and Use of Medical Cannabis for MS

The legal approval and usage of medical cannabis for multiple sclerosis vary across countries. The Food and Drug Administration (FDA) has not approved any marketing application for cannabis-based medicine for MS, and people with multiple sclerosis may experience symptoms that are not adequately controlled with FDA-approved medications. But it was recently asked to place cannabis-based therapy for progressive MS on the fast track (Reston 2019).

Medical cannabis can be prescribed to people with MS under strictly regulated conditions, but the eligibility of prescriber's cannabis-based medicines varies across countries, for instance, in the UK only specialist doctors with expertise in treating MS, are authorized to prescribe nabiximols. The latter has received approval and is currently available Canada, the United States and 21 European countries (Abuhasira 2018; Krcevski-Skvarc 2018).

Regarding Morocco, which is one of the world's top cannabis producers, and the substance's presence in the country dates back to the third century. The government approved Law on March 11, 2021, the legal use of cannabis. This would allow the marketing and use of cannabis for medical purposes like epilepsy.

The approval and legalization of cannabis for therapeutic use have led to extensive scientific studies focused on harnessing cannabinoids in clinical studies. Findings from these studies may further support the global legalization of medical cannabis.

VII. CONCLUSION

Multiple sclerosis is a neurological disorder characterized by inflammation and myelin degeneration that leads to complex and various neurological and non-neurological symptoms in teenagers and adults. Considering, a wide variety of symptoms involved, there is no single conventional treatment that works for all MS patients at all stages of the disease, but some treatments focus on slowing the progression and alleviating painful symptoms. Therefore, several studies have investigated the potential effect of Cannabinoids, especially anti-inflammatory, antiviral, and anticancer effects. Moreover, and since the 1990s, numerous scientific researches have demonstrated the potential role of cannabinoids in multiple sclerosis, via human clinical trials and animal models, studies have shown that cannabinoid-based drugs can effectively alleviate symptoms of multiple sclerosis. But these approved drugs have limited efficacy to manage the different symptoms due to their varying absorption, distribution and toxicity. Recently, Scientific research continues to study multiple sclerosis, and they are also focusing on the development and innovation of new treatments capable of minimizing symptoms and slowing or preventing the decline in function of multiple sclerosis.

The convergence of nanotechnology and cannabinoid drugs represents a groundbreaking advancement in modern medicine, offering innovative solutions to the challenges associated with conventional cannabis-derived treatments. The various types of nanoparticles encapsulating cannabinoids have shown remarkable results in enhancing the bioavailability of these molecules and minimizing their toxic effects on tissues. despite these advances in cannabinoid nanomedicine research, further studies are needed to guarantee the safety and efficacy of nano-cannabinoid medicine.

REFERENCES

Abd-Nikfarjam, B., Dolati-Somarin, A., Baradaran Rahimi, V., & Askari, V. R. (2023). Cannabinoids in neuroinflammatory disorders: Focusing on multiple sclerosis, Parkinsons, and Alzheimers diseases. *BioFactors (Oxford, England)*, *49*(3), 560–583.

Adam, K. C., Doss, M. K., Pabon, E., Vogel, E. K., & de Wit, H. (2020). Δ9-Tetrahydrocannabinol (THC) impairs visual working memory performance: A randomized crossover trial. *Neuropsychopharmacology: Official Publication of the American College of Neuropsychopharmacology*, *45*(11), 1807–1816.

Aghazadeh Tabrizi, M., Baraldi, P. G., Borea, P. A., & Varani, K. (2016). Medicinal chemistry, pharmacology, and potential therapeutic benefits of cannabinoid CB2 receptor agonists. *Chemical Reviews*, *116*(2), 519–560.

Albelo-Martínez, M., & Rizvi, S. (2025). Progressive multiple sclerosis: Evaluating current therapies and exploring future treatment strategies. Neurotherapeutics: the journal of the American Society for Experimental NeuroTherapeutics, e00601. Advance online publication. https://doi.org/DOI: 10.1016/j.neurot.2025.e00601

Alessandria, G., Meli, R., Infante, M. T., Vestito, L., Capello, E., & Bandini, F. (2020). Long-term assessment of the cognitive effects of nabiximols in patients with multiple sclerosis: A pilot study. *Clinical Neurology and Neurosurgery*, *196*, 105990.

Anil, S. M., Shalev, N., Vinayaka, A. C., Nadarajan, S., Namdar, D., Belausov, E., & Koltai, H. (2021). Cannabis compounds exhibit anti-inflammatory activity in vitro in COVID-19-related inflammation in lung epithelial cells and pro-inflammatory activity in macrophages. *Scientific Reports*, *11*(1), 1462.

Aparicio-Blanco, J., Sebastián, V., Benoit, J. P., & Torres-Suárez, A. I. (2019). Lipid nanocapsules decorated and loaded with cannabidiol as targeted prolonged release carriers for glioma therapy: In vitro screening of critical parameters. *European Journal of Pharmaceutics and Biopharmaceutics*, *134*, 126–137. DOI: 10.1016/j.ejpb.2018.11.020

Argenziano, M., Tortora, C., Bellini, G., Di Paola, A., Punzo, F., & Rossi, F. (2019). The endocannabinoid system in pediatric inflammatory and immune diseases. *International Journal of Molecular Sciences*, *20*(23), 5875.

Babayeva, M., & Loewy, Z. G. (2023). Cannabis pharmacogenomics: A path to personalized medicine. *Current Issues in Molecular Biology*, *45*(4), 3479–3514.

Balezina, O. P., Tarasova, E. O., & Bogacheva, P. O. (2024). Myogenic Classical Endocannabinoids, Their Targets and Activity. *Biochemistry (Moscow)*, *89*(10), 1759–1778.

Barker, H., & Ferraro, M. J. (2024). Exploring the versatile roles of the endocannabinoid system and phytocannabinoids in modulating bacterial infections. *Infection and Immunity*, *92*(6), e00020–e00024.

Belgers, V., Röttgering, J. G., Douw, L., Klein, M., Ket, J. C., van de Ven, P. M., & de Witt Hamer, P. C. (2023). Cannabinoids to improve health-related quality of life in patients with neurological or oncological disease: A meta-analysis. *Cannabis and Cannabinoid Research*, *8*(1), 41–55.

Bell, A. D., MacCallum, C., Margolese, S., Walsh, Z., Wright, P., & Daeninck, P. J. & External Review Panel. (2024). Clinical practice guidelines for cannabis and cannabinoid-based medicines in the management of chronic pain and co-occurring conditions. *Cannabis and Cannabinoid Research*, *9*(2), 669–687.

Berland, C., Castel, J., Terrasi, R., Montalban, E., Foppen, E., Martin, C., & Gangarossa, G. (2022). Identification of an endocannabinoid gut-brain vagal mechanism controlling food reward and energy homeostasis. *Molecular Psychiatry*, *27*(4), 2340–2354.

Bernerth, J. B., & Walker, H. J. (2020). Altered states or much to do about nothing? A study of when cannabis is used in relation to the impact it has on performance. *Group & Organization Management*, *45*(4), 459–478.

Bietar, B., Tanner, S., & Lehmann, C. (2023). Neuroprotection and beyond: The central role of CB1 and CB2 receptors in stroke recovery. *International Journal of Molecular Sciences*, *24*(23), 16728.

Bonn-Miller, M. O., Pollack, C. V.Jr, & Casarett, D.. (2019). Priority considerations for medicinal cannabis-related research. *Cannabis and Cannabinoid Research*, *4*(3), 139–157. DOI: 10.1089/can.2019.004531579832

Braile, M., Marcella, S., Marone, G., Galdiero, M. R., Varricchi, G., & Loffredo, S. (2021). The interplay between the immune and the endocannabinoid systems in cancer. *Cells*, *10*(6), 1282.

Broadley, S. A., Barnett, M. H., Boggild, M., Brew, B. J., Butzkueven, H., Heard, R., Hodgkinson, S., Kermode, A. G., Lechner-Scott, J., & Macdonell, R. A., (2014). Therapeutic approaches to disease modifying therapy for multiple sclerosis in adults: An Australian and New Zealand perspective Part 3 Treatment practicalities and recommendations. *Journal of Clinical Neuroscience*, *21*, 1857–1865. DOI: 10.1016/j.jocn.2014.01.017

Cabral, G. A., Raborn, E. S., Griffin, L., Dennis, J., & Marciano-Cabral, F. (2008). CB2 receptors in the brain: Role in central immune function. *British Journal of Pharmacology*, *153*(2), 240–251.

Calapai, F., Cardia, L., Esposito, E., Ammendolia, I., Mondello, C., Lo Giudice, R., & Mannucci, C. (2022). Pharmacological aspects and biological effects of cannabigerol and its synthetic derivatives. *Evidence-Based Complementary and Alternative Medicine : eCAM*, *2022*(1), 3336516.

Calapai, F., Cardia, L., Sorbara, E. E., Navarra, M., Gangemi, S., Calapai, G., & Mannucci, C. (2020). Cannabinoids, blood–brain barrier, and brain disposition. *Pharmaceutics*, *12*, 265. DOI: 10.3390/pharmaceutics12030265

Caruso, E. A., De Nunzio, V., Tutino, V., & Notarnicola, M. (2025). The Endocannabinoid System: Implications in Gastrointestinal Physiology and Pathology. *International Journal of Molecular Sciences*, *26*(3), 1306.

Ciuffreda, P., Xynomilakis, O., Casati, S., & Ottria, R. (2024). Fluorescence-Based Enzyme Activity Assay: Ascertaining the Activity and Inhibition of Endocannabinoid Hydrolytic Enzymes. *International Journal of Molecular Sciences*, *25*(14), 7693.

Collin, C., Ehler, E., Waberzinek, G., Alsindi, Z., Davies, P., Powell, K., & Ambler, Z. (2010). A double-blind, randomized, placebo-controlled, parallel-group study of Sativex, in subjects with symptoms of spasticity due to multiple sclerosis. *Neurological Research*, *32*(5), 451–459.

Contentti, E. C., López, P. A., Criniti, J., Pettinicchi, J. P., Tavolini, D., Mainella, C., & Alonso, R. (2021). Use of cannabis in patients with multiple sclerosis from Argentina. *Multiple Sclerosis and Related Disorders*, *51*, 102932.

de Fátima dos Santos Sampaio, M., de Paiva, Y. B., Sampaio, T. B., Pereira, M. G., & Coimbra, N. C. (2024). Therapeutic applicability of cannabidiol and other phytocannabinoids in epilepsy, multiple sclerosis and Parkinson's disease and in comorbidity with psychiatric disorders. *Basic & Clinical Pharmacology & Toxicology*, *134*(5), 574–601.

Durán-Lobato, M., Álvarez-Fuentes, J., Fernández-Arévalo, M., & Martín-Banderas, L. (2022). Receptor-targeted nanoparticles modulate cannabinoid anticancer activity through delayed cell internalization. *Scientific Reports*, *12*, 1–17. DOI: 10.1038/s41598-022-05301-z

Eddin, L. B., Meeran, M. N., Subramanya, S. B., Jha, N. K., & Ojha, S. (2024). Therapeutic potential of agents targeting cannabinoid type 2 receptors in organ fibrosis. *Pharmacology Research & Perspectives*, *12*(6), e1219.

El-Mansoury, B., Laadraoui, J., El Khiat, A., El Hiba, O., Jayakumar, A. R., Ferssiwi, A., & Bitar, A. (2023). *Ataxia in Multiple Sclerosis: From Current Understanding to Therapy*. Ataxia–Practice Essentials and Interventions.

Ergisi, M., Erridge, S., Harris, M., Kawka, M., Nimalan, D., Salazar, O., Loupasaki, K., Ali, R., Holvey, C., & Coomber, R.. (2022). An updated analysis of clinical outcome measures across patients from the UK Medical Cannabis Registry. *Cannabis and Cannabinoid Research*.

Fernández-Ruiz, J., Romero, J., & Ramos, J. A. (2015). Endocannabinoids and neurodegenerative disorders: Parkinson's disease, Huntington's chorea, Alzheimer's disease, and others. Endocannabinoids, 233-259.

Filippini, G., Minozzi, S., Borrelli, F., Cinquini, M., & Dwan, K. (2022). Cannabis and cannabinoids for symptomatic treatment for people with multiple sclerosis. *Cochrane Database of Systematic Reviews*, 5.

Fine, P. G., & Rosenfeld, M. J. (2013). The endocannabinoid system, cannabinoids, and pain. *Rambam Maimonides Medical Journal*, *4*(4), e0022.

Fraguas-Sánchez, A. I., Torres-Suárez, A. I., Cohen, M., Delie, F., Bastida-Ruiz, D., Yart, L., Martin-Sabroso, C., & Fernández-Carballido, A. (2020). PLGA Nanoparticles for the Intraperitoneal Administration of CBD in the Treatment of Ovarian Cancer: In Vitro and In Ovo Assessment. *Pharmaceutics*, *12*, 439.

Friedman, D., French, J. A., & Maccarrone, M. (2019). Safety, efficacy, and mechanisms of action of cannabinoids in neurological disorders. *Lancet Neurology, 18*(5), 504–512.

Gajofatto, A., & Benedetti, M. D. (2015). Treatment strategies for multiple sclerosis: When to start, when to change, when to stop? *World Journal of Clinical Cases: WJCC, 3*(7), 545.

García-Gutiérrez, M. S., Torregrosa, A. B., Navarrete, F., Navarro, D., & Manzanares, J. (2025). A comprehensive review of the multifaceted role of cannabinoid receptor type 2 in neurodegenerative and neuropsychiatric disorders. *Pharmacological Research, 213*, 107657. DOI: 10.1016/j.phrs.2025.107657

Gottschling, S., Ayonrinde, O., Bhaskar, A., Blockman, M., D'agnone, O., Schecter, D., & Cyr, C. (2020). Safety considerations in cannabinoid-based medicine. *International Journal of General Medicine*, 1317–1333.

Grabon, W., Rheims, S., Smith, J., Bodennec, J., Belmeguenai, A., & Bezin, L. (2023). CB2 receptor in the CNS: From immune and neuronal modulation to behavior. *Neuroscience and Biobehavioral Reviews, 150*, 105226.

Grotenhermen, F. (2003). Pharmacokinetics and pharmacodynamics of cannabinoids. *Clinical Pharmacokinetics, 42*(4), 327–360.

Gunasekaran, T., Haile, T., Nigusse, T., & Dhanaraju, M. D. (2014). Nanotechnology: An effective tool for enhancing bioavailability and bioactivity of phytomedicine. *Asian Pacific Journal of Tropical Biomedicine, 4*, S1–S7. DOI: 10.12980/APJTB.4.2014C980

Gustavsen, S., Søndergaard, H. B., Linnet, K., Thomsen, R., Rasmussen, B. S., Sorensen, P. S., & Oturai, A. B. (2021). Safety and efficacy of low-dose medical cannabis oils in multiple sclerosis. *Multiple Sclerosis and Related Disorders, 48*, 102708.

Haddad, F., Dokmak, G., & Karaman, R. (2022). The efficacy of cannabis on multiple sclerosis-related symptoms. *Life (Chicago, Ill.), 12*(5), 682.

Hakami, A. Y., & Alshehri, F. S. (2025). Therapeutic potential of cannabinoids in neurological conditions: A systematic review of clinical trials. *Frontiers in Pharmacology, 16*, 1521792.

Han, K. H., Lim, S., Ryu, J., Lee, C. W., Kim, Y., Kang, J. H., & Kim, J. J. (2009). CB1 and CB2 cannabinoid receptors differentially regulate the production of reactive oxygen species by macrophages. *Cardiovascular Research, 84*(3), 378–386.

Henshaw, F. R., Dewsbury, L. S., Lim, C. K., & Steiner, G. Z. (2021). The effects of cannabinoids on pro-and anti-inflammatory cytokines: A systematic review of in vivo studies. *Cannabis and Cannabinoid Research, 6*(3), 177–195.

Hill, A. (2024). Health, Wellness, and a Balanced Endocannabinoid System. *Caring for the Ages, 25*(1), 15.

Iannotti, F. A. (2023). Cannabinoids, endocannabinoids, and synthetic cannabimimetic molecules in neuromuscular disorders. *International Journal of Molecular Sciences, 25*(1), 238.

Jean-Gilles, L., Braitch, M., Latif, M. L., Aram, J., Fahey, A. J., Edwards, L. J., & Constantinescu, C. S. (2015). Effects of pro-inflammatory cytokines on cannabinoid CB 1 and CB 2 receptors in immune cells. *Acta Physiologica (Oxford, England)*, *214*(1), 63–74.

Johnson, B. W., Strand, N. H., Raynak, J. C., Jara, C., Habtegiorgis, K., Hand, B. A., & Maloney, J. A. (2025). Cannabinoids in Chronic Pain Management: A Review of the History, Efficacy, Applications, and Risks. *Biomedicines*, *13*(3), 530.

Jones, É., & Vlachou, S. (2020). A critical review of the role of the cannabinoid compounds δ9-tetrahydrocannabinol (δ9-THC) and cannabidiol (CBD) and their combination in multiple sclerosis treatment. *Molecules (Basel, Switzerland)*, *25*, 4930.

Kaplan, B. L. (2013). The role of CB1 in immune modulation by cannabinoids. *Pharmacology & Therapeutics*, *137*(3), 365–374.

Kaszyńska, A. A. (2024). Cannabinoids: Potential for modulation and enhancement when combined with vitamin B12 in case of neurodegenerative disorders. *Pharmaceuticals*, *17*(6), 813.

Khan, H., Ghori, F. K., Ghani, U., Javed, A., & Zahid, S. (2022). Cannabinoid and endocannabinoid system: A promising therapeutic intervention for multiple sclerosis. *Molecular Biology Reports*, *49*(6), 5117–5131.

Kitdumrongthum, S., & Trachootham, D. (2023). An individuality of response to cannabinoids: Challenges in safety and efficacy of cannabis products. *Molecules (Basel, Switzerland)*, *28*(6), 2791.

Koppo, K. (2025). current knowledge and future. Cognitive Stimulants: from Caffeine to Cannabinoids-Current and Future Perspectives, 46.

Kopustinskiene, D. M., Masteikova, R., Lazauskas, R., & Bernatoniene, J. (2022). Cannabis sativa L. bioactive compounds and their protective role in oxidative stress and inflammation. *Antioxidants*, *11*(4), 660.

Kumar, M. R. P., Vijayalakshmi, C., & Ramanathan, M. (2023). Isoflavones as Nutraceuticals in Stroke: Therapeutic Targets and Signaling Pathways. In Martin, C. R., Patel, V. B., & Preedy, V. R. (Eds.), *Preedy Nutraceuticals, Supplements, and Herbal Medicine in Neurological Disorders* (pp. 959–978). Academic Press.

Kuzumi, A., Yamashita, T., Fukasawa, T., Yoshizaki-Ogawa, A., Sato, S., & Yoshizaki, A. (2024). Cannabinoids for the treatment of autoimmune and inflammatory skin diseases: A systematic review. *Experimental Dermatology*, *33*(3), e15064.

Lefebvre, È., Tawil, N., & Yahia, L. H. (2024). Transdermal delivery of cannabidiol for the management of acute inflammatory pain: A comprehensive review of the literature. *International Journal of Molecular Sciences*, *25*(11), 5858.

Legare, C. A., Raup-Konsavage, W. M., & Vrana, K. E. (2022). Therapeutic Potential of Cannabis, Cannabidiol, and Cannabinoid-Based Pharmaceuticals. *Pharmacology*, *107*(3-4), 131–149. DOI: 10.1159/000521683

Longoria, V., Parcel, H., Toma, B., Minhas, A., & Zeine, R. (2022). Neurological benefits, clinical challenges, and neuropathologic promise of medical marijuana: A systematic review of cannabinoid effects in multiple sclerosis and experimental models of demyelination. *Biomedicines, 10*(3), 539.

Lowe, H., Toyang, N., Steele, B., Bryant, J., & Ngwa, W. (2021). The endocannabinoid system: A potential target for the treatment of various diseases. *International Journal of Molecular Sciences, 22*(17), 9472.

Lucas, C. J. (2018). Galettis, Schneider J. The pharmacokinetics and the pharmacodynamics of cannabinoids. *British Journal of Clinical Pharmacology, 84*, 2477. DOI: 10.1111/bcp.13710

Manera, C., & Bertini, S. (2021). Cannabinoid-Based Medicines and Multiple Sclerosis. *Advances in Experimental Medicine and Biology, 1264*, 111–129.

Markovà, J., Essner, U., Akmaz, B., Marinelli, M., Trompke, C., Lentschat, A., & Vila, C. (2019). Sativex® as add-on therapy vs. further optimized first-line ANTispastics (SAVANT) in resistant multiple sclerosis spasticity: A double-blind, placebo-controlled randomised clinical trial. *The International Journal of Neuroscience, 129*, 119–128.

Martínez-Aguilar, L. M., Ibarra-Sánchez, A., Guerrero-Morán, D. J., Macías-Silva, M., Muñoz-Bello, J. O., Padilla, A., & González-Espinosa, C. (2023). Lysophosphatidylinositol promotes chemotaxis and cytokine synthesis in mast cells with differential participation of GPR55 and CB2 receptors. *International Journal of Molecular Sciences, 24*(7), 6316.

Martínez-Torres, A. M., & Morán, J. (2024). CB1 Receptor Activation Provides Neuroprotection in an Animal Model of Glutamate-Induced Excitotoxicity Through a Reduction of NOX-2 Activity and Oxidative Stress. *CNS Neuroscience & Therapeutics, 30*(11), e70099.

Martins, A. M., Gomes, A. L., Vilas Boas, I., Marto, J., & Ribeiro, H. M. (2022). Cannabis-based products for the treatment of skin inflammatory diseases: A timely review. *Pharmaceuticals, 15*(2), 210.

Matheson, J., Zhou, X. M. M., Bourgault, Z., & Le Foll, B. (2021). Potential of fatty acid amide hydrolase (FAAH), monoacylglycerol lipase (MAGL), and diacylglycerol lipase (DAGL) enzymes as targets for obesity treatment: A narrative review. *Pharmaceuticals, 14*(12), 1316.

Millar, S. A., Stone, N. L., Bellman, Z. D., Yates, A. S., England, T. J., & O'Sullivan, S. E. (2019). A systematic review of cannabidiol dosing in clinical populations. *British Journal of Clinical Pharmacology, 85*(9), 1888–1900. DOI: 10.1111/bcp.1403831222854

Miralpeix, C., Reguera, A. C., Fosch, A., Zagmutt, S., Casals, N., Cota, D., & Rodríguez-Rodríguez, R. (2021). Hypothalamic endocannabinoids in obesity: An old story with new challenges. *Cellular and Molecular Life Sciences : CMLS*, 1–22.

Murphy, M., Kaur, V., Bui, H. L., Yang, T., Erridge, S., Holvey, C., & Sodergren, M. H. (2024). Clinical outcome analysis of patients with multiple sclerosis–Analysis from the UK Medical Cannabis Registry. *Multiple Sclerosis and Related Disorders, 87*, 105665.

Mustafa, W., Elgendy, N., Salama, S., Jawad, M., & Eltoukhy, K. (2021). The effect of cannabis on the clinical and cytokine profiles in patients with multiple sclerosis. *Multiple Sclerosis International, 2021*(1), 6611897.

Narouze, S. N. (2021). Cannabinoids and pain: mechanisms of action. In *Cannabinoids and Pain* (pp. 191–204). Springer International Publishing.

Nouh, R. A., Kamal, A., & Abdelnaser, A. (2023). Cannabinoids and multiple sclerosis: A critical analysis of therapeutic potentials and safety concerns. *Pharmaceutics, 15*(4), 1151.

Nouh, R. A., Kamal, A., Oyewole, O., Abbas, W. A., Abib, B., Omar, A., & Abdelnaser, A. (2024). Unveiling the potential of cannabinoids in multiple sclerosis and the dawn of nano-cannabinoid medicine. *Pharmaceutics, 16*(2), 241.

Nouh, R. A., Kamal, A., Oyewole, O., Abbas, W. A., Abib, B., Omar, A., Mansour, S. T., & Abdelnaser, A. (2024). Unveiling the Potential of Cannabinoids in Multiple Sclerosis and the Dawn of Nano-Cannabinoid Medicine. *Pharmaceutics, 16*, 241.

O'Brien, K. (2022). Cannabidiol (CBD) in cancer management. *Cancers (Basel), 14*(4), 885.

O'Brien, K., Blair, P., O'Brien, K., & Blair, P. (2021). Endocannabinoid system. Medicinal Cannabis and CBD in Mental Healthcare, 7-56.

Oddi, S., Fiorenza, M. T., & Maccarrone, M. (2023). Endocannabinoid signaling in adult hippocampal neurogenesis: A mechanistic and integrated perspective. *Progress in Lipid Research, 91*, 101239.

Ojha, S., & Kumar, B. (2018). A review on nanotechnology based innovations in diagnosis and treatment of multiple sclerosis. *J. Cell Immunother., 4*, 56–64. DOI: 10.1016/j.jocit.2017.12.001

Patravale, V. B., Date, A. A., & Kulkarni, R. M. (2004). Nanosuspensions: A promising drug delivery strategy. *The Journal of Pharmacy and Pharmacology, 56*, 827–840.

Pete, D. D., & Narouze, S. N. (2021). Endocannabinoids: Anandamide and 2 arachidonoylglycerol (2-AG). In *Cannabinoids and Pain* (pp. 63–69). Springer International Publishing.

Quintero, J. M., Diaz, L. E., Galve-Roperh, I., Bustos, R. H., Leon, M. X., Beltran, S., & Dodd, S. (2024). The endocannabinoid system as a therapeutic target in neuropathic pain: A review. *Expert Opinion on Therapeutic Targets, 28*(9), 739–755.

Rabinow, B. E. (2004). Nanosuspensions in drug delivery. *Nature Reviews. Drug Discovery, 3*, 785–796. DOI: 10.1038/nrd1494

Rahaman, O., & Ganguly, D. (2021). Endocannabinoids in immune regulation and immunopathologies. *Immunology, 164*(2), 242–252.

Rakotoarivelo, V., Mayer, T. Z., Simard, M., Flamand, N., & Di Marzo, V. (2024). The impact of the CB2 cannabinoid receptor in inflammatory diseases: An update. *Molecules (Basel, Switzerland), 29*(14), 3381.

Rakotoarivelo, V., Sihag, J., & Flamand, N. (2021). Role of the endocannabinoid system in the adipose tissue with focus on energy metabolism. *Cells*, *10*(6), 1279.

Rezende, B., Alencar, A. K. N., de Bem, G. F., Fontes-Dantas, F. L., & Montes, G. C. (2023). Endocannabinoid system: Chemical characteristics and biological activity. *Pharmaceuticals*, *16*(2), 148.

Rice, J., & Cameron, M. (2018). Cannabinoids for treatment of MS symptoms: State of the evidence. *Current Neurology and Neuroscience Reports*, *18*(8), 50.

Sacco, R., Riccitelli, G. C., Disanto, G., Bogousslavsky, J., Cavelti, A., Czell, D., & Zecca, C. (2024). Effectiveness, Safety and Patients' Satisfaction of Nabiximols (Sativex®) on Multiple Sclerosis Spasticity and Related Symptoms in a Swiss Multicenter Study. *Journal of Clinical Medicine*, *13*(10), 2907.

Sharon, N., Yarmolinsky, L., Khalfin, B., Fleisher-Berkovich, S., & Ben-Shabat, S. (2024). Cannabinoids' Role in Modulating Central and Peripheral Immunity in Neurodegenerative Diseases. *International Journal of Molecular Sciences*, *25*(12), 6402.

Shkodina, A. D., Bardhan, M., & Chopra, H., (2024). Pharmacological and Non-pharmacological Approaches for the Management of Neuropathic Pain in Multiple Sclerosis. *CNS Drugs*, *38*, 205–224.

Śmiarowska, M., Białecka, M., & Machoy-Mokrzyńska, A. (2022). Cannabis and cannabinoids: Pharmacology and therapeutic potential. *Neurologia i Neurochirurgia Polska*, *56*(1), 4–13.

Squintani, G., Donato, F., Turri, M., Deotto, L., Teatini, F., Moretto, G., & Erro, R. (2016). Cortical and spinal excitability in patients with multiple sclerosis and spasticity after oromucosal cannabinoid spray. *Journal of the Neurological Sciences*, *370*, 263–268.

Strand, N., D'souza, R. S., Karri, J., Kalia, H., Weisbein, J., Kassa, B. J., & Abd-Elsayed, A. (2023). Medical cannabis: A review from the American society of pain and neuroscience. *Journal of Pain Research*, •••, 4217–4228.

Tait, M. A., Costa, D. S., Campbell, R., Norman, R., Warne, L. N., Schug, S., & Rutherford, C. (2023). Health-related quality of life in patients accessing medicinal cannabis in Australia: The QUEST initiative results of a 3-month follow-up observational study. *PLoS One*, *18*(9), e0290549.

Tylawsky, D. E., Kiguchi, H., Vaynshteyn, J., Gerwin, J., Shah, J., Islam, T., Boyer, J. A., Boué, D. R., Snuderl, M., & Greenblatt, M. B.. (2023). P-selectin-targeted nanocarriers induce active crossing of the blood–brain barrier via caveolin-1-dependent transcytosis. *Nature Materials*, *22*, 391–399. DOI: 10.1038/s41563-023-01481-9

Vallée, A., Vallée, J. N., & Lecarpentier, Y. (2021). Potential role of cannabidiol in Parkinson's disease by targeting the WNT/β-catenin pathway, oxidative stress and inflammation. *Aging (Albany NY)*, *13*(7), 10796.

Vasincu, A., Rusu, R. N., Ababei, D. C., Larion, M., Bild, W., Stanciu, G. D., & Bild, V. (2022). Endocannabinoid modulation in neurodegenerative diseases: In pursuit of certainty. *Biology (Basel)*, *11*(3), 440.

Vuic, B., Milos, T., Tudor, L., Konjevod, M., Nikolac Perkovic, M., Jazvinscak Jembrek, M., & Svob Strac, D. (2022). Cannabinoid cb2 receptors in neurodegenerative proteinopathies: New insights and therapeutic potential. *Biomedicines*, *10*(12), 3000.

Wang, T., Collet, J. P., Shapiro, S., & Ware, M. A. (2008). Adverse effects of medical cannabinoids: A systematic review. *Canadian Medical Association Journal*, *178*(13), 1669–1678. DOI: 10.1503/cmaj.07117818559804

Yarar, E. (2021). Role and function of endocannabinoid system in major depressive disease. *Medical Cannabis and Cannabinoids*, *4*(1), 1–12.

Zajicek, J. P., Hobart, J. C., Slade, A., Barnes, D., & Mattison, P. G.MUSEC Research Group. (2012). Multiple sclerosis and extract of cannabis: Results of the MUSEC trial. *Journal of Neurology, Neurosurgery, and Psychiatry*, *83*(11), 1125–1132.

Zajicek, J. P., Sanders, H. P., Wright, D. E., Vickery, P. J., Ingram, W. M., Reilly, S. M., & Thompson, A. J. (2005). Cannabinoids in multiple sclerosis (CAMS) study: Safety and efficacy data for 12 months follow up. *Journal of Neurology, Neurosurgery, and Psychiatry*, *76*(12), 1664–1669.

Zeng, Y., Li, Z., Zhu, H., Gu, Z., Zhang, H., & Luo, K. (2020). Recent Advances in Nanomedicines for Multiple Sclerosis Therapy. *ACS Applied Bio Materials*, *3*, 6571–6597. DOI: 10.1021/acsabm.0c00953

Zeraatkar, D., Cooper, M. A., Agarwal, A., Vernooij, R. W., Leung, G., Loniewski, K., & Busse, J. W. (2022). Long-term and serious harms of medical cannabis and cannabinoids for chronic pain: A systematic review of non-randomised studies. *BMJ Open*, *12*(8), e054282.

Zhang, Q., Melchert, P. W., & Markowitz, J. S. (2024). Pharmacokinetic variability of oral cannabidiol and its major metabolites after short-term high-dose exposure in healthy subjects. *Medical Cannabis and Cannabinoids*, *7*(1), 1–9.

Chapter 9
Therapeutic Applications of Secondary Metabolites From Cannabis Sativa in Neurodegenerative Diseases

Hamid Kabdy
https://orcid.org/0009-0004-7438-3638
Laboratory of Pharmacology, Neurobiology, and Behavior, Semlalia Faculty of Sciences, Cadi Ayyad University, Morocco

Aitbaba Abdelfatah
https://orcid.org/0000-0002-5650-3868
Biology, Health, and Ecology Laboratory, Department of Biology, Faculty of Science, Abdelmalek Essaâdi University, Morocco

Azraida Hajar
Cadi Ayyad University, Morocco

Jaouhari Yasmina
https://orcid.org/0009-0007-4795-3051
Laboratory of Pharmacology, Neurobiology, Anthropobiology, and Environment, Morocco

Oufquir Sara
https://orcid.org/0000-0001-5014-7025
Cadi Ayyad University, Morocco

Agouram Fatimazahra
https://orcid.org/0000-0002-8073-2126
Cadi Ayyad University, Morocco

El Yazoulia Loubna
Cadi Ayyad University, Morocco

Chait Abderrahman
Laboratory of Pharmacology, Neurobiology, Anthropobiology, and Environment, Morocco

ABSTRACT

Neurodegenerative diseases (NDDs) are progressive disorders involving the loss of neuronal structure and function, leading to cognitive, behavioral, and motor decline. With curative therapies lacking, interest has turned to naturally derived compounds with neuroprotective potential. Plant secondary metabolites show anti-inflammatory, antioxidant, and neuroprotective effects, making them promising candidates for NDD management. Cannabis sativa produces cannabinoids such as cannabidiol (CBD) and tetrahydrocannabinol (THC), along with terpenes and flavonoids,

DOI: 10.4018/979-8-3373-1365-8.ch009

which act on the endocannabinoid system to regulate neuroinflammation, oxidative stress, and neuronal plasticity, showing benefits in multiple sclerosis, epilepsy, anxiety, insomnia, Alzheimer's, and Parkinson's disease. Likewise, constituents of Nicotiana tabacum, including nicotine and related alkaloids, can influence neuroinflammatory pathways and provide neuroprotection at controlled doses. This chapter examines their therapeutic potential, mechanisms, evidence, and translational challenges.

INTRODUCTION

The fight against neurodegenerative diseases remains one of the greatest challenges of modern medicine. While science has achieved remarkable breakthroughs in many fields, neurodegeneration remains a territory riddled with uncertainties (Ali, 2024). Diseases such as Alzheimer's, Parkinson's, and multiple sclerosis progressively strip millions of patients of their autonomy, affecting not only their quality of life but also that of their families. To date, no therapeutic intervention has been able to halt or reverse their progression, despite decades of intensive research. Existing treatments can alleviate symptoms and improve patients' quality of life, but they are unable to stop the advancement of these devastating conditions (Palanisamy et al., 2025).

In the face of this deadlock, a fresh perspective turns toward nature, and more specifically toward the secondary metabolites produced by medicinal plants with remarkable pharmacological properties (Vaou et al., 2025). These bioactive compounds, often overlooked, hold promise as innovative solutions to the limitations of conventional treatments (Vaou et al., 2025). Their diverse mechanisms of action, including anti-inflammatory (Laadraoui et al., 2025; Nunes et al., 2020), neuroprotective (Mohd Sairazi & Sirajudeen, 2020), and antioxidant effects (P. A. Silva et al., 2019), make them compelling candidates for addressing the complex pathways involved in neurodegeneration (Del Río & Segura-Carretero, 2023). Among these, this plants with contrasting profiles stand out: *Cannabis sativa (Cs)*, a symbol of controversy but also of scientific innovation, a paradoxical plant oscillating between health risk and therapeutic opportunity. this plant highlight the untapped therapeutic potential of secondary metabolites and invite further exploration into their role in mitigating the burden of neurodegenerative diseases (Nagarajan et al., 2025).

Cs, through its cannabinoids such as cannabidiol (CBD) and tetrahydrocannabinol (THC), represents a promising source of therapeutic solutions (Bukowska, 2024). CBD, being non-psychoactive, is distinguished by its anti-inflammatory, antioxidant, and analgesic properties (El-Mernissi et al., 2024), while THC plays a key role in pain modulation and the reduction of muscle spasms. These compounds, by interacting with the endocannabinoid system, target essential mechanisms such as neuroinflammation, oxidative stress, and neuronal plasticity (Bukowska, 2024).

Recent studies highlight their potential in treating neurodegenerative diseases such as Alzheimer's and Parkinson's, by reducing inflammation and protecting neurons (Stasiłowicz-Krzemień et al., 2024). However, further research is needed to elucidate their mechanisms of action and mitigate risks, particularly those associated with the psychoactive effects of THC, Cannabis presents promising potential for reducing the impact of neurodegenerative diseases (Stasiłowicz-Krzemień et al., 2024).

In this context, other metabolites present in *Cs* also play a key role in its therapeutic mechanisms, notably terpenoids and polyphenols. These bioactive compounds, with diverse pharmacological properties, are found in various types of plant extracts, such as essential oils and alcoholic extracts. Terpenoids are known for their anti-inflammatory, antioxidant, and neuroprotective effects, while polyphenols contribute to reducing oxidative stress and modulating cellular signaling pathways involved in neurodegenerative processes(Kumar et al., 2025). Together, these compounds provide *Cs* with a broadened therapeutic potential, warranting further investigation into their mechanisms of action and their synergy with major cannabinoids.

These discoveries open new avenues for the development of targeted therapies that harness the neuroprotective properties of secondary metabolites while minimizing adverse effects. This chapter offers an innovative approach to investigating the interactions between bioactive compounds derived from Cs and the complex mechanisms underlying neurodegenerative diseases. The objective is to conduct an in-depth analysis of the specific pharmacological properties of these molecules, with a particular focus on the molecular pathways involved in their therapeutic potential. This analysis will be grounded in a comprehensive body of data from both experimental and clinical studies. Finally, ethical and regulatory considerations often overlooked will be addressed in order to assess the challenges and opportunities associated with the integration of these metabolites into modern medical practice.

1. GEOGRAPHICAL ORIGINS

In nature, cannabis is an annual flowering plant, meaning it completes its life cycle from germination to seed production within a single year (Haggarty & Rabinak, 2025). This species originates from equatorial and subtropical regions, primarily in Central Asia (McPartland et al., 2019), where two locations are believed to be its cradle: the foothills of the Himalayas and the plains of the Pamir . However, the plant has a wide geographic distribution, growing in Canada, the United States of America, Europe, and Africa (Zimniewska, 2022). Cannabis is an ancient plant, but the enthusiasm it has generated, particularly since at least the past century has significantly transformed both the plant itself and the world. It is likely the first plant ever domesticated by humans (Li, 1974). The cultivation of *Cs* in Morocco has undergone a significant geographical expansion since the 1980s, initially concentrated in the historical Rif region. Driven by socioeconomic factors and the diffusion of agricultural knowledge among seasonal laborers, cannabis farming progressively spread beyond its traditional zones. During periods of relative tolerance from authorities, returning workers introduced cannabis cultivation to new rural areas. By the 1990s, increasing land scarcity and intensified forest controls prompted the search for new cultivable areas, particularly in the province of Taounate (Bachir et al., 2022). Collaborations with local farmers led to the implementation of modern agricultural techniques including the use of fertilizers, pesticides, mechanized tools, and advanced irrigation systems especially in less topographically constrained regions such as the irrigated valleys of Chefchaouen. As a result, cannabis cultivation expanded from a few dozen hectares in the 1950s to over 100,000 hectares by the early 2000s, mainly distributed across the provinces of Al Hoceima, Chefchaouen, Taounate, Tétouan, and Larache (Afsahi, 2017; Bachir et al., 2022).

2. SECONDARY METABOLITES FROM *CANNABIS SATIVA*

Cs, commonly referred to as "marijuana" or "cannabis," is a designation that encompasses all parts of the plant, whether growing or not, including the seeds, the resin extracted from any part of the plant, and any compound, manufactured product, salt, derivative, mixture, or preparation derived from the plant, its seeds, or its resin (Bouarfa et al., 2020). *Cs* is an herbaceous, flowering, and dioecious plant. From a taxonomic perspective, it is generally classified as a single species, *Cs*, comprising three principal varieties: Cannabis var. sativa, *Cannabis var. indica*, and *Cannabis var. ruderalis*. Each variety exhibits distinct morphological traits and is associated with specific phytochemical profiles and applications. Nevertheless, the three subspecies or varieties of cannabis can interbreed, resulting in blurred boundaries between them. Consequently, some researchers advocate for recognizing cannabis as a single species, *Cs* (Bouarfa et al., 2020). The two most common varieties, *C. sativa* var. *sativa* and *C. sativa* var. *indica*, are characterized by their high levels of THC, making them particularly sought after for both medicinal and recreational purposes (Kumar et al., 2025). This trait significantly influences their market demand and therapeutic applications. Furthermore, the term "Cannabis" is specifically defined as the drug derived from the herbaceous, dicotyledonous, dioecious, apetalous, annual plant *C.sativa* L., along with its variants within the Cannabaceae family. This plant exhibits a unique biosynthesis of terpenophenolic cannabinoids and phytocannabinoids, which primarily accumulate in the glandular trichomes of the plant (Bouarfa et al., 2020). 1. 2. 3.

2.1. Chemical Composition

To date, more than 500 compounds such as cannabinoids, flavonoids, terpenes, and fatty acids have been identified in the leaves and buds of the plant (Alves et al., 2020). Cannabinoids belong to a class consisting of 86 identified terpenophenolic secondary metabolites found in C. sativa. Another major class of compounds produced by this plant includes the 140 terpenes identified so far. Other classes include 50 identified hydrocarbons, 34 sugars and related compounds, 27 nitrogen-containing compounds, 25 non-cannabinoid phenols, 23 fatty acids, 23 flavonoids, 20 simple acids, 13 simple ketones, 13 simple esters and lactones, 12 simple aldehydes, 11 proteins, 11 steroids, 9 elements, 3 vitamins, and 2 pigments. It is important to note that no free flavonoids have been identified in C. sativa; instead, they are present in their corresponding glycoside forms. Given the wide array of molecules it contains, it is understandable that cannabis is a highly complex plant with numerous pharmacological effects and medical applications (Hartsel et al., 2016).

2.2. Phytocannabinoïds

Currently, 538 natural compounds have been identified in C. sativa. Among them, over 100 are classified as phytocannabinoids due to their shared chemical structure (Hanuš, 2009). From a chemical standpoint, phytocannabinoids have a lipid-based structure that includes alkylresorcinol and monoterpene components within their molecules (Hanuš et al., 2016). Moreover, they are predominantly found in the resin secreted by the trichomes of female plants, while male leaves possess fewer glandular trichomes and produce only small amounts of psychoactive compounds. Phytocannabinoids are categorized into neutral cannabinoids (lacking a carboxyl group) and cannabinoid acids (containing a carboxyl group) (Hanuš, 2009). In *Cs*, cannabinoids are biosynthesized and accumulate in the form of cannabinoid acids, which are then decarboxylated into their neutral forms. Specifically, the alkylation of olivetolic acid with geranyl pyrophosphate by a prenyltransferase yields cannabigerolic acid (CBGA). Through the action of cannabinoid synthase enzymes, CBGA leads to the formation of cannabidiol acid (CBDA), cannabichromenic acid (CBCA), and Δ-9-THCA (Sirikantaramas & Taura, 2017).

2.3. Terpenoids

More than 200 terpenoids, which contribute to the aroma of *C. sativa*, have been identified in the plant's flowers and leaves, potentially representing up to 10% of trichome content (Booth et al., 2017). Among these, limonene, myrcene, and pinene are the most prevalent and are highly volatile. These compounds act as insect repellents and possess anti-feedant properties against herbivorous animals. The combination of different terpenoids and phytocannabinoid acids in C. sativa creates a synergistic mechano-chemical defense strategy against various predators (Farag & Kayser, 2017). The production of terpenoids varies according to specific environmental conditions. For example, similar to phytocannabinoids, terpenoids are produced as a defense mechanism. Their concentration increases in response to light exposure but decreases with higher soil fertility (Farag & Kayser, 2017). As lipophilic molecules, terpenoids interact with animal cell membranes, both in the brain and in peripheral tissues (Russo, 2016). Notably, olfactory tests conducted on mice exposed to terpenoids for one hour revealed significant behavioral changes, suggesting a potential impact on brain function (Russo, 2016).

2.4. Compounds Derived From *Cannabis sativa*

The most commonly produced and consumed form of *Cs* worldwide is dried cannabis (Chouvy, 2019). However, this raw material can be processed into various formulations suited for multiple routes of administration. Among the most widely used are cannabis oils, which are typically administered orally or sublingually (Spindle et al., 2020). In addition, the consumption of food products infused with cannabis extracts is also widespread (Grof, 2018). These liquid extracts are currently receiving considerable attention in the medical field, as they allow for the development of more standardized derivatives that may be better suited for therapeutic applications (Sarma et al., 2020). *Cs L.* is well known for its richness in secondary metabolites, particularly cannabinoids, terpenes, and flavonoids, which give rise to a wide functional diversity in its derivatives. Among the approximately 120 identified cannabinoids, Δ-9-THC is the primary psychoactive compound, while CBD, which is non-euphoric, has been extensively studied for its anti-inflammatory, anxiolytic, antioxidant, and neuroprotective effects. These molecules are predominantly concentrated in the glandular trichomes located on the inflorescences of the female plant (Malabadi et al., 2023). The extraction of these active compounds can be achieved through various physico-chemical processes, such as supercritical CO_2 extraction, organic solvent extraction, or lipid infusion, resulting in a wide range of products including concentrated oils, resins, tinctures, crystalline isolates, and full-spectrum distillates (Sarma et al., 2020). Currently, the market offers a wide range of hemp oil-based products, including topical formulations that primarily contain CBD as the active ingredient. In contrast, the limited scientific data on the therapeutic potential of cannabis roots has discouraged major companies from investing in the development and commercialization of root-based products, unlike the more widely accepted hemp-derived oils (Booth et al., 2017). At present, various do-it-yourself recipes are available for preparing plant-based topical products. This suggests the possibility of formulating topicals using cannabis root extracts, particularly for the relief of localized inflammatory conditions, such as eczema, which is among the most frequently reported traditional uses. However, the main limitation remains the lack of comprehensive information regarding the chemical composition of the roots and the biological effects of their active constituents. Moreover, the industrial derivatives of Cs are widely utilized for their fibers, seeds, and biomass. Hemp fibers, known for their length and durability, are used in the production of technical textiles, fine papers, composite materials, and environmentally friendly insulation. Hemp seeds, which are rich in complete proteins, essential fatty acids (omega-3 and omega-6), and minerals, are highly valued in both human and animal nutrition. The oil extracted from these seeds, although devoid of cannabinoids, is highly sought after in the cosmetic and nutraceutical industries due to its balanced lipid profile. Additionally, the plant's lignocellulosic residues can be converted into bioplastics, biomaterials, or biofuel, thereby integrating Cs into a circular economy model (Mechoulam, 2019). The remarkable diversity of Cs extracts and derivatives highlights its multidisciplinary potential, positioned at the intersection of biotechnology, pharmacognosy, agro-industry, and green chemistry. The plant continues to attract growing scientific interest, particularly in light of recent advances in cannabinoid pharmacology.

3. METHODS OF CONSUMING CANNABIS

There are several scientifically recognized methods of administering cannabis products, including inhalation, ingestion, mucosal absorption, and topical application. Each route of administration influences how the active compounds primarily delta-9-THC and CBD are absorbed, distributed, metabolized, and ultimately how they exert their pharmacological effects (Goodman et al., 2019).

Inhalation is the most commonly used method of cannabis consumption. This route involves the absorption of active compounds through the respiratory system, allowing for rapid transfer of cannabinoids from the lungs into the bloodstream. Inhalation can be classified into two main categories: combustion-based and non-combustion-based methods (Goodman et al., 2019). Combustion involves burning the cannabis product (e.g., joints, blunts, hand pipes, or water pipes) to produce smoke (Borodovsky et al., 2016). In contrast, non-combustion methods heat cannabis below its combustion point to produce an aerosol rather than smoke. These include vaporization (using a vaporizer for dried cannabis or hashish), dabbing (which involves heating cannabis concentrates such as resin or distillate to high temperatures), and vaping (using electronic devices to aerosolize liquid cannabis extracts). Due to the rapid pulmonary absorption, the effects of inhaled cannabis are typically felt within seconds to minutes, with peak intensity occurring between 15 and 30 minutes post-inhalation (Borodovsky et al., 2016).

The second major route of administration is ingestion, which involves the oral intake of cannabis products that are subsequently metabolized via the gastrointestinal and hepatic systems. Edible cannabis products include infused beverages, oils (in liquid or capsule form), and food items such as dried fruit or crackers (Florimbio et al., 2023). Since the legalization of cannabis, oral ingestion has become increasingly popular in regions such as Morocco (Mzali et al., 2023). When cannabis is consumed orally, THC is metabolized by the liver into 11-hydroxy-THC (11-OH-THC), a metabolite known to be equally or even more potent than THC itself. As a result of this first-pass metabolism, the onset of effects is delayed typically occurring 30 minutes to over an hour after consumption—with peak effects observed between 2 and 4 hours (Florimbio et al., 2023).

Mucosal absorption is another route that utilizes the vascularized tissues of the oral, sublingual, vaginal, or rectal mucosa. Products designed for this route include suppositories, lozenges, oral sprays, and sublingual strips. In this case, cannabinoids are absorbed directly into the systemic circulation through the mucosal membranes, bypassing significant first-pass metabolism in the liver. Effects generally begin within approximately 30 minutes, with peak intensity occurring between 1 and 4 hours, depending on the specific product and site of application (Borodovsky et al., 2023).

Topical application involves applying cannabis-infused products directly to the skin, such as creams, balms, or exfoliants, typically for localized relief or cosmetic purposes. These formulations act locally at the site of application and generally do not result in systemic absorption of cannabinoids, meaning they do not produce psychoactive effects. However, transdermal patches designed to deliver cannabinoids into the bloodstream through the skin represent an exception.

It is essential to recognize that the choice of administration route significantly influences the pharmacokinetics and pharmacodynamics of cannabis. Each method varies in terms of onset time, peak effect, duration, and intensity, which in turn affects both the desired therapeutic outcomes and the risk of adverse effects. For instance, inhalation leads to a rapid onset and relatively short duration of action, while ingestion results in delayed but longer-lasting effects. Understanding these distinctions is critical for both clinical and recreational users to ensure safe and effective use of cannabis products.

Figure 1. Methods of consuming cannabis

4. CANNABIS AND NEURODEGENERATIVE DISEASES

4.1. Mode of Action

Cs is a medicinal plant with a remarkable phytochemical diversity, widely investigated for its therapeutic potential in various diseases, particularly neurodegenerative disorders. Its secondary metabolites-mainly phytocannabinoids, terpenes, and flavonoids-exert multifaceted biological activities through diverse cellular and molecular mechanisms. The endocannabinoid system (ECS) represents the principal target through which *Cs* modulates central nervous system function (Wu, 2019). This system consists of G protein–coupled receptors CB1 and CB2, their endogenous ligands anandamide (AEA) and 2-arachidonoylglycerol (2-AG), and enzymes responsible for their synthesis and degradation. Phytocannabinoids such as THC and CBD interact directly or indirectly with these receptors, influencing key physiological processes including neuroinflammation, neuronal plasticity, and cell survival (Atalay et al., 2019; Mechoulam & Parker, 2013)

THC, the primary psychoactive constituent of cannabis, acts as a partial agonist of CB1 and CB2 receptors, leading to inhibition of excitatory neurotransmitter release such as glutamate and dopamine. This inhibition contributes to the reduction of excitotoxicity, a central mechanism in the pathogenesis of several neurodegenerative diseases (Pertwee, 2008). THC also modulates intracellular signaling pathways by activating adenylyl cyclase and ERK1/2 kinases, thereby promoting neuronal plasticity and survival (Bockmann et al., 2023). However, its psychoactive effects limit clinical use, prompting greater interest in CBD, which lacks psychotropic properties and exhibits a safer pharmacological profile.

CBD exerts neuroprotective and anti-inflammatory actions through multiple mechanisms (Atalay et al., 2019). Although it shows low affinity for CB1 and CB2 receptors, CBD functions as a negative allosteric modulator of CB1, mitigating the adverse psychotropic effects of THC (Pertwee, 2008). Moreover, CBD interacts with several non-cannabinoid molecular targets, including peroxisome proliferator-activated receptor gamma (PPARγ), serotonin 5-HT1A receptors, transient receptor potential vanilloid 1 (TRPV1) channels, and adenosine A2A receptors (Bih et al., 2015; Fernández-Ruiz et al., 2013). Through PPARγ activation, CBD modulates transcriptional activity of genes implicated in oxidative stress and inflammation, downregulating pro-inflammatory cytokines such as TNF-α, IL-1β, and IL-6 (Campos et al., 2012). This anti-inflammatory action, together with its potent antioxidant properties, contributes to slowing the progression of neurodegeneration.

The antioxidant effects of CBD stem both from its ability to directly scavenge reactive oxygen species (ROS) and from its regulatory influence on cellular antioxidant defenses (Pereira et al., 2021). CBD enhances the activity of superoxide dismutase (SOD) and catalase while reducing the formation of lipid peroxidation products such as malondialdehyde (Atalay et al., 2019). It protects neurons from glutamate-induced cytotoxicity and restores mitochondrial integrity by improving respiratory function and ATP synthesis (Campos et al., 2012; Mehta et al., 2013). Furthermore, CBD stabilizes mitochondrial membranes, preventing cytochrome c release and subsequent activation of apoptotic pathways. These effects have been validated in experimental models of Alzheimer's and Parkinson's diseases, in which CBD improved cognitive performance, decreased neuroinflammation, and protected dopaminergic neurons (Martín-Moreno et al., 2011).

Beyond CBD and THC, other phytocannabinoids such as cannabigerol (CBG), cannabichromene (CBC), and cannabidivarin (CBDV) contribute to the overall pharmacological profile of *Cs*. CBG interacts with α2-adrenergic and 5-HT1A receptors, producing neuroprotective and anxiolytic effects (Cascio et al., 2010). CBC modulates TRPA1 and TRPV3 channels, thereby influencing calcium signaling and synaptic transmission (De Petrocellis et al., 2011). These combined interactions create a synergistic effect known as the "entourage effect," where the therapeutic activity of the whole plant extract surpasses that of isolated compounds (Russo, 2011). This synergy results from the cooperative actions of cannabinoids, terpenes, and flavonoids naturally coexisting in the plant.

Terpenes and essential oils from *Cs* also play crucial roles in its neuropharmacological profile. β-caryophyllene, one of the most abundant terpenes, acts as a selective CB2 receptor agonist devoid of psychoactive effects, conferring potent anti-inflammatory and analgesic properties (Kabdy, Baslam, et al., 2024). In animal models of neuroinflammation, β-caryophyllene reduced microglial activation and suppressed nitric oxide (NO) and cytokine production, thereby protecting neurons from degeneration. Other terpenes such as linalool, myrcene, limonene, and pinene exhibit anxiolytic, sedative, and antinociceptive effects through modulation of GABAergic and glutamatergic neurotransmission (Nuutinen, 2018; Russo, 2011). Linalool decreases neuronal hyperexcitability and enhances sleep quality, while myrcene induces muscle relaxation and potentiates the analgesic effects of THC (Al-Harrasi et al., 2022).

Essential oils derived from *Cs*, obtained through hydrodistillation or supercritical CO_2 extraction, contain complex mixtures of monoterpenes and sesquiterpenes responsible for diverse therapeutic properties (Kabdy, Baslam, et al., 2024). These volatile compounds exert significant antioxidant activity and can potentiate the actions of phytocannabinoids by modifying membrane permeability and enhancing cannabinoid bioavailability (Rahn & Hohmann, 2009). Recent studies have also demonstrated that *Cs* essential oils inhibit acetylcholinesterase and butyrylcholinesterase, enzymes involved in acetylcholine degradation, suggesting their potential in improving cholinergic transmission in Alzheimer's disease and related disorders (Nuutinen, 2018).

Recently, research has highlighted the significant anti-inflammatory and anti-arthritic effects of Cs essential oils (CSEO) (Kabdy, Azraida, et al., 2024). This study provides compelling evidence supporting their therapeutic potential in the management of arthritis and inflammatory diseases. The findings not only underscore the efficacy of CSEO in reducing inflammation but also emphasize the need for further in-depth investigations to elucidate the underlying molecular mechanisms and to confirm their safety and efficacy in clinical settings. Preliminary data from this research offer promising prospects for the development of new strategies for the prevention and treatment of inflammatory pathologies (Kabdy, Azraida, et al., 2024).

Current evidence indicates that *Cs* extracts and essential oils act through pleiotropic mechanisms on the nervous system, combining anti-inflammatory, antioxidant, neuromodulatory, and neuroprotective effects. These actions involve interactions with multiple molecular targets including CB1 and CB2 receptors, PPARγ, 5-HT1A, TRPV1, and A2A receptors, as well as modulation of oxidative stress enzymes (Kaszyńska, 2024). This complex pharmacological profile gives *Cs* great therapeutic potential in managing conditions such as multiple sclerosis, epilepsy, Alzheimer's and Parkinson's diseases, depression, and anxiety disorders (Cassano et al., 2020; Fernández-Ruiz et al., 2013). Nevertheless, despite these encouraging findings, challenges remain regarding extract standardization, chemical variability among cultivars, and long-term safety considerations. Regulatory and ethical barriers also continue to limit clinical translation in many regions.

Figure 2. Cannabis sativa and neurodegenerative diseases

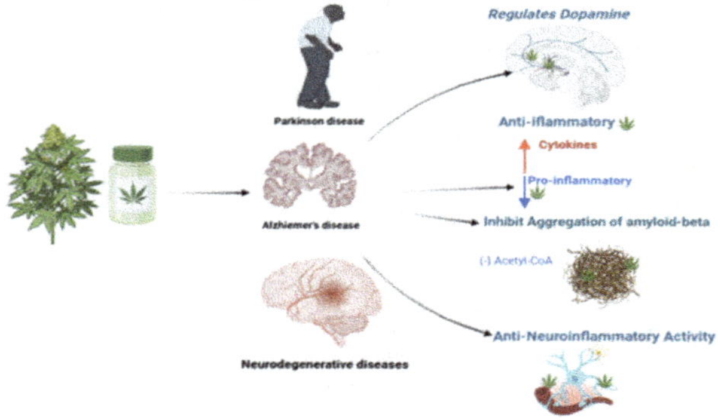

4.2. Cannabis and the Endocannabinoid System

The ECS principally consists of cannabinoid receptors (CB1 and CB2), endogenous cannabinoids (anandamide and 2-AG), and the enzymes that modulate their synthesis and degradation (O'Brien et al., 2021). Cannabis's main psychoactive compound, THC, binds to CB1 receptors in the brain, notably in areas such as the hippocampus, cortex, and basal ganglia, which are involved in memory, emotion, and motor function. Activating CB1 receptors reduces glutamate release, leading to decreased neuroexcitability and neuroinflammation, which are critical in the pathogenesis of neurodegenerative diseases like Alzheimer's and Parkinson's (Kumar, 2024).

Cannabidiol, however, has a less direct effect on CB1 receptors, but is thought to modulate the ECS in a more subtle way, potentiating the effects of endogenous cannabinoids by inhibiting their breakdown (Corroon and Felice, 2019). CBD possesses antioxidant and anti-inflammatory properties that may prevent the neurotoxic effects of chronic neurodegeneration (Jha et al., 2024). Recently, multi-directional biological actions have been reported in different preclinical models, such as the antioxidant and anti-inflammatory properties of CBD (Pellati et al., 2018;). It appears that CBD is more favored than other compounds in the phytocannabinoid group. Notwithstanding the advantageous pharmacological actions of CBD itself, the presence of this compound in the Δ9-THC environment reduces the undesirable side effects of 99-THC, thus improving its safety profile (Gundugurti et al., 2024).

4.3. Tobacco Metabolites and Nicotinic Receptors

Nicotinic acetylcholine receptor (nAChR) binding in the brain is the most studied alkaloid in tobacco. Such receptors are present in multiple regions involved in cognition, notably the hippocampus, cerebral cortex, and basal ganglia (Echeverria et al., 2023). Nicotine's binding to nAChRs modulates the release of neurotransmitters (dopamine, acetylcholine, etc.) that affect cognitive function, attentiveness, and memory. This is especially relevant in diseases such as Alzheimer's and Parkinson's where cholinergic dysfunction is a a characteristic feature (Crestini et al., 2024). Another neuroprotective agent is cotinine, a metabolite of nicotine. Cotinine binds to nAChRs and, by increasing the availability of neurotrophic factors such as brain-derived neurotrophic factor (BDNF), may promote neurogenesis, reduce oxidative stress, and improve cognitive function (Echeverria et al., 2023).

4.4. Synergistic Interaction Between the ECS and Nicotinic Systems

A promising target for synergy is the interaction between the endocannabinoid system and nicotinic receptors. Both systems are implicated in the control of neuroinflammation, neurotransmitter release, and cell survival. Potentially, cannabis and tobacco metabolites can jointly regulate these systems to improve therapeutic responses.

Co-Modulation of the Inflammation Process: Both cannabinoids and nicotine have been shown to decrease neuroinflammation, an important pathologic feature of neurodegenerative disorders. Cannabinoids such as THC and CBD reduce the release of pro-inflammatory cytokines (TNF-α, IL-6) via CB2 receptors on glial cells (microglia and astrocytes). CB1 receptor immunoreactivity is strongly associated with mRNA in GABAergic neurons, and cannabinoids affect the release of slower neurotransmitters such as opiates, acetylcholine, dopamine, and norepinephrine. CB1 receptors inhibit hippocampal synaptic plasticity, including LTP and LTD, via a receptor-dependent mechanism. Studies in CB1 receptor-deficient mice suggest that endocannabinoids consistently impair synaptic remodeling, possibly involving anandamide as an intrinsic messenger for cognitive regulation (Lutz et al., 2015). CB1 receptor mRNA, assessed by RT-PCR, is present in striatal and extra-striatal regions such as the globus pallidus and substantia nigra in the PD brain, with reduced levels in neurons of the caudate nucleus and dorsal putamen (Fernandez-Moncada et al., 2023). Research in neurological diseases reveals the unique presence of CB2 in microglial cells in diseases such as Alzheimer's, multiple sclerosis and ALS (Vuic et al., 2022). CB2 stimulation, like CB1, promotes adult synaptogenesis and may influence blood-brain barrier permeability. Evidence supports the involvement of the CB2 receptor in psychiatric and neurological conditions, with a distinct brain distribution mainly in postsynaptic neuronal somatodendritic regions, leading to distinct effects compared to CB1 receptors (Kumar; 2024). In contrast, CB2 receptors are predominantly located in the nuclei of postsynaptic neurons, leading to downstream activation that hyperpolarizes the membrane potential and inhibits neuronal firing (Cabañero et al., 2021). Reduction of neuronal excitability is achieved by CB2 receptor expression in a variety of ways, including modulation of K+ channels in VTA dopamine neurons and involvement in Gq11-PLC-IP3 pathways in frontal cortex neurons (Rathod et al., 2023). CB2 receptors also affect hunger, weight, depression, stress, and schizophrenia-like behavior through various mechanisms. Nicotine may also inhibit inflammatory cytokine release and promote cholinergic anti-inflammatory effects through its action on nAChRs (Zhang et al., 2022). In combination, cannabis and tobacco metabolites could synergistically attenuate neuroinflammation more successfully than either alone.

The regulation of oxidative stress: Oxidative stress contributes to neuronal damage in neurodegenerative diseases, caused by an imbalance between free radicals and antioxidants (Houldsworth, 2024). A number of studies indicate a critical role for ROX and OS in the pathogenesis of Alzheimer's disease through their deleterious effects on molecules, particularly proteins. Recent studies in the post-mortem AD brain have shown the presence of oxidative injury to key enzymes involved in energy metabolism, proteins related to neurotransmitters, mitochondrial proteins, and proteasomal components (Ganguly et al., 2017). Although whether increased ROS generation is a primary cause or a consequence of AD remains controversial, some studies have shown severe OS in early stages of AD, prior to significant Aβ accumulation (Zuo et al., 2019). THC and CBD both exhibit antioxidant properties. They scavenge free radicals and reduce lipid peroxidation. Nicotine, through the activation of nAChRs, also induces the release of glutathione and other antioxidants, thereby reducing oxidative damage. This combination may result in increased antioxidant activity that protects neurons from oxidative damage (Liu et al., 2024).

5. REGULATORY AND ETHICAL CHALLENGES

Cannabis and its derivatives have shown potential in managing symptoms and slowing the progression of neurodegenerative disorders. In Alzheimer's disease, research suggests that cannabinoids may reduce the accumulation of amyloid-beta plaques and mitigate neuroinflammation, two critical pathological hallmarks of the disease (Xiong & Lim, 2021). Studies indicate that both natural and synthetic cannabinoid agonists can activate CB1 and CB2 receptors, leading to decreased Aβ toxicity and tau phosphorylation key processes in Alzheimer's progression (Möller et al., 2025). Moreover, CBD's anxiolytic and anti-agitation properties can alleviate behavioral disturbances commonly observed in Alzheimer's patients, such as aggression and sleep disorders (Outen et al., 2021).

In computational modeling, a study demonstrates that the active component of marijuana, Δ9-THC, competitively inhibits the enzyme acetylcholinesterase (AChE) and prevents AChE-induced amyloid β-peptide (Aβ) aggregation, a key pathological marker of Alzheimer's disease (Vitale et al., 2024). THC binds to the enzyme's active site through interactions such as hydrogen bonding and Van der Waals forces with various amino acid residues (Suttithumsatid et al., 2023). This competitive inhibition could potentially enhance cholinergic neurotransmission by preventing the breakdown of acetylcholine, which is often diminished in Alzheimer's patients (Abdel-Salam & Khadrawy, 2019).

In Parkinson's disease (PD), cannabinoids have been explored for their potential to modulate dopamine signaling pathways and alleviate motor symptoms, including tremors, rigidity, and bradykinesia (Han et al., 2020). Additionally, evidence suggests that cannabis may improve non-motor symptoms such as pain, sleep disturbances, and depression, which significantly impact patients' quality of life (More & Choi, 2015). Despite these promising effects, clinical studies remain limited, and further research is needed to confirm the efficacy and safety of cannabinoids in PD management.

The endocannabinoid system (ECS), particularly its CB1 and CB2 receptors, plays a regulatory role in basal ganglia function and is closely involved in the pathogenesis of PD (Bioque et al., 2013). These receptors are abundantly expressed in the basal ganglia's neural circuits, where they interact with glutamatergic, GABAergic, and dopaminergic signaling systems, offering a unique therapeutic target (More & Choi, 2015). Preclinical studies have demonstrated that modulating ECS components can influence neuroinflammation, oxidative stress, and excitotoxicity key pathological processes in PD (Vasincu et al., 2022). The ECS also plays a significant role in regulating inflammatory responses in the central nervous system (Paes-Colli et al., 2022).

In PD, neuroinflammation is exacerbated by the accumulation of misfolded proteins, such as α-synuclein, which activate microglia and astrocytes. These activated glial cells release pro-inflammatory cytokines, creating a feedback loop that perpetuates inflammation and contributes to neuronal damage (Kam et al., 2020).

Cs has also gained significant attention for its potential therapeutic effects in other neurodegenerative conditions, such as Multiple Sclerosis (MS) and Amyotrophic Lateral Sclerosis (ALS). Cannabinoids interact with the body's ECS, which plays a crucial role in regulating various physiological processes (El-Mansoury et al., 2023; Lu & Mackie, 2021). For example, ALS is associated with altered endocannabinoid signaling, suggesting that restoring balance within this system could help alleviate symptoms (Pota et al., 2024). The activation of CB2 receptors has been linked to neuroprotective effects in ALS models (Rodríguez-Cueto et al., 2021). Selective CB2 agonists, for instance, have been shown to delay motor impairment and extend survival in mouse models of ALS. Recent metabolomics analyses have identified specific metabolites associated with ALS progression, including those related to the ECS, suggesting that enhancing ECS function could be a viable therapeutic strategy (Ito et al., 2024).

6. CONCLUSION

In conclusion, the therapeutic potential of secondary metabolites from *Cs* and *Nicotiana tabacum* represents a promising frontier in the fight against neurodegenerative diseases. Both plants, despite their distinct histories and reputations, harbor compounds with powerful bioactive properties that can influence key molecular pathways involved in neurodegeneration. The cannabinoids from cannabis, such as CBD and THC, offer promising neuroprotective effects by modulating inflammation, oxidative stress, and neuronal plasticity. Similarly, tobacco's nicotine, despite its controversial background, demonstrates a capacity to enhance cognitive function and provide neuroprotection through its interaction with nicotinic receptors. While preclinical studies and clinical trials show promising results, the therapeutic application of these plants must be approached with caution. The challenge lies not only in further validating the efficacy and safety of these compounds but also in navigating the complex ethical, regulatory, and societal issues surrounding their use. Moreover, the potential synergies between cannabis and tobacco metabolites could offer innovative treatment strategies, but their integration into clinical practice will require careful consideration and robust scientific evidence.

The future of neurodegenerative disease therapies may lie in the exploration of natural products, with cannabis and tobacco as powerful examples of plants offering therapeutic solutions. As research continues to uncover the mechanisms behind these bioactive compounds, it is crucial to refine our understanding of their full potential, while addressing the regulatory and ethical concerns that may arise. By bridging traditional knowledge with modern scientific inquiry, we may unlock novel approaches that transform the treatment and management of neurodegenerative diseases, providing hope for millions of patients worldwide.

REFERENCES

Abdel-Kader, M. S., Radwan, M. M., Metwaly, A. M., Eissa, I. H., Hazekamp, A., & ElSohly, M. A. (2024). Chemistry and pharmacology of delta-8-tetrahydrocannabinol. *Molecules (Basel, Switzerland)*, *29*(6), 1249.

Abdel-Salam, O. M. E., & Khadrawy, Y. A. (2019). The inhibition of serum cholinesterases by Cannabis sativa and/or tramadol. *J. Neurol. Forecast*, *2*, 1–4.

Abel, E. L. (1980). Cannabis in the ancient world. In *Marihuana : The first twelve thousand years* (pp. 3–35). Springer.

Ahmed, I., Rehman, S. U., Shahmohamadnejad, S., Zia, M. A., Ahmad, M., Saeed, M. M., & Liu, Q. (2021). Therapeutic attributes of endocannabinoid system against neuro-inflammatory autoimmune disorders. *Molecules (Basel, Switzerland)*, *26*(11), 3389.

Al-Harrasi, A., Bhatia, S., Behl, T., & Kaushik, D. (2022). Effects of essential oils on CNS. In *Role of Essential Oils in the Management of COVID-19* (pp. 269–297). CRC Press., https://www.taylorfrancis.com/chapters/edit/10.1201/9781003175933-20/effects-essential-oils-cns-ahmed-al-harrasi-saurabh-bhatia-tapan-behl-deepak-kaushik

Al-Khazaleh, A. K., Zhou, X., Bhuyan, D. J., Münch, G. W., Al-Dalabeeh, E. A., Jaye, K., & Chang, D. (2024). The Neurotherapeutic Arsenal in Cannabis sativa: Insights into Anti-Neuroinflammatory and Neuroprotective Activity and Potential Entourage Effects. *Molecules (Basel, Switzerland)*, *29*(2), 410.

Alhowail, A. (2021). Molecular insights into the benefits of nicotine on memory and cognition. *Molecular Medicine Reports*, *23*(6), 1–6.

Ali, M. (2024). Neurodegenerative Diseases: Insights into Early Diagnosis and Therapeutic Approaches. *Trends in Life Sciences and Biotechnology*, *1*(02), 98–106.

Alves, P., Amaral, C., Teixeira, N., & Correia-da-Silva, G. (2020). Cannabis sativa: Much more beyond Δ9-tetrahydrocannabinol. *Pharmacological Research*, *157*, 104822.

Alves, P., Amaral, C., Teixeira, N., & Correia-da-Silva, G. (2020). Cannabis sativa : Much more beyond Δ9-tetrahydrocannabinol. *Pharmacological Research*, *157*, 104822.

Atalay, S., Jarocka-Karpowicz, I., & Skrzydlewska, E. (2019). Antioxidative and anti-inflammatory properties of cannabidiol. *Antioxidants*, *9*(1), 21.

Atalay, S., Jarocka-Karpowicz, I., & Skrzydlewska, E. (2019). Antioxidative and anti-inflammatory properties of cannabidiol. *Antioxidants*, *9*(1), 21.

Bachir, F., Eddouks, M., Arahou, M., & Fekhaoui, M. (2022). Origin, Early History, Cultivation, and Characteristics of the Traditional Varieties of Moroccan Cannabis sativa L. *Cannabis and Cannabinoid Research*, *7*(5), 603–615. DOI: 10.1089/can.2021.0020

Bankole, H. A., Kanmodi, R. I., Kazeem, M. I., Wusu, A. D., Fatai, A. A., Oddiri, R. T., & Oyedele, A. Q. K. (2024). Interactive Properties of Alkaloids from Datura stramonium, Moringa oleifera, and Carica papaya with Human Receptor Proteins of Psychoactive Compounds from Cannabis sativa and Nicotiana tabacum. *Tropical Journal of Natural Product Research, 8*(7).

Banožić, M., Babić, J., & Jokić, S. (2020). Recent advances in extraction of bioactive compounds from tobacco industrial waste-a review. *Industrial Crops and Products, 144*, 112009.

Barratt, E. S., Beaver, W., & White, R. (1974). The effects of marijuana on human sleep patterns. Biological Psychiatry. https://psycnet.apa.org/record/1974-24717-001

Bendik, P. B., Rutt, S. M., Pine, B. N., Sosnoff, C. S., Blount, B. C., Zhu, W., & Wang, L. (2022). Anabasine and anatabine exposure attributable to cigarette smoking: National health and nutrition examination survey (NHANES) 2013–2014. *International Journal of Environmental Research and Public Health, 19*(15), 9744.

Bih, C. I., Chen, T., Nunn, A. V., Bazelot, M., Dallas, M., & Whalley, B. J. (2015). Molecular targets of cannabidiol in neurological disorders. *Neurotherapeutics; the Journal of the American Society for Experimental NeuroTherapeutics, 12*(4), 699–730.

Bioque, M., García-Bueno, B., MacDowell, K. S., Meseguer, A., Saiz, P. A., Parellada, M., Gonzalez-Pinto, A., Rodriguez-Jimenez, R., Lobo, A., Leza, J. C., & Bernardo, M. (2013). Peripheral Endocannabinoid System Dysregulation in First-Episode Psychosis. *Neuropsychopharmacology: Official Publication of the American College of Neuropsychopharmacology, 38*(13), 2568–2577. DOI: 10.1038/npp.2013.165

Biswas, S., & Bagchi, A. (2023). Study of the Effects of Nicotine and Caffeine for the Treatment of Parkinson's Disease. *Applied Biochemistry and Biotechnology, 195*(1), 639–654.

Bockmann, E. C., Brito, R., Madeira, L. F., Da Silva Sampaio, L., De Melo Reis, R. A., França, G. R., & Calaza, K. D. C. (2023). The Role of Cannabinoids in CNS Development: Focus on Proliferation and Cell Death. *Cellular and Molecular Neurobiology, 43*(4), 1469–1485. DOI: 10.1007/s10571-022-01263-y

Booth, J. K., Page, J. E., & Bohlmann, J. (2017). Terpene synthases from Cannabis sativa. *PLoS One, 12*(3), e0173911.

Borodovsky, J. T., Crosier, B. S., Lee, D. C., Sargent, J. D., & Budney, A. J. (2016). Smoking, vaping, eating : Is legalization impacting the way people use cannabis? *The International Journal on Drug Policy, 36*, 141–147.

Borodovsky, J. T., Struble, C. A., Habib, M. I., Hasin, D. S., Shmulewitz, D., Walsh, C., Livne, O., Aharonovich, E., & Budney, A. J. (2023). Exploring survey methods for measuring consumption quantities of cannabis flower and concentrate products. *The American Journal of Drug and Alcohol Abuse, 49*(6), 733–745. DOI: 10.1080/00952990.2023.2246635

Bouarfa, M., Lebtar, S., Boukhira, S., & Bousta, D. (2020). An ethnobotanical and ethnopharmacological survey of Cannabis sativa of Taounate Region in Northern Morocco. *International Journal of Pharmaceutical Sciences Review and Research, 64*(2), 116–122.

Bukowska, B. (2024). Current and potential use of biologically active compounds derived from Cannabis sativa L. in the treatment of selected diseases. *International Journal of Molecular Sciences*, *25*(23), 12738.

Cabañero, D., Martín-García, E., & Maldonado, R. (2021). The CB2 cannabinoid receptor as a therapeutic target in the central nervous system. *Expert Opinion on Therapeutic Targets*, *25*(8), 659–676.

Calik, M. W., & Carley, D. W. (2017). Effects of cannabinoid agonists and antagonists on sleep and breathing in Sprague-Dawley rats. *Sleep (Basel)*, *40*(9), zsx112.

Campos, A. C., Moreira, F. A., Gomes, F. V., Del Bel, E. A., & Guimarães, F. S. (2012). Multiple mechanisms involved in the large-spectrum therapeutic potential of cannabidiol in psychiatric disorders. *Philosophical Transactions of the Royal Society of London. Series B, Biological Sciences*, *367*(1607), 3364–3378. DOI: 10.1098/rstb.2011.0389

Cascio, M., Gauson, L., Stevenson, L., Ross, R., & Pertwee, R. (2010). Evidence that the plant cannabinoid cannabigerol is a highly potent α2 -adrenoceptor agonist and moderately potent 5HT1A receptor antagonist. *British Journal of Pharmacology*, *159*(1), 129–141. DOI: 10.1111/j.1476-5381.2009.00515.x

Cassano, T., Villani, R., Pace, L., Carbone, A., Bukke, V. N., Orkisz, S., Avolio, C., & Serviddio, G. (2020). From Cannabis sativa to cannabidiol : Promising therapeutic candidate for the treatment of neurodegenerative diseases. *Frontiers in Pharmacology*, *11*, 124.

Chouvy, P.-A. (2019). Cannabis cultivation in the world : Heritages, trends and challenges. *EchoGéo*, *48*. https://journals.openedition.org/echogeo/17591

Clarke, R. C., & Merlin, M. D. (2015). Small, Ernest. 2015. Evolution and classification of Cannabis sativa (Marijuana, Hemp) in relation to human utilization. [Springer.]. *Botanical Review*, *81*(3), 189–294.

Corroon, J., & Felice, J. F. (2019). The Endocannabinoid System and its Modulation by Cannabidiol (CBD). *Alternative Therapies in Health and Medicine*, 25.

Couch, D. G., Maudslay, H., Doleman, B., Lund, J. N., & O'Sullivan, S. E. (2018). The use of cannabinoids in colitis : A systematic review and meta-analysis. *Inflammatory Bowel Diseases*, *24*(4), 680–697.

Crestini, A., Carbone, E., Rivabene, R., Ancidoni, A., Rosa, P., Tata, A. M., & Piscopo, P. (2024). A Systematic Review on Drugs Acting as Nicotinic Acetylcholine Receptor Agonists in the Treatment of Dementia. *Cells*, *13*(3), 237.

Davis, M. P. (2014). Cannabinoids in pain management : CB1, CB2 and non-classic receptor ligands. *Expert Opinion on Investigational Drugs*, *23*(8), 1123–1140. DOI: 10.1517/13543784.2014.918603

De Petrocellis, L., Ligresti, A., Moriello, A. S., Allarà, M., Bisogno, T., Petrosino, S., Stott, C. G., & Di Marzo, V. (2011). Effects of cannabinoids and cannabinoid-enriched Cannabis extracts on TRP channels and endocannabinoid metabolic enzymes. *British Journal of Pharmacology*, *163*(7), 1479–1494. DOI: 10.1111/j.1476-5381.2010.01166.x

Del Río, C., & Segura-Carretero, A. (2023). Neuroprotection with bioactive compounds. In Nutrients (Vol. 15, Numéro 21, p. 4612). MDPI. https://www.mdpi.com/2072-6643/15/21/4612

Echeverria, V., Mendoza, C., & Iarkov, A. (2023). Nicotinic acetylcholine receptors and learning and memory deficits in Neuroinflammatory diseases. *Frontiers in Neuroscience*, *17*, 1179611.

El Khomsi, M., Dandani, Y., Chaachouay, N., & Hmouni, D. (2022). Ethnobotanical study of plants used for medicinal, cosmetic, and food purposes in the region of Moulay Yacoub, Northeast of Morocco. *Journal of Pharmacy & Pharmacognosy Research*, *10*, 13–29.

El-Mansoury, B., Laadraoui, J., El Khiat, A., El Hiba, O., Jayakumar, A. R., Ferssiwi, A., & Bitar, A. (2023). *Ataxia in Multiple Sclerosis: From Current Understanding to Therapy*. Ataxia–Practice Essentials and Interventions.

El-Mernissi, R., El Menyiy, N., Zouhri, A., El-Mernissi, Y., Diai, F., Siddique, F., Dabiellil, F., Almaary, K. S., Amhamdi, H., Abboussi, O., & Hajji, L. (2024). Phytochemical profiling and bioactivity evaluation of CBD- and THC-enriched Cannabis sativa extracts: In vitro and in silico investigation of antioxidant and anti-inflammatory effects. *Open Chemistry*, *22*(1), 20240119. DOI: 10.1515/chem-2024-0119

Farag, S., & Kayser, O. (2017). The Cannabis Plant: Botanical Aspects. In Preedy, V. R. (Ed.), *Handbook of Cannabis and Related Pathologies* (pp. 3–12). Academic Press., DOI: 10.1016/B978-0-12-800756-3.00001-6

Fernandez-Moncada, I., Eraso-Pichot, A., Dalla Tor, T., Fortunato-Marsol, B., & Marsicano, G. (2023). An enquiry to the role of CB1 receptors in neurodegeneration. *Neurobiology of Disease*, 106235.

Fernández-Ruiz, J., Sagredo, O., Pazos, M. R., García, C., Pertwee, R., Mechoulam, R., & Martínez-Orgado, J. (2013). Cannabidiol for neurodegenerative disorders: Important new clinical applications for this phytocannabinoid? *British Journal of Clinical Pharmacology*, *75*(2), 323–333. DOI: 10.1111/j.1365-2125.2012.04341.x

Florimbio, A. R., Walton, M. A., Coughlin, L. N., Lin, L. A., & Bonar, E. E. (2023). Perceived risk of harm for different methods of cannabis consumption: A brief report. *Drug and Alcohol Dependence*, *251*, 110915.

Ganguly, G., Chakrabarti, S., Chatterjee, U., & Saso, L. (2017). Proteinopathy, oxidative stress and mitochondrial dysfunction: Cross talk in Alzheimer's disease and Parkinson's disease. *Drug Design, Development and Therapy*, 797–810.

Gaston, T. E., & Szaflarski, J. P. (2018). Cannabis for the Treatment of Epilepsy : An Update. *Current Neurology and Neuroscience Reports*, *18*(11), 73. DOI: 10.1007/s11910-018-0882-y

Gérard, L., Kostrzewa, L., Legrand, K., & Tubiana-Potiez, A. (s. d.). Alcool, tabac et cannabis. Consulté 23 novembre 2023, à l'adresse https://www.researchgate.net/profile/Alexandra-Tubiana/publication/274248693_Alcool_tabac_et_cannabis_chez_les_jeunes_lorrains_en_2011_Etude_sur_la_cohorte_francaise_de_SEYLE/links/5559964f08ae980ca6106f6f/Alcool-tabac-et-cannabis-chez-les-jeunes-lorrains-en-2011-Etude-sur-la-cohorte-francaise-de-SEYLE.pdf

Goodman, S., Leos-Toro, C., & Hammond, D. (2019). Methods to Assess Cannabis Consumption in Population Surveys : Results of Cognitive Interviewing. *Qualitative Health Research*, *29*(10), 1474–1482. DOI: 10.1177/1049732318820523

Gundugurti, P. R., Banda, N., Yadlapalli, S. S. R., Narala, A., Thatikonda, R., Kocherlakota, C., & Kothapalli, K. S. (2024). Evaluation of the efficacy, safety, and pharmacokinetics of nanodispersible cannabidiol oral solution (150 mg/mL) versus placebo in mild to moderate anxiety subjects: A double blind multicenter randomized clinical trial. *Asian Journal of Psychiatry*, *97*, 104073.

Guo, F., Qin, X., Mao, J., Xu, Y., & Xie, J. (2024). Potential Protective Effects of Pungent Flavor Components in Neurodegenerative Diseases. *Molecules (Basel, Switzerland)*, *29*(23), 5700.

Haggarty, C. J., & Rabinak, C. (2025). Cannabis : A History. In Marusak, H. (Ed.), *Cannabis and the Developing Brain* (pp. 7–31). Springer Nature Switzerland., DOI: 10.1007/978-3-031-87990-6_2

Han, Q.-W., Yuan, Y.-H., & Chen, N.-H. (2020). The therapeutic role of cannabinoid receptors and its agonists or antagonists in Parkinson's disease. *Progress in Neuro-Psychopharmacology & Biological Psychiatry*, *96*, 109745. DOI: 10.1016/j.pnpbp.2019.109745

Hanuš, L. O. (2009). Pharmacological and therapeutic secrets of plant and brain (endo)cannabinoids. *Medicinal Research Reviews*, *29*(2), 213–271. DOI: 10.1002/med.20135

Hanuš, L. O., & Appendino, G. (2016). Natural Product Reports. *Natural Product Reports*, *33*, 1357–1392.

Hanuš, L. O., & Hod, Y. (2020). Terpenes/terpenoids in cannabis: Are they important? *Medical Cannabis and Cannabinoids*, *3*(1), 25–60.

Hanuš, L. O., Meyer, S. M., Muñoz, E., Taglialatela-Scafati, O., & Appendino, G. (2016). Phytocannabinoids : A unified critical inventory. *Natural Product Reports*, *33*(12), 1357–1392.

Hartsel, J. A., Eades, J., Hickory, B., & Makriyannis, A. (2016). Cannabis sativa and Hemp. In *Nutraceuticals* (pp. 735–754). Elsevier.

Hasenoehrl, C., Storr, M., & Schicho, R. (2017). Cannabinoids for treating inflammatory bowel diseases : Where are we and where do we go? *Expert Review of Gastroenterology & Hepatology*, *11*(4), 329–337. DOI: 10.1080/17474124.2017.1292851

Hosseini, L., Mahmoudi, J., Pashazadeh, F., Salehi-Pourmehr, H., & Sadigh-Eteghad, S. (2021). Protective effects of nicotinamide adenine dinucleotide and related precursors in alzheimer's disease: A systematic review of preclinical studies. *Journal of Molecular Neuroscience*, *71*, 1425–1435.

Houldsworth, A. (2024). Role of oxidative stress in neurodegenerative disorders: A review of reactive oxygen species and prevention by antioxidants. *Brain Communications*, *6*(1), fcad356.

Hwang, E. S., & Song, S. B. (2020). Possible adverse effects of high-dose nicotinamide: Mechanisms and safety assessment. *Biomolecules*, *10*(5), 687.

Iarkov, A., Mendoza, C., & Echeverria, V. (2021). Cholinergic receptor modulation as a target for preventing dementia in Parkinson's disease. *Frontiers in Neuroscience*, *15*, 665820.

Islam, M. R., Akash, S., Islam, M. M., Sarkar, N., Kumer, A., Chakraborty, S., & Khan, H. (2024). Alkaloids as drug leads in Alzheimer's treatment: Mechanistic and therapeutic insights. *Brain Research*, •••, 148886.

Ito, D., Iida, M., Iguchi, Y., Hashizume, A., Yamada, S., Kishimoto, Y., Komori, S., Shimamura, T., Takemoto, Y., Nakatochi, M., Akashi, T., Hinohara, K., Lee-Okada, H.-C., Niwa, J., Sobue, G., Tanaka, S., Takashina, K., Yokomizo, T., & Katsuno, M. (2024). Dysregulated expanded endocannabinoid system as therapeutic targets of amyotrophic lateral sclerosis (p. 2024.01.12.575341). *bioRxiv*. https://doi.org/DOI: 10.1101/2024.01.12.575341

Jha, S. K., Nelson, V. K., Suryadevara, P. R., Panda, S. P., Pullaiah, C. P., Nuli, M. V., & Jafari, S. M. (2024). Cannabidiol and neurodegeneration: From molecular mechanisms to clinical benefits. *Ageing Research Reviews*, *100*, 102386.

Kabdy, H., Azraida, H., Agouram, F., Oufquir, S., Laadraoui, J., Baslam, A., Aitbaba, A., Ouazzani, M. E., Elyazouli, L., Aboufatima, R., Garzoli, S., & Chait, A. (2024). Antiarthritic and Anti-Inflammatory Properties of Cannabis sativa Essential Oil in an Animal Model. *Pharmaceuticals*, *17*(1), 1. Advance online publication. DOI: 10.3390/ph17010020

Kabdy, H., Baslam, A., Belbachir, A., Garzoli, S., & Chait, A. (2024). Chemical Analysis, Antioxidant, Toxicology, and Antinociceptive Effects of Moroccan Cannabis sativa Essential Oil. *Journal of Biological Regulators and Homeostatic Agents*, *38*(3), 3. Advance online publication. DOI: 10.23812/j.biol.regul.homeost.agents.20243803.165

Kam, T.-I., Hinkle, J. T., Dawson, T. M., & Dawson, V. L. (2020). Microglia and astrocyte dysfunction in parkinson's disease. *Neurobiology of Disease*, *144*, 105028. DOI: 10.1016/j.nbd.2020.105028

Kaszyńska, A. A. (2024). Cannabinoids: Potential for modulation and enhancement when combined with vitamin B12 in case of neurodegenerative disorders. *Pharmaceuticals*, *17*(6), 813.

Kumar, A., Suradkar, N., Rawat, M., Yesansure, D., Dharmalingam, S., Kumar, D., Chauhan, A., Srivastav, Y., & Tandey, R. (2025). Simultaneous Quantification Of Terpenoids And Phenolic Compounds In Cannabis Sativa Using HPLC And LC-MS. *Eurasian Journal of Analytical Chemistry*, *20*, 147–152. DOI: 10.53555/ejac.v20i1.1174

Kumar, U. (2024). Cannabinoids: Role in Neurological Diseases and Psychiatric Disorders. *International Journal of Molecular Sciences*, *26*(1), 152.

Laadraoui, J., Kabdy, H., Oufquir, S., Aitlarradia, M., Agouram, F., Ait Baba, A., & El-Mansoury, B. (2025). The Therapeutic Role of Secondary Metabolites in Neurodegenerative and Inflammatory Diseases, Mechanisms, and Potential Applications. In *Secondary Metabolites and Their Applications in Various Diseases* (pp. 129–192). IGI Global Scientific Publishing.

Laws, J. S.III, & Smid, S. D. (2024). Characterizing cannabis-prevalent terpenes for neuroprotection reveal a role for α and β-pinenes in mitigating amyloid β-evoked neurotoxicity and aggregation in vitro. *Neurotoxicology*, *100*, 16–24.

Li, H.-L. (1974). The origin and use of Cannabis in eastern Asia linguistic-cultural implications. *Economic Botany*, *28*(3), 293–301.

Lim, K., See, Y. M., & Lee, J. (2017). A systematic review of the effectiveness of medical cannabis for psychiatric, movement and neurodegenerative disorders. *Clinical Psychopharmacology and Neuroscience : the Official Scientific Journal of the Korean College of Neuropsychopharmacology*, *15*(4), 301.

Lins, B. R., Anyaegbu, C. C., Hellewell, S. C., Papini, M., McGonigle, T., De Prato, L., & Fitzgerald, M. (2023). Cannabinoids in traumatic brain injury and related neuropathologies: Preclinical and clinical research on endogenous, plant-derived, and synthetic compounds. *Journal of Neuroinflammation*, *20*(1), 77.

Liu, Y., Chen, G., Cao, J., Zhang, J., Luo, C., & Huang, L. (2024). The "double-edged sword effect" of nicotine. *Fitoterapia*, 106102.

Lo, L. A., Christiansen, A. L., Strickland, J. C., Pistawka, C. A., Eadie, L., Vandrey, R., & MacCallum, C. A. (2024). Does acute cannabidiol (CBD) use impair performance? A meta-analysis and comparison with placebo and delta-9-tetrahydrocannabinol (THC). *Neuropsychopharmacology : Official Publication of the American College of Neuropsychopharmacology*, 1–12.

Lu, H.-C., & Mackie, K. (2021). Review of the Endocannabinoid System. *Biological Psychiatry: Cognitive Neuroscience and Neuroimaging*, *6*(6), 607–615. DOI: 10.1016/j.bpsc.2020.07.016

Lutz, B., Marsicano, G., Maldonado, R., & Hillard, C. J. (2015). The endocannabinoid system in guarding against fear, anxiety and stress. *Nature Reviews. Neuroscience*, *16*(12), 705–718.

Majdi, A., Kamari, F., Vafaee, M. S., & Sadigh-Eteghad, S. (2017). Revisiting nicotine's role in the ageing brain and cognitive impairment. *Reviews in the Neurosciences*, *28*(7), 767–781.

Malabadi, R. B., Kolkar, K. P., & Chalannavar, R. K. (2023). Cannabis sativa : Ethnobotany and Phytochemistry. *International Journal of Innovation Scientific Research and Review*, *5*(2), 3990–3998.

Martín-Moreno, A. M., Reigada, D., Ramírez, B. G., Mechoulam, R., Innamorato, N., Cuadrado, A., & de Ceballos, M. L. (2011). Cannabidiol and other cannabinoids reduce microglial activation in vitro and in vivo: Relevance to Alzheimer's disease. *Molecular Pharmacology*, *79*(6), 964–973.

Mastinu, A., Premoli, M., Ferrari-Toninelli, G., Tambaro, S., Maccarinelli, G., Memo, M., & Bonini, S. A. (2018). Cannabinoids in health and disease : Pharmacological potential in metabolic syndrome and neuroinflammation. *Hormone Molecular Biology and Clinical Investigation*, *36*(2), 20180013. DOI: 10.1515/hmbci-2018-0013

McPartland, J. M., Hegman, W., & Long, T. (2019). Cannabis in Asia: Its center of origin and early cultivation, based on a synthesis of subfossil pollen and archaeobotanical studies. *Vegetation History and Archaeobotany*, *28*(6), 691–702. DOI: 10.1007/s00334-019-00731-8

Mechoulam, R. (2019). The pharmacohistory of Cannabis sativa. Cannabinoids as therapeutic agents, 1-20.

Mehta, A., Prabhakar, M., Kumar, P., Deshmukh, R., & Sharma, P. L. (2013). Excitotoxicity : Bridge to various triggers in neurodegenerative disorders. *European Journal of Pharmacology*, *698*(1-3), 6–18.

Merghany, R. M., El-Sawi, S. A., Naser, A. F. A., Ezzat, S. M., Moustafa, S. F., & Meselhy, M. R. (2024). A comprehensive review of natural compounds and their structure–activity relationship in Parkinson's disease: Exploring potential mechanisms. *Naunyn-Schmiedeberg's Archives of Pharmacology*, 1–30.

Miller, G., Pareek, O., Penman, S. L., & Thanos, P. K. (2024). The Effects of Nicotine and Cannabinoids on Cytokines. *Current Pharmaceutical Design, 30*(31), 2468–2484.

Mohd Sairazi, N. S., & Sirajudeen, K. N. S. (2020). Natural Products and Their Bioactive Compounds : Neuroprotective Potentials against Neurodegenerative Diseases. *Evidence-Based Complementary and Alternative Medicine: eCAM, 2020*(1), 6565396. DOI: 10.1155/2020/6565396

Möller, J. E. L., Schmitt, F. W., Günther, D., Stöver, A., & Bouter, Y. (2025). The synthetic cannabinoid WIN 55,212-2 attenuates cognitive and motor deficits and reduces amyloid load in 5XFAD Alzheimer mice. *Pharmacology, Biochemistry, and Behavior, 247*, 173944. DOI: 10.1016/j.pbb.2024.173944

More, S. V., & Choi, D.-K. (2015). Promising cannabinoid-based therapies for Parkinson's disease : Motor symptoms to neuroprotection. *Molecular Neurodegeneration, 10*(1), 17. DOI: 10.1186/s13024-015-0012-0

Mukherjee, U., Sehar, U., Brownell, M., & Reddy, P. H. (2024). Mechanisms, consequences and role of interventions for sleep deprivation: Focus on mild cognitive impairment and Alzheimer's disease in elderly. *Ageing Research Reviews, 100*, 102457.

Mulpuri, Y., Marty, V. N., Munier, J. J., Mackie, K., Schmidt, B. L., Seltzman, H. H., & Spigelman, I. (2018). Synthetic peripherally-restricted cannabinoid suppresses chemotherapy-induced peripheral neuropathy pain symptoms by CB1 receptor activation. *Neuropharmacology, 139*, 85–97.

Mzali, A., Mahraz, A. M., Benlabchir, A. A., Benzeid, H., Hassani, A. C., Amalich, S., & Doukkali, A. (2023). Ethnobotanical and Ethnopharmacological Surveys of Cannabis sativa (Beldiya Species) Use in the Provinces of Taounate and Al Hoceïma. *Tropical Journal of Natural Product Research, 7*(12).

Nagarajan, P., Kannan, S., Raj, A., Louis, L. R. P., Bazeer, A. B., Kaliaperumal, R., Anand, M. A. V., & Patil, S. J. (2025). Future Prospects in Medicinal Plant-Derived Drug Development : Unveiling Opportunities and Challenges. In Sawicka, B., Messaoudi, M., & Rebiai, A. (Eds.), *Biotechnology and Phytochemical Prospects in Drug Discovery* (pp. 241–266). Springer Nature Singapore., DOI: 10.1007/978-981-96-2790-5_9

Nallathambi, R., Mazuz, M., Ion, A., Selvaraj, G., Weininger, S., Fridlender, M., Nasser, A., Sagee, O., Kumari, P., Nemichenizer, D., Mendelovitz, M., Firstein, N., Hanin, O., Konikoff, F., Kapulnik, Y., Naftali, T., & Koltai, H. (2017). Anti-Inflammatory Activity in Colon Models Is Derived from Δ9-Tetrahydrocannabinolic Acid That Interacts with Additional Compounds in Cannabis Extracts. *Cannabis and Cannabinoid Research, 2*(1), 167–182. DOI: 10.1089/can.2017.0027

Nasrin, S. (2022). *Potential Interactions of Cannabinoids and Their Metabolites on Major Drug Metabolizing Enzymes: Implication on Co-use of Tobacco and Cannabis*. Washington State University.

Nielsen, S., Germanos, R., Weier, M., Pollard, J., Degenhardt, L., Hall, W., Buckley, N., & Farrell, M. (2018). The Use of Cannabis and Cannabinoids in Treating Symptoms of Multiple Sclerosis: A Systematic Review of Reviews. *Current Neurology and Neuroscience Reports*, *18*(2), 8. DOI: 10.1007/s11910-018-0814-x

Nunes, C. dos R., Barreto Arantes, M., Menezes de Faria Pereira, S., Leandro da Cruz, L., de Souza Passos, M., Pereira de Moraes, L., Vieira, I. J. C., & Barros de Oliveira, D. (2020). Plants as sources of anti-inflammatory agents. *Molecules (Basel, Switzerland)*, *25*(16), 3726.

Nuutinen, T. (2018). Medicinal properties of terpenes found in Cannabis sativa and Humulus lupulus. *European Journal of Medicinal Chemistry*, *157*, 198–228.

O'Brien, K., Blair, P., O'Brien, K., & Blair, P. (2021). Endocannabinoid System. Medicinal Cannabis and CBD in Mental Healthcare, 7-56.

Ortiz, Y. T., McMahon, L. R., & Wilkerson, J. L. (2022). Medicinal cannabis and central nervous system disorders. *Frontiers in Pharmacology*, *13*, 881810.

Outen, J. D., Burhanullah, H., Vandrey, R., Amjad, H., Harper, D. G., Patrick, R. E., May, R. L., Agronin, M. E., Forester, B. P., & Rosenberg, P. B. (2021). Cannabinoids for Agitation in Alzheimer's Disease. The American journal of geriatric psychiatry: official journal of the American Association for Geriatric Psychiatry, 29(12), 1253-1263. https://doi.org/DOI: 10.1016/j.jagp.2021.01.015

Paes-Colli, Y., Aguiar, A. F. L., Isaac, A. R., Ferreira, B. K., Campos, R. M. P., Trindade, P. M. P., de Melo Reis, R. A., & Sampaio, L. S. (2022). Phytocannabinoids and Cannabis-Based Products as Alternative Pharmacotherapy in Neurodegenerative Diseases : From Hypothesis to Clinical Practice. *Frontiers in Cellular Neuroscience*, *16*. Advance online publication. DOI: 10.3389/fncel.2022.917164

Palanisamy, S., Balasubramanian, A., Kanchana, M., Sisubalan, N., & Selvakesavan, R. K. (2025). Curcumin-Loaded Nanoparticles in Neurodegenerative Diseases : Alzheimer's, Parkinson's, and Amyotrophic Lateral Sclerosis. In Sivamaruthi, B. S., Sisubalan, N., Kesika, P., & Varaprasad, K. (Eds.), *Nanoparticles in Modern Neurological Treatment* (pp. 225–237). Springer Nature Switzerland., DOI: 10.1007/978-3-031-80185-3_8

Pathak, S., Jeyabalan, J. B., Liu, K., Cook, P., Lange, B., Kim, S., & Dhanasekaran, M. (2024). Assessing effects of Cannabis on various neuropathologies: A systematic review. *Journal of Ayurveda and Integrative Medicine*, 100911.

Pellati, F., Borgonetti, V., Brighenti, V., Biagi, M., Benvenuti, S., & Corsi, L. (2018). Cannabis sativa L. and nonpsychoactive cannabinoids: Their chemistry and role against oxidative stress, inflammation, and cancer. *BioMed Research International*, *2018*(1), 1691428.

Pereira, S. R., Hackett, B., O'Driscoll, D. N., Sun, M. C., & Downer, E. J. (2021). Cannabidiol modulation of oxidative stress and signalling. *Neuronal Signaling*, *5*(3), NS20200080.

Pertwee, R. G. (2008). The diverse CB1 and CB2 receptor pharmacology of three plant cannabinoids : Δ9 -tetrahydrocannabinol, cannabidiol and Δ9 -tetrahydrocannabivarin. *British Journal of Pharmacology*, *153*(2), 199–215. DOI: 10.1038/sj.bjp.0707442

Pota, V., Sansone, P., De Sarno, S., Aurilio, C., Coppolino, F., Barbarisi, M., Barbato, F., Fiore, M., Cosenza, G., Passavanti, M. B., & Pace, M. C. (2024). Amyotrophic Lateral Sclerosis and Pain : A Narrative Review from Pain Assessment to Therapy. *Behavioural Neurology, 1228194*. Advance online publication. DOI: 10.1155/2024/1228194

Radosevich, S. R., Holt, J. S., & Ghersa, C. (1997). *Weed Ecology : Implications for Management*. John Wiley & Sons.

Rahn, E. J., & Hohmann, A. G. (2009). Cannabinoids as pharmacotherapies for neuropathic pain : From the bench to the bedside. *Neurotherapeutics; the Journal of the American Society for Experimental NeuroTherapeutics, 6*(4), 713–737.

Rathod, S. S., Agrawal, Y. O., Nakhate, K. T., Meeran, M. N., Ojha, S., & Goyal, S. N. (2023). Neuroinflammation in the Central Nervous System: Exploring the Evolving Influence of Endocannabinoid System. *Biomedicines, 11*(10), 2642.

Rice, J., & Cameron, M. (2018). Cannabinoids for Treatment of MS Symptoms : State of the Evidence. *Current Neurology and Neuroscience Reports, 18*(8), 50. DOI: 10.1007/s11910-018-0859-x

Rodríguez-Cueto, C., Gómez-Almería, M., García Toscano, L., Romero, J., Hillard, C. J., de Lago, E., & Fernández-Ruiz, J. (2021). Inactivation of the CB2 receptor accelerated the neuropathological deterioration in TDP-43 transgenic mice, a model of amyotrophic lateral sclerosis. *Brain Pathology (Zurich, Switzerland), 31*(6), e12972. DOI: 10.1111/bpa.12972

Russo, E. B. (2011). Taming THC : Potential cannabis synergy and phytocannabinoid-terpenoid entourage effects. *British Journal of Pharmacology, 163*(7), 1344–1364. DOI: 10.1111/j.1476-5381.2011.01238.x

Russo, E. B. (2016). Current therapeutic cannabis controversies and clinical trial design issues. *Frontiers in Pharmacology, 7*, 309.

Sarma, N. D., Waye, A., ElSohly, M. A., Brown, P. N., Elzinga, S., Johnson, H. E., Marles, R. J., Melanson, J. E., Russo, E., Deyton, L., Hudalla, C., Vrdoljak, G. A., Wurzer, J. H., Khan, I. A., Kim, N.-C., & Giancaspro, G. I. (2020). Cannabis Inflorescence for Medical Purposes : USP Considerations for Quality Attributes. *Journal of Natural Products, 83*(4), 1334–1351. DOI: 10.1021/acs.jnatprod.9b01200

Scherma, M., Masia, P., Deidda, M., Fratta, W., Tanda, G., & Fadda, P. (2018). New perspectives on the use of cannabis in the treatment of psychiatric disorders. *Medicines (Basel, Switzerland), 5*(4), 107.

Sharkey, K. A., Darmani, N. A., & Parker, L. A. (2014). Regulation of nausea and vomiting by cannabinoids and the endocannabinoid system. *European Journal of Pharmacology, 722*, 134–146.

Shoaib, S., Ansari, M. A., Fatease, A. A., Safhi, A. Y., Hani, U., Jahan, R., & Islam, N. (2023). Plant-derived bioactive compounds in the management of neurodegenerative disorders: Challenges, future directions and molecular mechanisms involved in neuroprotection. *Pharmaceutics, 15*(3), 749.

Silva, PA, Y., Borba, B. C., Pereira, V. A., Reis, M. G., Caliari, M., Brooks, M. S. L., & Ferreira, T. A. (2019). Characterization of tomato processing by-product for use as a potential functional food ingredient: Nutritional composition, antioxidant activity and bioactive compounds. *International Journal of Food Sciences and Nutrition*, *70*(2), 150–160. DOI: 10.1080/09637486.2018.1489530

Siqueira, L. M., Ryan, S. A., Gonzalez, P. K., Patrick, S. W., Quigley, J., & Walker, L. R. (2017). Nicotine and tobacco as substances of abuse in children and adolescents. *Pediatrics*, *139*(1).

Sirikantaramas, S., & Taura, F. (2017). Cannabinoids : Biosynthesis and biotechnological applications. Cannabis sativa L.-Botany and biotechnology, 183-206.

Śmiarowska, M., Białecka, M., & Machoy-Mokrzyńska, A. (2022). Cannabis and cannabinoids: Pharmacology and therapeutic potential. *Neurologia i Neurochirurgia Polska*, *56*(1), 4–13.

Solowij, N., & Battisti, R. (2008). The chronic effects of cannabis on memory in humans : A review. *Current Drug Abuse Reviews*, *1*(1), 81–98.

Soni, S., & Verma, L. (2024). Nicotine and Neurotransmitters an Update. *Research Journal of Pharmacy and Technology*, *17*(6), 2605–2612.

Spindle, T. R., Cone, E. J., Goffi, E., Weerts, E. M., Mitchell, J. M., Winecker, R. E., Bigelow, G. E., Flegel, R. R., & Vandrey, R. (2020). Pharmacodynamic effects of vaporized and oral cannabidiol (CBD) and vaporized CBD-dominant cannabis in infrequent cannabis users. *Drug and Alcohol Dependence*, *211*, 107937.

Starowicz, K., & Finn, D. P. (2017). Cannabinoids and pain : Sites and mechanisms of action. *Advances in Pharmacology*, *80*, 437–475.

Stasiłowicz-Krzemień, A., Nogalska, W., Maszewska, Z., Maleszka, M., Dobroń, M., Szary, A., Kępa, A., Żarowski, M., Hojan, K., & Lukowicz, M. (2024). The use of compounds derived from cannabis sativa in the treatment of epilepsy, painful conditions, and neuropsychiatric and neurodegenerative disorders. *International Journal of Molecular Sciences*, *25*(11), 5749.

Suttithumsatid, W., Kara, J., Lomlim, L., Nualsri, C., & Panichayupakaranant, P. (2023). Acetylcholinesterase Inhibitory Activity of Standardized Cannabinoids-rich Fractions. *Pharmaceutical and Biomedical Research*, *9*(3), 173–182.

Targa Dias Anastacio, H., Matosin, N., & Ooi, L. (2022). Neuronal hyperexcitability in Alzheimer's disease: What are the drivers behind this aberrant phenotype? *Translational Psychiatry*, *12*(1), 257.

Turner, B. D., Sloan, S. W., & Currell, G. R. (2019). Novel remediation of per-and polyfluoroalkyl substances (PFASs) from contaminated groundwater using Cannabis Sativa L.(hemp) protein powder. *Chemosphere*, *229*, 22–31.

Vaillancourt, R., Gallagher, S., Cameron, J. D., & Dhalla, R. (2022). Cannabis use in patients with insomnia and sleep disorders : Retrospective chart review. *Canadian Pharmacists Journal : CPJ*, *155*(3), 175–180. DOI: 10.1177/17151635221089617

Vaou, N., Voidarou, C., Rozos, G., Saldari, C., Stavropoulou, E., Vrioni, G., & Tsakris, A. (2025). Unraveling Nature's Pharmacy : Transforming Medicinal Plants into Modern Therapeutic Agents. *Pharmaceutics*, *17*(6), 754.

Vasincu, A., Rusu, R.-N., Ababei, D.-C., Larion, M., Bild, W., Stanciu, G. D., Solcan, C., & Bild, V. (2022). Endocannabinoid Modulation in Neurodegenerative Diseases : In Pursuit of Certainty. *Biology (Basel)*, *11*(3), 440. DOI: 10.3390/biology11030440

Vaucher, J., Keating, B. J., Lasserre, A. M., Gan, W., Lyall, D. M., Ward, J., Smith, D. J., Pell, J. P., Sattar, N., & Paré, G. (2018). Cannabis use and risk of schizophrenia : A Mendelian randomization study. *Molecular Psychiatry*, *23*(5), 1287–1292.

Viana, M. D. B., Aquino, P. E. A. D., Estadella, D., Ribeiro, D. A., & Viana, G. S. D. B. (2022). Cannabis sativa and cannabidiol: A therapeutic strategy for the treatment of neurodegenerative diseases? *Medical Cannabis and Cannabinoids*, *5*(1), 207–219.

Vitale, R. M., Morace, A. M., D'Errico, A., Ricciardi, F., Fusco, A., Boccella, S., Guida, F., Nasso, R., Rading, S., Karsak, M., Caprioglio, D., Iannotti, F. A., Arcone, R., Luongo, L., Masullo, M., Maione, S., & Amodeo, P. (2024). Identification of Cannabidiolic and Cannabigerolic Acids as MTDL AChE, BuChE, and BACE -1 Inhibitors Against Alzheimer's Disease by In Silico, In Vitro, and In Vivo Studies. *Phytotherapy Research*, *39*. Advance online publication. DOI: 10.1002/ptr.8369

Vuic, B., Milos, T., Tudor, L., Konjevod, M., Nikolac Perkovic, M., Jazvinscak Jembrek, M., & Svob Strac, D. (2022). Cannabinoid cb2 receptors in neurodegenerative proteinopathies: New insights and therapeutic potential. *Biomedicines*, *10*(12), 3000.

Wang, K., & Chen, X. (2024). Protective effect of flavonoids on oxidative stress injury in Alzheimer's disease. *Natural Product Research*, •••, 1–28.

White, C. M. (2019). A review of human studies assessing cannabidiol's (CBD) therapeutic actions and potential. *Journal of Clinical Pharmacology*, *59*(7), 923–934.

Wu, J. (2019). Cannabis, cannabinoid receptors, and endocannabinoid system : Yesterday, today, and tomorrow. *Acta Pharmacologica Sinica*, *40*(3), 297–299.

Xiong, Y., & Lim, C.-S. (2021). Understanding the Modulatory Effects of Cannabidiol on Alzheimer's Disease. *Brain Sciences*, *11*(9), 1211. DOI: 10.3390/brainsci11091211

Zeine, R. R., & Teasdale, B. W. (2023). *Medical Cannabis and the Effects of Cannabinoids on Fighting Cancer, Multiple Sclerosis, Epilepsy, Parkinson's, and Other Neurodegenerative Diseases*. IGI Global.

Zhang, W., Lin, H., Zou, M., Yuan, Q., Huang, Z., Pan, X., & Zhang, W. (2022). Nicotine in inflammatory diseases: Anti-inflammatory and pro-inflammatory effects. *Frontiers in Immunology*, *13*, 826889.

Zhang, W., Pan, X., Fu, J., Cheng, W., Lin, H., Zhang, W., & Huang, Z. (2024). Phytochemicals derived from Nicotiana tabacum L. plant contribute to pharmaceutical development. *Frontiers in Pharmacology*, *15*, 1372456.

Zimniewska, M. (2022). Hemp Fibre Properties and Processing Target Textile : A Review. *Materials (Basel), 15*(5), 1901. DOI: 10.3390/ma15051901

Zuo, L., Prather, E. R., Stetskiv, M., Garrison, D. E., Meade, J. R., Peace, T. I., & Zhou, T. (2019). Inflammaging and oxidative stress in human diseases: From molecular mechanisms to novel treatments. *International Journal of Molecular Sciences, 20*(18), 4472.

Chapter 10
Epigenetic Modulation by Gut Microbiota Metabolites:
A Therapeutic Avenue for Neurodegenerative Diseases

Hanane Ait Hammou
https://orcid.org/0009-0008-4806-494X
Chouaib Doukkali University, Morocco

Hanane Moummou
https://orcid.org/0009-0002-9252-4292
Private University of Marrakech, Morocco

Najoua Elhidar
University Cadi Ayyad, Morocco

Abderrahmane Achbani
ISPITS Marrakech, Morocco

Wafae Belokda
ISPITS Safi, Morocco

Soraia El Baz
Faculty of Sciences, University Cadi Ayyad, Morocco

Hicham Chatoui
https://orcid.org/0000-0002-4707-1645
ISPITS Marrakech, Morocco

Hassania Khalloufi EL
ISPITS Marrakech, Morocco

Abdellatif Ait Ougjij
ISPITS Marrakech, Morocco

Hanane Zine
Chouaib Doukkali University, Morocco

Abdelali Bitar
Chouaib Doukkali University, Morocco

ABSTRACT

The gut microbiota plays a pivotal role in maintaining human health. Composed of billions of microorganisms interacting with the digestive system, it is increasingly recognized for influencing pathophysiological processes beyond the digestive tract, particularly in neurodegenerative diseases. Secondary metabolites produced by the gut microbiota, such as short-chain fatty acids, polyphenols, and bile acids can modulate epigenetic processes in the brain. These epigenetic changes, including activating or inhibiting genes involved in inflammation, neuroprotection, and

DOI: 10.4018/979-8-3373-1365-8.ch010

neuronal repair, may offer new therapeutic perspectives for neurodegenerative diseases such as Alzheimer's and Parkinson's. Therefore, optimizing the gut microbiota and understanding these interactions provide promising avenues for innovative therapies for neurodegenerative diseases. This chapter provides a comprehensive overview of how microbial metabolites exert epigenetic effects and their potential for treating neurodegenerative diseases. It also includes references for further exploration of this exciting field.

INTRODUCTION

The gut-brain axis has recently emerged as a key area in the understanding of neurodegenerative diseases, highlighting the role of gut microbiota and its metabolites in modulating brain function (Morais et al., 2021; Rutsch et al., 2020). The human gastrointestinal (GI) tract contains the most diverse microbial community in the human body, including bacteria, fungi, viruses, and archaea (Hoegenauer et al., 2022; Miyauchi et al., 2023). Approximately 2000 bacterial species have been identified in the human gut, and the gut microbiota is estimated to contain nearly 150 times more genes than the human genome (Almeida et al., 2019; Qin et al., 2010).

The gut microbiome gradually reaches an adult configuration between the ages of 3 and 6 and remains stable throughout adulthood (Roswall et al., 2021; Yatsunenko et al., 2012). The biological functions of the adult gut microbiome include the regulation of nutrient harvest from the diet (Jumpertz et al., 2011), immune system modulation (Miyauchi et al., 2023; Schlute et al., 2020), cholesterol metabolism (Haghikia et al., 2022; Yao et al., 2022), bile acid (BA) transformation (Collins et al., 2023; Sato et al., 2021), production of antimicrobial peptides (Ma et al., 2022; Sberro et al., 2019), and drug metabolism (Javdan et al., 2020; Zimmermann et al., 2019).

Recent studies have revealed that microbiota-derived metabolites, such as short-chain fatty acids, can cross the blood-brain barrier and influence gene expression in the brain through epigenetic mechanisms. According to Hsiao et al. (2013), the gut microbiota can modulate central nervous system responses and influence neurological processes by producing biologically active molecules. For example, these metabolites can regulate DNA methylation and histone modifications, thereby affecting gene expression and neuronal function (Loh et al., 2024).

In the context of neurodegenerative diseases, studying the interaction between microbial metabolites and epigenetic mechanisms could lead to new therapeutic approaches for complex disorders such as Parkinson's and Alzheimer's disease (Loh et al., 2024). These findings open avenues for further exploration of interventions targeting the gut microbiota, offering promising prospects for slowing or preventing the progression of these devastating pathologies.

However, it is important to acknowledge the current limitations and open questions in this emerging field. Considerable interindividual variability in gut microbiota composition complicates the identification of universal microbial signatures associated with neurodegenerative diseases. Moreover, many mechanistic insights have been derived from animal models, and the translatability of these findings to humans remains challenging due to species differences and complex environmental influences. These aspects highlight the need for cautious interpretation of existing data and emphasize the importance of further longitudinal and clinical studies (Vogt et al., 2017).

This chapter provides an in-depth analysis of the gut microbiota, highlighting its main physiological functions. It then examines the mechanisms through which the gut–brain axis contributes to the development of neurodegenerative diseases. Particular attention is given to secondary metabolites produced by the microbiota and their role in the epigenetic regulation of these pathologies. Finally, the chapter explores therapeutic perspectives based on gut microbiota modulation, with a specific focus on Alzheimer's and Parkinson's diseases.

I. GENERAL OVERVIEW OF NEURODEGENERATIVE DISEASES

Neurodegenerative diseases encompass a group of chronic and progressive disorders of the central nervous system, characterized by neuronal degeneration and the accumulation of abnormal proteins, such as β-amyloid, tau, or α-synuclein (Soto & Pritzkow, 2018). The most common conditions include Alzheimer's disease, Parkinson's disease, amyotrophic lateral sclerosis (ALS), dementia with Lewy bodies, and Huntington's disease.

Although their clinical manifestations vary (cognitive, motor, or behavioral impairments), these disorders share common pathogenic mechanisms such as oxidative stress, chronic neuroinflammation, mitochondrial dysfunction, and epigenetic alterations (Heneka et al., 2015; Lin & Beal, 2006). Current therapeutic approaches remain mostly symptomatic, but emerging strategies-such as gene therapy, immunotherapy, modulation of the gut microbiota, and nanotechnology-are generating growing interest (Ciechanover & Kwon, 2015; Fei et al., 2022).

II. AN INSIGHT INTO GUT MICROBIOTA

The human body hosts a vast and diverse community of microorganisms, including bacteria, viruses, archaea, and unicellular eukaryotes. While these microbes colonize all body surfaces, the majority are located in the gastrointestinal tract (Backhed, 2005; Neish, 2009), which is anatomically divided into the stomach, small intestine, and large intestine. Each section presents a distinct microenvironment with specific physicochemical barriers that selectively support particular microbial communities.

It was long believed that the stomach was considered sterile due to its low pH, bactericidal secretions, bile reflux, mucus barrier, and active peristalsis. However, it is now known to host acid-tolerant bacteria such as *Streptococcus*, *Neisseria*, *Lactobacillus*, and *Campylobacter pyloridis* later renamed *Helicobacter pylori*. About 65% of gastric bacterial phylotypes originate from the oral cavity (Kazor et al., 2003). The dominant bacterial phyla in the healthy stomach include Firmicutes, Bacteroidetes, Actinobacteria, Fusobacteria, and Proteobacteria (Nardone & Compare, 2015).

The small intestine, essential for digestion and nutrient absorption, comprises the duodenum, jejunum, and ileum. In the duodenum, bile acids, pancreatic enzymes, and antimicrobial peptides limit microbial growth, resulting in low bacterial density (10^3–10^4 CFU/ml), dominated by Firmicutes and Actinobacteria (El Aidy et al., 2015). The jejunum shows increased microbial density and diversity (10^3–10^7 CFU/ml), favoring Gram-positive aerobes and facultative anaerobes such as *Lactobacillus*, *Enterococcus*, and *Streptococcus*. As contents move into the ileum, bacterial load rises further (up to 10^9 CFU/ml), with a shift toward anaerobic and Gram-negative species, especially near the ileocecal junction, where the composition resembles that of the colon.

The large intestine, composed of the cecum and the ascending, transverse, and descending colon, functions mainly in water reabsorption and fermentation of undigested substrates. Due to its slow transit time and anaerobic conditions, it harbors the highest microbial density (~10^{12} CFU/ml), predominantly Firmicutes and Bacteroidetes (Eckburg et al., 2005). The relative abundance of these phyla varies with age and health status, making their ratio a potential biomarker of disease risk (Mariat et al., 2009). The luminal microbiota primarily includes *Bacteroides*, *Bifidobacterium*, *Streptococcus*, *Enterobacteriaceae*, *Enterococcus*, *Clostridium*, *Lactobacillus*, and *Ruminococcus*, while the mucosal layer is enriched in *Clostridium*, *Lactobacillus*, *Enterococcus*, and *Akkermansia*. Some pathogens, such as *Campylobacter jejuni*, *Salmonella enterica*, *Vibrio cholerae*, *Escherichia coli*, and *Bacteroides fragilis*, may be present in small proportions (~0.1%) (Hollister et al., 2015).

Several factors influence microbial persistence along the gut, including gastric acidity, bile acids, digestive enzymes, and antimicrobial proteins. Additional determinants include chemical (pH, oxygen levels, redox potential), biological (mucus, bile, secretory IgA), and structural features (Seekatz et al., 2019; Leite et al., 2020).

Gut microbiota development begins early in life potentially in utero, as suggested by microbial detection in placental tissues (Aagaard et al., 2014; Rodriguez et al., 2015). The mode of delivery plays a critical role: vaginal birth exposes the newborn to maternal lactobacilli, while cesarean delivery delays colonization by *Bacteroides* and increases *Clostridium* abundance (Vershina et al., 2014; Jakobsson et al., 2014). In infancy, the microbiota is less diverse and dominated by Actinobacteria and Proteobacteria. Diversity expands rapidly during the first year, approaching an adult-like profile, though with individualized patterns (Rodriguez et al., 2015). In adults, the microbiota remains relatively stable but can be altered by diet, antibiotics, and illness (Eckburg et al., 2005). In older age, diversity tends to decline, with increased abundance of *Bacteroidetes* and *Clostridium* clusters, particularly in centenarians (Biagi et al., 2010; Claesson et al., 2011). These age-related changes, including reduced short-chain fatty acid production and elevated proteolysis, are thought to contribute to chronic low-grade inflammation known as "inflammaging" (Claesson et al., 2012; Biagi et al., 2013).

III. GUT MICROBIOTA FUNCTIONALITIES

Metagenomic studies have demonstrated that the gut microbiome possesses a coding capacity at least 150 times greater than the human genome, enabling biochemical pathways that are absent in human evolution (Kurokawa et al., 2007; Qin et al., 2010). Consequently, the gut microbiota plays a fundamental role in host physiology, contributing to key functions such as pathogen defense and immune system maturation. Additionally, it is involved in processes including the production of secondary bile acids (BAs), lipd metabolism, protein catabolism, xenobiotic degradation, and the synthesis of water-soluble vitamins (Figure 1). Beyond these roles, increasing attention is being given to its important impact on brain functions, leading to the concept of the brain gut-enteric microbiota axis (Figure 1) (Crudele et al., 2023).

Figure 1. Gut microbiota functions

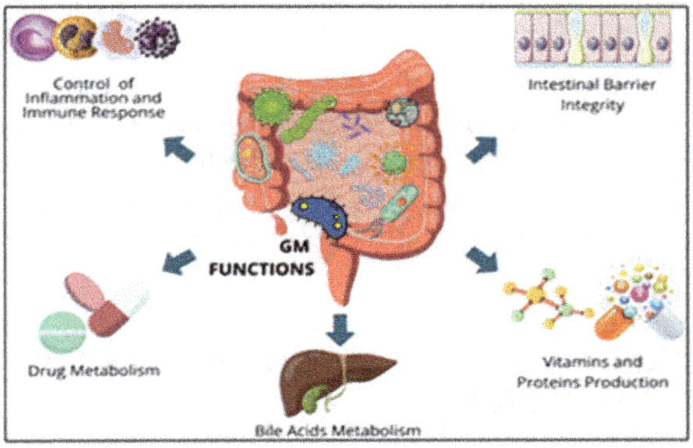

Source: Crudele et al. (2023)

1. **Vitamins** Humans cannot synthesize any vitamin, except D. However, it has been shown that members of the gut microbiota are able to synthesize vitamin C, K as well as most of the water-soluble B vitamins, such as biotin, cobalamin, folates, nicotinic acid, panthotenic acid, pyridoxine, riboflavin and thiamine (Hill et al., 1997). In contrast to dietary vitamins, which are adsorbed in the proximal tract of the small intestine, the predominant uptake of microbially produced vitamins occurs in the colon (Said & Mohammed 2006).
2. **Xenobiotics** Microorganisms residing in the human intestinal tract serve as a key site for xenobiotic metabolism. The gut microbiome, which encompasses the diverse microbial community in the gastrointestinal tract, can influence the metabolic fate of pharmaceuticals, environmental toxins, and heavy metals, thereby altering their pharmacokinetics. Through direct chemical modifications, whether occurring within the intestines or via re-entry through enterohepatic circulation, the gut microbiome can enhance metabolism or trigger bioactivation, depending on the enzymatic activity present in the microbial environment. (Koppel et al., 2017 ; Nelson, 2018)
3. **Metabolic functions** The gut microbiota plays a crucial nutritional role by fermenting nondigestible dietary fiber and carrying out anaerobic conversions of peptides and proteins, thereby enabling the host to recover metabolic energy (Acheson & Luccioli 2004). Notably, its capacity to process otherwise indigestible dietary components is essential, as this function not only supplies energy sources but also supports gut health maintenance (Guarner & Malagelada 2003).
4. **Barrier effect** In the intestinal lumen, pathogens and commensal bacteria compete for nutrients and epithelial adhesion sites. Additionally, the microbiota produces bacteriocins and can stimulate epithelial cells to produce antimicrobial peptides. It also induces the secretion of IgA and promotes the proper functioning of tight junctions between epithelial cells (Hooper, 2004), thereby reducing the invasion of pathogenic bacteria. (Hooper, 2004)
5. **Effects on the Immune System** Beyond its barrier function, the gut microbiota plays a crucial role in immune system development and maturation, as demonstrated by comparisons between germ-free and conventionally raised mice (Macpherson & Harris, 2004). Germ-free

mice exhibited significant immune deficiencies, including impaired lymphoid structures and reduced immunoglobulin and cytokine levels, which were restored upon microbiota colonization.

Beyond its overall impact, specific bacterial species modulate immune balance by influencing T cell populations. While some bacteria promote Th17 cell development (Gaboriau-Routhiau, 2009), others enhance Treg cells through short-chain fatty acid production, contributing to intestinal immune homeostasis (Atarashi et al., 2011; Smith, 2013).

IV. GUT-BRAIN AXIS: BIDIRECTIONAL COMMUNICATION AND MICROBIAL INFLUENCE

The bidirectional signaling between the (GI) tract and the brain plays a crucial role in maintaining homeostasis. This communication occurs at neural (central and enteric nervous systems), hormonal, and immunological levels. The gut microbiome influences the central nervous system (CNS) primarily through microbial-derived metabolites, including short-chain fatty acids (SCFAs), secondary bile acids (2BAs), and tryptophan metabolites (Tolhurst et al., 2012; Yano et al., 2015). Some of these intermediates interact directly with enteroendocrine and enterochromaffin cells, as well as the mucosal immune system. Others cross the intestinal barrier, enter systemic circulation, and may even penetrate the blood-brain barrier (Haghikia et al., 2015). However, it remains unclear whether these microbial-derived metabolites reach the brain in sufficient concentrations to modulate specific neural circuits. Alternatively, gut microbial signals may communicate with the brain via neural pathways, particularly through vagal and spinal afferents (Bravo et al., 2011).

Increasing evidence suggests that gut microbiota significantly impact brain function, leading to the concept of the brain-gut-enteric microbiota axis (Rhee et al., 2009) While much research has focused on brain-to-gut signaling, many studies are now exploring how gut microbiota influence brain activity and behavior. Experimental approaches to dissect this relationship include probiotic administration, antibiotic-induced dysbiosis, and pathogenic infections, each providing valuable insights into the role of the gut microbiome in neurobehavioral regulation 5 (Gareau et al., 2008; Denou et al., 2010).

For instance, Lyte et al. demonstrated that mice infected with *Campylobacter jejuni* exhibited anxiety-like behaviors within hours of infection, even before a significant immune response occurred. This finding suggests that the nervous system can directly detect changes in the gut microbiota and identify pathogenic threats (Lyte et al., 1998). Further studies involving *C. jejuni* and *Citrobacter rodentium* showed activation of specific brain regions linked to anxiety (Lyte et al., 2006; Goehler et al., 2008).

Likewise, chronic infection with *Helicobacter pylori* in mice led to alterations in stomach function and feeding behavior, potentially mimicking symptoms of functional dyspepsia observed in humans (Bercı'k et al., 2002; Bercik et al., 2009). These feeding abnormalities were associated with changes in hypothalamic neuropeptides, a brain region critical for appetite regulation (Bercik et al., 2009). Notably, the behavioral and biochemical changes persisted even after the eradication of *H. pylori*, indicating long-term effects (Bercik et al., 2009).

To further explore the relationship between gut microbiota and the brain, Research on the gut-brain connection has revealed significant differences between Germ-Free (GF) mice, which lack gut microbiota, and Specific Pathogen-Free (SPF) mice, which have a normal bacterial colonization. GF mice exhibited an abnormal stress response and lower brain-derived neurotrophic factor (BDNF) levels compared to SPF mice (Heijtz et al., 2011). Additionally, SPF mice demonstrated greater exploratory tendencies and reduced anxiety-like behavior relative to GF mice (Neufeld et al., 2011). Gene expression analysis highlighted notable differences in neurotransmitter-related, synaptic, and neurotrophic factor-associated genes between the two groups (Heijtz et al., 2011). Furthermore, introducing the commensal bacterium *Bacteroides thetaiotaomicron* into GF mice significantly influenced intestinal gene expression, particularly in pathways related to neurotransmitter metabolism (Hooper et al., 2001).

Disrupting the gut microbiota in adult mice using non-absorbable antibiotics further increased exploratory behavior and modified Brain-Derived Neurotrophic Factor (BDNF) levels in the hippocampus and amygdala (Bercik et al., 2011). Emerging evidence also suggests that the gut microbiota may influence memory and cognitive functions (Gareau et al., 2011).

Together, these animal studies provide robust evidence for a bidirectional communication axis between the gut microbiota and the central nervous system. They emphasize the significant role of gut bacteria in modulating behavior, brain function, and cognition.

V. SECONDARY METABOLITES FROM GUT MICROBIOTA

The human gut microbiota, composed of trillions of microorganisms, plays a critical role in human health by producing diverse secondary metabolites that influence many physiological processes. These metabolites, resulting from transforming dietary or endogenous compounds by gut bacteria, have significant effects on metabolism, the immune system, and protection against pathogens.

1. Short Chain Fatty Acids (SCFA)

The metabolic activity of bacteria is characterized by the production of bioproducts such as short-chain fatty acids (SCFAs). These metabolites also have immunomodulatory properties that impact chemotaxis, cell proliferation, cytokine production, and the expansion of immune populations. Among these SCFAs, butyrate has been described as particularly effective in the induction of Tregs in vivo and in vitro (Furusawa et al., 2013). Interestingly, the large amount of bacteria and associated antigens contained in the digestive tract do not generate spontaneous inflammatory responses in the intestine under physiological conditions. It is therefore necessary for recognition mechanisms to transmit tolerogenic signals for the establishment of a mutualistic relationship between the host and its microbiota.

They contribute to protection from infection and inflammation (Fukuda et al., 2011; Kim et al., 2013; Maslowski et al., 2009), recruitment and maturation of various subsets of immune cells (Sina et al., 2009; Smith et al. 2013), and metabolism (Bellahcene et al. 2013), and mediate host-microbe interactions.

Acetate, butyrate, and propionate are the major bacterial metabolites of the short-chain fatty acid (SCFA) class. They bind to GPCRs (G-protein-coupled receptors): GPR41 (G-protein regulator 41) or FFAR3 (free-fatty acid receptor 3), GPR43 or FFAR2, GPR109a, and OLFR78 (olfactory receptor 78). GPR43 can bind to all SCFAs, GPR41 to propionate and butyrate, GPR109a to butyrate, and OLFR78 to acetate and propionate.

2. Secondary Bile Acids

Bile acids, synthesized in the liver from cholesterol, are then conjugated to glycine or taurine. They are absorbed at the terminal ileum and then transported by the portal vein to the liver where they are excreted again in the bile (enterohepatic cycle). Approximately 5% of bile salts (0.2 to 0.3 g per day) escape this cycle and reach the colon where they are metabolized by the microbiota into so-called secondary bile acids as opposed to primary bile acids synthesized by the liver (Gérard, 2009; Ridlon et al., 2006). More than 20 secondary bile acids have been identified in human feces, demonstrating the wide variety of possible conversions of bile acids by the intestinal microbiota. These transformations include deconjugation, oxidation, and epimerization of hydroxyl groups, 7-dehydroxylation, formation of monoesters or polyesters, or desulfation (Ridlon et al., 2006). Deconjugation is catalyzed by a specific hydrolase, found in species belonging to the genera *Bacteroides, Clostridium, Lactobacillus, Listeria, Bifidobacterium, Streptococcus* and *Fusobacterium.* Oxidation of hydroxyl groups is carried out by hydroxysteroid dehydrogenases present in many bacterial genera. The concerted action of α and β-hydroxysteroid dehydrogenases results in the epimerization of bile acid (Ridlon et al., 2006). In humans, the two major secondary bile acids are deoxycholic and lithocholic, products of bacterial 7α-dehydroxylation of the primary bile acids, cholic and chenodeoxycholic acids.

3. Tryptophan Metabolism

Tryptophan, an essential amino acid for the body, can be metabolized through three main pathways: the kynurenine pathway, the indole pathway, and the serotonin pathway (Taleb, 2019). The serotonin pathway is a minor route and depends on the expression of the enzyme Tph1 (tryptophan hydroxylase 1) by intestinal epithelial cells, which is modulated by the intestinal microbiota (Reigstad et al., 2015; Yano et al., 2015). The kynurenine pathway leads to the production of kynurenine and several derivatives, including kynurenic acid, quinolinic acid, and picolinic acid (Dehhaghi et al., 2019; Wang et al., 2006). The indole pathway, mediated by bacterial tryptophanase, enables the production of indole and its derivatives, such as indole-3-aldehyde, indole-3-acetic acid, and indole-3-propionic acid (Taleb, 2019). Additionally, bacterial tryptophan decarboxylase facilitates the conversion of tryptophan into tryptamine (Williams et al., 2014) (Figure 2).

Figure 2. Tryptophan metabolic pathways

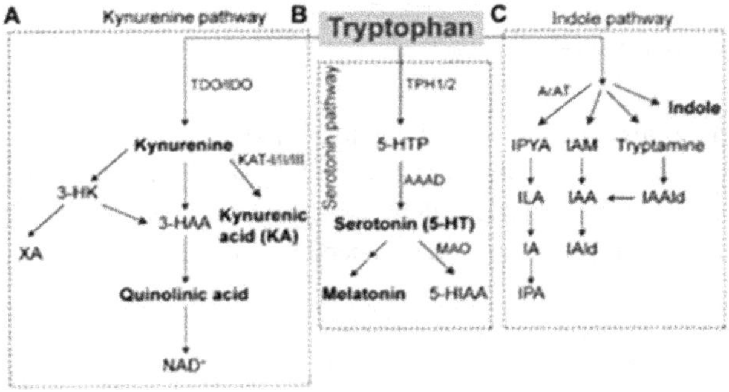

Source: Roth et al. (2021)

Tryptophan metabolism in the host involves three major pathways. (A) In the kynurenine pathway, which accounts for about 90% of tryptophan degradation, the enzyme TDO in the liver converts tryptophan into kynurenine, while IDO contributes in the brain, gastrointestinal tract, and liver. Kynurenine is further metabolized into kynurenic acid via KAT enzymes, or into quinolinic acid, which is then converted to NAD^+. Alternatively, it can be transformed into 3-hydroxykynurenine (3-HK) and then into xanthurenic acid (XA). (B) In the serotonin pathway, tryptophan is converted into 5-hydroxytryptophan (5-HTP) by TPH1 in enterochromaffin cells or TPH2 in neurons. 5-HTP is then decarboxylated into serotonin (5-HT), which can either be converted into melatonin, or degraded by MAO into 5-HIAA, a compound excreted in urine. (C) In the microbial pathway, gut microbes metabolize tryptophan into indole and various indole derivatives through specific bacterial enzymes. Tryptophan is first converted into indole-3-pyruvic acid (IPYA) via AAAD, then into indole-3-lactic acid (ILA) and anholocyclic acid (IA), and finally into indole-3-propionic acid (IPA). Other derivatives include indole-3-acetaldehyde (IAAld) and indole-3-aldehyde (IAld), some of which originate from tryptamine. Certain microbes can also convert tryptophan directly into indole. These three pathways collectively generate a broad spectrum of bioactive metabolites with significant roles in host physiology.

4. Microbial Neurotransmitters

The intestinal microbiota can produce neurotransmitters in the intestinal lumen that will have a local action. Many studies show that this intestinal microbiota is also able to modulate the levels of GABA, serotonin, catecholamines, and their receptors at the level of the Central Nervous System (CNS), thus modulating the major neurotransmission systems of the brain (Mazzoli and Pessione, 2016; Strandwitz, 2018). From a mechanistic perspective, the vagus nerve has many receptors that respond to the gut microbiota, such as TLR2, TLR4, GPR41, and GPR35 (Brun et al., 2013; Egerod et al., 2018; Hosoi et al., 2005; Nøhr et al., 2015), and gut hormones and cytokines such as CCK-A, GLP-1R, Y2R, H2R (H for histamine), H3R, IL-1R, and TNF-1/2R (Gray & Magnuson, 1987; Hansen, 2003; Steinberg et al., 2016). Nerve cells in the Enteric Nervous System (ENS) relay information from the gut microbiota to the vagus nerve via receptors that respond to viral RNAs – TLR3 and TLR7 – and bacterial components and metabolites – TLR4 and FFAR3 (Barajon et al., 2009; Nøhr et al., 2013).

5. Other Secondary Metabolites

Many other metabolites, such as trimethylamine and its precursors (choline and carnitine), polyphenols, arginine derivatives (putrescine, spermidine, spermine, citrulline, ornithine, agmatine), ATP, branched-chain fatty acids (isovalerate, isobutyrate, valerate), and histidine derivatives (histamine, imidazole propionate), are also produced by bacterial metabolism.

VI. EPIGENETIC EFFECT OF GUT MICROBIOTA SECONDARY METABOLITES ON NEURODEGENERATIVE DISEASES

The intestinal microbiota, composed of billions of microorganisms (bacteria, viruses, fungi), interacts with the brain via the gut-brain axis, a bidirectional communication system involving hormonal, immunological, and nervous mechanisms.

1. Epigenetic Mechanisms

Our body's genetic information is stored in DNA, which carries the instructions for producing proteins that perform various functions within the cell. Epigenetics is often defined as the study of phenotypic changes resulting from alterations in gene expression that do not stem directly from changes in the underlying DNA sequence. The modifications of epigenetics can be inherited across cell generations and are influenced by environmental factors such as diet, stress, toxins, and infections (Bird, 2007).

The main epigenetic mechanisms include:

- DNA methylation: Addition of methyl groups (CH3) to specific bases of DNA, usually on cytosines in sites rich in guanine and cytosine (CpG), which often represses gene expression.
- Histone modifications: These proteins around which DNA is wrapped can be modified by processes such as acetylation, methylation, or phosphorylation, which influences the accessibility of DNA and therefore gene expression.
- Non-coding RNAs: Molecules such as microRNAs and long non-coding RNAs regulate gene expression by inhibiting the translation of certain mRNAs or by modifying their stability.

2. Role of Microbial Metabolites in Epigenetic Mechanisms

The gut microbiota plays an important role in epigenetic modifications such as DNA methylation and histone modifications, which are known to influence the brain and behavior. Recent advances in multi-omics data integration and analysis provide an opportunity for a comprehensive investigation of gut microbiota and its contribution to host health (Tilocca et al., 2020). The impact of secondary metabolites produced by the gut microbiota on epigenetic regulation in neurodegenerative diseases has emerged as a key research area. These metabolites, which mainly include short-chain fatty acids (SCFAs), polyphenols, and neurotransmitters, can modulate gene expression in the brain, influence neuroinflammation mechanisms, and affect the progression of diseases such as Alzheimer's disease (Figure 3) and other neurodegenerative disorders.

Figure 3. Differences between physiological and pathological intestinal flora (a) Physiological microbial diversity is higher than that observed in Alzheimer's disease. Therefore, bacterial metabolites released into the systemic circulation have an anti-inflammatory effect. (b) While the abundance of anti-inflammatory bacteria decreased, the abundance of pro-inflammatory species increased in Alzheimer's disease. A large amount of pro-inflammatory microbial metabolites then passes into the blood and increase systemic inflammation

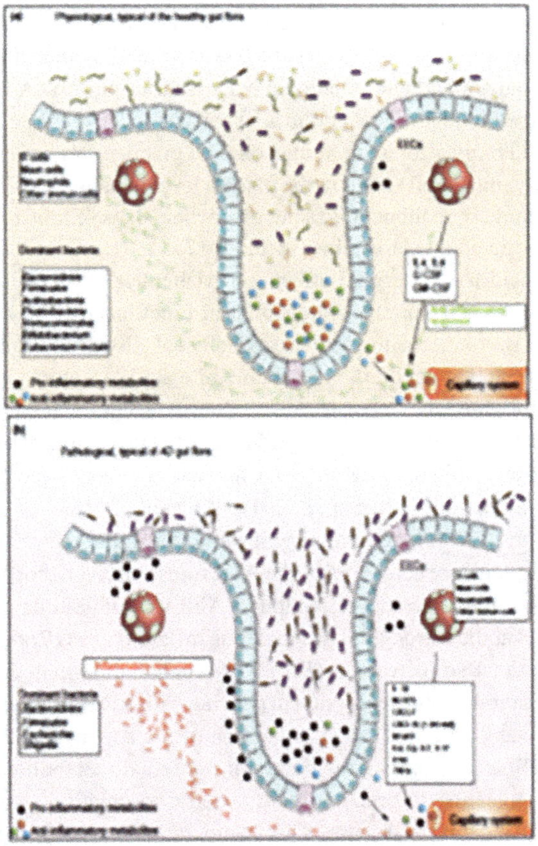

Source: Bostanciklioğlu (2019)

Butyrate is an important SCFA due to its function as a substrate for colonocyte proliferation and differentiation. It also modulates the inflammatory response by acting as an HDAC (Histone deacetylase) inhibitor, thereby decreasing the activation of the NF-κB pathway and the production of pro-inflammatory cytokines (Place et al., 2005), and increasing the expression of anti-inflammatory cytokines, such as IL-10 and IL-18, through the activation of GPCRs FFA2, FFA3, and GPR109a (Priyadarshini et al., 2018). Butyrate-producing bacteria mainly belong to the phylum Firmicutes, family Lachnospiraceae and the two species mainly responsible for this production are *Faecalibacterium prausnitizii* and *Eubacterium rectale*, whose abundance is decreased in patients with Alzheimer's disease, and is negatively correlated with the pro-inflammatory cytokine IL-6 and the NLRP3 inflammasome at the blood level (Marizzoni et al., 2020). Butyrate has also been shown to inhibit the formation of Aβ40 and Aβ42 peptide aggregates (Ho et al., 2018), and sodium butyrate protects against Aβ peptide-induced toxicity by acting on the GPR109a receptor in vitro (Sun et al., 2020). Furthermore, dietary butyrate administration reduces amyloid pathology and improves memory performance in the 5XFAD C57BL/6SJL mouse model (Fernando et al., 2020).

Cholesterol is metabolized by hepatocytes into bile acids such as cholic acid and chenodeoxycholic acid, which are then transformed into secondary bile acids (Ridlon et al., 2006) such as lithocholic acid, deoxycholic acid, and ursodeoxycholic acid by intestinal bacteria. They can cross the blood-brain barrier and enter neuronal cells via transporters. A recent study shows an association between low blood levels of cholic acid and chenodeoxycholic acid and the presence of amyloid deposits and brain atrophy in human clinical practice (Varma et al., 2021). A second study specifically analyzing 20 bile acids at the plasma level in patients with Alzheimer's disease showed a significant increase in lithocholic acid, glycochenodeoxycholic acid, glycodeoxycholic acid, and glycolithocholic acid (Marksteiner et al., 2017).

Ursodeoxycholic acid has beneficial effects on Alzheimer's disease. Indeed, it can inhibit NF-κB in microglial cells and improve mitochondrial functions of fibroblasts in patients with sporadic or familial forms of Alzheimer's disease (Bell et al., 2018; Joo et al., 2004). Other studies concerning treatments that act on the membrane bile acid receptors Gpbar1 (or TGR5) and nuclear FXR have highlighted the opposite effect of these receptors in AD pathology: a Gpbar1 receptor agonist significantly attenuates cognitive deficits and neuroinflammation in a mouse model of Aβ42 injection while an FXR receptor agonist aggravates the apoptosis of neuronal cells treated with Aβ42 peptides (Chen et al., 2019; Wu et al., 2018).

Tryptophan derivatives of the indole pathway, in particular indole-3-sulfate, indole-3-aldehyde, and indole-3-propionic acid, are activators of the AhR transcription factor (Hubbard et al., 2015) and can cross the blood-brain barrier. Activation of AhR will inhibit the activation and nuclear translocation of NF-κB at the astrocytic level and thus inflammation (Rothhammer et al., 2016). It has also been shown that an increase in dietary tryptophan in human clinical practice shows efficacy in reducing irritability and depression in patients with AD (Radulović & Minahan, 2004).

To enhance clarity and facilitate the visualization of the key mechanisms described above, the following table (Table 1) summarizes the main microbial metabolites involved in epigenetic regulation, their molecular targets, and their specific associations with neurodegenerative diseases, particularly Alzheimer's disease. This comparative overview highlights the complex interplay between gut-derived compounds and the host's epigenetic machinery, offering insights into potential therapeutic strategies.

Table 1. Microbial metabolites, their epigenetic effects, and associations with neurodegenerative diseases

Microbial Metabolite	Source / Associated Bacteria	Epigenetic Effects / Mechanisms	Receptors / Targets	Link with Neurological Diseases	References
Butyrate (SCFA)	Firmicutes, Lachnospiraceae, *Faecalibacterium prausnitzii*, *Eubacterium rectale*	HDAC inhibition → increased histone acetylation ↓ NF-κB activation (inflammatory pathway) ↑ expression of IL-10, IL-18 (anti-inflammatory cytokines)	GPR109a, FFA2, FFA3	↓ Aggregation of Aβ40/Aβ42 peptides ↓ Pro-inflammatory cytokines (IL-6, NLRP3) Memory improvement in Alzheimer's mouse models	Fernando et al., 2020 ; Ho et al., 2018 ; Marizzoni et al., 2020 ; Place et al., 2005 ; Priyadarshini et al., 2018 ; Sun et al., 2020
Secondary Bile Acids	Bacterial metabolism of primary bile acids (cholic acid, CDCA)	Modulation of nuclear receptors (FXR, TGR5) Mitochondrial and inflammatory effects	FXR (negative) TGR5/Gpbar1 (positive)	↓ Blood levels of cholic acid/CDCA associated with amyloid deposits and brain atrophy ↑ Lithocholic acid and derivatives in AD patients	Bell et al., 2018 ; Chen et al., 2019 ; Joo et al., 2004 ; Marksteiner et al., 2017 ; Ridlon et al., 2006 ; Varma et al., 2021 ; Wu et al., 2018
Ursodeoxycholic Acid (UDCA)	Secondary bile acid derivative	NF-κB inhibition Improved mitochondrial function	NF-κB, mitochondria	Reduced microglial inflammation Mitochondrial protection in sporadic/familial Alzheimer's forms	Bell et al., 2018 ; Joo et al., 2004
Tryptophan Derivatives (Indole Pathway)	Bacterial metabolites: indole-3-sulfate, indole-3-aldehyde, indole-3-propionic acid	Activation of AhR transcription factor Inhibition of NF-κB nuclear translocation in astrocytes	AhR (Aryl hydrocarbon receptor)	↓ Neuroinflammation Reduced irritability and depression in Alzheimer's patients	Hubbard et al., 2015 ; Radulović & Minahan, 2004 ; Rothhammer et al., 2016

VI. MICROBIOTA-BASED THERAPIES

Therapeutic perspectives involving the gut microbiota for neurodegenerative diseases have emerged as a particularly fascinating field of research. The gut microbiota, comprising billions of microorganisms, plays a crucial role in modulating various bodily functions, including brain health. Epigenetic modulation by microbiota metabolites paves the way for new therapeutic strategies for neurodegenerative pathologies such as Alzheimer's, Parkinson's, and multiple sclerosis. In parallel, innovative therapeutic approaches based on microbiota manipulation, drugs inspired by microbial metabolites, and personalized treatments are emerging as promising avenues.

1. Modulation of the Microbiota Through Prebiotics and probiotics

Therapies aimed at modulating the gut microbiota, such as probiotics, prebiotics, and dietary modifications, have shown significant potential for the prevention and treatment of neurodegenerative diseases (Ansari et al., 2020; Liu et al., 2015). Probiotics, which provide beneficial bacteria to the gut, can influence the production of protective metabolites, reducing systemic inflammation and altering neuronal gene expression. Similarly, prebiotics, which specifically feed beneficial bacteria, can stimulate the production of metabolites such as SCFAs, contributing to anti-inflammatory effects in the brain (Morgan, 1991; Radjabzadeh et al., 2022).

Notably, a meta-analysis, including a total of 12 studies comprising 852 patients with mild cognitive impairment or Alzheimer's disease), showed that probiotic supplementation significantly improves global cognitive function (SMD = 0.67), as well as memory, attention, and visuospatial abilities (Mo et al., 2024). The effect was particularly notable with interventions lasting several weeks.

2. Fecal Microbiota Transplantation (FMT)

Fecal microbiota transplantation (FMT) is a potentially therapeutic approach in which a healthy gut microbiota is transferred to a sick patient. It has shown promising results in patients with neurodegenerative diseases (langovan et al., 2019). This technique has shown beneficial effects in the treatment of Parkinson's disease, by modifying the composition of the gut microbiota and improving motor function in some patients (mouse model) (Sampson et al., 2016). FMT may be more effective, as it involves a broader range of microbiome modifications than prebiotics, probiotics or antibiotics (Smits et al., 2013). It has been shown that symptoms of Alzheimer's disease can be improved by FMT. Compared with mice from the same background and conventional breeding, FMT can regulate the disruption of gut microbiota and improve their cognitive function in mice with Alzheimer's disease (Yang, 2018; Sun et al., 2019b).

3. Nutrition and Modulation of the Microbiota

Diets play a crucial role in shaping the composition of the gut microbiota. Harnessing the power of the gut microbiota through dietary modifications could be a valuable approach to reducing the risk and symptoms of neurodegenerative diseases. Consumption of specific dietary components can shift the prevalence of bacterial communities in the gut toward a healthy enterotype, which may influence the production of beneficial metabolites by the microbiota. Several studies have shown that fermented foods, including kefir and foods rich in bioactive polyphenols and complex carbohydrates, such as grapes, pomegranate, and seaweed, may be effective in reducing neuroinflammation, oxidative stress, neurotransmitter dysfunction, and neuronal death associated with Alzheimer's and Parkinson's diseases (Gate et al., 2022; Turner et al., 2015; Yılmaz et al., 2019). Li et al. (2009) demonstrated that mice fed a diet containing 50% lean ground beef have a greater diversity of gut bacteria than those fed a standard rodent chow, and exhibit increased physical activity, reference memory, and less anxious behavior.

VII. LIMITATIONS OF MICROBIOTA-BASED THERAPIES

Despite the growing enthusiasm surrounding microbiota-based therapies including FMT, probiotics, prebiotics, synbiotics, and postbiotics several limitations still hinder their clinical translation and efficacy. First, there is substantial inter-individual variability in gut microbiota composition, driven by genetics, diet, environment, and medication use, which complicates the prediction of treatment outcomes and standardization of interventions (Lloyd-Price et al., 2019). Second, many current interventions lack specificity and mechanistic clarity; for example, probiotic strains may not colonize or persist in the host gut or may interact unpredictably with native microbiota or the immune system (Zmora et al., 2018). Furthermore, FMT, though effective in Clostridioides difficile infections, carries risks of pathogen transmission, and its long-term safety and regulatory oversight remain controversial (DeFilipp et al., 2019). Lastly, ethical and logistical issues related to donor screening, microbiota banking, and product stability further complicate the clinical application of these therapies. Thus, while promising, microbiota-based therapies require more precise characterization, rigorous clinical trials, and regulatory frameworks to be safely and effectively integrated into routine care.

CONCLUSION

Epigenetic modulation by gut microbiota metabolites represents a promising therapeutic avenue for neurodegenerative diseases. Recent findings highlight the importance of the gut-brain axis in regulating neuronal function and the influence of microbial metabolites on epigenetic mechanisms. By altering gene expression and modulating inflammation, oxidative stress, and the immune response, the gut microbiota could play a key role in preventing or slowing the progression of diseases such as Alzheimer's, Parkinson's, and other neurodegenerative disorders. Although further research is needed to better understand these complex mechanisms, interventions aimed at modulating the microbiota (such as the use of probiotics, prebiotics, or dietary adjustments) offer revolutionary therapeutic potential for the future of neurological care.

REFERENCES

Aagaard, K., Ma, J., Antony, K. M., Ganu, R., Petrosino, J., & Versalovic, J. (2014). The placenta harbors a unique microbiome. *Science Translational Medicine, 6*, 237ra65. PMID: 24848255

Almeida, A. A., Mitchell, L., Miguel, B., Samuel, C. F., & Gregory, B. G., (2019). A new genomic blueprint of the human gut microbiota. *Nature, 568*, 499–504. PMID: 30745586

Ansari, F., Pourjafar, H., Tabrizi, A., & Homayouni, A. (2020). The effects of probiotics and prebiotics on mental disorders: A review on depression, anxiety, Alzheimer, and autism spectrum disorders. *Current Pharmaceutical Biotechnology, 21*, 555–565. PMID: 31914909

Atarashi, K., Tanoue, T., Shima, T., Imaoka, A., Kuwahara, T., Momose, Y., Cheng, G., Yamasaki, S., & Saito, T.. (2011). Induction of colonic regulatory T cells by indigenous Clostridium species. *Science, 331*, 337–341. PMID: 21205640

Backhed, F. (2005). Host-bacterial mutualism in the human intestine. *Science, 307*, 1915–1920. PMID: 15790844

Barajon, I., Serrao, G., Arnaboldi, F., Opizzi, E., Ripamonti, G., Balsari, A., & Rumio, C. (2009). Toll-like receptors 3, 4, and 7 are expressed in the enteric nervous system and dorsal root ganglia. *The Journal of Histochemistry and Cytochemistry, 57*, 1013–1023. PMID: 19546475

Bell, S. M., Barnes, K., Clemmens, H., Al-Rafiah, A. R., Al-ofi, E. A., Leech, V., & Bandmann, O., (2018). Ursodeoxycholic Acid Improves Mitochondrial Function and Redistributes Drp1 in Fibroblasts from Patients with Either Sporadic or Familial Alzheimer's Disease. *Journal of Molecular Biology, 430*, 3942. PMID: 30171839

Bellahcene, M., O'Dowd, J. F., Wargent, E. T., Zaibi, M. S., & Hislop, D. C., (2013). Male mice that lack the G-protein-coupled receptor GPR41 have low energy expenditure and increased body fat content. *British Journal of Nutrition, 109*, 1755–1764. PMID: 23110765

Bercı'k, P., De Giorgio, R., Blennerhassett, P., Verdú, E. F., Barbara, G., & Collins, S. M. (2002). Immune-mediated neural dysfunction in a murine model of chronic Helicobacter pylori infection. *Gastroenterology, 123*, 1205–1215. PMID: 12360482

Bercik, P., Denou, E., Collins, J., Jackson, W., Lu, J., Jury, J., Deng, Y., Blennerhassett, P., Macri, J., McCoy, K. D., Verdu, E. F., & Collins, S. M. (2011). The intestinal microbiota affect central levels of brain-derived neurotropic factor and behavior in mice. *Gastroenterology, 141*, 599–609. PMID: 21683077

Bercik, P., Verdu, E. F., Foster, J. A., Lu, J., Scharringa, A., Kean, L., Wang, L., Blennerhassett, P., & Collins, S. M. (2009). Role of gut-brain axis in persistent abnormal feeding behavior in mice following eradication of Helicobacter pylori infection. *American Journal of Physiology. Regulatory, Integrative and Comparative Physiology, 296*, R587–R594. PMID: 19129375

Biagi, E., Candela, M., Turroni, S., Garagnani, P., Franceschi, C., & Brigidi, P. (2013). Ageing and gut microbes: Perspectives for health maintenance and longevity. *Pharmacological Research, 69*, 11–20. PMID: 23079287

Bird, A. (2007). Perceptions of epigenetics. *Nature, 447,* 396–398. PMID: 17522671

Bostanciklioğlu, M. (2019). The role of gut microbiota in pathogenesis of Alzheimer's disease. *Journal of Applied Microbiology, 127,* 954–967. PMID: 30920075

Bravo, J. A., Forsythe, P., Chew, M. V., Escaravage, E., Savignac, H. M., Dinan, T. G., Bienenstock, J., & Cryan, J. F. (2011). Ingestion of Lactobacillus strain regulates emotional behavior and central GABA receptor expression in a mouse via the vagus nerve. *Proceedings of the National Academy of Sciences of the United States of America, 108,* 16050–16055. PMID: 21876150

Brun, P., Giron, M. C., Qesari, M., Porzionato, A., Caputi, V., Zoppellaro, C., Banzato, S., Grillo, A. R., Spagnol, L., De Caro, R., & Pizzuti, D., (2013). Toll-like receptor 2 regulates intestinal inflammation by controlling integrity of the enteric nervous system. *Gastroenterology, 145,* 1323–1333. PMID: 23994200

Chen, Q., Ma, H., Guo, X., Liu, J., Gui, T., & Gai, Z. (2019). Farnesoid X Receptor (FXR) 212 Aggravates Amyloid-β-Triggered Apoptosis by Modulating the cAMP-Response ElementBinding Protein (CREB)/Brain-Derived Neurotrophic Factor (BDNF) Pathway In Vitro. *Medical Science Monitor, 25,* 9335–9345. PMID: 31812977

Ciechanover, A., & Kwon, Y. T. (2015). Degradation of misfolded proteins in neurodegenerative diseases: Therapeutic targets and strategies. *Experimental & Molecular Medicine, 47,* e147. PMID: 25766616

Claesson, M. J., Cusack, S., O'Sullivan, O., Greene-Diniz, R., de Weerd, H., & Flannery, E., (2011). Composition, variability, and temporal stability of the intestinal microbiota of the elderly. *Proceedings of the National Academy of Sciences of the United States of America, 108,* 4586–4591. PMID: 20571116

Claesson, M. J., Jeffery, I. B., Conde, S., Power, S. E., O'Connor, E. M., & O'Toole, P. W. (2012). Gut microbiota composition correlates with diet and health in the elderly. *Nature, 488,* 178–184. PMID: 22797518

Collins, S. L., Stine, J. G., Bisanz, J. E., Okafor, C. D., & Patterson, A. D. (2023). Bile acids and the gut microbiota: Metabolic interactions and impacts on disease. *Nature Reviews. Microbiology, 21,* 236–247. PMID: 36253479

Crudele, L., Gadaleta, R. M., Cariello, M., & Moschetta, A., (2023). Gut microbiota in the pathogenesis and therapeutic approaches of diabetes. The lancet,, 97: 104821 eBioMedicine 2023;97: 104821.DOI: 10.1016/j.ebiom.2023.104821

DeFilipp, Z., Bloom, P. P., & Torres Soto, M., (2019). Drug-Resistant E. coli Bacteremia Transmitted by Fecal Microbiota Transplant. *The New England Journal of Medicine, 381,* 2043–2050. PMID: 31665575

Dehhaghi, M., Kazemi Shariat Panahi, H., & Guillemin, G. J. (2019). Microorganisms, Tryptophan Metabolism, and Kynurenine Pathway: A Complex Interconnected Loop Influencing Human Health Status. *International Journal of Tryptophan Research : IJTR, 12,* 1178646919852996. PMID: 31258331

Denou, E., Bercik, P., & Collins, S. M. (2010). Perturbation of the intestinal microbiota alters behavior in mice. *Gastroenterology*, *136*(Suppl. 1), A-776–A-7.

Eckburg, P. B., Bik, E. M., Bernstein, C. N., Purdom, E., Dethlefsen, L., & Sargent, M., (2005). Diversity of the human intestinal microbial flora. *Science*, *308*(5728), 1635–1638. PMID: 15831718

Egerod, K. L., Petersen, N., Timshel, P. N., Rekling, J. C., & Wang, Y., (2018). Profiling of G protein-coupled receptors in vagal afferents reveals novel gut-to-brain sensing mechanisms. *Molecular Metabolism*, *12*, 62–75. PMID: 29673577

El Aidy, S., Bogert, V. D. B., & Kleerebezem, M. (2015). The small intestine microbiota, nutritional modulation and relevance for health. *Current Opinion in Biotechnology*, *32*, 14–20. PMID: 25308830

Fei, Y., Ma, Y., Zhang, H., Li, H., Feng, G., & Fang, J. (2022). Nanotechnology for research and treatment of the intestine. *Journal of Nanobiotechnology*, *20*, 430. PMID: 36175955

Fernando, W. M. A. D. B., Martins, I. J., Morici, M., Bharadwaj, P., Rainey-Smith, S. R., Lim, W. L. F., & Martins, R. N. (2020). Sodium Butyrate Reduces Brain Amyloid-β Levels and Improves Cognitive Memory Performance in an Alzheimer's Disease Transgenic Mouse Model at an Early Disease Stage. *Journal of Alzheimer's Disease*, *74*, 91–99. PMID: 31958090

Fukuda, S., Toh, H., Hase, K., Oshima, K., & Nakanishi, Y., (2011). Bifidobacteria can protect from enteropathogenic infection through production of acetate. *Nature*, *469*, 543–547. PMID: 21270894

Furusawa, Y., Obata, Y., Fukuda, S., Takaho, A. E., Gaku, N., & Daisuke, T., (2013). Commensal microbe-derived butyrate induces the differentiation of colonic regulatory T cells. *Nature*, *504*, 446–450. PMID: 24226770

Gaboriau-Routhiau, V., Rakotobe, S., Lecuyer, E., Mulder, I., Lan, A., Bridonneau, C., Rochet, V., Pisi, A., Paepe, M. D., Brandi, G., Eberl, G., Snel, J., Kelly, D., & Cerf-Bensussan, N. (2009). The key role of segmented filamentous bacteria in the coordinated maturation of gut helper T cell responses. *Immunity*, *31*, 677–689. PMID: 19833089

Gareau, M. G., Silva, M. A., & Perdue, M. H. (2008). Pathophysiological mechanisms of stress-induced intestinal damage. *Current Molecular Medicine*, *8*, 274–281. PMID: 18537635

Gareau, M. G., Wine, E., Rodrigues, D. M., Cho, J. H., Whary, M. T., Philpott, D. J., Macqueen, G., & Sherman, P. M. (2011). Bacterial infection causes stress-induced memory dysfunction in mice. *Gut*, *60*, 307–317. PMID: 20966022

Gérard, P. (2009). Gastrointestinal tract: microbial metabolism of steroids. In Timmis, K. N. (Ed.), *Handbook of Hydrocarbon and Lipid Microbiology*. Springer Heidelberg.

Goehler, L. E., Park, S. M., Opitz, N., Lyte, M., & Gaykema, R. P. A. (2008). Campylobacter jejuni infection increases anxiety-like behavior in the holeboard: Possible anatomical substrates for viscerosensory modulation of exploratory behavior. *Brain, Behavior, and Immunity*, *22*, 354–366. PMID: 17920243

Gray, T. S., & Magnuson, D. J. (1987). Neuropeptide neuronal efferents from the bed nucleus of the stria terminalis and central amygdaloid nucleus to the dorsal vagal complex in the rat. *The Journal of Comparative Neurology*, *262*, 365–374. PMID: 3655018

Guarner, F., & Malagelada, J. R. (2003). Gut flora in health and disease. *Lancet*, *361*, 512–519. PMID: 12583961

Haghikia, A., Jorg, S., Duscha, A., Berg, J., Manzel, A., Waschbisch, A., Hammer, A., & Lee, D., (2015). Dietary fatty acids directly impact central nervous system autoimmunity via the small intestine. *Immunity*, *43*, 817–829. PMID: 26488817

Haghikia, A., Zimmermann, F., Schumann, P., Jasina, A., Roessler, J., & Schmidt, D., (2022). Propionate attenuates atherosclerosis by immune-dependent regulation of intestinal cholesterol metabolism. *European Heart Journal*, *43*, 518–533. PMID: 34597388

Hansen, M. B. (2003). The Enteric Nervous System I: Organisation and Classification. *Pharmacology & Toxicology*, *92*, 105–113. PMID: 12753424

Heijtz, R. D., Wang, S., & Anuar, F. (2011). Pattersson, S. Normal gut microbiota modulates brain development and behavior. *Proceedings of the National Academy of Sciences of the United States of America*, *108*, 3047–3052. PMID: 21282636

Heneka, M. T., Carson, M. J., El Khoury, J., & Landreth, G. E., (2015). Neuroinflammation in Alzheimer's disease. *Lancet Neurology*, *14*, 388–405. PMID: 25792098

Hill, M. J. (1997). Intestinal flora and endogenous vitamin synthesis. *European Journal of Cancer Prevention*, *6*(Suppl 1), S43–S45. PMID: 9167138

Ho, L., Ono, K., Tsuji, M., Mazzola, P., Singh, R., & Pasinetti, G. M. (2018). Protective roles of intestinal microbiota derived short chain fatty acids in Alzheimer's disease-type betaamyloid neuropathological mechanisms. *Expert Review of Neurotherapeutics*, *18*, 83–90. PMID: 29095058

Hoegenauer, C., Hammer, H. F., Mahnert, A., & Moissl-Eichinger, C. (2022). Methanogenic archaea in the human gastrointestinal tract. *Nature Reviews. Gastroenterology & Hepatology*, *19*, 805–813. PMID: 36050385

Hollister, E. B., Gao, C., & Versalovic, J. (2015). Compositional and functional features of the gastrointestinal microbiome and their effects on human health. *Gastroenterology*, *146*(6), 1449–1458. DOI: 10.1053/j.gastro.2014.01.052 PMID: 24486050

Hooper, L. V. (2004). Bacterial contributions to mammalian gut development. *Trends in Microbiology*, *12*, 129–134. PMID: 15001189

Hooper, L. V., Wong, M. H., Thelin, A., Hansson, L., Falk, P. G., & Gordon, J. I. (2001). Molecular analysis of commensal host-microbial relationships in the intestine. *Science*, *291*, 881–884. PMID: 11157169

Hosoi, T., Okuma, Y., Matsuda, T., & Nomura, Y. (2005). Novel pathway for LPS-induced afferent vagus nerve activation: Possible role of nodose ganglion. *Autonomic Neuroscience*, *120*, 104–107. PMID: 15919243

Hsiao, E. Y., McBride, S. W., & Hsien, S., (2013). Microbiota modulate behavioral and physiological abnormalities associated with neurodevelopmental disorders. *Cell, 155*, 1451–1463. PMID: 24315484

Hubbard, T. D., Murray, I. A., & Perdew, G. H. (2015). Special section on drug metabolism and the microbiome - Minireview indole and tryptophan metabolism: Endogenous and dietary routes to ah receptor activation. In Drug Metabolism and Disposition. American Society for Pharmacology and Experimental Therapy, 43, 1522–1535.

Jakobsson, H. E., Abrahamsson, T. R., Jenmalm, M. C., Harris, K., Quince, C., Jernberg, C., Björkstén, B., Engstrand, L., & Andersson, A. F. (2014). Decreased gut microbiota diversity, delayed Bacteroidetes colonisation and reduced Th1 responses in infants delivered by caesarean section. *Gut, 63*, 559–566. DOI: 10.1136/gutjnl-2012-303249 PMID: 23926244

Javdan, B. Lopez J.G, Chankhamjon P., Lee Y.J, et al. (2020). Personalized Mapping of Drug Metabolism by the Human Gut Microbiome, Cell 181, 1661–1679.e1622.

Joo, S. S., Won, T. J., & Lee, D. I. (2004). Potential role of ursodeoxycholic acid in suppression of Nuclear factor kappa B in microglial cell line (BV-2). *Archives of Pharmacal Research, 9*, 954–960. PMID: 15473667

Jumpertz, R., Le, D. S., Turnbaugh, P. J., Trinidad, C., Bogardus, C., Gordon, J. I., & Krakoff, J. (2011). Energy-balance studies reveal associations between gut microbes, caloric load, and nutrient absorption in humans. *The American Journal of Clinical Nutrition, 94*, 58–65. PMID: 21543530

Kazor, C. E., Mitchell, P. M., Lee, A. M., Stokes, L. N., Loesche, W. J., Dewhirst, F. E., & Paster, B. J. (2003). Diversity of bacterial populations on the tongue dorsa of patients with halitosis and healthy patients. *Journal of Clinical Microbiology, 41*(2), 558–563. PMID: 12574246

Kim M,H., Kang, S.G, Park, J.H, Yanagisawa, M., Kim, C.H. (2013). Short-chain fatty acids activate GPR41 and GPR43 on intestinal epithelial cells to promote inflammatory responses in mice. Gastroenterology, 145, 396–406. e10.

Koppel, N., Maini, R. V., & Balskus, E. P. (2017). Chemical transformation of xenobiotics by the human gut microbiota. *Science, 356*(6344), eaag2770. PMID: 28642381

Kurokawa, K., Itoh, T., Kuwahara, T., Oshima, K., Toh, H., Toyoda, A., Takami, H., & Morita, H.. (2007). Comparativemetagenomics revealed commonly enriched gene sets in human gutmicrobiomes. *DNA Research : an International Journal for Rapid Publication of Reports on Genes and Genomes, 14*, 169–181. PMID: 17916580

Leite, G. G. S., Weitsman, S., Parodi, G., Shreya, C., Rashin, S., Maritza, S., Walter, M., Maria, J. V. M., Gillian, M. B., Ruchi, M., Simon, K. L., Laith, H. J., Shirley, P., Ali, R., & Mark, P. (2020). Mapping the segmental Microbiomes in the human mall bowel in comparison with stool: A REIMAGINE study. *Digestive Diseases and Sciences, 65*, 2595–2604. PMID: 32140945

Li, W., Dowd, S. E., Scurlock, B., Acosta-Martinez, V., & Lyte, M. (2009). Memory and learning behavior in mice is temporally associated with dietinduced alterations in gut bacteria. *Physiology & Behavior, 96*, 557–567. PMID: 19135464

Lin, M. T., & Beal, M. F. (2006). Mitochondrial dysfunction and oxidative stress in neurodegenerative diseases. *Nature*, *443*, 787–795. PMID: 17051205

Liu, X., Cao, S., & Zhang, X. (2015). Modulation of gut microbiota–brain Axis by probiotics, prebiotics, and diet. *Journal of Agricultural and Food Chemistry*, *63*, 7885–7895. PMID: 26306709

Loh, J. S., Mak, W. Q., Tan, L. K. S., Ng, C. X., & Chan, H. H., (2024). Microbiota–gut–brain axis and its therapeutic applications in neurodegenerative diseases. *Signal Transduction and Targeted Therapy*, *9*, 37. PMID: 38360862

Lyte, M., Li, W., Opitz, N., Gaykema, R. P. A., & Goehler, L. E. (2006). Induction of anxiety-like behavior in mice during the initial stages of infection with the agent of murine colonic hyperplasia Citrobacter rodentium. *Physiology & Behavior*, *89*, 350–357. PMID: 16887154

Lyte, M., Varcoe, J. J., & Bailey, M. T. (1998). Anxiogenic effect of subclinical bacterial infection in mice in the absence of overt immune activation. *Physiology & Behavior*, *65*, 63–68. PMID: 9811366

Ma, Y., Guo, Z., Xia, B., Zhang, Y., Liu, X., & Yu, Y., (2022). Identification of antimicrobial peptides from the human gut microbiome using deep learning. *Nature Biotechnology*, *40*, 921–931. PMID: 35241840

Macpherson, A. J., & Harris, N. L. (2004). Interactions between commensal intestinal bacteria and the immune system. *Nature Reviews. Immunology*, *4*, 478–485. PMID: 15173836

Mariat, D., Firmesse, O., Levenez, F., Guimaraes, V., Sokol, H., Doré, J., Corthier, G., & Furet, J. P. (2009). The Firmicutes/Bacteroidetes ratio of the human microbiota changes with age. *BMC Microbiology*, *9*(1), 123. PMID: 19508720

Marizzoni, M., Cattaneo, A., Mirabelli, P., Festari, C., Lopizzo, N., Nicolosi, V., & Mombelli, E.. (2020). Short-Chain Fatty Acids and Lipopolysaccharide as Mediators Between Gut Dysbiosis and Amyloid Pathology in Alzheimer's Disease. *Journal of Alzheimer's Disease*, *78*, 683–697. PMID: 33074224

Marksteiner, J., Blasko, I., Kemmler, G., Koal, T., & Humpel, C. (2017). Bile acid quantification of 20 plasma metabolites identifies lithocholic acid as a putative biomarker in Alzheimer's disease. *Metabolomics*, *14*, 1–10. PMID: 29249916

Maslowski, K. M., Vieira, A. T., Ng, A., Kranich, J., & Sierro, F., (2009). Regulation of inflammatory responses by gut microbiota and chemoattractant receptor GPR43. *Nature*, *461*, 1282–1286. PMID: 19865172

Mazzoli, R., & Pessione, E. (2016). The neuro-endocrinological role of microbial glutamate and GABA signaling. *Frontiers in Microbiology*, *7*, 1934. PMID: 27965654

Miyauchi, E., Shimokawa, C., Steimle, A., Desai, M. S., & Ohno, H. (2023). The impact of the gut microbiome on extra-intestinal autoimmune diseases. *Nature Reviews. Immunology*, *23*, 9–23. PMID: 35534624

Mo, R., Jiang, M., Xu, H., & Jia, R. (2024). Effect of probiotics on cognitive function in adults with mild cognitive impairment or Alzheimer's disease: A meta-analysis of randomized controlled trials. *Medicina Clínica (English Ed.)*, *162*, 565–573. PMID: 38641508

Morais, L. H., Schreiber, I., & Mazmanian, S. K. (2021). The gut microbiota–brain axis in behaviour and brain disorders. *Nature Reviews. Microbiology*, *19*, 241–255. PMID: 33093662

Morgan, M. Y. (1991). The treatment of chronic hepatic encephalopathy. *Hepato-Gastroenterology*, *38*, 377–387. PMID: 1662661

Nardone, G., & Compare, D. (2015). The human gastric microbiota: Is it time to rethink the pathogenesis of stomach diseases? *United European Gastroenterology Journal*, *3*(3), 255–260. PMID: 26137299

Neish, A. S. (2009). Microbes in gastrointestinal health and disease. *Gastroenterology*, *136*, 65–80. DOI: 10.1053/j.gastro.2008.10.080 PMID: 19026645

Nelson, D. R. (2018). Cytochrome P450 diversity in the tree of life. *Biochimica et Biophysica Acta*, *1866*, 141–154. DOI: 10.1016/j.bbapap.2017.05.003 PMID: 28502748

Neufeld, K. M., Kang, N., Bienenstock, J., & Foster, J. A. (2011). Reduced anxiety-like behavior and central neurochemical change in germ-free mice. *Neurogastroenterology and Motility*, *23*, 255–264. PMID: 21054680

Nøhr, M. K., Egerod, K. L., Christiansen, S. H., Gille, A., Offermanns, S., Schwartz, T. W., & Møller, M. (2015). Expression of the short chain fatty acid receptor GPR41/FFAR3 in autonomic and somatic sensory ganglia. *Neuroscience*, *290*, 126–137. PMID: 25637492

Nøhr, M. K., Pedersen, M. H., Gille, A., Egerod, K. L., Engelstoft, M. S., Husted, A. S., & Sichlau, R. M.. (2013). GPR41/FFAR3 and GPR43/FFAR2 as cosensors for short-chain fatty acids in enteroendocrine cells vs FFAR3 in enteric neurons and FFAR2 in enteric leukocytes. *Endocrinology*, *154*, 3552–3564. PMID: 23885020

Place, R. F., Noonan, E. J., & Giardina, C. (2005). HDAC inhibition prevents NF-κB activation by suppressing proteasome activity: Down-regulation of proteasome subunit expression stabilizes IκBα. *Biochemical Pharmacology*, *70*, 394–406. PMID: 15950952

Priyadarshini, M., Kotlo, K. U., Dudeja, P. K., & Layden, B. T. (2018). Role of Short Chain Fatty Acid Receptors in Intestinal Physiology and Pathophysiology. *Comprehensive Physiology*, *8*, 1091. PMID: 29978895

Qin, J., Li, R., Raes, J., Arumugam, M., & Burgdorf, K. S., (2010). A human gut microbial gene catalogue established by metagenomic sequencing. *Nature*, *464*, 59–65. PMID: 20203603

Radjabzadeh, D. Bosch, J. A., Uitterlinden, A. G., Zwinderman, A. H., Ikram, M. A., van Meurs et al. (2022). Gut microbiome-wide association study of depressive symptoms. *Nature Communications*, *13*, 7128. PMID: 36473852

Radulović, M., & Minahan, N. (2004). Tryptophan in Daily Diets and Residents' Mood States at Bayshore Health Center, Alzheimer's Unit, 105-119.

Reigstad, C. S., Salmonson, C. E., Rainey, J. F.III, Szurszewski, J. H., Linden, D. R., Sonnenburg, J. L., Farrugia, G., & Kashyap, P. C. (2015). Gut microbes promote colonic serotonin production through an effect of short-chain fatty acids on enterochromaffin cells. The FASEB Journal, 29, 1395. - Ren, J., & Bertrand, P. P. (2008). Purinergic receptors and synaptic transmission in enteric neurons. *Purinergic Signalling*, *4*, 255–266.

Ridlon, J. M., Kang, D. J., & Hylemon, P. B. (2006). Bile salt biotransformations by human intestinal bacteria. *Journal of Lipid Research*, *47*, 241–259. PMID: 16299351

Rodriguez, J. M., Murphy, K., Catherine, S., Ross, R. P., Kober, L. O., Juge, N., Avershina, E., Rudi, K., Narbad, A., Jenmalm, M. C., Marchesi, J. R., & Collado, M. C. (2015). The composition of the gut microbiota throughout life, with an emphasis on early life. *Microbial Ecology in Health and Disease*, *26*, 26050. PMID: 25651996

Roswall, J., Olsson, L. M., Kovatcheva-Datchary, P., Nilsson, S., Tremaroli, V., & Simon, M. C.. (2021). Developmental trajectory of the healthy human gut microbiota during the first 5 years of life. *Cell Host & Microbe*, *29*, 765–776.e763. PMID: 33794185

Roth, W., Zadeh, K., Vekariya, R., Ge, Y., & Mohamadzadeh, M. (2021). Tryptophan metabolism and gut-brain homeostasis. *International Journal of Molecular Sciences*, *22*, 2973. PMID: 33804088

Rothhammer, V., & Quintana, F. G. (2016). Environmental control of autoimmune inflammation in the central nervous system. *Current Opinion in Immunology*, *43*, 46–53. PMID: 27710839

Rutsch, A., Kantsjö, J. B., & Ronchi, F. (2020). The gut-brain axis: How microbiota and host inflammasome influence brain physiology and pathology. *Frontiers in Immunology*, *11*, 604179. PMID: 33362788

Said, H. M., & Mohammed, Z. M. (2006). Intestinal absorption of watersoluble vitamins: An update. Current Opinion in Gastroenterology, 22:140-146. 5.

Sampson, T. R., Debelius, J. W., Thron, T., Janssen, S., Shastri, G. G., & Ilhan, Z. E., (2016). Gut microbiota regulate motor deficits and neuroinflammation in a model of Parkinson's disease. *Cell*, *167*, 1469–1480. PMID: 27912057

Sato, Y., Atarashi, K., Plichta, D. R., Arai, Y., Sasajima, S., & Kearney, S. M., (2021). Novel bile acid biosynthetic pathways are enriched in the microbiome of centenarians. *Nature*, *599*, 458–464. PMID: 34325466

Sberro, H., Brayon, J. F., Zlitni, S., Edfors, F., & Greenfield, N.. (2019). Large-scale analyses of human microbiomes reveal thousands of small, novel genes. *Cell*, *178*, 1245–1259.e1214. PMID: 31402174

Seekatz, A. M., Schnizlein, M. K., Koenigsknecht, M. J., Baker, J. R., William, L. H., Barry, E. B., Vincent, B. Y., & Duxin, S. (2019). Spatial and temporal analysis of the stomach and small-intestinal microbiota in fasted healthy humans. *MSphere*, *4*, e00126–e00129. PMID: 30867328

Sina, C., Gavrilova, O., Förster, M., Till, A., Derer, S., & Hildebrand, F., (2009). G protein-coupled receptor 43 is essential for neutrophil recruitment during intestinal inflammation. *The Journal of Immunology : Official Journal of the American Association of Immunologists, 183*, 7514–7522. PMID: 19917676

Smith, P. M., Howitt, M. R., Panikov, N., Michaud, M., & Gallini, C. A., (2013b). The microbial metabolites, short-chain fatty acids, regulate colonic Treg cell homeostasis. *Science, 341*, 569–573. PMID: 23828891

Smits, L. P., Bouter, K. E., de Vos, W. M., Borody, T. J., & Nieuwdorp, M. (2013). Therapeutic potential of fecal microbiota transplantation. *Gastroenterology, 145*, 946–953. PMID: 24018052

Soto, C., & Pritzkow, S. (2018). Protein misfolding, aggregation, and conformational strains in neurodegenerative diseases. *Nature Neuroscience, 21*, 1332–1340. PMID: 30250260

Steinberg, B. E., Silverman, H. A., Robbiati, S., Gunasekaran, M. K., Tsaava, T., & Battinelli, E.. (2016). Cytokinespecific Neurograms in the Sensory Vagus Nerve. *Bioelectronic Medicine, 3*, 7–17. PMID: 30003120

Strandwitz, P. (2018). Neurotransmitter modulation by the gut microbiota. *Brain Research, 1693*, 128–133. PMID: 29903615

Sun, J., Xu, J., Ling, Y., Wang, F., Gong, T., & Yang, C., (2019b). Fecal microbiota transplantation alleviated Alzheimer's disease-like pathogenesis in APP/PS1 transgenic mice. *Translational Psychiatry, 9*, 189. PMID: 31383855

Sun, J., Yuan, B., Wu, Y., Gong, Y., Guo, W., Fu, S., Luan, Y., & Wang, W. (2020). Sodium Butyrate Protects N2a Cells against A β Toxicity in Vitro. *Mediators of Inflammation, 1*, 7605160. PMID: 32377164

Taleb, S. (2019). Tryptophan Dietary Impacts Gut Barrier and Metabolic Diseases. *Frontiers in Immunology, 10*, 2113. PMID: 31552046

Tilocca, B., Pieroni, L., Soggiu, A., Britti, D., Bonizzi, L., & Roncada, P., (2020). Gut–brain axis and neurodegeneration: State-of-the-art of meta-omics sciences for microbiota characterization. *International Journal of Molecular Sciences, 21*, 4045. PMID: 32516966

Tolhurst, G., Heffron, H., Lam, Y. S., Parker, H. E., Habib, A. M., Diakogiannaki, E., & Cameron, J., (2012). Short-chain fatty acids stimulate glucagon-like peptide-1 secretion via the G-proteincoupled receptor FFAR2. *Diabetes, 61*, 364–371. PMID: 22190648

Turner, R. S., Thomas, R. G., Craft, S., van Dyck, C. H., Mintzer, J., Reynolds, B. A., Brewer, J. B., Rissman, R. A., Raman, R., & Aisen, P. S. (2015). A randomized, double-blind, placebocontrolled trial of resveratrol for Alzheimer disease. *Neurology, 85*, 1383–1391. PMID: 26362286

Varma, V. R., Wang, Y., An, Y., Varma, S., Bilgel, M., Doshi, J., Legido-Quigley, C., & Delgado, J. C., (2021). Bile acid synthesis, modulation, and dementia: A metabolomic, transcriptomic, and pharmacoepidemiologic study. *PLoS Medicine, 18*, e1003615. PMID: 34043628

Vershina, E., Storrø, O., Øien, T., Johnsen, R., Pope, P., & Rudi, K. (2014). Major faecal microbiota shifts in composition and diversity with age in a geographically restricted cohort of mothers and their children. *FEMS Microbiol*, *87*, 280–290. PMID: 24112053

Vogt, N. M., Kerby, R. L., Dill-McFarland, K. A., Harding, S. J., Merluzzi, A. P., Johnson, S. C., Carlsson, C. M., Asthana, S., Zetterberg, H., Blennow, K., Bendlin, B. B., & Rey, F. E. (2017). Gut microbiome alterations in Alzheimer's disease. *Scientific Reports*, *7*, 13537. PMID: 29051531

Wang, J., Simonavicius, N., Wu, X., Swaminath, G., Reagan, J., Tian, H., & Ling, L. (2006). Kynurenic acid as a ligand for orphan G protein-coupled receptor GPR35. *The Journal of Biological Chemistry*, *281*, 22021–22028. PMID: 16754668

Williams, B. B., Van Benschoten, A. H., Cimermancic, P., Donia, M. S., & Zimmermann, M., (2014). Discovery and characterization of gut microbiota decarboxylases that can produce the neurotransmitter tryptamine. *Cell Host & Microbe*, *16*, 495–503. PMID: 25263219

Wu, X., Lv, Y. G., Du, Y. F., Chen, F., Reed, M. N., Hu, M., Suppiramaniam, V., Tang, S. S., & Hong, H. (2018). Neuroprotective effects of INT-777 against Aβ1–42-induced cognitive impairment, neuroinflammation, apoptosis, and synaptic dysfunction in mice. *Brain, Behavior, and Immunity*, *73*, 533–545. PMID: 29935310

Yang, L. (2018). *Effects of Fecal Microbiota Transplantation on Alzheimer's Disease Mice and its Molecular Mechanisms*. Zhengzhou University.

Yano, J. M., Yu, K., Donaldson, G. P., Shastri, G. G., Ann, P., Ma, L., Nagler, C. R., Ismagilov, R. F., Mazmanian, S. K., & Hsiao, E. Y. (2015). Indigenous bacteria from the gut microbiota regulate host serotonin biosynthesis. *Cell*, *161*, 264–276. PMID: 25860609

Yao, L., D'Agostino, G. D., Park, J., Hang, S., & Adhikari, A.. (2022). A biosynthetic pathway for the selective sulfonation of steroidal metabolites by human gut bacteria. *Nature Microbiology*, *7*, 1404–1418. PMID: 35982310

Yatsunenko, T., Rey, F. E., Manary, M. J., Trehan, I., Dominguez-Bello, M. G., & Contreras, M., (2012). Human gut microbiome viewed across age and geography. *Nature*, *486*, 222–227. PMID: 22699611

Yılmaz, İ., Enver Dolar, M., & Özpınar, H. (2019). Effect of administering kefir on the changes in fecal microbiota and symptoms of inflammatory bowel disease: A randomized controlled trial. *The Turkish Journal of Gastroenterology : the Official Journal of Turkish Society of Gastroenterology*, *30*, 242–253. PMID: 30662004

Zimmermann, M., Zimmermann-Kogadeeva, M., Wegmann, R., & Goodman, A. L. (2019). Mapping human microbiome drug metabolism by gut bacteria and their genes. *Nature*, *570*, 462–467. PMID: 31158845

Zmora, N., Zilberman-Schapira, G., & Suez, J., (2018). Personalized Gut Mucosal Colonization Resistance to Empiric Probiotics Is Associated with Unique Host and Microbiome Features. *Cell*, *174*, 1388–1405.e21. PMID: 30193112

Chapter 11
Integration of Nanotechnology and Secondary Metabolites for Neurodegenerative Disease

Vivek Pazhamalai
https://orcid.org/0000-0003-2088-7545
Department of BioEngineering, School of Engineering, Vels Institute of Science, Technology, and Advanced Studies, India

Subha Lakshmi S.
Department of BioEngineering, School of Engineering, Vels Institute of Science, Technology, and Advanced Studies, India

S. Vignesh
https://orcid.org/0009-0007-0827-9797
Department of BioEngineering, School of Engineering, Vels Institute of Science, Technology, and Advanced Studies, India

E. Sreeram
https://orcid.org/0009-0001-7797-6824
Department of BioEngineering, School of Engineering, Vels Institute of Science, Technology, and Advanced Studies, India

Ivo Romauld S.
https://orcid.org/0000-0003-0610-0646
Department of BioEngineering, School of Engineering, Vels Institute of Science, Technology, and Advanced Studies, India

Meenambiga S. S.
https://orcid.org/0000-0002-5445-525X
Department of BioEngineering, School of Engineering, Vels Institute of Science, Technology, and Advanced Studies, India

K. Rajakumari
Department of BioEngineering, School of Engineering, Vels Institute of Science, Technology, and Advanced Studies, India

ABSTRACT

Neurodegenerative disorders (NDDs), including Alzheimer's disease (AD), Parkinson's disease (PD), Prion disease (PrD), and amyotrophic lateral sclerosis (ALS), present huge difficulties on account of the consistently advancing loss of neurons. Nanotechnology offers a promising solution to these difficulties. Designed nanomaterials can cross the BBB, upgrading prescription conveyance to brain tissues. Nanoparticles further develop drug solubility, stability, bioavail-

DOI: 10.4018/979-8-3373-1365-8.ch011

ability, and designated conveyance while diminishing side effects. Generally used nanocarriers incorporate liposomes, emulsions, solid lipid nanoparticles, and polymeric nanoparticles. These nanocarriers enhance drug conveyance, bioavailability, and stability, effectively conveying phytomedicine and manufactured drugs. Nanomedicine progressions show potential for treating neurodegenerative infections by further developing medication transport across the BBB and limiting fringe side effects. Research needs to be focused on biocompatibility, Environmental effects, and administrative obstacles to the drug development.

1. INTRODUCTION

Neurodegenerative disorders (NDDs) represent one of the critical challenges in contemporary medicine. States characterized by continuous damage to the nervous tissue and irreversible cellular failure greatly hamper mental, muscular, and sensory expertise and, thus, reduce the overall wellness of billions of entities. With over 55 million individuals affected worldwide and an anticipated rise to 139 million by 2050, driven by the global aging trend (Alzheimer's Association, 2024), Alzheimer's alone affects a burgeoning population. Ultramodern exploration has stressed the diversified and complex etiology of NDDs. Common pathological features include proteotoxic stress due to the aggregation of misfolded proteins, oxidative stress due to the excessive production of free radicals, and habitual neuroinflammation caused by a disabled vulnerable response (Jadoon et al., 2024). Importantly, these molecular differences frequently occur before symptoms have shown up, and therefore, early opinion and treatment aren't possible. The available treatments are symptomatic and seldom address the underlying disease process. Furthermore, clinical efficacy of the treatment is also mostly restricted due to an important role played by the BBB in the protection of neural tissue, while that same reality hinders speedy delivery of drugs to the brain (Mohapatra et al., 2024). These pitfalls dictate the immediate need for new and robust therapeutic approaches.

Table 1. Concise overview of major NDDs, summarizing their definitions, clinical manifestations, prevalence patterns, and associated therapeutic challenges. This comparative framework emphasizes the heterogeneity of disease presentation while also revealing the convergence of treatment limitations, underscoring the rationale for novel therapeutic exploration

Disorder	Definition	Symptoms	Prevalence	Treatment Challenges
Alzheimer's Disease	Progressive neurodegenerative disorder	Memory loss, confusion	High	Limited treatment options, difficulty in early diagnosis
Parkinson's Disease	Disorder affecting motor coordination	Tremors, rigidity, bradykinesia	Moderate	Symptom management, minimal disease-modifying therapies
Prion Disease	Rare, fatal neurodegenerative condition	Rapid cognitive decline	Low	Lack of effective treatments
Amyotrophic Lateral Sclerosis (ALS)	Degeneration of motor neurons	Muscle weakness, paralysis	Low–moderate	No cure, symptomatic management only

To overcome these patient challenges, nanotechnology has surfaced as a transformative strategy. Engineered nanoparticles offer the eventuality to access the BBB, enabling precise and targeted delivery of remedial moieties to affected brain regions (Silindir Gunay et al., 2016). Of particular interest is the encapsulation of secondary metabolites - bioactive composites deduced from natural sources known for their antioxidant, anti-inflammatory, and neuroprotective parcels (Sharma et al., 2021). Classes similar to flavonoids, terpenoids, and alkaloids have shown strong eventuality in mollifying oxidative stress and neuroinflammation. still, their clinical restatement remains limited due to poor solubility, rapid-fire declination, and minimum brain penetration.

Research Gap and Objectives

Even though nanotechnology has really improved how we deliver drugs, not much research has looked into using natural compounds called secondary metabolites in nanocarrier systems to treat brain diseases like neurodegenerative disorders (NDDs). The most current exploration studies either look at nanocarriers on their own or natural metabolites independently. This means we don't really understand how well they could work together, especially when it comes to getting past the blood-brain barrier and guarding brain cells. This chapter aims to fill that gap by looking at how important secondary metabolites can be in helping with issues like cell damage from stress, protein buildup, and brain inflammation. It also examines how nanotechnology can help these mixes by making them more stable, easier for the body to use, and better at reaching the brain. Plus, it suggests ways to combine these natural treatments with smart nanotech approaches. The chapter also talks about the problems with current NDD treatments, explains the basics of how nanotechnology works for delivering drugs to the brain, describes how different kinds of secondary metabolites work and protect the brain, and analyzes how they're put into nanocarriers. It also points out the hurdles we face, what the future might hold, and how these ideas could actually be used. By bringing together nanotechnology and the healing power of natural metabolites, this chapter offers a hopeful new approach that could not only ease symptoms but also change how these diseases progress, giving new hope to people with neurodegenerative disorders.

2. THE BLOOD–BRAIN BARRIER

2.1. Structural Description

The BBB is a technical interface that protects the brain while regulating the exchange of molecules between blood and neural tissue. It's composed of endothelial cells joined by tight junctions, supported by the rudimentary lamella, pericytes, and astrocyte end bases. Endothelial cells form the core of the hedge, confining most motes, while the rudimentary lamella stabilizes the structure and mediates cell relations. Pericytes regulate blood inflow and maintain tight junctions, and astrocytes cache growth factors that strengthen hedge integrity and control nutrient and waste transport (Chen et al., 2024; Schiera et al., 2024; Liu et al., 2024; Lau et al., 2024). Figure 1 illustrates the factors inclusively maintain BBB integrity and highlights remedial access to the brain is confined.

Figure 1. Structure of the Blood-Brain Barrier (BBB)

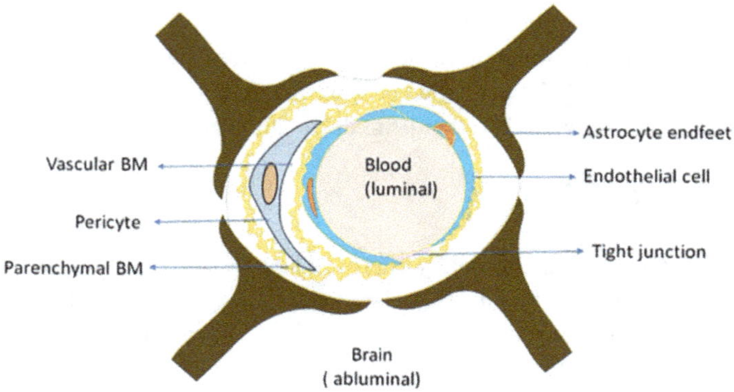

2.2. Functional Places

The BBB has three essential places to play. First out, it acts as a defensive guard, keeping out contagions, poisons, and colourful specifics (Chen et al., 2024). Next, it ensures picky permeability, allowing nutrients like glucose and amino acids to pass through while blocking larger or water-loving molecules (Schiera et al., 2024). Eventually, it helps maintain the neural terrain necessary for proper signaling by regulating ion attention and clearing out metabolic waste (Liu et al., 2024).

Limitations

While vital for brain protection, the BBB significantly restricts remedial access. Tight junctions help the entry of most medicines, enzymatic degradation degrades therapeutic molecules, and efflux transporters laboriously pump medicines back into circulation, reducing their concentration in the brain (Chen et al., 2024; Daraban et al., 2024; Narsinh et al., 2024). These mechanisms explain why conventional curatives frequently fail to achieve effective medicine delivery.

NDD Specific Impact

The restrictive nature of the BBB has direct consequences for neurodegenerative diseases (NDDs). In Alzheimer's disease, amyloid β accumulation further disrupts BBB integrity, worsening medicine delivery challenges. In Parkinson's disease, dopaminergic curatives are limited by poor penetration, reducing long-term efficacy. In amyotrophic side sclerosis (ALS), motor neuron–targeted medicines fail to reach sufficient attention due to efflux mechanisms. These exemplifications punctuate how BBB limitations directly hamper remedial progress in NDDs.

Exploration Gap and Transition

Although nanocarriers have been explored for synthetic drugs, their operation in encapsulating secondary metabolites for NDD remedy remains underdeveloped. Flavonoids, terpenoids, and alkaloids are examples of natural composites that retain neuroprotective, antioxidant, and anti-inflammatory properties (Sharma et al., 2021). However, their limited BBB penetration and poor bioavailability make clinical restatement difficult. Innovative approaches that combine nanotechnology with secondary metabolites to improve remedial efficacy are needed to close this gap. These challenges establish the explanation for exploring nanotechnology-predicated delivery systems. The coming section will examine how nanoparticles can be finagled to cross the BBB, summarize secondary metabolites, and deliver them precisely to affected brain regions, thereby advancing remedial issues in neurodegenerative conditions (Mohapatra et al., 2024).

3. NANOTECHNOLOGY: AN INNOVATIVE APPROACH

3.1. Concept and Scope

Nanotechnology refers to the manipulation of matter at the nanoscale, typically between 1 and 100 nanometers, where materials exhibit unique physical, chemical, and biological properties that differ significantly from their bulk forms. In the context of neurodegenerative disorders (NDDs), nanotechnology provides innovative solutions to long-standing therapeutic challenges. These include crossing the blood–brain barrier (BBB), protecting fragile therapeutic agents from degradation, and ensuring precise release within diseased neural environments. By tailoring nanomaterials in terms of composition, size, shape, and surface chemistry, researchers can exploit natural BBB transport pathways and design delivery systems that address disease-specific needs (Kainat et al., 2024; Nguyen et al., 2021; Ullah et al., 2024; Mohapatra et al., 2024).

3.2. Classes of Nanomaterials and Their Relevance

Nanomaterials can be broadly categorized into inorganic, organic, carbon-based, and composite systems, each offering distinct advantages for biomedical applications. Inorganic nanoparticles, such as gold or iron oxide, are valued for their stability, imaging contrast, and magnetic guidance. Organic systems, including liposomes, solid lipid nanoparticles, polymeric micelles, and dendrimers, provide biocompatibility and biodegradability while allowing versatile drug loading. Carbon-based nanomaterials, such as graphene derivatives and carbon nanotubes, contribute mechanical strength and electrical conductivity, though they require careful functionalization to ensure safety. Composite or hybrid systems combine multiple materials, such as lipid–polymer hybrids or core–shell architectures, to integrate complementary properties like stability, high loading capacity, and targeted delivery (Kainat et al., 2024).

These categories are particularly relevant to NDDs because they offer flexibility in targeting, compatibility with hydrophobic or sensitive secondary metabolites, and theranostic potential through optical or magnetic tracking. In this way, nanomaterials serve not only as carriers but also as multifunctional platforms that can diagnose, monitor, and treat disease simultaneously.

3.3. Physico-Chemical Features and Design Choices

The physicochemical features of nanoparticles are central to their therapeutic utility. Size is a critical determinant, with particles between 20 and 80 nm often striking the balance between efficient BBB transcytosis, tissue penetration, and circulation time.

Shape also influences cellular uptake, with spherical particles generally favored for broad applicability, while rods or discs can alter endothelial interactions. Surface charge plays a dual role: near-neutral or PEGylated surfaces extend circulation and reduce immune clearance, whereas mildly cationic surfaces enhance adsorptive transcytosis but risk increased protein binding. Surface chemistry further refines delivery, as ligand functionalization with transferrin, insulin, or ApoE motifs enables receptor-mediated transcytosis, while stealth coatings minimize immune recognition. Stability and release mechanisms are equally important, with core–shell designs preventing premature leakage and stimuli-responsive systems—triggered by pH, redox conditions, enzymes, or external stimuli such as ultrasound or magnetic fields—ensuring site-specific delivery (Mohapatra et al., 2024; Nguyen et al., 2021).

Together, these design choices allow nanoparticles to engage BBB pathways effectively, fit the oxidative and inflammatory microenvironments characteristic of NDDs, and maintain safety through controlled release and biodegradability.

3.4. Overcoming BBB Challenges with Nanotechnology

Nanotechnology provides several strategies to overcome the formidable challenges posed by the BBB. Lipid-based carriers, such as liposomes and solid lipid nanoparticles, exploit membrane affinity and can be decorated with ligands to facilitate receptor-mediated transcytosis, while also improving the solubility of hydrophobic secondary metabolites. Polymeric nanoparticles, including PLGA and PEG-PLA systems, offer sustained release and enzymatic protection, with tunable surfaces for receptor targeting and endosomal escape. Receptor-mediated delivery is particularly promising, as ligand-decorated nanoparticles can engage transferrin, insulin, or LRP1 receptors to cross endothelial cells via vesicular transport. Adsorptive transcytosis provides an alternative route, where mildly cationic surfaces interact with the negatively charged endothelial glycocalyx to promote uptake. Physical adjuncts such as focused ultrasound can transiently loosen tight junctions, magnetic guidance can concentrate iron oxide carriers, and gene-editing approaches can enhance receptor expression, all of which synergize with nanoparticle design to improve brain delivery (Mohapatra et al., 2024). Figure 2 illustrates the mechanisms, mapping endothelial structures, tight junctions, and efflux systems to nanoparticle features and transport routes, thereby linking biological barriers to engineering solutions.

Figure 2. Mechanism of nanoparticle transport across the BBB

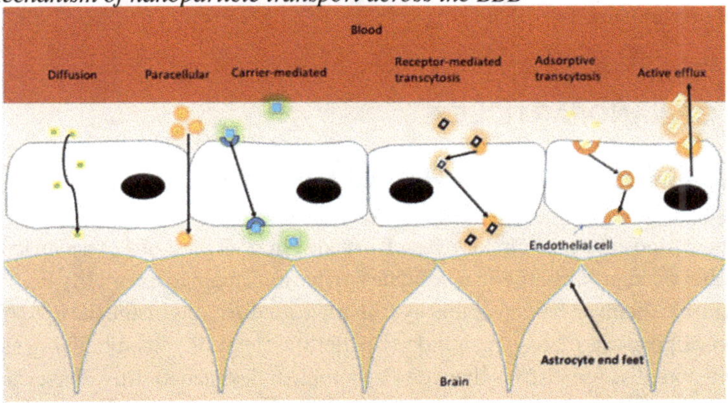

3.5. NDD-Specific Impact and Integrated Use Cases

The impact of nanotechnology becomes most evident when applied to specific neurodegenerative disorders. In Alzheimer's disease, nanoparticles can be engineered to target LRP1 or RAGE receptors, encapsulate antioxidants or anti-amyloid agents, and release them in response to oxidative stress, while theranostic tracking verifies delivery to plaque regions. In Parkinson's disease, transferrin- or lactoferrin-liganded nanoparticles can deliver dopamine precursors or neuroprotective metabolites, with pH-responsive release ensuring endosomal escape. In amyotrophic lateral sclerosis, long-circulating, efflux-evasive carriers, including cell membrane–coated nanoparticles, can home to inflamed vasculature and co-deliver anti-inflammatory metabolites to motor neurons. For prion diseases, stable, enzyme-resistant cores with targeted ligands provide rapid delivery of fast-acting payloads to counter widespread neurotoxicity. These examples demonstrate how nanotechnology design choices interconnect with disease-specific pathology, translating engineering principles into therapeutic outcomes.

3.6. Research Gap, Novelty, and Objectives

Despite these advances, a critical research gap remains: the underuse of secondary metabolites in nanocarrier systems. Compounds such as flavonoids, terpenoids, and alkaloids possess strong neuroprotective, antioxidant, and anti-inflammatory properties, yet their poor solubility, instability, and limited BBB penetration have hindered clinical translation (Sharma et al., 2021). The novelty of this chapter lies in addressing this gap by proposing the synergistic integration of secondary metabolites into BBB-targeted, stimuli-responsive nanoparticles. This approach enhances brain delivery, protects sensitive payloads, and synchronizes release with the oxidative and inflammatory signals characteristic of NDD microenvironments.

The objectives of this section are therefore to define the classes and properties of nanomaterials relevant to NDD therapy, explain how engineered features map onto BBB transport mechanisms, demonstrate disease-anchored design examples that interconnect nanotechnology decisions with

NDD pathology, and prepare the ground for a detailed discussion of secondary metabolites and their optimal carrier pairings.

4. TYPES OF NANOPARTICLES

4.1. Liposomes, Emulsions, and Solid Lipid Nanoparticles

Liposomes: Liposomes are circular vesicles made out of at least one concentric lipid bilayer encasing a watery center. This one-of-a-kind design permits them to exemplify both hydrophilic medications inside their watery center and hydrophobic medications inside the lipid bilayer itself. Key benefits of liposomes include their biocompatibility, flexibility in embodying assorted drug types, and the potential for surface change to accomplish designated drug conveyance (Patel et al., 2024).

Emulsions: Emulsions are colloidal scatterings of two immiscible liquids, typically oil and water, stabilized by the presence of surfactants. They are portrayed considering their stage piece: oil-in-water (O/W) emulsions, where oil drops are dispersed in an endless water stage, and water-in-oil (W/O) emulsions, where water drops are dispersed in an oil stage. Emulsions offer a couple of advantages, including the ability to encapsulate hydrophobic drugs inside the oil stage and the relative ease of preparation (Patel et al., 2024).

Strong Lipid Nanoparticles (SLNs): SLNs are submicron-sized particles made out of strong lipids. Contrasted with liposomes, SLNs offer a few benefits, including further developed stability, controlled drug delivery, and security of encapsulated drugs from degradation. This improved soundness is ascribed to the strong idea of the lipid center, which limits drug spillage and forestalls oxidation. SLNs have shown promise in effectively water-soluble medications and have applications in different drug details (Atapour-Mashhad et al., 2024).

4.2. Nanostructured Lipid Carriers and Micelles

Nanostructured Lipid Carriers (NLCs): NLCs address a high-level type of strong lipid nanoparticles (SLNs), integrating both strong and fluid lipids inside their construction. This one-of-a-kind creation offers a few benefits over conventional SLNs, including fundamentally upgraded drug stacking limit and further developed controlled discharge profiles. The presence of fluid lipids inside the strong lipid network disturbs the profoundly requested precious stone design, making voids that can accommodate a larger amount of medication while keeping up with the, generally speaking, primary honesty of the transporter (Atapour-Mashhad et al., 2024).

Micelles: Micelles are spherical totals shaped by amphiphilic particles, like surfactants, in fluid conditions. These atoms have both hydrophilic (water-loving) and hydrophobic (water-hating) moieties. In watery arrangements, they self-gather into round structures with a hydrophobic center, equipped for solubilizing and typifying hydrophobic medications, and a hydrophilic shell that guarantees their stability in the fluid environment. This one-of-a-kind design empowers micelles to work on the dissolvability and bioavailability of ineffectively water-soluble medications, while likewise shielding them from degradation and upgrading their safety (Singh et al., 2024).

4.3. Polylacticcoglycolic Acid (PLGA) Nanoparticles

PLGA, a biodegradable and biocompatible copolymer made out of lactic corrosive and glycolic corrosive monomers, has emerged as a promising stage for drug delivery applications. Its exceptional properties, including controlled and supported drug delivery, biodegradability, and negligible harmfulness, make it an appealing contender for different restorative applications. PLGA nanoparticles can really embody a great many restorative specialists, including small particles, proteins, and nucleic acids. Additionally, their adaptability reaches out to different organizational courses, like oral, intravenous, and intramuscular, offering adaptability in treatment systems (Chakraborty et al., 2024).

5. NANO PHYTOMEDICINE

5.1. Integration of Nanotechnology and Phytomedicine

PLGA, a biodegradable and biocompatible copolymer made out of lactic corrosive and glycolic corrosive monomers, has emerged as an ideal stage for drug delivery applications. Its remarkable properties, including controlled and supported drug delivery, biodegradability, and insignificant toxicity, make it an alluring possibility for different helpful applications. PLGA nanoparticles can really epitomize many remedial specialists, including small atoms, proteins, and nucleic acids. In addition, their adaptability stretches out to different administration routes, like oral, intravenous, and intramuscular, offering adaptability in treatment systems (Chakraborty et al., 2024).

This flexibility, joined with their biocompatibility and biodegradability, makes PLGA nanoparticles a promising methodology for the conveyance of optional metabolites. Numerous optional metabolites, obtained from plants, have intense restorative properties however frequently experience the ill effects of poor solubility, fast degradation, and restricted bioavailability.

Epitome inside PLGA nanoparticles can defeat these constraints by:

- Further developing dissolvability and bioavailability: PLGA nanoparticles can improve the solubility and bioavailability of inadequately soluble auxiliary metabolites, in this way expanding their therapeutic efficacy (Patel et al., 2024).
- Safeguarding from corruption: The PLGA network can shield embodied optional metabolites from degradation by catalysts and unforgiving environmental circumstances, extending their shelf life and improving their stability (Patel et al., 2024).
- Empowering controlled discharge: PLGA nanoparticles can be designed to deliver optional metabolites in a controlled and supported way, giving delayed remedial impacts and decreasing the recurrence of organization (Dan et al., 2024).
- Working with designated conveyance: PLGA nanoparticles can be surface-altered with a focus on ligands to guide them to explicit cells or tissues, improving therapeutic efficacy and limiting incidental effects (Singh et al., 2024).

By utilizing the extraordinary properties of PLGA nanoparticles, scientists can foster imaginative medication delivery systems for therapeutic agents, making ready for additional effective and targeted treatments for various diseases.

5.2. Benefits of Nano Phytomedicine in NDD Treatment

Neurodegenerative problems (NDDs) present a huge challenge because of the constraints of current helpful choices. The blood-brain barrier (BBB) represents a significant impediment to medication conveyance, limiting the penetration of numerous therapeutic agents into the cerebrum. In any case, nano phytomedicine offers a few key benefits that address these difficulties and hold huge potential for the treatment of NDDs:

- **Upgraded Mind Infiltration**: Nanoparticles, because of their small size and remarkable surface properties, can all the more actually cross the BBB contrasted with conventional drug delivery systems. This improved mind infiltration permits phytochemicals to arrive at their objectives inside the cerebrum and apply their restorative impacts more effectively (Patel et al., 2024).
- **Upgraded Cancer Prevention Agent and Mitigating Properties**: Numerous phytochemicals have powerful cell reinforcement and calming properties. These properties are vital in NDDs, where oxidative stress and neuroinflammation assume critical roles in sickness movement. By encapsulating these phytochemicals inside nanoparticles, their cell reinforcement and mitigating exercises can be improved and saddled all the more (Ojha et al., 2022).
- **Neuroprotection**: A couple of phytochemicals, for instance, curcumin and resveratrol, have shown neuroprotective properties, including the ability to protect neurons from oxidative damage, advance neuronal resilience, and support neuronal regeneration. Nano phytomedicine can work on the movement of these neuroprotective phytochemicals to the frontal cortex, thus expanding their supportive potential (Dan et al., 2024).
- **Diminished Aftereffects**: Regular foundational organization of phytochemicals can frequently prompt askew impacts and fundamental secondary effects. By using nanocarriers to target specific cerebrum regions, the foundational openness to phytochemicals can be limited, decreasing the risk of adverse incidental effects (Patel et al., 2024).
- **Synergistic Impacts**: Many plant separates contain a mind-boggling combination of phytochemicals with corresponding or synergistic helpful impacts. Nano phytomedicine empowers the co-conveyance of different phytochemicals inside a solitary nanoparticle detailing, possibly upgrading their general therapeutic efficacy (Chakraborty et al., 2024).
- **Supported and Controlled Delivery**: Nanocarriers can be intended to maintain and control the arrival of phytochemicals inside the cerebrum. This guarantees a steady restorative impact over the long run, lessening the requirement for incessant dosing and improving quiet consistency (Chakraborty et al., 2024).

By and large, the incorporation of nanotechnology with phytomedicine offers a novel and promising methodology for the treatment of NDDs. By upgrading the conveyance and adequacy of plant-based therapeutics, nano phytomedicine can possibly work on persistent results and personal satisfaction. Figure 3 represents the application of nanotechnology in phytomedicine delivery for gastrointestinal disorders. Table 2 overviews the Nano-Phytomedicine in NDD Therapy.

Figure 3. Nano-phytomedicine in drug delivery

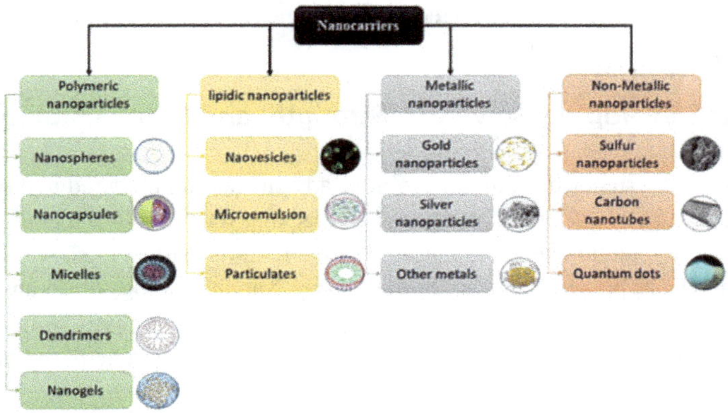

Table 2. Nano-phytomedicine in NDD therapy

Nano-Phytomedicine	Mechanism of Action	Benefits
Curcumin Nanoparticles	Enhanced brain penetration and antioxidant properties	Improved cognitive function in Alzheimer's disease
Resveratrol Nanoparticles	Neuroprotection and anti-inflammatory effects	Reduced neuronal damage in Parkinson's disease
Ginkgo Biloba Extracts	Enhanced stability and bioavailability	Improved cognitive function in dementia
Quercetin Nanoparticles	Antioxidant and anti-inflammatory properties	Reduced oxidative stress and inflammation in neurodegenerative disorders

6. ENHANCING DRUG DELIVERY WITH NANOPARTICLES

Nanomaterials will change drug delivery through the designated, effective, and controlled organization of drugs. Conventional strategies frequently face difficulties, for example, poor solubility and quick breakdown, which can restrict the viability of many drugs. However, nanoparticles address these issues with customizable sizes, unique surface properties, and multifunctional features. They improve drug steadiness, help bioavailability, and consider the exact focus of diseased tissues or cells. The utilization of nanoparticles in drug delivery is noteworthy; they can cross biological barriers, for example, the blood-brain barrier (BBB), provide sustained drug discharge, and decrease side effects on the body. Hydrophilic and hydrophobic medications can be integrated into nanoparticle (NP) formulations, shielding drugs from degradation before arriving at their objective locales. Normal sorts of nanoparticles include liposomes, polymeric nanoparticles, dendrimers, and metallic nanoparticles. These advancements are broadly used to treat malignant growth, neurodegenerative illnesses, and diseases. Nanoparticles address a critical progression in personalized medicine, with the possibility to change drug delivery and treatment strategies for different illnesses. However, inquiries concerning well-being and adequacy remain. Further examinations are expected to understand these progressions completely (Sell et al., 2023; Al-Zoubi et al., 2022).

6.1. Further Developing Solvency, Strength, and Bioavailability

Phytochemicals, esteemed for their restorative advantages, are hindered in their clinical use by constraints like poor solubility, instability, and lacking retention in the body. These issues lead to unfortunate retention and deficient restorative fixations at the objective site. Moreover, phytochemicals can be broken down rapidly in the body. Nanotechnology offers answers for these difficulties, guaranteeing effective delivery and improved execution of phytochemicals (Grigore et al., 2017). Table 3 overviews these aspects of developing solvency, strength, and bioavailability:

6.1.1. Further Developing Dissolvability

A huge test for most phytochemicals is their low solubility in water. Poor solubility keeps these mixtures from blending in with biological liquids, lessening their restorative potential. Nanotechnology resolves this issue by decreasing phytochemicals to nanometer sizes, expanding the surface region, and upgrading disintegration rates. Nanoparticles can be treated with exceptional coatings, surfactants, or polymers to further develop water dissolvability and similarity with biological frameworks. Moreover, phytochemicals can be integrated into lipid-based carriers, like liposomes and Nano-emulsions, making a hydrophilic shell around a hydrophobic core. This approach further develops solvency and copies the body's natural transport systems (Sutradhar et al., 2014).

6.1.2. Further Developing Strength

Therapeutic agents are defenceless against corruption because of natural variables like light, heat, oxygen, dampness, and pH changes. These awarenesses lessen their adequacy, entangling their utilization of medicines. Nanotechnology answers by embodying phytochemicals in nanoparticle frameworks, safeguarding them from external damage, and upgrading stability. Methods like exemplification in liposomes, polymeric nanoparticles, and strong lipid nanoparticles establish defensive conditions that forestall degradation. These particles can be designed to oppose oxidation, hydrolysis, and other breakdown processes by adding stabilizers or making powerful defensive layers. Imaginative techniques, similar to co-polymer blends or multilayer coatings, have shown superior stability for delicate mixtures under harsh circumstances (Chakraborty et al., 2022).

6.1.3. Target Specificity

Nanoparticles offer a novel drug delivery system, empowering designated organizations to explicit cells, tissues, or organs. This explicitness is accomplished by modifying nanoparticle surfaces with ligands, antibodies, peptides, or particles that bind to receptors at the objective site. This guarantees the therapeutic agent is conveyed precisely where required, upgrading efficacy while lessening exposure to healthy tissues. Such accuracy is beneficial for sicknesses requiring entrance of biological barriers, similar to the BBB in neurodegenerative problems, or targeting specific tumor microenvironments in cancer. For example, nanoparticles modified with transferrin, a ligand for transferrin receptors, facilitate the effective delivery of therapeutic agents to the brain or tumors. Essentially, nanoparticles modified with folic acid target folate receptors, profoundly expressed in numerous growths, adding one more layer of explicitness. Progressed focusing on methodologies include double or multi-focusing, where nanoparticles draw in numerous receptors, expanding take-up by target cells, and overcoming drug resistance mechanisms (Huang et al., 2024).

6.2. Reduced Toxicity

Phytochemical-based treatments frequently convey the dangers of fundamental toxicity at high doses. Numerous phytochemicals have adverse impacts, causing unwanted secondary effects and restricting clinical use. Nanocarriers altogether decrease foundational toxicity by restricting helpful specialist accessibility to non-target tissues. Exemplifying phytochemicals in nanoparticles shields them from untimely delivery and conveys them directly to target locales. This designated conveyance technique upgrades wellbeing profiles, taking into account higher helpful dosages without significant incidental effects. Nanocarriers can likewise be intended for improved responsive delivery, delivering their freight at explicit locations in response to pH, temperature, or the presence of specific catalysts. In malignant growth treatment, pH-responsive nanoparticles can deliver their payload in the acidic climate of cancers, saving nonpartisan pH tissues (Al-Zoubi et al., 2022).

Table 3. Advantages of nanocarriers over conventional drug delivery

Aspect	Conventional Drug Delivery	Nanocarriers
Solubility	Poor solubility for many drugs	Enhanced solubility through Nano formulation
Stability	Susceptible to degradation	Improved stability and protection from environmental factors
Bioavailability	Limited bioavailability	Enhanced absorption and bioavailability
Target Specificity	Low target specificity	High target specificity through surface modification
Toxicity	Higher systemic toxicity	Reduced toxicity through targeted delivery
Controlled Release	Limited control over the release	Controlled and sustained release of therapeutic agents

Understanding the advantages of nano-phytomedicine and the mechanisms by which nanoparticles enhance drug delivery is crucial for developing effective treatments for neurodegenerative disorders.

7. TYPES OF NANOCARRIERS FOR DRUG DELIVERY

Nanocarriers are particular conveyance frameworks intended to ship restorative agents to specific objective sites in the body. They come in different structures, each with remarkable properties and applications. Figure 4 provides an overview of properties of nanocarriers in Table 4.

Figure 4. Types of nanocarriers used in drug delivery

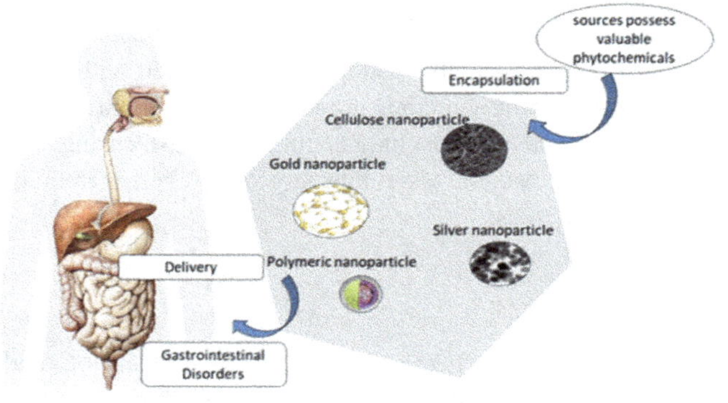

Polymeric Nanoparticles: A Versatile Tool

These tiny particles have many useful features, which we'll explore in this section. Polymeric nanoparticles (NPs) are solid structures in which bioactive chemicals can be encapsulated or absorbed. Polymeric NPs, given their structure, namely, a nanocapsule or NP, are manufactured using a variety of processes, which mainly fall into two categories: top-down methods, in which NPs are initiated from a solution and gathered into a unified structure, and bottom-up methods, in which NPs are prepared from the monomer level to the final structure. Regardless of the final structure or preparation process, there are some finishing stages, such as drying to remove excess solvents and additives. There are numerous drying processes available, the most notable of which being air freeze drying, spray freeze drying, vacuum freeze drying, and lyophilization. Top-down fabrication methods include solvent emulsification-evaporation, solvent emulsification-diffusion, nanoprecipitation, salting out, dialysis, and supercritical fluid technology (SCF). Furthermore, the most widely utilized bottom-up approaches include emulsion polymerization and recombinant DNA technology (Zeynalzadeh et al., 2024).

Micelles

Polymeric micelles are the missing link in efforts to utilize poorly soluble drugs. Polymeric micelles are core-shell structures with hydrophilic and hydrophobic regions that each serve a specific purpose. The hydrophilic shell is responsible for pharmacokinetics and interactions with proteins and cells, while the hydrophobic core is primarily important for regulating drug release. Polymeric micelles are self-assembly structures that arise when amphiphilic block copolymers aggregate in aqueous solutions above the critical micellar concentration. Generally, there are three types of polymeric micelles: micelle-forming polymer-drug conjugates, polymeric micellar nanocontainers, and polyion complex micelles, each with its own procedure for fabrication. The methods for preparing micelles fall into four main categories: dialysis, solvent evaporation, nanoprecipitation, and freeze-drying (Wu et al., 2025).

Nanocomposites

A composite material combines the properties of several different materials. It contains a matrix and perhaps a period of reinforcement. One-dimensional items include tubes and threads, two-dimensional objects are clay and other layered materials, and three-dimensional objects are spheres. Composites are often classified based on the following attributes: laminated composites, fibrous composites, and particulate composites. When polymers are used as the matrix, these composites are called "polymer-based composites." Recent laboratory and clinical research has concentrated on several biodegradable and bioresorbable materials, along with enhancements to the design of biomedical devices. Polymer-based composites have shown promise across multiple areas, including drug delivery, tissue engineering, and applications that bear loads. Polymer-based composites can combine the best qualities of their constituents, creating materials with unique properties (Jin et al., 2024).

Dendrimers

In biomedicine, nanodrugs (NDs) are addressed with dendrimers, which are the smallest and most precisely defined nanoformulations. Dendrimer size, core-shell mix, and surface functional groups can be varied to create different types of nanocarriers that aid in the delivery of medications and genes to the brain. Dendrimers are used to treat Alzheimer's, Parkinson's, and prion disorders because of their powerful anti-amyloidogenic properties. PAMAM dendrimers have been employed to develop novel anti-amyloidogenic compounds. PAMAM dendrimers in three generations (G3, G4, and G5) disaggregated existing fibrils into amorphous forms, preventing amyloid formation of the Aβ peptide (Aβ 1-28). The larger PAMAM dendrimer generations were more effective at interfering with amyloid aggregation. Cationic Carbosilane (CBS) dendrimers are effective therapeutic agents in a rotenone-induced mouse model of Parkinson's disease, suppressing α-syn fibrillation while also encapsulating and complexing rotenone, providing dual neuroprotective benefits (Mahjubin-Tehran et al., 2024).

Table 4. Properties of common nanocarriers

Nanocarrier	Properties	Applications
Polymeric Nanoparticles	Versatile tool, solid structures, encapsulate or absorb bioactive chemicals	Drug delivery, tissue engineering
Micelles	Core-shell structures, hydrophilic and hydrophobic regions	Solubility enhancement, drug delivery
Nanocomposites	Composite material with matrix and reinforcement	Drug delivery, tissue engineering
Dendrimers	Smallest, most precisely defined nanoformulations	Drug delivery, gene therapy, treatment of Alzheimer's, Parkinson's, and prion disorders

These nanocarriers offer diverse properties and applications, making them valuable tools in the field of drug delivery.

7.1. Latest Advancements in Nanocarrier Technology

As of late, a few huge advances have been made in nanocarrier innovation to foster medications that are focused on, successful, and safe. Procedures are being created to further develop the essential focusing of explicit cells and tissues to limit any accidental results of helpful specialists. For example, nanocarriers have been developed that can specifically target endothelial cells of the vascular system; these cells are critical to maintaining vascular health and regulating inflammatory responses (Dan et al., 2024). The surface of these carriers has also been changed to boost biocompatibility and lower immune detection. Polymer coating and mimetic camouflage have proven to be good ways for reducing immune recognition, while at the same time increasing medicine administration (Patel et al., 2024). However, improvements in polymeric micelles, liposomes, and magnetic nanoparticles have greatly enhanced the performance, precision, and efficacy of drug delivery systems compared to traditional methods. Although these improvements represent a big leap forward in nanocarrier tech, they have the potential to improve patient outcomes and therapeutic efficacy, which is something to be excited about (Patel et al., 2024).

7.2. Case Studies of Nanocarrier Applications

- **Cancer Therapy:** The use of nanocarriers in malignant growth treatment has shown huge commitment lately. Liposomes and micelles, specifically, have been utilized to convey chemotherapeutic medications directly to cancer cells, consequently diminishing systemic toxicity. For example, a review demonstrated the use of stealth nanocarriers to deliver drugs to cancer cells while avoiding off-target effects (Ojha et al., 2022). This approach has been shown to further develop therapy results and diminish incidental effects, making it a reasonable procedure for disease treatment.
- **Cardiovascular Illnesses:** Nanocarriers have also been examined for their actual limit in treating cardiovascular afflictions, similar to atherosclerosis and disturbance. These nanocarriers can be planned to target unequivocal cells or tissues inside the cardiovascular system, including endothelial cells, which play an essential part in the development of atherosclerosis. By conveying restorative specialists, like mitigating medications or cell reinforcements, directly to the objective site, nanocarriers can further improve treatment results and decrease the risk of side effects. A review has exhibited the utilization of nanocarriers to target endothelial cells and treat cardiovascular diseases, featuring the capability of this methodology (Patel et al., 2024).
- **Breast Cancer Detection and Treatment:**

The recognition and therapy of bosom malignant growth have been worked on using lipid and attractive nanoparticles. These nanocarriers can be intended to target explicit malignant growth cells, like breast cancer cells, and deliver restorative agents, for example, chemotherapy medications or antibodies, directly to the cancer site. Furthermore, attractive nanoparticles can be utilized for imaging, taking into account early discovery of bosom malignant growth. A review has shown the utilization of lipid and attractive nanoparticles in the designated treatment of breast cancer, featuring the capability of this approach to achieve further developed treatment results and diminish secondary effects (Wu et al., 2025).

7.3. Potential Risks and Challenges of Using Nanocarriers

Nanocarriers provide numerous benefits, but they also provide significant risks and challenges:

- **Biocompatibility and Safety:** It's important to make sure that nanocarriers don't trigger harmful immune reactions or toxicity. Biocompatibility means that a material can create the right response in a living organism when it's implanted. Safety deals with the negative effects that these nanocarriers might have on our biological tissues and organs (Sell et al., 2023).
- **Environmental Impact:** We need to explore the long-term effects of nanocarriers on ecosystems and biodiversity. Although nanocarriers can be beneficial, they might also bring unexpected environmental issues (Al-Zoubi et al., 2022).
- **Regulatory Obstacles:** There are many complicated rules we must navigate to gain approval for clinical use. The research and use of nanocarriers raise big regulatory concerns, including safety, effectiveness, and compliance with the rules. However, these challenges are crucial because they ensure that new technologies are safe for everyone (Al-Zoubi et al., 2022).

Each type of nanocarrier has well-defined advantages and can be selected based on the drug requirements and the target site in the body.

8. CHARACTERIZATION TECHNIQUES

In Vitro, Ex Vivo, In Situ, and In Vivo Methods

Understanding the way of behaving and collaborations of nanomaterials, especially those conveying optional metabolites, requires a thorough methodology using a range of portrayal procedures. An overview of the advantages and the characterization techniques is provided in Tables 5 and 6. These methods can be extensively classified into four primary sorts:

In Vitro Strategies: These methods include concentrating on organic cycles outside their normal setting, regularly in a controlled lab environment, utilizing cells or natural particles. Models incorporate cell culture tests to survey cytotoxicity and cell uptake, biochemical measures to assess the delivery energy of exemplified compounds, and atomic science procedures to examine interactions with cell parts. In vitro techniques offer high command over exploratory circumstances, empowering reproducibility and cost-effectiveness. Notwithstanding, they miss the mark on the physiological intricacy of living creatures, possibly restricting their prescient insight for in vivo results (Hamzah et al., 2024).

Ex Vivo Strategies: These methods include concentrating on tissues or organs outside the body while keeping up with some of their physiological properties. Models incorporate organ culture, tissue cuts, and perfusion frameworks. Ex vivo strategies overcome any barrier between in vitro and in vivo investigations by considering the examination of cell cooperations inside an all more physiologically important setting. Be that as it may, impediments include the restricted life expectancy of tissues and the trouble in keeping up with local tissue design and capability (Mascolini et al., 2024).

In Situ Strategies: These methods permit the investigation of organic cycles inside their normal setting, regularly inside the organic entity's own tissue or organ. Models remembered for situ hybridization, immunohistochemistry, and in situ protein examines. In situ strategies give important data about the spatial and transient dissemination of nanomaterials and their connections within the life form. In any case, they can be in fact testing, may include obtrusive methodology, and might upset the regular condition of the tissue (Mukalel et al., 2024).

In Vivo Strategies: These methods include concentrating on natural cycles inside living organisms, giving the most physiologically applicable information. Models incorporate creature models, clinical studies, and entire-body imaging methods. In vivo examinations give vital data on the pharmacokinetics, biodistribution, and helpful adequacy of nanomaterial-based drug delivery systems. In any case, they include moral contemplations, can be costly, and may display changeability between people (Ibrahiem et al., 2024).

A mix of these procedures is often fundamental for an exhaustive comprehension of the way of behaving and the organic impacts of nanomaterials. In vitro investigations can give starting bits of knowledge into essential collaborations, while ex vivo and in situ examinations can give more physiologically important data. At last, in vivo examinations are critical for assessing the well-being and viability of nanomaterial-based drug delivery systems in a living organism.

Table 5. Advantages and limitations of each technique

Method	Advantages	Limitations
In Vitro	High control, reproducible, cost-effective, less ethical concerns	Lack of physiological context, limited complexity
Ex Vivo	Maintains tissue architecture, and bridges gap between in vitro and in vivo	Limited tissue lifespan, difficulty in maintaining conditions
In Situ	Preserves tissue context, spatial and temporal resolution	Technically challenging, limited accessibility
In Vivo	Physiologically relevant, comprehensive data	Ethical concerns, high cost, longer duration, potential variability

Table 6. Characterization methods for nanocarriers

Method	Description	Application
Dynamic Light Scattering (DLS)	Measures size distribution and zeta potential of nanoparticles	Assessing nanoparticle size, stability, and dispersion properties
Transmission Electron Microscopy (TEM)	Provides high-resolution images of nanoparticle morphology	Examining the detailed structure and shape of nanoparticles
Scanning Electron Microscopy (SEM)	Produces images with surface topography information	Analyzing surface morphology and particle size
Atomic Force Microscopy (AFM)	Uses a cantilever to measure surface properties	Assessing surface characteristics and mechanical properties
Fourier-Transform Infrared Spectroscopy (FTIR)	Identifies functional groups and chemical bonds	Characterizing chemical composition and interactions
X-ray Diffraction (XRD)	Determines crystalline structure	Analyzing crystallinity and phase identification

continued on following page

Table 6. Continued

Method	Description	Application
Nuclear Magnetic Resonance (NMR)	Provides information on molecular structure and dynamics	Investigating chemical structure and nanoparticle interactions
Thermogravimetric Analysis (TGA)	Measures weight change in response to temperature	Assessing thermal stability and composition
Differential Scanning Calorimetry (DSC)	Measures heat flow associated with phase transitions	Studying thermal behavior and stability

These characterization methods are essential for understanding the properties and behavior of nanocarriers, which is crucial for their development and application in drug delivery and other fields.

9. REGULATORY AND MANUFACTURING CHALLENGES IN NANOMEDICINE

9.1. Security and Administrative Contemplations in Nanomedicine

Guaranteeing that nanomaterials and nanomedicines are ok for human use and the climate is a fundamental part of nanomedicine. This includes evaluating their likely toxicological effects, biocompatibility, and environmental effects (Souto et al., 2024; Caputo et al., 2024). Poisonousness assessment is basic, as it includes deciding the potential harm that nanomaterials and their degradation products might pose to people and the climate. This is accomplished through different strategies, including in vitro and in vivo examinations, as well as computational modeling, to distinguish toxic effects like irritation, oxidative stress, or DNA damage. For example, a study in 2024 found that nanoparticles utilized in disease therapy caused negligible toxicity in healthy cells but were profoundly harmful to malignant cancer cells (Souto et al., 2024). Biocompatibility is another critical variable, alluding to the capacity of a material to cooperate with living tissues without causing unfriendly responses. Techniques like in vitro and in vivo examinations, and histological and immunological examinations assist with recognizing issues like aggravation or resistant reactions. A study in 2024 showed that nanoparticles utilized in drug delivery were biocompatible and didn't cause huge tissue harm or adverse reactions (Caputo et al., 2024). Also, surveying the natural effect is indispensable to guarantee ecological well-being, including the assessment of expected impacts on biological systems through lab and field studies, and computational modeling to identify risks with like toxicity to amphibian organisms, soil contamination, or disturbance of established pecking orders. Administrative bodies, for example, the FDA and EMA, give rules to nanomedicine; nonetheless, particular guidelines custom-made to the remarkable properties of nanotechnology are as yet to address difficulties presented by nanomaterials' small size, high surface area, and interaction with biological systems in novel ways.

9.2. Hurdles in Clinical Translation of Nanocarriers

The clinical interpretation of nanocarriers faces several challenges, including:

Organic Challenges

Guaranteeing the protected and powerful conveyance of prescriptions to the objective site is a critical test in the clinical translation of nanocarriers (Coey et al., 2024). Nanocarriers should be intended to explore the complex natural climate and arrive at the objective site without causing unfriendly impacts. This requires a profound comprehension of the communications among nanocarriers and organic frameworks, as well as the improvement of techniques to relieve the expected harmfulness.

Manufacturing Challenges

Increasing the creation of nanocarriers while keeping up with consistency and quality is a huge assembling challenge (Caputo et al., 2024). Nanocarriers are complicated frameworks that require exact command over their structure, size, and surface properties. Guaranteeing that these properties are predictable across huge groups of nanocarriers is fundamental for their protected and powerful use in clinical settings.

Regulatory Obstacles

Exploring the complex administrative scene and getting endorsement for clinical use is a critical test in the improvement of nanocarriers (Wang et al., 2024). Administrative organizations require thorough testing and assessment of nanocarriers to guarantee their security and adequacy. Nonetheless, the interesting properties of nanocarriers can make it hard to apply customary administrative systems, featuring the requirement for new rules and principles.

Cost-Effectiveness Challenges

Guaranteeing that nanomedicines are superior compared with existing treatments is a huge test in their clinical understanding (Havelikar et al., 2024; Aggarwal et al., 2024). Nanocarriers are in many cases, more costly to create than customary drugs, and their utilization might require critical investments in infrastructure and preparation. Nonetheless, the expected advantages of nanocarriers, including further developed efficacy and diminished toxicity, may legitimize these expenses. Further exploration is expected to completely assess the expense adequacy of nanocarriers and to recognize techniques for decreasing their expenses.

Table 7. Regulatory and manufacturing challenges in nanomedicine

Challenge	Description
Safety	Evaluating potential toxicological effects and biocompatibility.
Regulatory	Navigating complex regulatory guidelines and obtaining approvals.
Manufacturing	Scaling up production while maintaining quality and consistency.
Cost-Effectiveness	Ensuring nanomedicines are cost-effective compared to existing therapies

10. POTENTIAL OF NANOMEDICINE IN NEURODEGENERATIVE DISORDERS

Nanomedicine in the Treatment of Neurodegenerative Diseases: An Assigned Methodology

Neurodegenerative ailments like Alzheimer's infection and Parkinson's illness are among the most inciting issues to treat as a result of the multifaceted design of the frontal cortex and the presence of the blood-brain barrier (BBB). The improvement of nanomedicine has become an ever-evolving technique that beats these troubles by passing medicinal experts directly onto the influenced mind regions (Ferrão & Rai, 2024; Yang et al., 2024; Al-Muhanna et al., 2024).

Challenges in Treating Neurodegenerative Illnesses

Quite possibly one of the greatest obstacles in treating cerebrum problems is the blood-brain barrier (BBB). This normal safeguard instrument shields the mind from hurtful substances, yet in addition keeps most medications from reaching the cerebrum. Traditional medicines for Alzheimer's and Parkinson's infections battle with low medication delivery efficiency, prompting restricted beneficial effects and increased side effects (Krsek & Baticic, 2024; Xu et al., 2024).

Role of Nanomedicine in Overcoming These Difficulties

Nanomedicine includes the utilization of nanoparticles, which are tiny particles intended to convey medications and cross the BBB successfully. Because of their small size and surface adjustments, nanoparticles can enter the barrier and convey drugs precisely to the impacted synapses. This designated drug conveyance fundamentally further develops treatment adequacy and decreases side effects (Palaniyappan et al., 2024).

10.1. Applications in Alzheimer's and Parkinson's Diseases

Alzheimer's Disease: In Alzheimer's disease, the gathering of amyloid-beta plaques is a fundamental process contributing to mental deterioration. Nanoparticles can be designed to convey drugs that forestall or separate these plaques, easing back or in any event, turning around sickness movement. Also, they can be utilized to move calming specialists to decrease neuroinflammation, which is one more key factor in Alzheimer's pathology (Yang et al., 2024; Al-Muhanna et al., 2024).

Parkinson's Disease: Parkinson's disease is characterized by the deficiency of dopamine-creating neurons, prompting development problems like tremors and stiffness. Conventional dopamine substitution treatments frequently have restricted viability because of unfortunate cerebrum entrance. Nanoparticles can be utilized to convey dopamine or dopamine-delivering forerunners directly to the affected neurons, giving a more maintained and effective therapeutic impact (Xu et al., 2024).

10.2. Targeted Delivery of Medications to the CNS

Treating Central Nervous System disorders is difficult due to the blood-brain barrier, which blocks most therapeutic agents from reaching the brain. Nanocarrier-based designated drug conveyance has arisen as a promising solution for overcoming this challenge. This technique guarantees exact medication conveyance to neurons, improving treatment adequacy while limiting fundamental aftereffects (Madadi et al., 2024; McLoughlin et al., 2024; Roghani et al., 2024). A visual representation of this targeted delivery is illustrated in Figure 5.

Designated Medication Conveyance to the CNS

Step 1: Plan and Synthesis of Nanocarriers

The first step involves designing and synthesizing nanocarriers, such as:

- Liposomes (lipid-based transporters)
- Polymeric nanoparticles (biodegradable polymer-based frameworks)
- Dendrimers (spread polymeric designs)
- Inorganic nanoparticles (such as gold or silica nanoparticles)

These nanocarriers are designed to have explicit size, charge, and surface science that work with their capacity to cross the BBB. Surface changes, like the connection of ligands, further improve their focus on ability (Jean-Baptiste et al., 2024).

Step 2: Loading of Therapeutic Agents

When orchestrated, the nanocarriers are stacked with restorative specialists, including:

- Drugs (e.g., neuroprotective mixtures)
- Genes (for gene therapy applications)
- Proteins or Peptides, (for example, neurotrophic factors)

Embodiment within the nanocarrier safeguards the remedial specialists from corruption and guarantees controlled discharge at the objective site (Mendake et al., 2024).

Step 3: Targeting and Crossing the BBB

The nanocarriers must successfully traverse the BBB to reach the CNS. This can be achieved through:

- Receptor-mediated transport: Utilizing ligands or antibodies that bind receptors on endothelial cells of the BBB (e.g., transferrin or insulin receptors).
- Adsorptive-mediated transport: Using positively charged nanocarriers that interact with negatively charged components of the BBB.
- Nanoparticle-mediated brief BBB disruption: Certain nanoparticles can temporarily and reversibly disturb the BBB, allowing drug passage (Youssef et al., 2024).

Step 4: Delivery to Neurons

When inside the CNS, nanocarriers navigate to the specific site of action, such as neurons affected by neurodegenerative illnesses (Alzheimer's, Parkinson's). The therapeutic agents are delivered through mechanisms such as:

- pH-sensitive release (responding to changes in pH within the CNS).
- Enzyme-mediated degradation (triggered by CNS-specific enzymes).
- External stimuli-responsive release (such as heat, light, or magnetic fields).

Step 5: Therapeutic Effects and Clearance

After successful drug delivery, the therapeutic agents interact with neurons, producing beneficial effects like:

- Reducing inflammation and oxidative stress
- Promoting neuronal regeneration
- Restoring neurotransmitter balance

The nanocarriers and therapeutic agents are then metabolized and cleared from the body through normal excretion pathways, reducing toxicity risks.

Key Advantages of Designated Medication Conveyance to the CNS

1. **Enhanced Drug Efficacy** – Ensures precise delivery to the affected brain regions, increasing treatment effectiveness.
2. **Reduced Systemic Toxicity** – Minimizes drug exposure to non-target organs, reducing side effects.
3. **Improved Patient Compliance** – Controlled and prolonged drug release reduces dosing frequency, enhancing patient comfort and adherence (Madadi et al., 2024; McLoughlin et al., 2024; Roghani et al., 2024).

Designated drug conveyance to the CNS is a significant advancement in nanomedicine. Nanocarriers, like liposomes and polymeric nanoparticles, can be designed to cross the BBB and deliver drugs directly to neurons. This technique ensures that the therapeutic agents reach the intended site of action, enhancing treatment efficacy and minimizing systemic toxicity.

Figure 5. Targeted nanocarrier delivery to neuronal cells

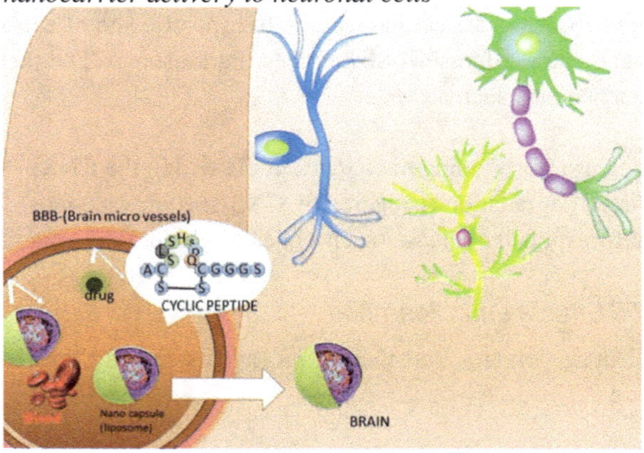

Table 8. Applications of Nanocarriers in Neurodegenerative Disorders (NDDs)

Application	Description
Alzheimer's Disease	Delivery of drugs to reduce amyloid plaques.
Parkinson's Disease	Delivery of dopamine or other therapeutic agents.
Neuroprotection	Delivery of antioxidants and anti-inflammatory agents.
Neural Stem Cell Therapy	Delivery of factors that promote neural stem cell differentiation

11. CONCLUSION

Emerging as a transforming approach to treating neurodegenerative disorders (NDDs), nanomedicine offers fresh therapeutic possibilities where traditional therapies have long faced limits. One of the main barriers in this industry, the blood–brain barrier (BBB) limits the access of most therapeutic medications into the brain. By means of specific carriers including liposomes, emulsions, and solid lipid nanoparticles, nanotechnology helps to enable targeted drug delivery to impacted brain areas. By lowering negative effects, these nanocarriers boost general therapeutic efficacy via enhancing drug solubility, stability, and accuracy. Combining technical advancement with nature's active substances, nanotechnology's incorporation with phytomedicine especially secondary metabolites with antioxidant and neuroprotective qualities, further enhances its therapeutic potential. Still, difficulties remain even with these improvements. Large-scale manufacturing obstacles, safety concerns, and regulatory obstacles all slow clinical translation. We have to provide them with assurance of their biocompatibility, repeatability, and long-term safety - all essential criteria for practical medical use of nanocarriers. Addressing this issue requires not only rigorous scientific validation, but also intentional collaboration among academic, industry, and regulatory practitioners to establish prescriptive standards and pathways to clearance. Going forward, nanomedicine appears to be extremely promising. Personalized medicine strategies matched to certain patient profiles and smart nanoparticles able to react to disease-specific signals will likely change treatment strategies for NDDs. Turning lab inventions into clinical practice calls for cooperative research, ongoing investment, and interdisciplinary collaboration. Nanomedicine has the potential to totally revolutionize the treatment of neurodegenerative diseases via constant learning and enhancement. Offering more precise, useful, and accessible therapeutic options may greatly improve patient outcomes and quality of life, therefore signifying a big shift in modern medicine's approach to some of the most challenging neurological disorders of our time.

REFERENCES

Aggarwal, D., (2024). Liposome-based nanomedicine for advanced combinatorial cancer therapy. *Biochemical and Cellular Archives, 24*(1).

Al-Muhanna, M. K., (2024). Surface-engineered quantum dot nanocomposites for neurodegenerative disorder remediation and avenue for neuroimaging. *Nanotechnology Reviews, 13*(1), 20240128.

Al-Zoubi, M. S., (2022). Nanomedicine tactics in cancer treatment: Challenge and hope. *Critical Reviews in Oncology/Hematology, 174*, 103677. PMID: 35385774

Atapour-Mashhad, H., (2024). Preparation and characterization of novel nanostructured lipid carriers (NLC) and solid lipid nanoparticles (SLN) containing coenzyme Q10 as potent antioxidants and antityrosinase agents. *Heliyon, 10*(11). PMID: 38882272

Caputo, F., (2024). Toward an international standardisation roadmap for nanomedicine. *Drug Delivery and Translational Research.* PMID: 38865038

Chakraborty, B., (2024). Evolution of lipid nanoparticles as charioteers of Alzheimer's disease therapeutics. *Applied Materials Today, 41*, 102442.

Chakraborty, K., (2022). Emerging concepts in designing next-generation multifunctional nanomedicine for cancer treatment. *Bioscience Reports, 42*(7), BSR20212051. PMID: 35638450

Chen, T., (2024). Cellular and molecular mechanisms of the blood–brain barrier dysfunction in neurodegenerative diseases. *Fluids and Barriers of the CNS, 21*(1), 60. PMID: 39030617

Coey, T. (2024). Nanomedicine in oncology: A critical review on epidemiology, health impacts, challenges, and future outlook. *Progress in Drug Discovery & Biomedical Science, 7*(1).

Dan, S., (2024). Synergistic approach of nanocomposite film in cancer: A new paradigm for diagnosis and therapy in cancer. In *Anticancer Therapeutics* (p. 317). Bionanotechnology, Nanomedicine and Phytochemicals.

Daraban, B. S., (2024). Latest perspectives on Alzheimer's disease treatment: The role of blood–brain barrier and antioxidant-based drug delivery systems. *Molecules (Basel, Switzerland), 29*(17), 4056. PMID: 39274904

Ferrão, R., (2024). Advanced polymeric nanoparticles for the treatment of neurodegenerative diseases. In *Chemical Physics of Polymer Nanocomposites* (pp. 843–885).

Grigore, M. E. (2017). Organic and inorganic nano-systems used in cancer treatment. *Journal of Medical Research and Health Education, 1*(3).

Hamzah, M. L., (2024). Formulation and development of frovatriptan succinate in situ gel for nasal drug delivery: In vitro and ex vivo evaluation. *Pakistan Journal of Pharmaceutical Sciences, 37*(3). PMID: 39340843

Havelikar, U., (2024). Comprehensive insights into mechanism of nanotoxicity, assessment methods and regulatory challenges of nanomedicines. *Discover Nano, 19*(1), 165. PMID: 39365367

Huang, X., (2024). Advances and applications of nanoparticles in cancer therapy. *MedComm. Oncology, 3*(1), e67.

Ibrahiem, B., (2024). Magnetic targeting of lornoxicam/SPION bilosomes loaded in a thermosensitive in situ hydrogel system for the management of osteoarthritis. *Drug Delivery and Translational Research, 14*(7), 1982–2002. PMID: 38158473

Jadoon, L., (2024). Nano-elicitation and hydroponics: A synergism to enhance plant productivity and secondary metabolism. *Planta, 259*(4), 80. PMID: 38436711

Jean-Baptiste, Z., Pathak, Y., & Sneed, K. B. (2024, February). Nanomedicine-Based Drug Delivery Systems and the Treatment of Autism Spectrum Disorders: A Review. In *Micro* (Vol. 4, No. 1, pp. 132-141). MDPI.

Jin, X., (2024). Nanoengineered polymers and other organic materials in lung cancer treatment: Bridging the gap between research and clinical applications. *European Polymer Journal*, 112891.

Kainat, S., (2024). Physicochemical attributes, structural characterization, and catalytic properties of nanomaterials. In *Nanomaterials in Biomass Conversion* (pp. 143–167). Woodhead Publishing.

Krsek, A., (2024). Nanotechnology-driven therapeutic innovations in neurodegenerative disorders: A focus on Alzheimer's and Parkinson's disease. *Future Pharmacology, 4*(2), 352–379.

Lau, K., (2024). Blood–brain barrier alterations and their impact on Parkinson's disease pathogenesis and therapy. *Translational Neurodegeneration, 13*(1), 37. PMID: 39075566

Liu, D., (2024). Regulation of blood–brain barrier integrity by Dmp1-expressing astrocytes through mitochondrial transfer. *Science Advances, 10*(26), eadk2913. PMID: 38941455

Madadi, A. K., & Sohn, M. J. (2024). Advances in intrathecal nanoparticle delivery: Targeting the blood–cerebrospinal fluid barrier for enhanced CNS drug delivery. *Pharmaceuticals, 17*(8), 1070. PMID: 39204177

Mahjubin-Tehran, M., (2024). Dendrimer-encapsulated curcumin: Illuminating pathways to cancer treatment excellence. *European Polymer Journal*, 113358.

Mascolini, M. V., (2024). Ex vivo, in vivo and in silico studies of corneal biomechanics: A systematic review. *Physical and Engineering Sciences in Medicine, 10*, 1–39. PMID: 38598066

McLoughlin, C. D., (2024). Overcoming the blood–brain barrier: Multifunctional nanomaterial-based strategies for targeted drug delivery in neurological disorders. *Small Science, 4*(12), 2400232. PMID: 40213484

Mohapatra, P., (2024). How precise are nanomedicines in overcoming the blood–brain barrier? A comprehensive review of the literature. *International Journal of Nanomedicine*, ●●●, 2441–2467. PMID: 38482521

Mukalel, A. J., (2024). Oxidized mRNA lipid nanoparticles for in situ chimeric antigen receptor monocyte engineering. *Advanced Functional Materials*, 2312038. PMID: 39628840

Narsinh, K. H., (2024). Strategies to improve drug delivery across the blood–brain barrier for glioblastoma. *Current Neurology and Neuroscience Reports, 24*(5), 123–139. PMID: 38578405

Nguyen, T. T., (2021). Nanotechnology-based drug delivery for central nervous system disorders. *Biomedicine and Pharmacotherapy, 143*, 112117. PMID: 34479020

Ojha, A.. (2022). Nanoparticles and nanomaterials-based recent approaches in upgraded targeting and management of cancer: A review. *Cancers (Basel), 15*(1), 162. PMID: 36612158

Palaniyappan, R., (2024). Nanotechnology-driven approaches in overcoming drug delivery challenges for neurodegenerative diseases. *Indian Journal of Pharmaceutical Education and Research, 58*(4s), S1185–S1200.

Palaniyappan, R., (2024). Nanotechnology-driven approaches in overcoming drug delivery challenges for neurodegenerative diseases. *Indian Journal of Pharmaceutical Education and Research, 58*(4s), S1185–S1200.

Patel, P., (2024). Lipid-based nanoparticles in delivering bioactive compounds for improving therapeutic efficacy. *Pharmaceuticals, 17*(3), 329. PMID: 38543115

Patel, S., (2024). Latest developments in biomaterial interfaces and drug delivery: Challenges, innovations, and future outlook. *Zeitschrift für Naturforschung C, 21*(0). PMID: 39566511

Roghani, A. K., (2024). Treating Alzheimer's disease using nanoparticle-mediated drug delivery strategies/systems. *Ageing Research Reviews*, 102291. PMID: 38614367

Schiera, G., (2024). Involvement of astrocytes in the formation, maintenance, and function of the blood–brain barrier. *Cells, 13*(2), 150. PMID: 38247841

Sell, M., (2023). Application of nanoparticles in cancer treatment: A concise review. *Nanomaterials (Basel, Switzerland), 13*(21), 2887. PMID: 37947732

Silindir Gunay, M., (2016). Drug delivery systems for imaging and therapy of Parkinson's disease. *Current Neuropharmacology, 14*(4), 376–391. PMID: 26714584

Singh, A., (2024). Nanoparticle formulations: A sustainable approach to biodegradable and non-biodegradable products. In *Nanocarrier Vaccines* (pp. 95–151). Biopharmaceutics-Based Fast Track Development.

Souto, E. B., (2024). Regulatory insights into nanomedicine and gene vaccine innovation: Safety assessment, challenges, and regulatory perspectives. *Acta Biomaterialia*. PMID: 38604468

Sutradhar, K. B., (2014). Nanotechnology in cancer drug delivery and selective targeting. *International Scholarly Research Notices, 2014*(1), 939378.

Ullah, Q., (2024). Exploring the influence of nanoparticles and PGPRs on the physico-chemical characteristics of wheat plants: A review. *EPH-International Journal of Agriculture and Environmental Research, 10*(1), 1–9.

Wang, J., (2024). A journey of challenges and victories: A bibliometric worldview of nanomedicine since the 21st century. *Advanced Materials, 36*(15), 2308915. PMID: 38229552

Wu, Y., (2025). Micelles as nanocarriers of fluorescent and magnetic dendrimers for potential bimodal imaging applications. *Dyes and Pigments*, *232*, 112482.

Xu, Q., et al. (2024). Potential applications of nanomedicine in the treatment of Parkinson's disease. *Human Brain, 3*(1).

Yang, H., (2024). Recent progress in nanomedicine for the diagnosis and treatment of Alzheimer's diseases. *ACS Nano*, *18*(50), 33792–33826. PMID: 39625718

Youssef, F. S., Fouad, O. A., Ismail, S. H., & Mohamed, G. G. (2024). Therapeutic and toxicological aspects of some metal nanoparticles on the central nervous system: A review. *Egyptian Journal of Veterinary Science*, *55*(3), 733–745.

Zeynalzadeh, E., (2024). Navigating the neurological frontier: Macromolecular marvels in overcoming blood–brain barrier challenges for advanced drug delivery. *Heliyon*, *10*(15). PMID: 39170552

Chapter 12
Nanotechnology and Secondary Metabolites: A Synergistic Approach to Combat Neurodegeneration

Kranthi Kumar Kotha
https://orcid.org/0000-0002-2935-4262
Dayananda Sagar University, Bangalore, India

Aruga Chandrakala
S.V.U. College of Pharmaceutical Sciences, S.V. University, Tirupati, India

Rajasekhar Komarla Kumarachari
https://orcid.org/0000-0001-5611-1410
Meenakshi College of Pharmacy, Meenakshi Academy of Higher Education and Research, Chennai, India

Sarad Pawar Naik Bukke
Kampala International University, Ishaka-Bushenyi, Uganda

Pugazhendhi Sarangapani
Meenakshi College of Pharmacy, Meenakshi Academy of Higher Education and Research, Chennai, India

L. Karpagavalli
Meenakshi College of Pharmacy, Meenakshi Academy of Higher Education and Research, Chennai, India

ABSTRACT

Neurodegenerative diseases, including Alzheimer's, Parkinson's, and Huntington's, are pressing global health challenges characterized by complex pathologies and limited treatment options. This chapter explores the synergistic potential of nanotechnology and secondary metabolites to address these challenges. Secondary metabolites, such as polyphenols, alkaloids, and terpenoids, exhibit antioxidant, anti-inflammatory, and neuroregenerative properties, making them promising candidates for therapeutic intervention. However, their clinical application is hindered by low bioavailability and poor blood-brain barrier (BBB) penetration. Nanotechnology offers innovative solutions by enabling precise drug delivery, enhancing stability, and facilitating BBB crossing through advanced nanocarriers like liposomes, nanoparticles, and dendrimers. Case studies on nano-enabled formulations, including curcumin, piperine, resveratrol, etc, demonstrate significant efficacy in preclinical neurodegenerative models.

DOI: 10.4018/979-8-3373-1365-8.ch012

1. INTRODUCTION

Neurodegenerative diseases represent a complex and multifaceted challenge in modern medicine. Despite advances in our understanding of their underlying mechanisms, effective therapeutic solutions remain elusive. This chapter aims to explore how the synergy between nanotechnology and secondary metabolites can provide novel strategies to combat neurodegeneration.

1.1. Background on Neurodegenerative Diseases

Neurodegenerative diseases, such as Alzheimer's disease (AD), Parkinson's disease (PD), and Huntington's disease (HD), represent an escalating global health concern. The aging population is a major contributing factor to their increasing prevalence, with these disorders projected to affect millions more in the coming decades (Hou, Y., et al. 2019). AD alone is estimated to impact over 50 million individuals worldwide, a figure expected to triple by 2050. These diseases impose significant socioeconomic burdens, including the cost of care, loss of productivity, and emotional distress for patients and caregivers (WHO, 2017).

The pathophysiology of neurodegenerative diseases is multifaceted, involving protein misfolding, oxidative stress, mitochondrial dysfunction, neuroinflammation, and neuronal death (Dash et al.2024). Despite extensive research, current therapeutic strategies predominantly focus on symptomatic relief, such as improving cognitive function in AD (Peng, Y., et al. 2023) or alleviating motor symptoms in PD, rather than addressing the underlying disease mechanisms (Satapathy and Kumar, 2024). Furthermore, the presence of the BBB, a selective physiological barrier that restricts the entry of most therapeutic agents into the brain, significantly hampers drug delivery. The pressing need for innovative solutions to address these challenges has catalyzed interest in exploring novel therapeutic avenues (Teleanu, R. I., et al. 2022).

1.2. Role of Nanotechnology in Drug Delivery

Nanotechnology has emerged as a transformative tool in medicine, particularly in addressing the challenges associated with brain-targeted therapies. By employing nanoscale carriers, researchers can enhance the delivery and efficacy of therapeutic agents, including secondary metabolites, across the BBB.

Nanocarriers such as liposomes, nanoparticles, dendrimers, and micelles are engineered to encapsulate bioactive compounds, protecting them from enzymatic degradation and improving their pharmacokinetics. Key advantages of nanotechnology in drug delivery include enhanced bioavailability, targeted delivery, and controlled release. Nanoformulations improve the solubility and stability of hydrophobic compounds, such as many secondary metabolites, ensuring better absorption and effectiveness. Surface functionalization of nanocarriers with ligands enables targeted delivery to specific brain regions, thereby minimizing systemic toxicity and enhancing therapeutic outcomes. Furthermore, nanocarriers can be engineered to release drugs in a sustained manner, maintaining therapeutic concentrations over extended periods and reducing the frequency of administration.

For instance, polymeric nanoparticles loaded with curcumin have demonstrated enhanced brain uptake and neuroprotective effects in preclinical models of AD. Similarly, liposomal formulations of quercetin have shown promising results in reducing oxidative stress and inflammation in PD models (Nayab, D.E., et al. 2023).

1.3. Therapeutic Potential of Secondary Metabolites

Secondary metabolites, or phytochemicals, are bioactive compounds synthesized by plants as part of their defense mechanisms against environmental stressors. These compounds have emerged as promising candidates for the treatment of neurodegenerative diseases due to their diverse pharmacological properties, including antioxidant, anti-inflammatory, anti-apoptotic, and neuroprotective effects. Unlike conventional drugs, secondary metabolites often exert multitarget effects, which is particularly advantageous in addressing the complex pathology of neurodegeneration.

Mechanistically, secondary metabolites can modulate critical pathways involved in neurodegeneration. For example, polyphenols such as resveratrol and curcumin reduce oxidative stress by neutralizing reactive oxygen species (ROS) and upregulating endogenous antioxidant systems. Flavonoids like quercetin exhibit anti-inflammatory effects by inhibiting the activation of microglia, the brain's resident immune cells, thereby reducing neuroinflammation (Kumari, N., et al. 2023). Additionally, alkaloids such as huperzine A promote neuroregeneration by enhancing synaptic plasticity and neuronal survival through the modulation of cholinergic signaling and inhibition of acetylcholinesterase (Friedli, M. J., & Inestrosa, N. C., 2021).

Despite their therapeutic potential, the clinical translation of secondary metabolites faces challenges, including poor water solubility, low bioavailability, and rapid metabolism. Overcoming these limitations is critical to harnessing their full potential for treating neurodegenerative diseases (Yakubu, J., & Pandey, A. V., 2024).

The synergy between nanotechnology and secondary metabolites offers a unique opportunity to overcome the limitations of conventional therapies and develop innovative solutions for neurodegenerative diseases. This chapter explores the intersection of these two fields, emphasizing their combined potential to revolutionize the treatment landscape for neurodegeneration.

2. NANOTECHNOLOGY FOR ENHANCED PHYTOCHEMICAL DELIVERY

Nanotechnology offers a promising approach to overcoming the challenges associated with the delivery of phytochemicals, particularly for the treatment of neurodegenerative diseases. By improving the pharmacokinetics and pharmacodynamics of secondary metabolites, nanocarriers can significantly enhance their therapeutic potential. This synergy not only optimizes drug delivery but also provides innovative strategies to address limitations such as poor solubility, rapid metabolism, and low bioavailability of phytochemicals. In this context, the integration of nanotechnology and phytochemistry has the potential to reshape the therapeutic landscape for neurodegeneration.

2.1. Types of Nanocarriers

Nanocarriers are engineered to enhance the delivery of phytochemicals, offering innovative solutions to overcome the limitations of conventional drug delivery systems. Important types of nanocarriers include (Figure 1):

Liposomes: Liposomes are biocompatible and biodegradable vesicles composed of phospholipid bilayers, capable of encapsulating both hydrophilic and lipophilic agents. These carriers protect phytochemicals from enzymatic degradation and facilitate their controlled release. For example, liposomal formulations of curcumin have been shown to enhance its stability and bioavailability, making them highly effective for neuroprotective applications (Rudzińska, M., et al. 2024).

Nanoparticles: Nanoparticles, including polymeric and metallic types, are versatile carriers with tunable surface properties. They can be functionalized with targeting ligands to cross the BBB and deliver phytochemicals directly to affected brain regions. Polymeric nanoparticles loaded with resveratrol have demonstrated significant neuroprotective effects in preclinical models of AD by reducing oxidative stress and inflammation (Ahlawat, et al. 2020).

Dendrimers: Dendrimers are highly branched, tree-like polymers offering precise control over drug loading and release. Their unique architecture allows for the encapsulation or conjugation of multiple therapeutic agents, including secondary metabolites. Dendrimers functionalized with quercetin have shown promise in improving its solubility and targeting capability for neurodegenerative disease treatment (Goswami, et al. 2024).

Polymeric Micelles: Polymeric micelles are amphiphilic structures that self-assemble in aqueous environments, forming a hydrophobic core and a hydrophilic shell. These carriers are particularly suited for poorly soluble phytochemicals, such as curcumin and flavonoids. Polymeric micelles enhance the solubility and stability of these compounds, improving their therapeutic efficacy in neurodegeneration (Li, L., et al. 2022).

Nanoemulsions: Nanoemulsions are stable colloidal systems consisting of oil, water, and surfactants. They are designed to enhance the solubility, absorption, and bioavailability of hydrophobic phytochemicals. Nanoemulsions containing resveratrol have demonstrated improved pharmacokinetic profiles and brain-targeting efficiency, making them valuable for combating neurodegenerative diseases (Jalili, C., et al. 2023).

Figure 1. Nanocarriers for enhanced phytochemical delivery

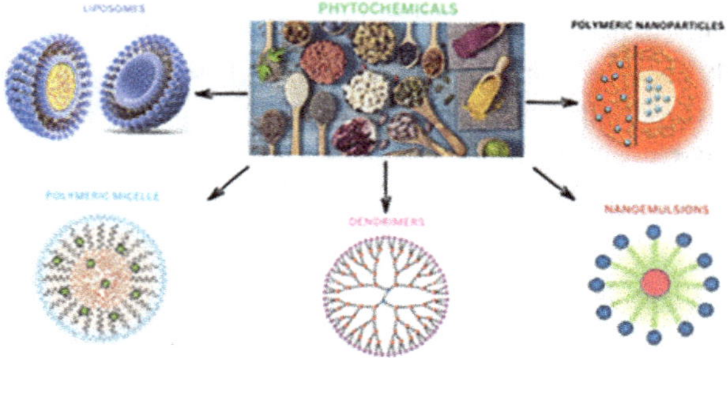

These nanocarriers provide a robust platform for optimizing the delivery of secondary metabolites, addressing challenges such as poor bioavailability and limited BBB permeability. By leveraging their unique properties, nanotechnology paves the way for innovative therapeutic strategies in the fight against neurodegeneration.

2.2. Advantages of Nanotechnology in Drug Delivery

a. Improved Stability of Phytochemicals: Many phytochemicals are prone to degradation due to environmental factors such as light, heat, and oxygen. Nanocarriers protect these compounds, preserving their therapeutic activity and extending their shelf life. This stability ensures that the bioactive agents maintain their efficacy during storage and delivery (Rezagholizade-Shirvan, A., et al. 2024).

b. Prolonged Circulation Time: Nanocarriers can be engineered to evade rapid clearance by the immune system, extending their presence in the bloodstream. This prolonged circulation allows for sustained therapeutic action, reducing the frequency of dosing and improving patient compliance (Bhattacharya, T., et al. 2022).

c. Enhanced Permeability and Retention (EPR) Effect: The nanoscale size of carriers enables preferential accumulation in diseased tissues with leaky vasculature, such as the brain in neurodegenerative conditions. The EPR effect facilitates targeted delivery of therapeutic agents, increasing their concentration at the site of action while minimizing systemic side effects (Rahman, et al. 2022).

d. Integration of Multifunctional Properties: Modern nanocarriers can be designed to integrate multiple functionalities, such as combining therapeutic and diagnostic capabilities. This "theranostic" approach enables real-time monitoring of treatment efficacy while delivering therapeutic agents. Additionally, nanocarriers can be functionalized with ligands or antibodies to achieve active targeting, further enhancing their specificity for diseased tissues (Vashist, et al. 2023).

e. Potential for Reduced Dosage: By improving the pharmacokinetics and targeting capabilities of therapeutic agents, nanotechnology allows for lower drug dosages to achieve desired therapeutic effects. This reduction minimizes the risk of adverse effects and increases the safety profile of the treatment (Mittal, K. R., et al. 2022).

Collectively, these advantages underscore the transformative potential of nanotechnology in optimizing phytochemical delivery for neurodegenerative diseases. By addressing the limitations of conventional therapies, nanotechnology holds the promise of advancing more effective and patient-friendly treatment options.

2.3. Strategies for Brain Targeting

Surface Modification: Nanocarriers can be functionalized with ligands such as peptides, antibodies, or aptamers that specifically target receptors overexpressed on the endothelial cells of the BBB. This approach enhances the specificity and efficacy of drug delivery by facilitating receptor-mediated transcytosis across the BBB, ensuring that therapeutic agents reach their intended brain targets (Asimakidou, E., et al. 2024).

Stimuli-Responsive Systems: These advanced systems are designed to release drugs in response to specific stimuli, such as pH changes, temperature variations, or the presence of certain enzymes. For instance, pH-sensitive nanocarriers can release their payload in the slightly acidic environment of the brain's diseased regions, ensuring localized drug delivery and minimizing off-target effects. Similarly, enzyme-sensitive carriers respond to overexpressed enzymes in neurodegenerative conditions, providing precise control over drug release (Cao, et al. 2019).

Intranasal Delivery: This non-invasive strategy bypasses the BBB entirely by utilizing the olfactory and trigeminal nerve pathways for direct brain access. Intranasal delivery systems exploit the nasal epithelium's permeability to deliver drugs rapidly and effectively to the brain. This approach not only enhances drug bioavailability but also reduces systemic exposure, lowering the risk of side effects. Nanocarriers optimized for intranasal delivery further improve the stability and absorption of phytochemicals, making them a promising avenue for neurodegenerative therapies (Rajput A, et al. 2022).

By employing these strategies, nanotechnology offers innovative solutions to overcome the challenges of brain-targeted drug delivery, paving the way for more effective and targeted treatments for neurodegenerative diseases.

3. SECONDARY METABOLITES IN NEURODEGENERATIVE DISEASES

3.1. Neuroprotective Mechanisms

Secondary metabolites hold immense potential for addressing the complex pathology of neurodegenerative diseases. Their diverse bioactivities provide a multi-pronged approach to combating disease progression. This section delves into the neuroprotective mechanisms of these bioactive compounds, highlighting their potential to mitigate key pathological processes in neurodegeneration.

a. Antioxidant Activity

One of the primary mechanisms by which secondary metabolites exert neuroprotection is through their robust antioxidant properties. Neurodegenerative diseases are often associated with excessive production of ROS, which leads to oxidative stress and damage to cellular structures, including lipids, proteins, and DNA. Polyphenols such as curcumin, resveratrol, and epigallocatechin gallate (EGCG) act as powerful scavengers of ROS. By neutralizing these harmful species, they prevent the cascade of oxidative damage. Additionally, these compounds upregulate endogenous antioxidant defense systems, such as superoxide dismutase (SOD) and catalase, further enhancing the brain's resilience against oxidative stress. This dual action of direct ROS scavenging and endogenous system activation makes secondary metabolites invaluable in combating the oxidative component of neurodegenerative diseases (Rudrapal, M., et al. 2022).

b. Anti-inflammatory Effects

Chronic neuroinflammation is a hallmark of many neurodegenerative diseases, driven by the activation of microglia and the release of pro-inflammatory cytokines, such as tumor necrosis factor-alpha (TNF-α) and interleukin-6 (IL-6). Secondary metabolites such as flavonoids, alkaloids, and terpenoids exhibit potent anti-inflammatory properties. For example, quercetin and luteolin inhibit microglial activation and suppress the production of pro-inflammatory mediators, thereby reducing neuronal damage. These compounds also modulate key inflammatory signaling pathways, such as nuclear factor-kappa B (NF-κB) and mitogen-activated protein kinases (MAPKs), effectively dampening the inflammatory response. By targeting neuroinflammation, secondary metabolites not only protect existing neurons but also create a favorable environment for neural repair and regeneration (Yadav, et al. 2024).

c. Neuroregeneration

The ability to stimulate neurogenesis and support neuronal repair is another critical attribute of secondary metabolites in neurodegenerative diseases. Alkaloids such as huperzine A and galantamine have demonstrated efficacy in promoting synaptic plasticity and enhancing neuronal survival. These compounds often achieve this by modulating neurotransmitter systems, such as the cholinergic and dopaminergic pathways, which are typically impaired in diseases like Alzheimer's and Parkinson's (Chen, Z.-R., et al. 2022). Additionally, secondary metabolites can activate growth factors such as brain-derived neurotrophic factor (BDNF) and nerve growth factor (NGF), which are essential for the survival, growth, and differentiation of neurons. For instance, curcumin has been shown to increase BDNF levels, thereby promoting neural plasticity and repair (El Ouaamari, Y., et al. 2023).

Through their multifaceted actions, neutralizing oxidative stress, attenuating inflammation, and fostering neuroregeneration, secondary metabolites provide a holistic approach to mitigating neurodegenerative disease progression. Their potential to act on multiple pathological pathways simultaneously underscores their significance as promising candidates for the development of innovative neurotherapeutic strategies (Figure 2).

Figure 2. Multifaceted actions of secondary metabolites in neurodegenerative disease progression

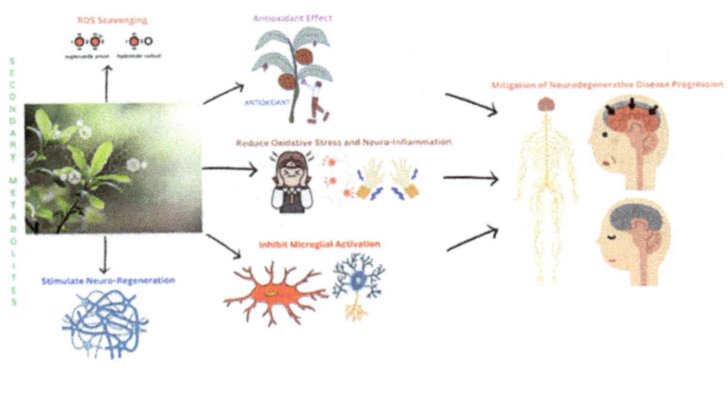

3.2. Secondary Metabolites with Neuroprotective Therapeutic Potential

Secondary metabolites are chemically diverse bioactive compounds with significant potential for combating neurodegeneration. Among them, polyphenols, alkaloids, and terpenoids have been extensively studied for their unique neuroprotective properties. This subsection explores the therapeutic relevance of these classes, emphasizing their roles in mitigating the complex pathology of neurodegenerative diseases.

3.2.1. Polyphenols

Polyphenols are a diverse group of plant-derived compounds known for their potent antioxidant, anti-inflammatory, and neuroprotective properties.

Flavonoids: Quercetin, a well-researched flavonoid, exhibits remarkable neuroprotective effects by scavenging ROS and inhibiting the activation of microglia, the brain's immune cells. Quercetin also regulates inflammatory pathways, such as nuclear factor-kappa B (NF-κB), thereby reducing neuroinflammation and promoting neuronal survival (Li Zhang, et al. 2024). Similarly, resveratrol, a stilbene polyphenol, has shown promise in neurodegenerative models by activating the sirtuin 1 (SIRT1) pathway, which enhances mitochondrial function and reduces oxidative stress. Resveratrol's ability to cross the BBB further enhances its therapeutic potential for targeting central nervous system (CNS) disorders (Azargoonjahromi, A. and Abutalebian, F., 2024).

Curcumin: Another notable polyphenol, curcumin, found in turmeric, exhibits multitarget neuroprotective effects. It neutralizes oxidative stress, inhibits amyloid-beta aggregation, a hallmark of Alzheimer's disease, and modulates neuroinflammatory pathways. Despite its low bioavailability, nanotechnology-based delivery systems have significantly enhanced curcumin's efficacy in preclinical models of neurodegeneration (da Costa, I. M., et al. 2019).

3.2.2. Alkaloids

Alkaloids are nitrogen-containing compounds with diverse pharmacological properties that make them effective candidates for neurodegenerative therapies.

Huperzine A: Derived from the Chinese club moss *Huperzia serrata*, huperzine A is a potent acetylcholinesterase inhibitor, which enhances cholinergic signaling critical for learning and memory. This makes it particularly effective in addressing cognitive decline in Alzheimer's disease. Huperzine A also exhibits antioxidant and anti-apoptotic effects, further supporting its neuroprotective role (Melgarejo da Rosa et al. 2022).

Berberine: Found in various medicinal plants, berberine demonstrates neuroprotective effects through its antioxidant and anti-inflammatory properties. Berberine enhances mitochondrial function and reduces the accumulation of misfolded proteins, such as amyloid-beta and alpha-synuclein, which are implicated in Alzheimer's and Parkinson's diseases. Additionally, its ability to activate AMP-activated protein kinase (AMPK) supports cellular energy homeostasis and neuronal survival (Cheng, Z., et al. 2022).

3.2.3. Terpenoids

Terpenoids are a large class of naturally occurring compounds with significant neuroprotective potential due to their antioxidant, anti-inflammatory, and neuroregenerative properties.

Rosmarinic Acid: This phenolic terpenoid, found in herbs like rosemary and mint, has demonstrated neuroprotective effects in preclinical studies. Rosmarinic acid reduces oxidative damage by neutralizing ROS and enhances mitochondrial function. It also modulates inflammatory mediators such as prostaglandins and cytokines, protecting neurons from chronic neuroinflammation (Cai, G., et al. 2022).

Ginkgolides: Derived from Ginkgo biloba, ginkgolides are diterpenoids known for their ability to improve cerebral blood flow and protect against neuronal damage. They act by modulating platelet-activating factor (PAF) receptors and reducing excitotoxicity, a process linked to neuronal death in conditions like ischemia and Alzheimer's disease (Noor-E-Tabassum, et al. 2022).

Each of these secondary metabolite classes offers unique mechanisms of action to address the multifactorial pathology of neurodegenerative diseases. By targeting oxidative stress, inflammation, protein misfolding, and neuronal repair, these compounds represent a promising foundation for developing novel therapeutics. Furthermore, integrating nanotechnology to optimize their delivery and bioavailability holds the potential to enhance their efficacy in clinical applications.

4. CASE STUDIES: NANOFORMULATIONS IN NEURODEGENERATIVE MODELS

4.1. Curcumin-Loaded Nanoparticles

Curcumin, a natural polyphenolic compound derived from *Curcuma longa*, has garnered significant attention for its therapeutic potential in managing AD. However, its clinical application is limited due to poor solubility, rapid degradation, and low bioavailability, which hinder its ability to cross the BBB and exert sustained pharmacological effects. Nanotechnology offers a transformative solution to these challenges by encapsulating curcumin within nanoparticles, thereby improving its stability, solubility, and targeted delivery (Fuloria, S., et al. 2022).

Several studies have shown that curcumin-loaded nanoparticles significantly enhance curcumin's bioavailability, allowing higher concentrations to reach the brain. These nanoformulations utilize various carriers, including polymeric nanoparticles, lipid-based systems, and solid lipid nanoparticles, which are designed to protect curcumin from degradation while enabling controlled and sustained release. For instance, polymeric nanoparticles fabricated using biocompatible materials such as PLGA (poly(lactic-co-glycolic acid)) have demonstrated the ability to transport curcumin across the BBB effectively (Figure 3) (Del Prado-Audelo, M. L., et al. 2019).

Figure 3. Curcumin loaded PLGA solid lipid nanoparticles

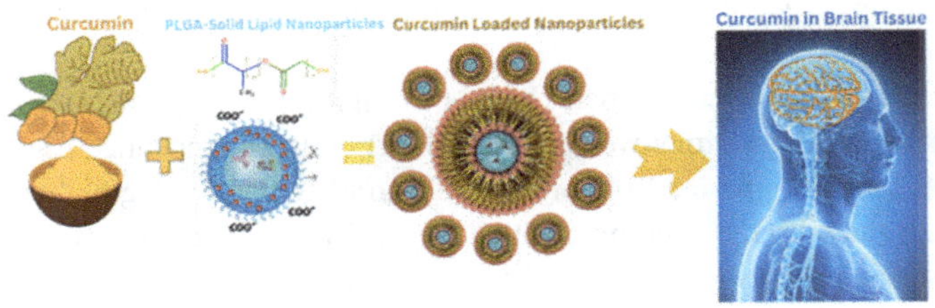

The therapeutic benefits of curcumin-loaded nanoparticles are particularly evident in their capacity to modulate critical pathological hallmarks of AD. By targeting amyloid-beta plaques, these formulations help inhibit aggregation and promote clearance, addressing one of the disease's primary drivers. Furthermore, curcumin's potent antioxidant and anti-inflammatory properties are amplified through nanoencapsulation, leading to a significant reduction in oxidative stress and neuroinflammation—two processes heavily implicated in neuronal damage and cognitive decline (Pei, J., et al. 2024).

Preclinical studies in animal models of Alzheimer's disease have provided compelling evidence of these advantages. Mice treated with curcumin-loaded nanoparticles exhibited improved cognitive performance, reduced amyloid plaque burden, and enhanced synaptic plasticity compared to those receiving free curcumin. These findings underscore the potential of nanotechnology to unlock curcumin's full therapeutic potential, transforming it into a viable neuroprotective agent for AD management (Yavarpour-Bali, H., et al. 2019).

In addition to enhancing efficacy, curcumin-loaded nanoparticles hold promise for minimizing potential side effects by enabling targeted delivery and reducing systemic exposure. This precision in drug delivery not only maximizes therapeutic benefits but also aligns with the broader goals of sustainability and safety in pharmacological interventions (Genchi, G., et al. 2024).

By addressing the limitations of traditional curcumin formulations and leveraging the unique capabilities of nanotechnology, curcumin-loaded nanoparticles exemplify the synergistic interplay between natural compounds and advanced drug delivery systems. This approach paves the way for innovative and effective strategies to combat neurodegenerative disorders like Alzheimer's disease.

4.2. Piperine-Loaded Chitosan Nanoparticles

Piperine, an alkaloid derived from *Piper nigrum* (black pepper), has shown considerable neuroprotective potential in addressing AD. Its ability to modulate amyloid-beta aggregation, attenuate neuroinflammation, and protect against oxidative stress makes it a promising candidate for therapeutic development. However, piperine's clinical utility is significantly limited by its poor water solubility, low bioavailability, and restricted ability to cross the BBB (Balakrishnan, R., et al. 2023).

Nanotechnology provides an innovative platform to overcome these barriers. Encapsulating piperine in chitosan-based nanoparticles offers a dual advantage: enhancing solubility and facilitating targeted delivery to the brain. Chitosan, a biodegradable and biocompatible polysaccharide, is widely recognized for its mucoadhesive properties and ability to open tight junctions in the BBB. These features make chitosan an ideal carrier for brain-targeted drug delivery systems (Omidian, H., et al. 2024).

Piperine-loaded chitosan nanoparticles have been engineered to optimize stability and ensure sustained release, resulting in improved pharmacokinetics and therapeutic efficacy. The encapsulation process protects piperine from premature degradation, allowing higher concentrations to reach the brain and exert their neuroprotective effects. This enhanced bioavailability translates into a more effective modulation of AD pathophysiology, particularly in reducing amyloid-beta aggregation, a hallmark of the disease. By inhibiting the formation of amyloid plaques and promoting their clearance, these formulations address a critical aspect of AD progression (Figure 4) (Manek, E., et al. 2020).

Figure 4. Advantages of piperine loaded chitosan nanoparticles

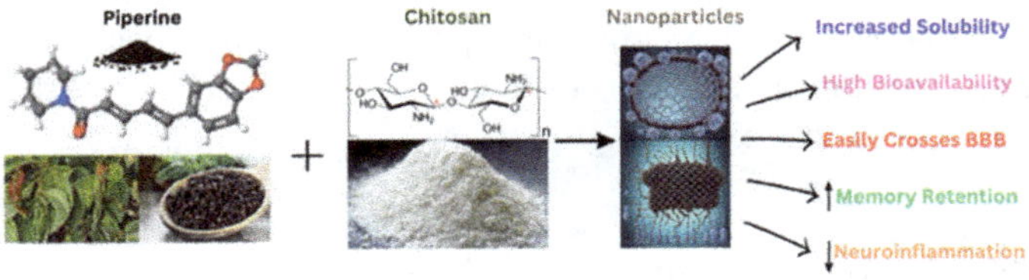

Moreover, piperine's anti-inflammatory and antioxidant properties are amplified through nanoformulation. Encapsulation not only enhances piperine's stability but also ensures its efficient uptake by neuronal cells. This targeted action mitigates neuroinflammation and oxidative damage, both of which are key contributors to neuronal degeneration and cognitive decline in AD. Preclinical studies in AD models have demonstrated that treatment with piperine-loaded chitosan nanoparticles results in improved memory retention, reduced neuroinflammatory markers, and better overall neuronal health compared to unencapsulated piperine (Coco, J. C., et al. 2023; Elnaggar, Y. S., et al. 2015).

The advantages of this nanotechnological approach extend beyond efficacy. Chitosan nanoparticles are inherently non-toxic and biodegradable, aligning with the principles of safe and sustainable pharmacological interventions. Furthermore, their ability to enable controlled drug release minimizes systemic side effects, making this delivery system not only effective but also patient-friendly (Fatma Kurul, et al. 2025).

Piperine-loaded chitosan nanoparticles exemplify the synergy between natural metabolites and nanotechnology, showcasing how traditional compounds can be transformed into advanced therapeutic agents. By addressing the limitations of piperine and enhancing its neuroprotective potential, this approach offers a promising pathway for developing innovative treatments for Alzheimer's disease (Chopra, H., et al. 2022).

4.3. Berberine-Loaded Carbon Nanotubes

Berberine, a bioactive alkaloid derived from plants such as *Berberis vulgaris*, has demonstrated significant neuroprotective potential due to its anti-inflammatory, antioxidant, and anti-amyloidogenic properties. Despite these promising attributes, berberine's clinical application in Alzheimer's disease (AD) is constrained by its limited bioavailability and inability to efficiently cross the BBB. Encapsulation in carbon nanotubes (CNTs) offers a cutting-edge solution to these challenges, enhancing berberine's therapeutic potential in managing AD (Elsori, D., et al. 2023).

Carbon nanotubes, renowned for their unique physicochemical properties, serve as exceptional carriers for drug delivery. Their high surface area, tunable functionalization, and ability to penetrate biological barriers make them ideal for transporting berberine across the BBB. By encapsulating berberine within functionalized CNTs, researchers have achieved significant improvements in its stability, bioavailability, and brain-targeting capabilities (Farhan, et al. 2025; Lohan, et al. 2017).

Berberine-loaded carbon nanotubes enable effective modulation of two critical aspects of AD pathology: amyloid-beta aggregation and oxidative stress. Amyloid-beta plaques, a hallmark of Alzheimer's disease, contribute to neurotoxicity and cognitive decline. Studies have shown that CNT-based delivery enhances berberine's ability to inhibit amyloid-beta aggregation and promote plaque clearance, thereby mitigating neuronal damage. Furthermore, the antioxidant properties of berberine are amplified through encapsulation, reducing oxidative stress and preventing further deterioration of neuronal cells (Figure 5) (Bai et al. 2025; Akshatha P. Kamath et al. 2024; Shao, X., et al. 2022).

Figure 5. Berberine loaded carbon nanotubes for Alzheimers disease

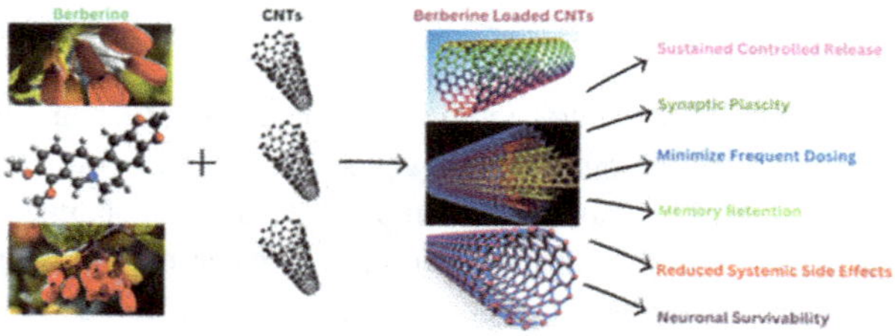

Another critical advantage of CNT-based delivery systems is their capacity for sustained and controlled drug release. This feature ensures a consistent therapeutic effect over extended periods, minimizing the need for frequent dosing and reducing potential systemic side effects. Preclinical studies in AD models have reported improvements in memory retention, synaptic plasticity, and neuronal survival following treatment with berberine-loaded CNTs (Preethy, H, et al. 2024).

While carbon nanotubes offer promising advancements, their biocompatibility and safety remain subjects of ongoing investigation. Functionalization strategies, such as coating CNTs with biocompatible polymers, are being employed to address potential toxicity and enhance their biodegradability. These advancements pave the way for safer and more effective applications of CNTs in neurodegenerative disease treatment (Al-zharani, et al. 2024).

By leveraging the synergistic interaction between berberine and carbon nanotubes, this innovative approach exemplifies the potential of nanotechnology to transform natural metabolites into robust therapeutic agents. The successful integration of berberine-loaded CNTs highlights a forward-thinking strategy for combating Alzheimer's disease, offering hope for more effective and targeted interventions in neurodegenerative care (Dey, A., et al. 2024).

4.4. Huperzine A Nanoparticles

Huperzine A, a natural alkaloid extracted from *Huperzia serrata*, is widely recognized for its potent acetylcholinesterase (AChE) inhibitory properties, which enhance cholinergic neurotransmission. This mechanism makes it a promising therapeutic agent for symptomatic relief in AD, particularly in addressing memory and cognitive decline. However, its therapeutic potential is limited by poor bioavailability, rapid metabolism, and challenges in crossing the BBB. Nanoparticle-based delivery systems have emerged as a transformative approach to overcoming these limitations and optimizing Huperzine A's efficacy in AD management (Zorkina, Y., et al. 2020).

Nanoparticles offer multiple advantages for delivering Huperzine A, including improved stability, enhanced solubility, and targeted delivery to the brain. By encapsulating Huperzine A in biocompatible and biodegradable nanoparticles, such as those based on lipids or polymers, researchers have significantly enhanced its pharmacokinetic profile. These formulations protect Huperzine A from degradation in the gastrointestinal tract and bloodstream, ensuring that a higher fraction of the compound reaches the brain (Dighe, S., et al. 2024).

The targeted delivery facilitated by nanoparticles ensures a concentrated therapeutic effect within the CNS. This precision minimizes systemic side effects while amplifying the compound's ability to inhibit AChE activity. By reducing AChE levels, Huperzine A nanoparticles enhance acetylcholine availability, thereby improving cholinergic signaling. This mechanism is critical for alleviating cognitive symptoms in AD patients, as cholinergic deficits are a hallmark of the disease (Gabriella, et al. 2021).

Beyond symptomatic relief, Huperzine A nanoparticles exhibit additional neuroprotective benefits. Studies have shown that these formulations can attenuate oxidative stress and reduce neuronal damage, further contributing to their therapeutic efficacy. Preclinical models of AD treated with Huperzine A nanoparticles have demonstrated improved learning and memory functions, reduced neuronal apoptosis, and enhanced synaptic plasticity (Meng, Q., et al. 2018).

Huperzine A nanoparticles exemplify the power of nanotechnology to harness and amplify the therapeutic potential of natural metabolites. By addressing the pharmacological and logistical challenges associated with traditional Huperzine A administration, these advanced delivery systems offer a promising pathway for more effective and targeted treatment strategies in Alzheimer's disease (Mir Najib, S. N., et al. 2023). This synergy between natural compounds and nanotechnology represents a significant step forward in the quest for innovative solutions to combat neurodegeneration.

4.5. Resveratrol Nanoemulsions

Resveratrol, a polyphenolic compound found in grapes, berries, and other plants, has garnered attention for its potent antioxidative and neuroprotective properties. In PD, where oxidative stress and mitochondrial dysfunction play central roles in neuronal degeneration, resveratrol offers significant therapeutic potential. However, its clinical application is limited by poor water solubility, low bioavailability, and rapid metabolism, which hinder its ability to reach and act effectively on the CNS. Nanoemulsion-based delivery systems have emerged as a promising solution to these challenges, amplifying resveratrol's efficacy in managing PD (Chinthu V. Saji, et al. 2024).

Nanoemulsions are colloidal systems that incorporate resveratrol into nanometer-sized droplets stabilized by surfactants. These formulations improve resveratrol's solubility and stability, enabling efficient absorption and distribution throughout the body, including the brain. Importantly, nanoemulsions can facilitate crossing the BBB, ensuring that resveratrol reaches the target sites of neurodegeneration (Bonferoni, M. C., et al. 2019).

In the context of PD, resveratrol nanoemulsions effectively combat oxidative stress by neutralizing ROS and reducing lipid peroxidation, processes that contribute to dopaminergic neuron loss. Furthermore, resveratrol supports mitochondrial health by restoring mitochondrial membrane potential and enhancing ATP production, which are critical for maintaining neuronal function.

These effects collectively protect neurons from the progressive degeneration that characterizes PD (Kim, S., et al. 2024).

Preclinical studies in PD models have demonstrated the therapeutic benefits of resveratrol nanoemulsions. Animals treated with these formulations exhibit improved motor function, reduced neuroinflammation, and decreased neuronal apoptosis compared to those receiving unformulated resveratrol. Additionally, the enhanced bioavailability provided by nanoemulsions prolongs resveratrol's action, offering sustained neuroprotection and mitigating disease progression (Burns, et al. 2023).

The versatility of nanoemulsion systems also allows for customization, such as incorporating additional neuroprotective agents or tuning release profiles to match the disease's progression. This adaptability makes nanoemulsions an attractive platform for developing more comprehensive treatment strategies (Petrovic, S., et al. 2024).

By addressing the pharmacological limitations of resveratrol and enhancing its therapeutic potential, nanoemulsions exemplify the synergistic integration of nanotechnology and natural metabolites. These advanced delivery systems not only unlock new possibilities for treating PD but also serve as a blueprint for optimizing other bioactive compounds in the fight against neurodegeneration (Mall, et al. 2024).

4.6. Quercetin Nanocapsules

Quercetin, a flavonoid found in various fruits and vegetables, is well-known for its antioxidative, anti-inflammatory, and neuroprotective properties. In PD, where oxidative stress, neuroinflammation, and dopaminergic neuron degeneration are key pathological features, quercetin holds considerable therapeutic promise. However, its poor water solubility, rapid metabolism, and limited ability to cross the BBB restrict its clinical application. Nanocapsulation offers a transformative approach to overcoming these challenges, enhancing quercetin's therapeutic efficacy in PD management (Das, S. S., et al. 2023).

Nanocapsules, which consist of a core-shell structure encapsulating quercetin, provide improved stability and protect the compound from premature degradation. This advanced formulation ensures efficient delivery of quercetin to the brain by facilitating its transport across the BBB. By enhancing bioavailability, nanocapsules allow therapeutic concentrations of quercetin to reach the CNS, where it can exert its neuroprotective effects (Hawthorne, et al. 2022).

In PD models, quercetin nanocapsules have demonstrated robust antioxidative properties by scavenging ROS and reducing oxidative damage, which are major contributors to dopaminergic neuron loss. Additionally, quercetin's anti-inflammatory effects are amplified through nanocapsulation, as it suppresses the activation of microglia and pro-inflammatory cytokines that exacerbate neuronal damage. Together, these actions help preserve dopaminergic neurons and mitigate the progression of PD (Tomou, E. M., et al. 2023).

Preclinical studies have shown that treatment with quercetin nanocapsules leads to significant improvements in motor function, a hallmark symptom of PD. Animals receiving these formulations exhibit reduced neuronal apoptosis and enhanced neuronal survival in the substantia nigra, the brain region most affected in PD. The sustained release capabilities of nanocapsules further extend quercetin's therapeutic action, providing prolonged neuroprotection with fewer dosing requirements (Camila de Oliveira, et al. 2024).

Moreover, nanocapsules are versatile and can be engineered using biocompatible and biodegradable materials, ensuring safety and reducing potential side effects. This aligns with the principles of sustainable and patient-friendly therapeutic approaches, making nanocapsules an ideal delivery system for neurodegenerative disorders like PD (Silva, D. F., et al. 2023).

The integration of quercetin and nanotechnology exemplifies a powerful synergy that unlocks the full potential of this natural compound. By addressing the pharmacokinetic and delivery challenges, quercetin nanocapsules represent a promising avenue for innovative and effective PD therapies, showcasing the vital role of nanotechnology in advancing treatments for neurodegeneration (Zhuo, Y., et al 2024).

4.7. Rosmarinic Acid Solid Lipid Nanoparticles

Rosmarinic acid, a natural phenolic compound derived from herbs like Rosmarinus officinalis and Salvia officinalis, is renowned for its potent antioxidative and anti-inflammatory properties. These attributes make it a promising candidate for mitigating the neurodegenerative processes characteristic of Huntington's disease (HD). However, the therapeutic potential of rosmarinic acid is hindered by its limited stability, rapid metabolism, and difficulty in crossing the BBB. Solid lipid nanoparticles (SLNs) offer an innovative solution to these challenges, enhancing the delivery and efficacy of rosmarinic acid in HD therapy (Bhatt, R., et al. 2014).

SLNs are submicron-sized carriers composed of biocompatible and biodegradable lipids that encapsulate therapeutic compounds, providing protection against degradation and controlled release. Encapsulating rosmarinic acid in SLNs significantly improves its stability, prolonging its bioactivity and ensuring sustained therapeutic action. Furthermore, the lipid-based composition of SLNs facilitates efficient transport of rosmarinic acid across the BBB, enabling targeted delivery to affected regions of the brain (Ashfaq, R., et al. 2023).

In preclinical models of Huntington's disease, rosmarinic acid-loaded SLNs have shown promising results. These formulations mitigate oxidative stress by neutralizing ROS and preventing lipid peroxidation, thereby reducing cellular damage. Additionally, the anti-inflammatory properties of rosmarinic acid are enhanced through nanocarrier-based delivery, effectively suppressing the activation of microglia and the release of pro-inflammatory cytokines. Together, these actions contribute to a reduction in neurodegeneration and promote neuronal survival in regions of the brain affected by HD (Fachel, F. N. S., et al. 2019).

Another critical advantage of SLNs is their ability to provide controlled and sustained release of rosmarinic acid, ensuring prolonged therapeutic effects with fewer dosing requirements. This feature not only improves treatment efficacy but also enhances patient compliance, particularly in chronic conditions like Huntington's disease (Gandhi, S., et al. 2024).

Preclinical studies have also demonstrated the potential of rosmarinic acid-loaded SLNs to improve motor function and cognitive performance in HD models. By addressing both the oxidative and inflammatory components of HD pathology, these formulations provide a multifaceted approach to neuroprotection (Lamptey, R. N. L., et al. 2022).

The development of rosmarinic acid-loaded SLNs highlights the synergistic potential of nanotechnology and natural metabolites in neurodegenerative disease therapy. By overcoming the pharmacokinetic limitations of rosmarinic acid, this innovative approach underscores the transformative role of nanocarriers in advancing treatments for Huntington's disease and other neurodegenerative disorders (Azhar, M. K., et al. 2023).

4.8. Other Promising Nano-enabled Phytochemicals

Several other phytochemicals, including gallic acid, epigallocatechin-3-gallate (EGCG), ferulic acid, coumarin, naringenin, and sesamol, have demonstrated significant potential in addressing neurodegenerative disorders. When combined with nanotechnology-based delivery systems, their therapeutic efficacy is substantially enhanced by improving bioavailability, stability, and brain-targeted delivery. Below, we explore the unique attributes and nanoformulations of these phytochemicals and their relevance to combating neurodegeneration (Figure 6).

a. Gallic Acid

Gallic acid, a polyphenolic compound found in berries, tea, and grapes, is celebrated for its antioxidative, anti-inflammatory, and anti-apoptotic effects. These properties make it particularly valuable in mitigating the oxidative damage and neuronal loss associated with diseases like Alzheimer's and Parkinson's. Nanocarriers such as polymeric nanoparticles and solid lipid nanoparticles encapsulating gallic acid enhance its bioavailability and enable precise delivery to the brain. Preclinical studies report reduced oxidative stress, improved neuronal survival, and restoration of cognitive and motor functions following treatment with gallic acid nanoformulations (Hadidi, M., et al. 2024).

b. Epigallocatechin-3-Gallate (EGCG)

EGCG, a catechin derived from green tea, is known for its neuroprotective effects, including its ability to modulate amyloid-beta aggregation, reduce oxidative stress, and suppress neuroinflammation. Nano-enabled delivery systems, such as liposomes and polymeric nanoparticles, enhance the stability and BBB permeability of EGCG, overcoming its rapid degradation and poor bioavailability. These advanced formulations have shown promising results in preclinical models of Alzheimer's and Parkinson's, demonstrating improved memory retention, reduced neuronal damage, and decreased inflammatory markers (Afzal, O., et al. 2022).

c. Ferulic Acid

Ferulic acid, abundant in grains, fruits, and vegetables, exhibits strong antioxidative and anti-inflammatory effects, making it a candidate for addressing neurodegeneration. Encapsulation in nanocarriers, such as chitosan nanoparticles and nanostructured lipid carriers, improves its solubility and facilitates sustained release. These nanoformulations have shown effectiveness in reducing oxidative stress and neuroinflammation in preclinical models of neurodegenerative diseases, supporting neuronal health and function (Purushothaman, J. R., and Rizwanullah, M. et al. 2024).

d. Coumarin

Coumarin, a phytochemical found in cinnamon, tonka beans, and certain plants, offers neuroprotective benefits by reducing oxidative stress, preventing apoptosis, and modulating neuroinflammation. However, its clinical application has been hampered by limited solubility and bioavailability. Nano-enabled systems, including polymeric and lipid-based nanoparticles, enhance its delivery to the brain and prolong its therapeutic effects. Coumarin-loaded nanoparticles have shown potential in alleviating symptoms and slowing disease progression in models of neurodegenerative disorders (Baloghová, J., et al. 2023).

e. Naringenin

Naringenin, a flavonoid derived from citrus fruits, is recognized for its antioxidative, anti-inflammatory, and neuroprotective properties. Despite its therapeutic promise, its poor water solubility and low systemic bioavailability restrict its utility. Nanoformulations such as lipid nanoparticles and nanoemulsions have been developed to overcome these limitations, enabling efficient brain delivery. Studies demonstrate that naringenin-loaded nanocarriers reduce oxidative damage, improve neuronal survival, and enhance cognitive and motor functions in neurodegenerative disease models (Jiang, H., et al. 2023).

f. Sesamol

Sesamol, a lignan found in sesame seeds, exhibits antioxidative and anti-apoptotic properties, making it an attractive compound for neuroprotection. Encapsulation in nanostructured lipid carriers and polymeric nanoparticles enhances its solubility, stability, and BBB permeability. Nano-enabled sesamol formulations have shown significant promise in reducing oxidative stress, neuroinflammation, and neuronal apoptosis in preclinical models of Alzheimer's and PD (Nair, A. B., et al. 2023).

The integration of nanotechnology with phytochemicals like gallic acid, EGCG, ferulic acid, coumarin, naringenin, and sesamol exemplifies the transformative potential of this synergistic approach in combating neurodegeneration. These nanoformulations address the pharmacokinetic limitations of natural metabolites, enhancing their therapeutic impact while reducing side effects. Collectively, they underscore the vital role of nanotechnology in unlocking the full potential of phytochemicals for effective and targeted interventions in neurodegenerative diseases.

Figure 6. Chemical structures of some promising nano-enabled secondary metabolites used in the management of neurodegenerative diseases

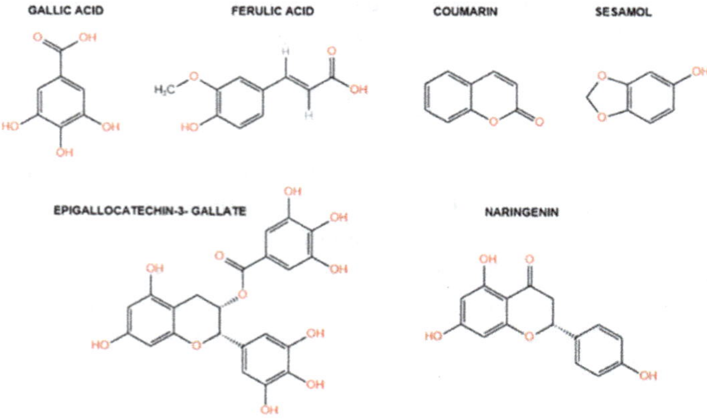

5. SYNERGISTIC EFFECTS

5.1. Enhancing Bioavailability and Stability

One of the key challenges in utilizing phytochemicals for therapeutic purposes lies in their inherent limitations, such as poor solubility, rapid degradation, and limited bioavailability. Nanotechnology offers a transformative solution to these issues, ensuring that phytochemicals retain their potency and therapeutic potential throughout their journey to the target site.

Nanocarriers, including liposomes, solid lipid nanoparticles, polymeric nanoparticles, and nanoemulsions, act as protective vessels for phytochemicals. These nanostructures shield the bioactive compounds from environmental factors such as heat, light, and oxygen, which can lead to degradation. They also guard against metabolic breakdown within the gastrointestinal tract or liver, ensuring that a larger proportion of the therapeutic agent remains intact for systemic absorption.

By enhancing the solubility of poorly water-soluble phytochemicals, nanocarriers significantly improve their absorption in the bloodstream. These systems often employ surface modifications, such as the addition of hydrophilic coatings or targeting ligands, to further facilitate cellular uptake and ensure efficient transport across biological barriers, including the BBB. This targeted delivery mechanism is especially critical in neurodegenerative disorders, where access to brain tissue is a significant challenge (Zazuli, Z., et al. 2025; Lu, H., et al. 2021).

In addition to protection, nanocarriers can provide controlled and sustained release of phytochemicals. This feature ensures a consistent therapeutic effect over time, reducing the need for frequent dosing and minimizing potential side effects. For instance, the encapsulation of curcumin, known for its poor stability and rapid metabolism, in solid lipid nanoparticles has demonstrated a marked improvement in its bioavailability, allowing higher concentrations to reach the brain and exert antioxidative and anti-inflammatory effects critical for neuroprotection (Qing, et al. 2024).

Preclinical studies further support the efficacy of nanotechnology-enhanced phytochemicals in combating neurodegenerative diseases. Phytochemicals like quercetin, rosmarinic acid, and epigallocatechin-3-gallate, when delivered via nanocarriers, exhibit amplified therapeutic effects, including reduced oxidative stress, improved neuronal survival, and alleviation of disease symptoms. These advancements are directly attributable to the enhanced stability and bioavailability afforded by nanotechnology (Fakhri, S., et al. 2022).

The integration of nanocarriers with phytochemicals addresses fundamental pharmacokinetic challenges, unlocking their full therapeutic potential. By enhancing bioavailability and stability, this synergistic approach ensures that natural compounds can effectively target and mitigate the pathological processes underlying neurodegeneration, paving the way for more effective and sustainable treatments.

5.2. Overcoming the Blood-Brain Barrier

The BBB, a selective and tightly regulated interface, plays a critical role in maintaining the brain's microenvironment but poses a significant obstacle to delivering therapeutic agents for neurodegenerative diseases. Many promising phytochemicals, despite their potent neuroprotective properties, struggle to cross the BBB due to their hydrophilic nature, molecular size, or rapid metabolism. Nanotechnology provides groundbreaking solutions to overcome these barriers, enabling efficient and targeted delivery of therapeutic compounds to the brain.

Nanocarriers such as liposomes, polymeric nanoparticles, solid lipid nanoparticles, and dendrimers are engineered with properties that enhance their ability to traverse the BBB. Surface functionalization plays a pivotal role in this process. By incorporating specific ligands, such as transferrin, lactoferrin, or apolipoproteins, nanocarriers can engage with receptors expressed on the BBB, facilitating receptor-mediated transcytosis. This targeted approach ensures the selective and efficient delivery of encapsulated phytochemicals to neural tissues, minimizing off-target effects.

Another innovative strategy involves the use of nanocarriers coated with surfactants or PEGylation (polyethylene glycol modification), which helps nanoparticles evade the immune system and prolong their circulation time in the bloodstream. This increases the likelihood of their interaction with the BBB and subsequent uptake into the CNS.

For instance, curcumin-loaded nanoparticles have demonstrated improved BBB penetration compared to free curcumin, allowing higher concentrations to reach affected brain regions. These formulations effectively reduce amyloid-beta plaques and oxidative stress, hallmarks of Alzheimer's disease. Similarly, chitosan nanoparticles encapsulating piperine have shown enhanced transport across the BBB, improving the compound's anti-inflammatory and neuroprotective effects in preclinical studies (Maria Inês, et al. 2023).

Nanocarriers also provide the advantage of controlled and sustained drug release, ensuring that therapeutic levels of phytochemicals are maintained in the brain over extended periods. This not only improves efficacy but also reduces the frequency of administration, enhancing patient compliance and reducing systemic side effects.

Preclinical research underscores the transformative potential of nanotechnology in crossing the BBB and improving treatment outcomes for neurodegenerative conditions such as Alzheimer's, Parkinson's, and Huntington's diseases. By enabling the brain-targeted delivery of phytochemicals, these advanced formulations address one of the most significant challenges in neuropharmacology (Agosti, E., et. al. 2024).

In conclusion, nanotechnology represents a paradigm shift in overcoming the BBB, empowering natural phytochemicals with the capacity to reach and treat neural tissues effectively. This synergistic approach holds immense promise for developing innovative therapies for neurodegenerative diseases, advancing both scientific discovery and clinical applications.

5.3. Multifunctional Nanocarriers

The recent evolution of multifunctional nanocarriers, which combine therapeutic and diagnostic functions, marks a significant breakthrough in the treatment of neurodegenerative diseases. Known as theranostics, this approach integrates drug delivery with real-time monitoring, enabling the precise targeting, assessment, and management of therapeutic interventions. In the context of phytochemicals, multifunctional nanocarriers are particularly promising, as they combine the neuroprotective benefits of natural compounds with advanced diagnostic capabilities, offering a more tailored and dynamic treatment strategy (Ramanathan S, et al. 2018).

Multifunctional nanocarriers are designed to perform multiple roles simultaneously, enhancing both the delivery and monitoring of therapeutic agents. By incorporating imaging agents such as fluorescent dyes, magnetic nanoparticles, or radiolabeled molecules into the nanocarrier system, it becomes possible to track the distribution of phytochemicals *in vivo*. These imaging capabilities provide crucial feedback on the biodistribution, brain penetration, and accumulation of drugs, enabling real-time assessment of therapeutic efficacy (Stiufiuc, G. F., & Stiufiuc, R. I., 2024; Chaparro, C. I. P., et al. 2023).

For example, nanocarriers encapsulating curcumin or resveratrol can be functionalized with magnetic nanoparticles, allowing magnetic resonance imaging (MRI) or positron emission tomography (PET) to monitor the pharmacokinetics of the compound. This dual functionality not only ensures that the drug reaches the intended site of action in the brain but also helps track the success of the therapy, facilitating dose adjustments and providing insights into patient-specific responses. Such systems are particularly valuable in neurodegenerative diseases, where subtle changes in disease progression can be challenging to monitor through conventional methods (Yallapu, M. M., et al. 2015).

Additionally, multifunctional nanocarriers can be engineered to incorporate both therapeutic phytochemicals and molecular probes that respond to specific biomarkers of neurodegeneration. For instance, nanoparticles loaded with phytochemicals such as epigallocatechin-3-gallate (EGCG) or quercetin can be coupled with diagnostic agents that target amyloid-beta plaques or tau protein tangles, common markers of Alzheimer's disease. These combined systems enable simultaneous drug delivery and biomarker imaging, allowing for more accurate diagnosis and monitoring of disease progression (Zhu, F. D., et al. 2021).

The precision offered by multifunctional nanocarriers also enhances personalized medicine approaches, as they enable the tailoring of treatments based on real-time data. By monitoring the pharmacokinetics and therapeutic effects of phytochemicals in individual patients, clinicians can adjust treatment plans to optimize outcomes. Moreover, the ability to visualize therapeutic agents in the brain offers invaluable insights into the mechanisms underlying disease and the patient's response to treatment.

In addition to their diagnostic and therapeutic roles, multifunctional nanocarriers also provide sustained and controlled release of phytochemicals, ensuring that therapeutic levels are maintained over time. This reduces the need for frequent dosing, minimizes side effects, and improves overall treatment adherence, particularly in the context of chronic neurodegenerative conditions (Das, A., et al. 2023).

In conclusion, the integration of therapeutic and diagnostic functions in multifunctional nanocarriers represents a significant step forward in treating neurodegenerative diseases. By combining phytochemicals with advanced imaging and biomarker-targeting technologies, these theranostic systems provide a powerful tool for personalized, precise, and effective treatment strategies. This innovative approach has the potential to revolutionize the management of neurodegenerative diseases, offering new opportunities for both research and clinical practice.

6. CHALLENGES AND FUTURE DIRECTIONS

6.1. Technical and Manufacturing Challenges

Scalability of Nanoformulations: While many nanoformulations show promise at the laboratory scale, scaling up production to meet clinical demand remains a significant hurdle. Issues such as batch-to-batch consistency, cost-effectiveness, and maintaining the functional integrity of nanocarriers during large-scale manufacturing must be addressed.

Standardization of Production Methods: The lack of standardized protocols for the synthesis and characterization of nanocarriers contributes to variability in their quality and performance. Establishing robust, reproducible manufacturing processes is critical for ensuring the reliability and efficacy of nanoformulations in clinical settings (Yedi H., et al. 2022).

6.2. Regulatory and Safety Considerations

Long-Term Toxicity and Biocompatibility Concerns: Nanocarriers often interact with biological systems in complex ways, raising questions about their long-term safety. Studies must evaluate their potential for inducing immune responses, bioaccumulation, and off-target effects to ensure patient safety.

Regulatory Frameworks for Clinical Translation: Regulatory agencies face challenges in evaluating nanotechnology-based drugs due to their unique properties. Developing comprehensive guidelines for preclinical testing, safety assessment, and efficacy evaluation is essential for advancing these therapies toward clinical use (Havelikar, U., et al. 2024).

6.3. Emerging Trends and Innovations

Smart Nanocarriers: These dynamic systems are designed for controlled drug release, responding to specific triggers such as temperature, pH, or light. Smart nanocarriers enable precise timing and localization of therapeutic action, reducing systemic side effects and improving treatment efficacy (Sun, L., et al. 2023).

Stimuli-Responsive Delivery Systems: Building on the principles of smart nanocarriers, stimuli-responsive systems are engineered to respond to changes in the disease microenvironment. For example, enzyme-sensitive carriers can release drugs in response to overexpressed enzymes in neurodegenerative conditions, ensuring targeted therapy (Liu, G., et al. 2020).

Integration with Other Technologies: Combining nanotechnology with cutting-edge tools like AI and CRISPR gene-editing technology opens new avenues for personalized medicine. AI algorithms can optimize nanocarrier design, while CRISPR can be integrated into nanocarrier systems for precise genetic modifications, offering transformative potential in treating neurodegenerative diseases (Anurogo D., 2023).

7. CONCLUSION

This chapter underscores the transformative potential of integrating nanotechnology and phytochemistry in addressing the complex challenges of neurodegenerative disorders. Neurodegeneration, characterized by multifactorial pathophysiology, demands innovative approaches that can simultaneously address bioavailability limitations, delivery barriers, and therapeutic efficacy. Nanotechnology emerges as a powerful enabler, offering the ability to enhance the stability and bioavailability of secondary metabolites while ensuring targeted and controlled delivery to affected neural tissues.

The use of multifunctional nanocarriers, such as liposomes, dendrimers, and polymeric nanoparticles, provides unprecedented opportunities to harness the therapeutic potential of bioactive compounds. These carriers not only protect phytochemicals from degradation but also enable site-specific delivery, minimizing off-target effects and enhancing therapeutic outcomes. Advances in stimuli-responsive nanocarriers further add a layer of precision, responding to the unique microenvironments of neurodegenerative tissues, such as oxidative stress or altered pH, to release their cargo in a controlled manner.

The synergy between nanotechnology and secondary metabolites extends beyond therapeutic delivery. Nanocarriers can be designed to simultaneously deliver multiple therapeutic agents, enabling a combinatorial approach to combat the multifactorial nature of neurodegeneration. Moreover, the potential to incorporate diagnostic functionalities within these systems paves the way for theranostic applications, where treatment and monitoring can occur concurrently, thereby streamlining the management of neurodegenerative diseases.

Future research directions emphasize the need for interdisciplinary collaboration between chemists, biologists, material scientists, and clinicians to optimize the design and functionality of nanocarriers. Efforts must also focus on bridging the gap between preclinical findings and clinical implementation by addressing scalability, manufacturing challenges, and long-term safety. Additionally, regulatory frameworks must evolve to accommodate the unique complexities of nanoformulations, ensuring that these innovations meet safety standards without stifling progress.

In conclusion, the integration of nanotechnology and secondary metabolites holds immense promise in redefining therapeutic paradigms for neurodegenerative disorders. By leveraging the complementary strengths of these fields, it is possible to develop holistic, targeted, and sustainable solutions. The road ahead demands innovation, collaboration, and perseverance to translate these laboratory breakthroughs into real-world clinical interventions that can transform patient outcomes and quality of life.

REFERENCES

Afzal, O., Dalhat, M. H., & Altamimi, A. S. A., (2022). Green tea catechins attenuate neurodegenerative diseases and cognitive deficits. *Molecules (Basel, Switzerland), 27*(21), 7604. DOI: 10.3390/molecules27217604 PMID: 36364431

Agosti, E., Zeppieri, M., & Antonietti, S., (2024). Navigating the nose-to-brain route: A systematic review on lipid-based nanocarriers for central nervous system disorders. *Pharmaceutics, 16*(3), 329. DOI: 10.3390/pharmaceutics16030329 PMID: 38543223

Ahlawat, J., Guillama Barroso, G., & Masoudi Asil, S., (2020). Nanocarriers as potential drug delivery candidates for overcoming the blood–brain barrier: Challenges and possibilities. *ACS Omega, 5*(22), 12583–12595. DOI: 10.1021/acsomega.0c01592 PMID: 32548442

Akshatha, P. K., Nayak, P. G., & John, J., (2024). Revolutionizing neurotherapeutics: Nanocarriers unveiling the potential of phytochemicals in Alzheimer's disease. *Neuropharmacology, 259*, 110096. DOI: 10.1016/j.neuropharm.2024.110096 PMID: 39084596

Al-zharani, M., Hasnain, M. S., Al-Eissa, M. S., & Alqahtani, R. A. (2024). Novel drug delivery to the brain for neurodegenerative disorder treatment using carbon nanotubes. *Journal of King Saud University. Science, 36*(11), 103513. DOI: 10.1016/j.jksus.2024.103513

Anurogo, D. (2023). The art of nanoimmunobiotechnomedicine in stem cells and gene editing. *Journal of Biomedical Research & Environmental Sciences, 4*(9), 1348–1369. DOI: 10.37871/jbres1757

Ashfaq, R., Rasul, A., & Asghar, S., (2023). Lipid nanoparticles: An effective tool to improve the bioavailability of nutraceuticals. *International Journal of Molecular Sciences, 24*(21), 15764. DOI: 10.3390/ijms242115764 PMID: 37958750

Asimakidou, E., Tan, J. K. S., Zeng, J., & Lo, C. H. (2024). Blood-brain barrier-targeting nanoparticles: Biomaterial properties and biomedical applications in translational neuroscience. *Pharmaceuticals (Basel, Switzerland), 17*(5), 612. DOI: 10.3390/ph17050612 PMID: 38794182

Azargoonjahromi, A., & Abutalebian, F. (2024). Unraveling the therapeutic efficacy of resveratrol in Alzheimer's disease: An umbrella review of systematic evidence. *Nutrition & Metabolism, 21*, 15. DOI: 10.1186/s12986-024-00792-1 PMID: 38504306

Azhar, M. K., Anwar, S., & Hasan, G. M., (2023). Comprehensive insights into biological roles of rosmarinic acid: Implications in diabetes, cancer and neurodegenerative diseases. *Nutrients, 15*(19), 4297. DOI: 10.3390/nu15194297 PMID: 37836581

Bai, M., Shao, X., & Wang, C., (2025). Application of carbon-based nanomaterials in Alzheimer's disease. *Materials Horizons*. Advance online publication. DOI: 10.1039/D4MH01256A PMID: 39526325

Balakrishnan, R., Azam, S., Kim, I. S., & Choi, D. K. (2023). Neuroprotective effects of black pepper and its bioactive compounds in age-related neurological disorders. *Aging and Disease, 14*(3), 750–777. DOI: 10.14336/AD.2022.1022 PMID: 37191428

Baloghová, J., Michalková, R., & Baranová, Z., (2023). Spice-derived phenolic compounds: Potential for skin cancer prevention and therapy. *Molecules (Basel, Switzerland)*, *28*(17), 6251. DOI: 10.3390/molecules28176251 PMID: 37687080

Bhatt, R., Singh, D., Prakash, A., & Mishra, N. (2014). Development, characterization, and nasal delivery of rosmarinic acid-loaded solid lipid nanoparticles for the effective management of Huntington's disease. *Drug Delivery*, *22*(7), 931–939. DOI: 10.3109/10717544.2014.880860 PMID: 24512295

Bhattacharya, T., Soares, G. A. B. E., & Chopra, H., (2022). Applications of phyto-nanotechnology for the treatment of neurodegenerative disorders. *Materials (Basel, Switzerland)*, *15*(3), 804. DOI: 10.3390/ma15030804 PMID: 35160749

Bonferoni, M. C., Rossi, S., & Sandri, G., (2019). Nanoemulsions for "nose-to-brain" drug delivery. *Pharmaceutics*, *11*(2), 84. DOI: 10.3390/pharmaceutics11020084 PMID: 30781585

Burns, J., Buck, A. C., D'Souza, S., Dube, A., & Bardien, S. (2023). Nanophytomedicines as therapeutic agents for Parkinson's disease. *ACS Omega*, *8*(45), 42045–42061. DOI: 10.1021/acsomega.3c04862 PMID: 38024675

Cai, G., Lin, F., & Wu, D., (2022). Rosmarinic acid inhibits mitochondrial damage by alleviating unfolded protein response. *Frontiers in Pharmacology*, *13*, 859978. DOI: 10.3389/fphar.2022.859978 PMID: 35652041

Cao, Z., Li, W., & Liu, R., (2019). pH- and enzyme-triggered drug release as an important process in the design of anti-tumor drug delivery systems. *Biomedicine and Pharmacotherapy*, *118*, 109340. DOI: 10.1016/j.biopha.2019.109340 PMID: 31545284

Chaparro, C. I. P., Simões, B. T., & Borges, J. P., (2023). A promising approach: Magnetic nanosystems for Alzheimer's disease theranostics. *Pharmaceutics*, *15*(9), 2316. DOI: 10.3390/pharmaceutics15092316 PMID: 37765284

Chen, Z.-R., Huang, J.-B., Yang, S.-L., & Hong, F.-F. (2022). Role of cholinergic signaling in Alzheimer's disease. *Molecules (Basel, Switzerland)*, *27*(6), 1816. DOI: 10.3390/molecules27061816 PMID: 35335180

Cheng, Z., Kang, C., & Che, S., (2022). Berberine: A promising treatment for neurodegenerative diseases. *Frontiers in Pharmacology*, *13*, 845591. DOI: 10.3389/fphar.2022.845591 PMID: 35668943

Chopra, H., Bibi, S., & Singh, I.,. (2022). Nanomedicines in the management of Alzheimer's disease: Current view and future prospects. *Frontiers in Aging Neuroscience*, *14*, 879114. DOI: 10.3389/fnagi.2022.879114 PMID: 35875806

Coco, J. C., Silvério, L. A. L., & Santos, É. M. d., (2023). Piperine extraction and encapsulation in polycaprolactone nanoparticles. *Cosmetics*, *10*(6), 152. DOI: 10.3390/cosmetics10060152 da Costa, I. M., Freire, M. A. M., & de Paiva Cavalcanti, J. R. L., (2019). Supplementation with Curcuma longa reverses neurotoxic and behavioral damage in models of Alzheimer's disease: A systematic review. *Current Neuropharmacology*, *17*(5), 406–421. DOI: 10.2174/09298673256 66180117112610 PMID: 29338678

Das, A., Adhikari, S., Deka, D., et al. (2023). An updated review on the role of nanoformulated phytochemicals in colorectal cancer. Medicina (Kaunas, Lithuania, 59(4), 685. https://doi.org/ .DOI: 10.3390/medicina59040685

Das, S. S., Jha, N. K., & Jha, S. K., (2023). Neuroprotective role of quercetin against alpha-synuclein-associated hallmarks in Parkinson's disease. *Current Neuropharmacology*, *21*(7), 1464–1466. DOI: 10.2174/1570159X21666221221092250 PMID: 36545726

Dash, U. C., Bhol, N. K., & Swain, S. K., (2024). Oxidative stress and inflammation in the pathogenesis of neurological disorders: Mechanisms and implications. *Acta Pharmaceutica Sinica. B*. Advance online publication. DOI: 10.1016/j.apsb.2024.10.004 PMID: 40041912

Del Prado-Audelo, M. L., Caballero-Florán, I. H., & Meza-Toledo, J. A., (2019). Formulations of curcumin nanoparticles for brain diseases. *Biomolecules*, *9*(2), 56. DOI: 10.3390/biom9020056 PMID: 30743984

Dey, A., Ghosh, S., & Rajendran, R. L., (2024). Alzheimer's disease pathology and assistive nanotheranostic approaches for its therapeutic interventions. *International Journal of Molecular Sciences*, *25*(17), 9690. DOI: 10.3390/ijms25179690 PMID: 39273645

Dighe, S., Jog, S., Momin, M., Sawarkar, S., & Omri, A. (2024). Intranasal drug delivery by nanotechnology: Advances in and challenges for Alzheimer's disease management. *Pharmaceutics*, *16*(1), 58. DOI: 10.3390/pharmaceutics16010058 PMID: 38258068

El Ouaamari, Y., Van den Bos, J., Willekens, B., Cools, N., & Wens, I. (2023). Neurotrophic factors as regenerative therapy for neurodegenerative diseases: Current status, challenges and future perspectives. *International Journal of Molecular Sciences*, *24*(4), 3866. DOI: 10.3390/ijms24043866 PMID: 36835277

Elnaggar, Y. S., Etman, S. M., Abdelmonsif, D. A., & Abdallah, O. Y. (2015). Intranasal piperine-loaded chitosan nanoparticles as brain-targeted therapy in Alzheimer's disease: Optimization, biological efficacy, and potential toxicity. *Journal of Pharmaceutical Sciences*, *104*(10), 3544–3556. DOI: 10.1002/jps.24557

Elsori, D., Rashid, G., & Khan, N. A., (2023). Nanotube breakthroughs: Unveiling the potential of carbon nanotubes as a dual therapeutic arsenal for Alzheimer's disease and brain tumors. *Frontiers in Oncology*, *13*, 1265347. DOI: 10.3389/fonc.2023.1265347 PMID: 37799472

Fachel, F. N. S., Schuh, R. S., & Veras, K. S., (2019). An overview of the neuroprotective potential of rosmarinic acid and its association with nanotechnology-based delivery systems: A novel approach to treating neurodegenerative disorders. *Neurochemistry International*, *122*, 47–58. DOI: 10.1016/j.neuint.2018.11.003 PMID: 30439384

Fakhri, S., Abdian, S., & Zarneshan, S. N., (2022). Nanoparticles in combating neuronal dysregulated signaling pathways: Recent approaches to the nanoformulations of phytochemicals and synthetic drugs against neurodegenerative diseases. *International Journal of Nanomedicine*, *17*, 299–331. DOI: 10.2147/IJN.S347187 PMID: 35095273

Friedli, M. J., & Inestrosa, N. C. (2021). Huperzine A and its neuroprotective molecular signaling in Alzheimer's disease. *Molecules (Basel, Switzerland)*, *26*(21), 6531. DOI: 10.3390/molecules26216531 PMID: 34770940

Fuloria, S., Mehta, J., & Chandel, A., (2022). A comprehensive review on the therapeutic potential of Curcuma longa Linn. in relation to its major active constituent curcumin. *Frontiers in Pharmacology*, *13*, 820806. DOI: 10.3389/fphar.2022.820806 PMID: 35401176

Gabriella, M., Buccioni, M., & Dal Ben, D., (2021). Efficacy of acetylcholinesterase inhibitors in Alzheimer's disease. *Neuropharmacology*, *190*, 108352. DOI: 10.1016/j.neuropharm.2020.108352 PMID: 33035532

Gandhi, S., Shastri, D. H., Shah, J., Nair, A. B., & Jacob, S. (2024). Nasal delivery to the brain: Harnessing nanoparticles for effective drug transport. *Pharmaceutics*, *16*(4), 481. DOI: 10.3390/pharmaceutics16040481 PMID: 38675142

Genchi, G., Lauria, G., Catalano, A., Carocci, A., & Sinicropi, M. S. (2024). Neuroprotective effects of curcumin in neurodegenerative diseases. *Foods*, *13*(11), 1774. DOI: 10.3390/foods13111774 PMID: 38891002

Goswami, S., Isaev, N. K., & Salmina, A. B., (2024). Recent developments in selective therapeutic targeting of functionalized nanomaterials to neurovascular units in overcoming the gaps in neurovascular therapy. *Brain Disorders (Amsterdam, Netherlands)*, *15*, 100162. DOI: 10.1016/j.dscb.2024.100162

Hadidi, M., Liñán-Atero, R., Tarahi, M., Christodoulou, M. C., & Aghababaei, F. (2024). The potential health benefits of gallic acid: Therapeutic and food applications. *Antioxidants*, *13*(8), 1001. DOI: 10.3390/antiox13081001 PMID: 39199245

Havelikar, U., Ghorpade, K. B., & Kumar, A., (2024). Comprehensive insights into the mechanism of nanotoxicity, assessment methods, and regulatory challenges of nanomedicines. *Discover Nano*, *19*, 165. DOI: 10.1186/s11671-024-04118-1 PMID: 39365367

Hawthorne, D., Pannala, A., Sandeman, S., & Lloyd, A. (2022). Sustained and targeted delivery of hydrophilic drug compounds: A review of existing and novel technologies from bench to bedside. *Journal of Drug Delivery Science and Technology*, *78*, 103936. DOI: 10.1016/j.jddst.2022.103936

Hou, Y., Dan, X., & Babbar, M., (2019). Ageing as a risk factor for neurodegenerative disease. *Nature Reviews. Neurology*, *15*(10), 565–581. DOI: 10.1038/s41582-019-0244-7 PMID: 31501588

Jalili, C., Kiani, A., & Gholami, M., (2023). Brain targeting-based nanocarriers loaded with resveratrol in Alzheimer's disease: A review. *IET Nanobiotechnology / IET*, *17*(3), 154–170. DOI: 10.1049/nbt2.12127 PMID: 36949020

Jiang, H., Zhang, M., & Lin, X., (2023). Biological activities and solubilization methodologies of naringin. *Foods*, *12*(12), 2327. DOI: 10.3390/foods12122327 PMID: 37372538

Kim, S., Jung, U. J., & Kim, S. R. (2024). Role of oxidative stress in blood–brain barrier disruption and neurodegenerative diseases. *Antioxidants*, *13*(12), 1462. DOI: 10.3390/antiox13121462 PMID: 39765790

Kumari, N., Anand, S., & Shah, K., (2023). Emerging role of plant-based bioactive compounds as therapeutics in Parkinson's disease. *Molecules (Basel, Switzerland)*, *28*(22), 7588. DOI: 10.3390/molecules28227588 PMID: 38005310

Kurul, Fatma, Turkmen, H., Cetin, A. E., & Topkaya, S. N. (2025). Nanomedicine: How nanomaterials are transforming drug delivery, bio-imaging, and diagnosis. *Next Nanotechnology*, *7*, 100129. DOI: 10.1016/j.nxnano.2024.100129

Lamptey, R. N. L., Chaulagain, B., & Trivedi, R., (2022). A review of the common neurodegenerative disorders: Current therapeutic approaches and the potential role of nanotherapeutics. *International Journal of Molecular Sciences*, *23*(3), 1851. DOI: 10.3390/ijms23031851 PMID: 35163773

Li, L., Zeng, Y., Chen, M., & Liu, G. (2022). Application of nanomicelles in enhancing bioavailability and biological efficacy of bioactive nutrients. *Polymers*, *14*(16), 3278. DOI: 10.3390/polym14163278 PMID: 36015535

Liu, G., Lovell, J. F., Zhang, L., & Zhang, Y. (2020). Stimulus-responsive nanomedicines for disease diagnosis and treatment. *International Journal of Molecular Sciences*, *21*(17), 6380. DOI: 10.3390/ijms21176380 PMID: 32887466

Lohan, S., Raza, K., & Mehta, S. K., (2017). Anti-Alzheimer's potential of berberine using surface-decorated multi-walled carbon nanotubes: A preclinical evidence. *International Journal of Pharmaceutics*, *530*(1–2), 263–278. DOI: 10.1016/j.ijpharm.2017.07.080 PMID: 28774853

Lu, H., Zhang, S., Wang, J., & Chen, Q. (2021). A review on polymer and lipid-based nanocarriers and its application to nano-pharmaceutical and food-based systems. *Frontiers in Nutrition*, *8*, 783831. DOI: 10.3389/fnut.2021.783831 PMID: 34926557

Mall, J., Naseem, N., & Haider, M. F., (2024). *Nanostructured lipid carriers as a drug delivery system: A comprehensive review with therapeutic applications*. Intelligent Pharmacy., DOI: 10.1016/j.ipha.2024.09.005

Manek, E., Darvas, F., & Petroianu, G. A. (2020). Use of biodegradable, chitosan-based nanoparticles in the treatment of Alzheimer's disease. *Molecules (Basel, Switzerland)*, *25*(20), 4866. DOI: 10.3390/molecules25204866 PMID: 33096898

Melgarejo da Rosa, M., Cassimiro de Amorim, L., & de Oliveira Alves, J. V., (2022). The promising role of natural products in Alzheimer's disease. *Brain Disorders (Amsterdam, Netherlands)*, *7*, 100049. DOI: 10.1016/j.dscb.2022.100049

Meng, Q., Wang, A., & Hua, H.. (2018). Intranasal delivery of huperzine A to the brain using lactoferrin-conjugated N-trimethylated chitosan surface-modified PLGA nanoparticles for treatment of Alzheimer's disease. *International Journal of Nanomedicine*, *13*, 705–718. DOI: 10.2147/IJN.S151474 PMID: 29440896

Mittal, K. R., Pharasi, N., & Sarna, B., (2022). Nanotechnology-based drug delivery for the treatment of CNS disorders. *Translational Neuroscience*, *13*(1), 527–546. DOI: 10.1515/tnsci-2022-0258 PMID: 36741545

Nair, A. B., Dalal, P., & Kadian, V., (2023). Formulation strategies for enhancing pharmaceutical and nutraceutical potential of sesamol: A natural phenolic bioactive. *Plants*, *12*(5), 1168. DOI: 10.3390/plants12051168 PMID: 36904028

Nayab, D. E., & Din, F. u., Ali, H., & others. (2023). Nano biomaterials-based strategies for enhanced brain targeting in the treatment of neurodegenerative diseases: An up-to-date perspective. *Journal of Nanobiotechnology*, *21*, 477. DOI: 10.1186/s12951-023-02250-1 PMID: 38087359

Noor-E-Tabassum, Das, R., Lami, M. S., Chakraborty, A. J., et al. (2022). Ginkgo biloba: A treasure of functional phytochemicals with multimedicinal applications. *Evidence-Based Complementary and Alternative Medicine : eCAM, 8288818*. Advance online publication. DOI: 10.1155/2022/8288818

Omidian, H., Gill, E. J., Dey Chowdhury, S., & Cubeddu, L. X. (2024). Chitosan nanoparticles for intranasal drug delivery. *Pharmaceutics*, *16*(6), 746. DOI: 10.3390/pharmaceutics16060746 PMID: 38931868

Pei, J., Palanisamy, C. P., & Natarajan, P. M., (2024). Curcumin-loaded polymeric nanomaterials as a novel therapeutic strategy for Alzheimer's disease: A comprehensive review. *Ageing Research Reviews*, *99*, 102393. DOI: 10.1016/j.arr.2024.102393 PMID: 38925479

Peng, Y., Jin, H., & Xue, Y. H., (2023). Current and future therapeutic strategies for Alzheimer's disease: An overview of drug development bottlenecks. *Frontiers in Aging Neuroscience*, *15*, 1206572. DOI: 10.3389/fnagi.2023.1206572 PMID: 37600514

Petrovic, S., Bita, B., & Barbinta-Patrascu, M.-E. (2024). Nanoformulations in pharmaceutical and biomedical applications: Green perspectives. *International Journal of Molecular Sciences*, *25*(11), 5842. DOI: 10.3390/ijms25115842 PMID: 38892030

Preethy, H. A., Rajendran, K., Sukumar, A. J., & Krishnan, U. M. (2024). Emerging paradigms in Alzheimer's therapy. *European Journal of Pharmacology*, *981*, 176872. DOI: 10.1016/j.ejphar.2024.176872 PMID: 39117266

Purushothaman, J. R., & Rizwanullah, M. (2024). Ferulic acid: A comprehensive review. *Cureus*, *16*(8), e68063. DOI: 10.7759/cureus.68063 PMID: 39347187

Rahman, M. M., Islam, M. R., & Akash, S., (2022). Recent advancements of nanoparticles application in cancer and neurodegenerative disorders: At a glance. *Biomedicine and Pharmacotherapy*, *153*, 113305. DOI: 10.1016/j.biopha.2022.113305 PMID: 35717779

Rajput, A., Pingale, P., & Dhapte-Pawar, V. (2022). Nasal delivery of neurotherapeutics via nanocarriers: Facets, aspects, and prospects. *Frontiers in Pharmacology, 13*, 979682. DOI: 10.3389/fphar.2022.979682 PMID: 36176429

Ramanathan, S., Archunan, G., & Sivakumar, M., (2018). Theranostic applications of nanoparticles in neurodegenerative disorders. *International Journal of Nanomedicine, 13*, 5561–5576. DOI: 10.2147/IJN.S149022 PMID: 30271147

Rezagholizade-Shirvan, A., Soltani, M., & Shokri, S., (2024). Bioactive compound encapsulation: Characteristics, applications in food systems, and implications for human health. *Food Chemistry: X, 24*, 101953. DOI: 10.1016/j.fochx.2024.101953 PMID: 39582652

Rudrapal, M., Khairnar, S. J., & Khan, J., (2022). Dietary polyphenols and their role in oxidative stress-induced human diseases: Insights into protective effects, antioxidant potentials, and mechanism(s) of action. *Frontiers in Pharmacology, 13*, 806470. DOI: 10.3389/fphar.2022.806470 PMID: 35237163

Rudzińska, M., Grygier, A., Knight, G., & Kmiecik, D. (2024). Liposomes as carriers of bioactive compounds in human nutrition. *Foods, 13*(12), 1814. DOI: 10.3390/foods13121814 PMID: 38928757

Saji, C. V., Mohanakumar, K. P., & Shenoi, R. A. (2024). Phytomedicine nanoformulations for Parkinson's disease. *Phytomedicine Plus: International Journal of Phytotherapy and Phytopharmacology*. Advance online publication. DOI: 10.1016/j.phyplu.2024.100704

Satapathy, T., & Kumar, D. (2024). A comprehensive review and recent advancement in the application of tannins for treating Parkinson's disease. *Pharmacological Research. Modern Chinese Medicine, 12*, 100499. DOI: 10.1016/j.prmcm.2024.100499

Shao, X., Yan, C., & Wang, C., (2022). Advanced nanomaterials for modulating Alzheimer's-related amyloid aggregation. *Nanoscale Advances, 5*(1), 46–80. DOI: 10.1039/d2na00625a PMID: 36605800

Silva, D. F., Melo, A. L. P., & Uchôa, A. F. C., (2023). Biomedical approach of nanotechnology and biological risks: A mini-review. *International Journal of Molecular Sciences, 24*(23), 16719. DOI: 10.3390/ijms242316719 PMID: 38069043

Stiufiuc, G. F., & Stiufiuc, R. I. (2024). Magnetic nanoparticles: Synthesis, characterization, and their use in the biomedical field. *Applied Sciences (Basel, Switzerland), 14*(4), 1623. DOI: 10.3390/app14041623

Sun, L., Liu, H., & Ye, Y. (2023). Smart nanoparticles for cancer therapy. *Signal Transduction and Targeted Therapy, 8*, 418. DOI: 10.1038/s41392-023-01642-x PMID: 37919282

Sun, Q., Mei, Lv., & Li, Y. (2024). Nanotechnology-based drug delivery systems for curcumin and its derivatives in the treatment of cardiovascular diseases. *Journal of Functional Foods, 122*, 106476. DOI: 10.1016/j.jff.2024.106476

Teleanu, R. I., Preda, M. D., & Niculescu, A. G., (2022). Current strategies to enhance delivery of drugs across the blood-brain barrier. *Pharmaceutics, 14*(5), 987. DOI: 10.3390/pharmaceutics14050987 PMID: 35631573

Tomou, E. M., Papakyriakopoulou, P., & Saitani, E. M., (2023). Recent advances in nanoformulations for quercetin delivery. *Pharmaceutics, 15*(6), 1656. DOI: 10.3390/pharmaceutics15061656 PMID: 37376104

Vashist, A., Manickam, P., & Raymond, A. D., (2023). Recent advances in nanotherapeutics for neurological disorders. *ACS Applied Bio Materials, 6*(7), 2614–2621. DOI: 10.1021/acsabm.3c00254 PMID: 37368486

Vashist, V., Mythri, C., & Kumarasamy, M. (2024). Natural products as potential modulators of pro-inflammatory cytokines signaling in Alzheimer's disease. *Brain Behavior and Immunity Integrative, 5*, 100048. DOI: 10.1016/j.bbii.2024.100048

WHO. (World Health Organization). (2017). Dementia: Number of people affected to triple in next 30 years. World Health Organization. https://www.who.int/news/item/07-12-2017-dementia-number-of-people-affected-to-triple-in-next-30-years#:~:text=As%20the%20global%20population%20ages,%2C%20 Director%2 DGeneral%20of%20WHO.

Yadav, A., Sharma, V., & Singh, G. (2024). Anti-Inflammatory Potential of Chalcone Related Compounds: An Updated Review. *ChemistrySelect, 9*(26), e202401321.

Yakubu, J., & Pandey, A. V. (2024). Innovative delivery systems for curcumin: Exploring nanosized and conventional formulations. *Pharmaceutics, 16*(5), 637. DOI: 10.3390/pharmaceutics16050637 PMID: 38794299

Yallapu, M. M., Nagesh, P. K., Jaggi, M., & Chauhan, S. C. (2015). Therapeutic applications of curcumin nanoformulations. *The AAPS Journal, 17*(6), 1341–1356. DOI: 10.1208/s12248-015-9811-z PMID: 26335307

Yavarpour-Bali, H., Ghasemi-Kasman, M., & Pirzadeh, M. (2019). Curcumin-loaded nanoparticles: A novel therapeutic strategy in treatment of central nervous system disorders. *International Journal of Nanomedicine, 14*, 4449–4460. DOI: 10.2147/IJN.S208332 PMID: 31417253

Zazuli, Z., Hartati, R., Rowa, C. R., & Asyarie, S., & Satrialdi. (2025). The potential application of nanocarriers in delivering topical antioxidants. *Pharmaceuticals, 18*(1), 56. DOI: 10.3390/ph18010056 PMID: 39861119

Zhang, L., Xu, L.-Y., & Tang, F., (2024). New perspectives on the therapeutic potential of quercetin in non-communicable diseases: Targeting Nrf2 to counteract oxidative stress and inflammation. *Journal of Pharmaceutical Analysis, 14*(6), 100930. DOI: 10.1016/j.jpha.2023.12.020 PMID: 39005843

Zhu, F. D., Hu, Y. J., & Yu, L., (2021). Nanoparticles: A hope for the treatment of inflammation in CNS. *Frontiers in Pharmacology, 12*, 683935. DOI: 10.3389/fphar.2021.683935 PMID: 34122112

Zhuo, Y., Zhao, Y.-G., & Zhang, Y. (2024). Enhancing drug solubility, bioavailability, and targeted therapeutic applications through magnetic nanoparticles. *Molecules (Basel, Switzerland), 29*(20), 4854. DOI: 10.3390/molecules29204854 PMID: 39459222

Zorkina, Y., Abramova, O., & Ushakova, V., (2020). Nano carrier drug delivery systems for the treatment of neuropsychiatric disorders: Advantages and limitations. *Molecules (Basel, Switzerland), 25*(22), 5294. DOI: 10.3390/molecules25225294 PMID: 33202839

Chapter 13
Integrative Nanomedicine Approaches and Secondary Metabolites in Neurodegenerative Diseases

Burak Çakar
https://orcid.org/0000-0002-4334-9320
Faculty of Medicine, Istinye University, Turkey

Cansu Çadirci
Health Sciences Institute, Aydın Adnan Menderes University, Turkey

Kubilay Dogan Kilic
https://orcid.org/0000-0002-9484-0777
Faculty of Medicine, Ege University, Turkey

ABSTRACT

The potential of nanoparticle-based drug delivery systems and bioactive secondary metabolites offers promising avenues for the treatment of neurodegenerative diseases. Neurodegeneration, driven by oxidative stress, protein aggregation, mitochondrial dysfunction, and neuroinflammation, presents complex therapeutic challenges. Nanocarriers such as polymeric, lipid-based, and metallic nanoparticles enable targeted delivery, improved blood–brain barrier penetration, and enhanced bioavailability of neuroprotective compounds. Meanwhile, secondary metabolites ncluding polyphenols, flavonoids, and alkaloids exhibit antioxidant, anti-inflammatory, and autophagy-regulating activities, positioning them as valuable agents for neuroprotection. The molecular mechanisms underlying these effects involve modulation of key cellular signaling pathways. Ongoing challenges in clinical translation, alongside regulatory considerations and future directions for integrating nanotechnology with natural bioactive compounds, are also critically assessed.

DOI: 10.4018/979-8-3373-1365-8.ch013

INTRODUCTION

With the rise in life expectancy and the ageing population, Neurodegenerative diseases (NDs) have also increased. This situation has drawn greater attention to new drug discoveries and therapeutic approaches for age-related diseases. NDs pose significant challenges due to their complex mechanisms and lack of effective treatments. While drugs and therapies are already available, extensive research continues to expand and reshape our current understanding. In this chapter, NDs are discussed, focusing on the application of nanotechnology in neurological research by identifying NPs and secondary metabolites that are increasingly used for their diagnosis and treatment. NPs, recognized as innovative drug delivery systems, offer unique advantages in their ability to cross the blood-brain barrier and deliver therapeutic agents directly to affected neuronal sites. The promising neuroprotective and anti-inflammatory properties of secondary metabolites derived from natural sources make them therapeutic candidates. Their synergistic use, their role in targeting molecular and cellular pathways of neurodegeneration, their clinical applications, and their accompanying conveniences and challenges are discussed. By reviewing current developments and highlighting prospects, this chapter aims to provide insights into how ongoing resarch can be further improved and directed.

1. COMMON NEURODEGENERATIVE DISEASES AND GENERAL CHARACTERISTICS

NDs are progressive and usually age-related diseases in which certain populations of neurons and connections of neurons in the brain and spinal cord gradually lose function These diseases are associated with molecular abnormalities such as the accumulation of protein aggregates (Heemels, 2016; Dugger & Dickson, 2017; Ruffini et al., 2022) They have an increasing prevalence in old age and include several diseases such as Alzheimer's disease (AD), Parkinson's disease (PD), Huntington's disease (HD), and Amyotrophic lateral sclerosis (ALS) (Heemels, 2016; Dugger & Dickson, 2017; Ciurea et al., 2023). These resulting diseases cause many neurological and psychiatric symptoms, such as motor dysfunction and cognitive impairments, depending on the parts of the brain that are damaged. New and effective treatment modalities have been used to slow and prevent the onset and progression of NDs (Figure 1), but have not been identified to date.

Figure 1. Illustration of new and effective treatment modalities for NDs

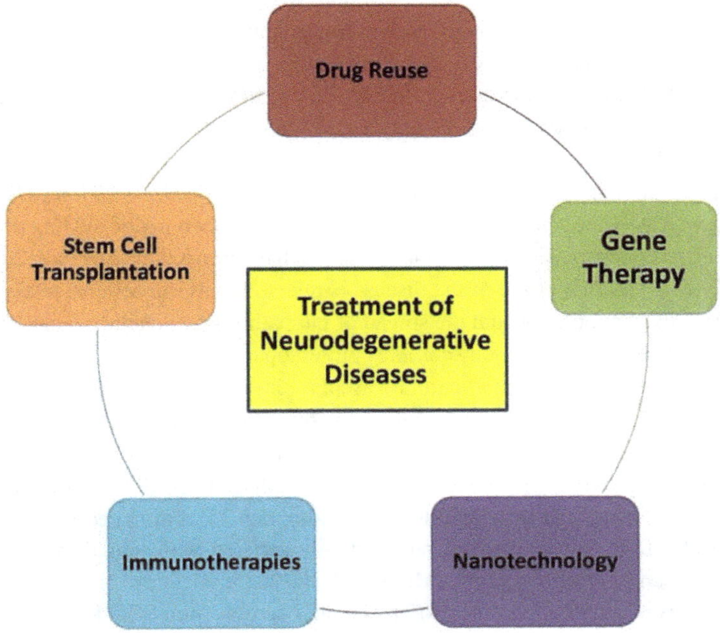

Control of inflammation and prevention of protein accumulation are among the potential treatment strategies for neurodegenerative dieases. With the rapid increase in the number of patients worldwide and the associated rise in diagnostic and treatment costs, there is growing interest and strong demand for the development of innovations in both diagnostic techniques and therapeutic approaches (Takeuchi, 2021). The main features of NDs are associated with the abnormal and progressive accumulation of proteins such as amyloid-β, tau, and α-synuclein, which aggregate in the brain and spinal cord and impair the function of neurons (Dugger & Dickson, 2017; Kovacs, 2016; Kovacs, 2019). Furthermore, chronic and uncontrolled inflammation plays an important role in disease progression, as overactivation of microgllial cells can exacerbate neuronal damage (Gao & Hong, 2008; Tanaka et al., 2020). The diagnosis of NDs are usually revealed by clinical symptoms, neuroimaging, and biomarker analysis. Current research focuses on developing new treatment methods that can slow or stop the progression of these diseases (Erkkinen et al., 2018; Kovacs, 2019; Slanzi et al., 2020). NDs are consequently increasing with aging and are highly complex disorders requiring deeper molecular research and more innovative therapeutic strategies. The pathology and molecular mechanisms of NDs are often associated with the inter- or intracellular accumulation of misfolded proteins. Examples include proteins such as amyloid-β and tau proteins in AD, α-synuclein in PD, and TAR DNA binding protein 43 in ALS (Dugger & Dickson, 2017; Erkkinen et al., 2018; Peng et al., 2020). The amplification of all these proteins behind cell-cell adhesion and transduction plays a key role in the initiation and progression of diseases (Peng et al., 2020).

1.1. Alzheimer's Disease

Worldwide, AD is recognized as the most common cause of dementia, representing a major global health problem and a progressive neurodegenerative disorder whose prevalence increases with the aging population (Erkkinen et al., 2018; Reith, 2018). The disease is characterized by distinct pathological hallmarks, includings amyloid plaques and neurofibrillary tangles composed of hyperphoshorylated tau proteins (Knopman et al., 2021; Blennow et al., 2006; DeTure & Dickson, 2019). The pathogenesis of AD is associated with the accumulation of amyloid-beta (Aβ) and tau proteins. The accumulation of these proteins leads to synaptic dysfunction, neuronal loss, and impaired endosomal-lysosomal clearance mechanisms (Knopman et al., 2021; DeTure & Dickson, 2019). The first phase of the disease is referred to as the cellular phase and occurs in parallel with Aβ accumulation and the spread of tau pathology (Scheltens et al., 2021; Fan et al., 2020). AD is recognized as a highly complex proteinopathy and is commonly associated with other age-related pathologies, such as cerebrovascular disease and Lewy body disease (DeTure & Dickson, 2019). The risk of AD is estimated to be 60-80% due to hereditary factors, although this estimate remains uncertain. APOE alleles are the strongest genetic risk factors associated with the disease (Scheltens et al., 2021; Blennow et al., 2006). Sporadic AD, on the other hand, generally has a later onset and is thought to affect more than 15 million people worldwide. The familial form of AD is early onset and rare and is associated with mutations in the amyloid precursor protein and presenilin genes (Masters et al., 2015; Blennow et al., 2006). Methods such as PET scans of amyloid and tau biomarkers and cerebrospinal fluid analysis are used to diagnose AD (Scheltens et al., 2021; Villain & Dubois, 2019). Current treatment methods are aimed at alleviating the symptoms of the disease and effective therapies targeting the underlying mechanisms have not yet been discovered (Masters et al., 2015; Ballard et al., 2011). Therapeutic agent development efforts for novel pharmacological treatments focus on anti-amyloid, anti-tau, and anti-inflammatory strategies and are in clinical trial stages (Scheltens et al., 2021). Some theories on the pathogenesis of AD include mechanisms such as gamma oscillations dysfunction, prion-like propagation and infection (Fan et al., 2020). Lifestyle-based prevention trials designed to prevent the disease aim to provide cognitive benefits to participants at increased risk of dementia (Scheltens et al., 2021). In conclusion, multidisciplinary research integrsting molecular biology, genetics and clinical neuroscianca will be essential for developing effective and mechanism-based treatment strategies for AD.

1.2. Parkinson's Disease

PD is the most common neurodegenerative movement disorder of the central nervous system. It usually occurs in older people and is more common in men than in women. Although the exact cause of the disease is unknown, it is believed to result from a combination of genetic predisposition and environmental influences (Balestrino & Schapira, 2019; Bloem et al., 2021; Kalia & Lang, 2015). PD affects approximately 1% of individuals over the age of 60 worldwide, with prevalence rates ranging from 108 to 257 per 100,000 people across Europe (Balestrino & Schapira, 2019; Samii et al., 2004). The main symptoms include classic motor symptoms such as tremors, muscle rigidity, slowness of movement (bradykinesia), and postural instability (Balestrino & Schapira, 2019; De Virgilio et al., 2016; Fahn, 2003). Genetic factors account for approximately 5-15% of PD cases anda re associated with several identified genes (De Virgilio et al., 2016; Fahn, 2003). Environmental factors such as exposure to neurotoxic substances, pesticides or heavy metals, as well as aging, also play a role in the development of the disease (Bhat et al., 2018; Ball et al., 2019). Recent studies provide increasing evidence that the immune responce and shronic inflammation are important factors in the pathogenesis of the disease (De Virgilio et al., 2016). In short, PD is a complex neurodegenerative disorder characterized by motor and non-motor symptoms and driven by the interplay of genetic, environmental and immunological mechanisms. While treatment options focus on alleviating symptoms, no current therapy has been proven to halt or reserve disease proression.

1.3. Huntington's Disease

HD, like other NDs, is an autosomal dominant neurodegenerative disease that affects the central nervous system. The disease is mainly caused by the expansion of CAG trinucleotide repeats in the huntingtin (HTT) gene and usually begins to manifest between the ages of 30-50 (McColgan & Tabrizi, 2018; Roos, 2010; Jimenez-Sanchez et al., 2017). This expansion leads to the production of a mutant huntingtin protein that acquires toxic gain-of-function properties and induces progressive neurodegeneration (Finkbeiner, 2011; Jimenez-Sanchez et al., 2017). HD is characterized by involuntary choreiform movements, psychiatric disorders, and dementia. As the disease progresses, choreic movements spread to all muscles, and severe psychiatric symptoms such as depression, irritability and aggression are often observed. Patients experience progressive cognitive decline and psychiatric disturbances (Roos, 2010; Finkbeiner, 2011; Martin, 1984). As the disease progresses, it leads to complete dependence on daily life and eventually death (Roos, 2010). Identified pathological processes of the disease include disruption of proteostasis, transcriptional dysregulation, mitochondrial dysfunnction and direct cytotoxixity of the mutant huntingtin protein (McColgan & Tabrizi, 2018; Rüb et al., 2016). Currently, there is no curative treatment for HD. Existing treatments aim to reduce symptoms and improve patients quality of life. Pharmacological and supportive interventions may be required to manage psychiatric symptoms such as depression and aggression (Roos, 2010; Walker, 2007). In recent years, therapies aimed at lowering mutant huntingtin expression have emerged as promising treatment approaches, particularly antisense oligonucleotide (ASO) therapy (McColgan & Tabrizi, 2018; Ross & Tabrizi, 2011). This means that studies to develop realistic approaches to delay or prevent the onset of the disease will be promising for patients (Gusella et al., 2014).

1.4. Amyotrofik Lateral Skleroz (ALS)

This neurodegenerative disease is characterized by the progressive degeneration of motor neurons, leading to muscle weakness, atrophy and paralysis (Heemels, 2016; Erkkinen et al., 2018) Although the exact pathogenesis of ALS is not fully understood, both genetic factors and various molecular mechanisms are known to play a role ALS is typically characterized by simultaneous involvement upper and lower motor neurons, although atypical variants may selectively affect one motor neuron type. This phenotypic diversity suggests that the disease may have distinct biological subtypes. The same gene mutation may result in varying clinical phenotypes, whisch likly contributes to the heterogeneity observed in ALS patients. (Grad et al., 2017; Feldman et al., 2018) Approximately 10% of current ALS cases are familial, and several pathogenic gene mutations have been identified in these cases In particular, hexanucleotide repeat expansions in the C9orf72 gene are found in a substantial proportion of both familial and sporadic ALS cases. These genetic discoveries highlight the molecular overlap between ALS and ALS and frontotemporal dementia (FTD) (Ghasemi & Brown, 2018; Masrori & Van Damme, 2020; Al-Chalabi et al., 2017) The pathogenesis of ALS remains largely unknown, but some mechanisms have been proposed, such as disturbances in protein homeostasis, changes in the biology of RNA-binding proteins and disturbances in cytoskeletal dynamics The accumulation of TDP43 protein in neurons has been shown to be a marker of the pathological process of ALS (Morgan & Orrell, 2016; Chiò et al., 2020) . Current therapeutic options are limited, with riluzole and edaravone being the only FDA-approved drugs that modestly slow disease progression. Overall, management is primarily supportive and multidisciplinary, focusing on symptom control and improving quality of life. Emerging therapeutic research includes stem cell therapy, gene silencing techniques, and neuroprotective strategies (Oskarsson et al., 2018; Chiò et al., 2020; Hardiman et al., 2011).

A better understanding of the genetic, molecular, and cellular basis of ALS will allow the development of personalized and targeted therapeutic approaches. Large-scale genomic and epidemiological studies will be crucial for identifying risk factors, pathogenic triggers, and disease modifiers (Morgan & Orrell, 2016; Al-Chalabi et al., 2017). Although the complex nature of ALS presents significant challenges in both clinical management and experimental research, continued advances in genetic technologies and molecular neuroscience are expected to drive the development of more effective and disease-modifying treatment strategies.

1.5. Molecular Mechanisms of Neurodegenerative Diseases and the Impact of Common Pathological Mechanisms

Molecular mechanisms of NDs are shaped by various pathways such as oxidative stress, protein aggregation, mitochondrial dysfunction, and genetic factors. A better understanding of these mechanisms will enable the development of novel therapeutic and preventive strategies for these disorders. Research efforts continue to intensify in order to elucidate these mechanisms and develop effective treatment approaches. In general, the molecular and pathological alterations in NDs can be summarized under the following key mechanisms:

- ***Oxidative stress and redox imbalance:*** Neurons are particularly vulnerable to oxidative damage due to their high oxygen consumption and limited antioxidant capacity. Overproduction of reactive oxygen species (ROS) is strongly implicated in the development of Alzheimer's, Parkinson's, and other neurodegenerative disorders (Liu et al., 2017; Sanghai & Tranmer, 2023; Tan et al., 2019).
- ***Protein aggregation*** Intracellular or extracellular accumulation of misfolded proteins is a common feature of NDs These proteins can be transported between cells and contribute to disease progression (Peng et al., 2020; Ross & Poirier, 2004)
- ***Mitochondrial dysfunction*** Mitochondrial dysfunction is linked to decreased energy production and increased cellular stress This makes it difficult for neurons to survive (Tan et al., 2019; Bossy-Wetzel et al., 2004)
- ***Ubiquitin-Proteasome system dysfunction*** The inability to properly degrade proteins can lead to cellular toxicity and neurodegeneration (Tan et al., 2019; Bossy-Wetzel et al., 2004)

- ***MicroRNA dysregulation*** MicroRNAs are involved in post-transcriptional regulation of gene expression and are commonly dysregulated in NDs (Juźwik et al., 2019)
- ***Neuroinflammation*** Dysfunction of glial cells and inflammatory responses play an important role in neurodegenerative processes (Sanghai & Tranmer, 2023; Guo et al., 2020)
- ***Genetic and environmental influences*** Genetic mutations and environmental factors can increase disease risk and influence pathogenesis (Santiago et al., 2017; Gan et al., 2018)
- ***Chronic inflammation immune responses in NDs*** are associated with the activation of microglia cells, the resident macrophages of the central nervous system These immune responses play a critical role in both disease pathogenesis and potential neuroregeneration (Amor et al., 2010; Stephenson et al., 2018) Molecules such as inflammatory cytokines and biologically active kynurenines play an important role in the development of NDs (Tanaka et al., 2020; Slanzi et al., 2020)
- ***Aging in NDs*** A primary risk factor for NDs Biological processes associated with aging play an important role in the pathogenesis of these diseases Biomarkers of aging, such as genomic instability, telomere shortening, epigenetic changes, and loss of proteostasis, are linked to susceptibility to NDs (Hou et al., 2019)
- ***Clinical diagnosis of NDs*** is usually based on symptomatology, neuroimaging, and laboratory findings diseases such as AD, frontotemporal dementia, PD, and HD are distinguished by different clinical symptoms and diagnostic criteria (Erkkinen et al., 2018) The pathogenesis of these diseases is related to processes such as the accumulation and intercellular migration of protein aggregates Inflammation and immune responses play an important role in both disease progression and potential treatment strategies Aging is a major risk factor for these diseases, and targeting biological processes is leading to the development of new therapeutic strategies.

2. NANOTECHNOLOGY

Nanotechnology is a multidisciplinary science that deals with the design, synthesis, and development of nanomaterials, as well as the control and manipulation of matter at the atomic and molecular level. Nanotechnology has its origins in Richard Feynman's pioneering concept of manipulating matter at the atomic scale. In the early 21st century, nanoscience emerged as a national research priority, leading to substantial global investments (Hulla et al., 2015). This technology plays a transformative role across multiple industries, offering revolutionary advancements in sectors such as agriculture, food, medicine, textiles, cosmetics, and packaging. In these fields, the unique physical, chemical, and biological properties of nanomaterials enable the development of more effective, durable, and sustainable products (Malik et al., 2023; Singh, 2017). Especially, nanotechnology is thought to optimize production processes and enhance efficiency in large-scale industrial areas such as the automotive, construction, and environmental management sectors (Malik et al., 2023). In medicine, nanotechnology holds immense promise (Figure 2). NPs are currently being used for diagnostic and screening purposes, advanced drug delivery systems, gene therapy, and tissue engineering applications (Emerich & Thanos, 2003; Ferrari & Downing, 2005). Moreover, nanotechnology has the potential to revolutionize biomedical research and clinical sciences (Ferrari & Downing, 2005). In addition, nanotechnology is making important contributions to the cosmetics industry, one of the fastest-growing sectors today. Nanosystems enhance skin penetration, improve ingredient stability, and provide controlled release profiles, resulting in superior cosmetic efficacy. However, research on nanotoxicology, human exposure, and environmental safety is still ongoing (Santos et al., 2019). Looking ahead, nanotechnology is expected to evolve toward more sustainable, biocompatible, and eco-friendly innovations (Malik et al., 2023). Overall, it has the potential to revolutionize a wide spectrum of scientific and industrial fields, and continued investment in research and innovation will further expand its applications in the coming decades.

Figure 2. Illustration of the uses of nanotechnology in medicine

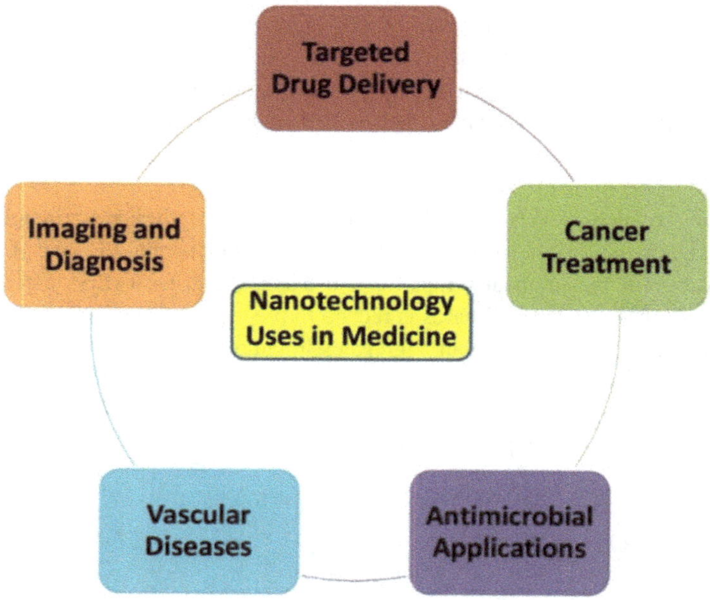

2.1. Nanoparticle Properties and Types

NPs are defined as very small particles ranging in size from 1 to 100 nanometers and are used in various applications. They attract attention because of their properties such as high surface area, unique optics and reactivity (McNamara & Tofail, 2017; Khan et al., 2017; Ealias & Saravanakumar, 2017). (McNamara & Tofail, 2017; Khan et al., 2017; Ealias & Saravanakumar, 2017). NPs offer a wide range of applications with various types and unique properties. The different types of NPs, such as metal, metal oxide, carbon-based and polymeric, offer significant advantages in fields such as medicine, environment and industry. The high surface area and reactivity offered by these particles make them ideal for many innovative applications. NPs offer a promising approach to the treatment of NDs. NPs are used as biosensors for the early diagnosis of NDs, which allows for more sensitive detection (Bilal et al., 2020; Cao & Zhang, 2022). When we look at the types of NPs, they are classified as follows (Figure 3);

Figure 3. Illustration of nanoparticle types

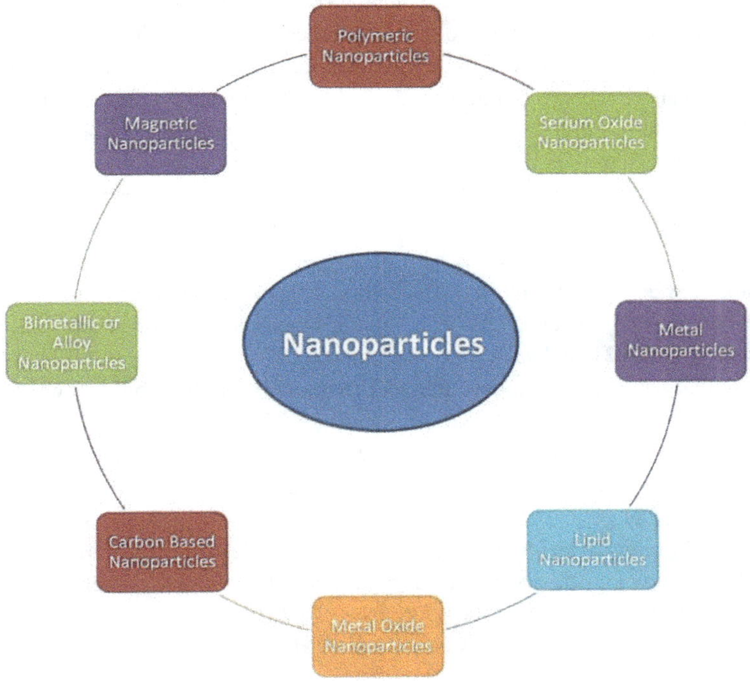

- ***Polymeric Nanoparticles:*** These types of NPs can target brain regions by releasing drugs in a highly controlled way and can be used in the treatment of NDs (Cano et al., 2020; Vissers et al., 2019). These NPs have shown promising results in preclinical models of NDs (Cano et al., 2020; Cunha et al., 2021). These NPs are used in Alzheimer's treatment by improving targeted distribution (Gupta et al., 2019; Fonseca-Santos et al., 2015). These NPs made of polymers are used in drug delivery systems and offer controlled release properties (Khan et al., 2017; Wang & Wang, 2014).
- ***Cerium Oxide Nanoparticles:*** Thanks to their antioxidant properties, cerium oxide NPs stand out as a potential therapy for the treatment of Alzheimer's, Parkinson's, and other NDs (Naz et al., 2017).
- ***Metal Nanoparticles*:** These NPs have been developed to improve brain drug delivery and for use in the diagnosis and treatment of NDs (Asefy et al., 2021; Vissers et al., 2019; Poovaiah et al., 2018). NPs made of metals such as gold (Au) and silver (Ag) are widely used in biomedical applications. These types have properties such as high chemical stability and biocompatibility (McNamara & Tofail, 2017; Hasan, 2015; Harilal et al., 2019).
- ***Lipid Nanoparticles:*** These types of NPs are composed of lipid fragments and are effectively used in many current biomedical applications. They usually have a diameter ranging from 10 to 1000 nm and are spherical in shape. Like the polymeric type, they

have a solid core of lipids and a matrix of soluble lipophilic molecules. Surfactants or emulsifiers stabilize the outer core of these NPs (Rawat et al., 2011). Lipid-based NPs have become a specialized field focused on the design and synthesis of NPs for various applications and research, such as drug delivery and delivery (Puri et al., 2009) and RNA release in cancer therapy (Gujrati et al., 2014). Due to these transport capacities, they enable drugs to reach the brain more effectively and increase bioavailability (Zeng et al., 2021; Fonseca-Santos et al., 2015).

- ***Metal Oxide Nanoparticles:*** NPs composed of metal oxides such as titanium dioxide (TiO2) and zinc oxide (ZnO) are used in catalysis and environmental applications (McNamara & Tofail, 2017; Khan et al., 2017).
- ***Carbon Based Nanoparticles:*** NPs composed of carbon atoms such as fullerene, graphene, and carbon nanotubes are notable for their electrical and thermal conductivity (Rizwan et al., 2021; Ealias & Saravanakumar, 2017). These nanomaterials are used for drug delivery and biomarker detection and are notable for their ability to cross the blood-brain barrier (Bilal et al., 2020; Karthivashan et al., 2018).
- ***Bimetallic or Alloy Nanoparticles:*** These NPs, which are a combination of two different metals, generally offer superior magnetic and catalytic properties (McNamara & Tofail, 2017; Hasan, 2015).
- ***Magnetic Nanoparticles*****:** Superparamagnetic iron oxide NPs (SPIONs), in particular, are used for both diagnostic and therapeutic purposes, functioning as theranostic agents for magnetic resonance imaging (MRI) (Amiri et al., 2013). These NPs can also be used for medical imaging and diagnosis (Vissers et al., 2019; Mukherjee et al., 2020).

2.2. Role and Types of Nanoparticles in Neurodegenerative Diseases Applications

The treatment of NDs is challenging due to biological barriers such as the blood-brain barrier (BBB). NPs offer an innovative approach capable of overcoming these barriers and delivering drugs to the brain. In particular, when we look at the effects of NPs on AD;

- ***Drug delivery and targeting:*** NPs increase the bioavailability of drugs by enabling them to cross the blood-brain barrier and reach targeted sites in the brain. This can improve the efficacy of drugs used in treatment (Cano et al., 2019; Zheng et al., 2017; Gupta et al., 2019).
- ***Reduction of amyloid beta (Aβ) plaques*****:** NPs can slow the progression of AD by reducing the accumulation of Aβ plaques and inhibiting Aβ fibrillation. For example, gold NPs and molybdenum disulfide NPs have been found effective in inhibiting Aβ aggregation (Han et al., 2017; Sanati et al., 2019; Hou et al., 2020).
- ***Neuroprotective effects:*** NPs may exert neuroprotective effects by reducing oxidative stress and protecting synapses. This may contribute to improving cognitive functions in Alzheimer's model animals (Cano et al., 2019; Han et al., 2017; Sun et al., 2016).
- ***Multifunctional therapeutic approaches:*** NPs can carry multiple therapeutic agents and thus target different pathological processes of AD simultaneously. For example, GM1-modified lipoprotein-like NPs can protect neurons while accelerating Aβ clearance (Huang et al., 2015). Reduction of Aβ plaques, neuroprotective effects, and multifunc-

tional therapeutic approaches suggest that NPs may be a promising tool in Alzheimer's treatment.

- ***Crossing the Blood-Brain Barrier:*** The blood-brain barrier is an important structure that maintains brain homeostasis and controls molecule movement. However, disruption of this barrier can lead to NDs. NPs have the capacity to transport drugs to the brain by crossing the blood-brain barrier and thus show promise in the treatment of NDs (Saraiva et al., 2016; Mukherjee et al., 2020; Asil et al., 2020).
- ***Types and applications of nanoparticles;*** Various types of NPs such as metal, polymeric, lipid-based, and hydrogel NPs are used in the diagnosis and treatment of NDs. These NPs offer functions such as drug delivery to brain tissue, regulating the cellular microenvironment, and increasing the efficacy of stem cell therapies (Vissers et al., 2019; Asefy et al., 2021). Cerium oxide NPs (CeONPs) stand out as a potential nanomedical approach in the treatment of NDs thanks to their antioxidant properties (Naz et al., 2017). The challenges faced in the transition of NPs to clinical applications include safety, toxicity, and efficacy. Further research and development is required to overcome these challenges (Mukherjee et al., 2020; Asil et al., 2020).
- ***Nanoparticle applications in AD*** offer significant potential in the diagnosis and treatment of the disease. NPs enable drugs to reach the brain effectively by crossing the blood-brain barrier and thus increase the distribution of therapeutic compounds (Gupta et al., 2019; Bilal et al., 2020; Karthivashan et al., 2018; Harilal et al., 2019). In its treatment; it shows promise in Alzheimer's treatment by increasing the effectiveness of drugs and reducing side effects. Furthermore, reconstituting existing drugs and delivering them via nanoparticle platforms may improve treatment efficacy (Chakraborty et al., 2024; Liu & Shen, 2023). NPs play an important role in the diagnosis and treatment of AD. Different types of NPs are used in both diagnosis and treatment processes, increasing the effectiveness of drugs by crossing the blood-brain barrier. Advances in this field may contribute to the development of more effective treatment methods in the future.
- ***Nanoparticle applications in PD are*** used to ensure effective delivery of drugs to brain tissue and alleviate the symptoms of the disease. Some examples of nanoparticle applications in PD: Puerarin-loaded NPs puerarin (PU)-loaded NPs increase the bioavailability of drugs, allowing them to reach the brain better. These NPs have been found effective in improving PD symptoms (Chen et al., 2019). Cu xO nanoparticle clusters reduce neurotoxicity by scavenging reactive oxygen species (ROS) and preventing memory loss in mice with PD (Hao et al., 2019). Iron chelation NPs; chelation NPs targeting iron accumulation help protect dopaminergic neurons in PD and prevent loss of motor control (Wang et al., 2017). Pramipexole dihydrochloride-loaded chitosan NPs improve PD symptoms by providing direct drug delivery to the brain via the nose (Raj et al., 2018). Biomimetic NPs Biomimetic (Cu2-xSe-PVP-Qe) NPs reduce neuroinflammation and improve PD symptoms by targeting microglia cells (Liu et al., 2020). Polymeric NPs are used as brain-targeted drug and gene delivery systems and enhance therapeutic efficacy in PD (Virameteekul et al., 2024; Torres-Ortega et al., 2019). Nanoparticle technology offers promising strategies to alleviate the symptoms of the disease and prevent neurodegeneration (Baskin et al., 2020; Leyva-Gómez et al., 2015; Paul & Yadav, 2020). Nanoparticle applications increase treatment efficacy and alleviate the symptoms of PD by ensuring effective delivery of drugs to the brain.

2.3. Which Signaling Pathways Nanoparticles Target in Neurodegenerative Diseases

In NDs, NPs are thought to be able to slow or halt disease progression by targeting various signaling pathways. These signaling pathways include mechanisms such as inflammation, apoptosis, oxidative stress, and protein misfolding, which often play an important role in neurodegenerative processes. When we examine some of the targeted signaling pathways;

- ***PI3K/Akt/mTOR pathway:*** This signaling pathway plays a critical role in apoptosis, inflammation, and oxidative stress processes. By targeting this pathway, NPs such as exosomes can deliver drugs to the central nervous system and offer potential therapeutic effects in the treatment of NDs (Iranpanah et al., 2023).
- ***NO/cGMP/CREB pathway***: Multifunctional NPs used in the treatment of AD support synaptic plasticity and memory improvement by activating the NO/cGMP/CREB pathway (Liu et al., 2021).
- ***Wnt/β-catenin pathway:*** In a model of AD, Curcumin-loaded NPs activate this pathway, enhancing neurogenesis and reversing significant cognitive impairments (Tiwari et al., 2014).
- ***ROS and Ca2+ signaling pathways:*** Photothermally conjugated polymer NPs, which are effective in modulating reactive oxygen species (ROS) and calcium (Ca2+) signaling pathways, reduce neuroinflammation (Li et al., 2022).
- ***BDNF pathway:*** Ceramic oxide NPs, which protect against free radicals, promote neuronal survival by modulating the brain-derived neurotrophic factor (BDNF) signaling pathway (Cimini et al., 2012).
- ***NF-κB/JNK/GSK3β Pathway:*** Anthocyanin-loaded NPs reduce neuroinflammation and neuroapoptosis by inhibiting this pathway (Kim et al., 2017).

2.4. Positive and Negative Effects, Challenges, and Advantages of Using Nanoparticles

NPs have significant potential in the treatment of NDs. Their ability to cross the blood-brain barrier and their capacity for targeted drug delivery make this technology promising. However, challenges such as safety and toxicity need to be overcome. Theranostic applications of NPs could revolutionize the management of NDs by combining both therapeutic and diagnostic capabilities (Ramanathan et al., 2018). NPs offer an innovative approach to the treatment of NDs, providing significant advantages in crossing the blood-brain barrier and targeted drug delivery.

- ***The blood-brain barrier (BBB)*** largely prevents the entry of foreign substances into the brain, making it difficult for therapeutic drugs to reach it effectively. This is one of the biggest challenges in the treatment of NDs (Lamptey et al., 2022; Ramanathan et al., 2018; Asil et al., 2020). Although NPs have the potential to overcome this barrier and enable drugs to reach the brain, there are still significant technical and biological barriers (Cano et al., 2020; Cunha et al., 2021; Mukherjee et al., 2020). Successful results obtained in animal models are difficult to transfer to human clinical applications. Further research is required to prove the clinical efficacy and safety of NPs (Cano et al., 2020; Benfenati & Lanzani, 2020).

- ***Targeted drug delivery;*** It can increase treatment efficacy by enabling direct delivery of drugs to targeted brain regions. It can play an important role in the treatment of Parkinson's, Alzheimer's, and other NDs (Cano et al., 2020; Ramanathan et al., 2018; Rafe, 2024). The targeting accuracy of NPs must be optimized to improve treatment efficacy. This is critical to ensure that drugs reach the right cells and brain regions (Vissers et al., 2019; Pirhaghi et al., 2024).
- ***Antioxidant properties;*** Especially cerium oxide NPs can be used in the treatment of NDs by scavenging reactive oxygen species. Thanks to these properties, these NPs have the potential to revolutionize existing treatments (Naz et al., 2017).
- ***Safety and toxicity:*** Safety and toxicity issues are among the major challenges in the transition of NPs to clinical applications. Solving these issues is critical for NPs to be used more widely (Mukherjee et al., 2020; Asil et al., 2020). Transferring the successes achieved in preclinical models to clinical applications involves technological and scientific challenges. Achieving this transition could revolutionize the treatment of NDs (Cano et al., 2020; Mukherjee et al., 2020). It is used in areas such as drug delivery, imaging, biosensors, and tissue engineering (McNamara & Tofail, 2017; Harish et al., 2022; Wang & Wang, 2014). Environmental applications offer environmental solutions such as pollution control and water treatment (Liu, 2006; Ealias & Saravanakumar, 2017). Industrial applications include catalysis, energy storage, and electronic devices (Khan et al., 2017; Hasan, 2015). Different types such as metal, polymeric, lipid-based, and hydrogel NPs can be used in the diagnosis and treatment of NDs (Vissers et al., 2019; Mukherjee et al., 2020; Rafe, 2024). Whether NPs damage nerve cells varies depending on the type of nanoparticle used and the route of administration. The biocompatibility and long-term effects of different types of NPs are not fully understood (Mukherjee et al., 2020; Guo & Yi, 2023). In general, it has been reported that NPs can have both positive and negative effects on nerve cells.
- ***Adverse effects;*** Aluminum oxide NPs (Nano-Al2O3) can cause oxidative damage, inflammatory responses, and histopathological damage by reaching the brain. Furthermore, adverse effects such as mitochondrial dysfunction and apoptosis have been observed in dopaminergic neurons (Liu et al., 2020). Zinc oxide NPs (Nano-ZnO) can damage nervous system functions by inducing oxidative damage and inflammatory responses in the brain (Liu et al., 2019). Silver NPs (AgNPs) can cause neurotoxicity in nerve cells, especially at low concentrations These NPs can lead to oxidative stress and glutamate excitotoxicity (Begum et al., 2016). The general mechanisms of neurotoxicity can lead to neurotoxicity by damaging the cell membrane, disrupting the cell cycle, and causing the formation of reactive oxygen species (ROS) (Lojk et al., 2020).
- ***Positive effects;*** Serum oxide anoparticles can protect nerve cells by protecting against oxidative stress (Schubert et al., 2006). Bioactive composite NPs can show positive effects in the treatment of brain injuries by promoting the survival and differentiation of nerve cells (Yuan et al., 2022). The effect of NPs on nerve cells varies depending on the particle type, coating material, and application dose. While some NPs have neurotoxic effects, others may show neuroprotective properties. Therefore, the use of NPs needs to

be carefully evaluated. Various treatment approaches have been developed for these diseases, but most of them only alleviate symptoms.

- ***Blood circulation and targeting:*** NPs need to overcome biological barriers to remain in the bloodstream for a long time and reach the target tissue. In this process, there are risks of NPs being eliminated by the immune system (Thomas & Weber, 2019; Barua & Mitragotri, 2014).
- ***Tissue and cell passage:*** Effective penetration of NPs into target tissues and cells is very difficult. It involves steps such as crossing the endothelial layer and then spreading in the tissue interstitium and entering the cell (Barua & Mitragotri, 2014).
- ***Technical and manufacturing challenges in design and production:*** The complex nature of NPs requires meticulous design and engineering knowledge. Scaling and manufacturing processes must be carefully controlled to ensure consistency in the production process (Desai, 2012). Many factors such as the safety and efficacy of NPs, bioavailability, targeting, and potential immune toxicities have the potential to be affected by small changes (Desai, 2012).
- ***Clinical application and approval processes;*** Most nanoparticle-based drug designs are still limited to in vivo and in vitro research. Clinical approval processes are more complex compared to traditional drug development stages (Gavas et al., 2021; Desai, 2012).
- ***Regulatory shortcomings***; The lack of regulatory standards for the demonstration and evaluation of NPs as a unique therapeutic agent candidate slows the development and approval processes considerably (Desai, 2012).

2.5. Utilization of Nanoparticles in Neurodegenerative Diseases and Various Innovative Therapeutic Modalities

NPs offer innovative solutions in many areas of medicine such as drug delivery, imaging, and cancer and vascular disease treatment. These particles have the potential to reduce side effects while increasing treatment efficacy, making them a step ahead of other options. However, issues such as toxicity and environmental impacts need to be carefully considered. NPs can promote neuronal tissue regeneration and neuroprotection by enhancing the transport of neuroprotective drugs into the brain (Cunha et al., 2021; Mukherjee et al., 2020). Although NDs are on the rise with the aging world population and present with symptoms affecting memory, cognition, and movement, studies on various innovative treatment modalities are ongoing. Antioxidant therapies, mitochondrial function and oxidative oxidative damage play an important role in NDs. Antioxidant molecules are used to combat this stress Antioxidant molecules are used to combat this stress. Antioxidants such as polyphenols, carotenes, vitamins, and melatonin are used as neuroprotective agents in conditions such as Alzheimer's, Parkinson's, and Huntington's diseases (Morén et al., 2022; Rekatsina et al., 2020). Natural pharmaceutical ingredients are growing in popular as an alternative to synthetic drugs due to their neuroprotective properties and biocompatibility. However, clinical trials of these compounds are limited (Andrade et al., 2023). Therapeutic approaches for NDs have not gone beyond alleviating symptoms. Various strategies such as antioxidants, nanotechnology, drug repositioning, gene therapy, and natural compounds have shown promise in the treatment of these diseases. However, more research is needed to improve the effectiveness of these approaches. NPs are innovative materials that offer a variety of applications in medicine. In particular, these particles play an important role in drug delivery, imaging, cancer therapy, and treatment of vascular diseases. NPs enable the delivery of drugs directly to targeted sites, increasing treatment efficacy and reducing side effects. This is particularly important in cancer treatment (McNamara & Tofail, 2017; Aghebati-Maleki et al., 2019; Anselmo & Mitragotri, 2016; Joseph et al., 2023; Flores et al., 2019; Yetisgin et al., 2020)

In imaging and diagnosis, NPs are used as contrast agents in diagnostic methods such as magnetic resonance imaging (MRI). This provides clearer and more accurate imaging (McNamara & Tofail, 2017; Anselmo & Mitragotri, 2016; Sanvicens & Marco, 2008).

In cancer treatment, NPs are used in cancer therapy by methods such as gene and drug delivery to cancer cells and photodynamic therapy. This reduces the risk of damage to healthy cells while increasing treatment efficacy (Aghebati-Maleki et al., 2019; Joseph et al., 2023; Sanvicens & Marco, 2008; Ziental et al., 2020).

In vascular diseases, NPs are used to treat vascular diseases such as atherosclerosis, thrombosis, and restenosis. This allows drugs to reach the diseased area directly, increasing treatment efficacy (Flores et al., 2019).

The repurposing of existing drugs for NDs is advantageous as the pharmacokinetic and pharmacodynamic profiles of these drugs are already known. This strategy offers potential treatments for diseases such as Alzheimer's, Parkinson's, Huntington's, Multiple Sclerosis, and ALS (Durães et al., 2018).

Gene therapy can provide therapeutic benefits through direct correction of pathogenic mechanisms of disease, neuroprotection, and symptom control. Studies on novel vectors and gene delivery routes are ongoing (Sudhakar & Richardson, 2019; Chen et al., 2020; Sun & Roy, 2021). Viral vector-based gene therapies, which have shown success in diseases such as spinal muscular atrophy, are also promising for other NDs (Sun & Roy, 2021; Chen et al., 2020; Sudhakar & Richardson, 2019; O'Connor & Boulis, 2015).

Neural stem cell transplantation may provide benefits against NDs through mechanisms such as neurotrophic factor production, reduction of neuroinflammation and cell regeneration. This treatment strategy focuses on the integration of stem cells in the central nervous systems and their potential to enhance neuroplasticity (De Gioia et al., 2020; Bonaventura et al., 2021).

Immunotherapies involve the use of the immune system or its derivatives to fight misfolded and accumulated proteins. Monoclonal antibodies and specific antigens trigger adaptive immune responses and studies are ongoing (Mortada et al., 2021). Furthermore, antimicrobial applications of metal oxide NPs such as titanium dioxide are used in the photodynamic inactivation of antibiotic-resistant bacteria (Ziental et al., 2020).

3. CHARACTERISTICS AND TYPES OF SECONDARY METABOLITES

Secondary metabolites are known as small organic molecules produced by organisms that are generally not essential for growth or survival, but provide advantages under certain conditions. These metabolites are particularly common in organisms such as fungi, bacteria, and plants and exhibit a variety of biological activities (Scharf et al., 2014; Bills & Gloer, 2016; Pagare et al., 2015).Secondary metabolites are often implicated in the organism's interaction with its environment. In plants, they provide protection against biotic or abiotic stresses (Pagare et al., 2015; Hadacek, 2002). Fungi can use these metabolites in pathogenic interactions or communication with other microorganisms (Pusztahelyi et al., 2015; Macheleidt et al., 2016). Secondary metabolites show a wide variety of chemical structures and biological activities. This diversity offers great potential for drug discovery and synthetic biology (Bills & Gloer, 2016; O'Connor, 2015).

Secondary metabolites are used in applications as drugs, food preservatives, biopesticides, and research tools. However, some of them can be toxic and therefore their use may be limited (Madariaga-Mazón et al., 2019). Fungal antibiotics such as penicillin are important examples of secondary metabolites (Bills & Gloer, 2016). In plants, secondary metabolites function as defense mechanisms against insects and microbes (Hadacek, 2002; Pusztahelyi et al., 2015).

Mycotoxins, toxic secondary metabolites produced by some fungi, can cause contamination in food and feed (Bills & Gloer, 2016). Secondary metabolites are compounds with a wide chemical and biological diversity that play important roles in the interactions of organisms with their environment. These metabolites are characterized by both beneficial and harmful effects and are used in different industrial and medical applications.

Food-derived secondary metabolites, plant foods such as fruits and vegetables, contain secondary metabolites such as terpenoids, polyphenols and phytosterols. These compounds provide inhibitory effects on the development of chronic diseases associated with oxidative stress (Gomes et al., 2024). Further investigation of these plant-derived compounds contributes to the development of new therapeutic strategies.

3.1. Role and Impact of Secondary Metabolites in Neurodegenerative Diseases

Secondary metabolites are being explored as possible new therapeutic agents in the treatment of NDs. Secondary metabolites used in NDs are various compounds derived from plants, fungi, and marine sponges. These metabolites offer potential benefits in the treatment of NDs. We briefly look at the role and impact of the secondary metabolites used in NDs;

Alkaloids; and alkaloid derivatives such as galantamine and huperzine A are used in the symptomatic management of NDs (Kong et al., 2021).

Polyphenols have attracted much attention for their antioxidant properties in neurodegenerative disorders such as AD (Vicente-Zurdo et al., 2024). Furthermore, consumption of polyphenol-rich foods plays a role in the protection against chronic diseases (Vicente-Zurdo et al., 2024)

Terpenoids, polyphenols, glucosinolates, glucosinolates, phytosterols, and capsaicinoids are found in foods of plant origin and are considered as therapeutic and preventive agents for the prevention of NDs (Gomes et al., 2024).

Biflavonoids; Biflavonoids, which have the potential to alter tau and amyloid beta aggregation in AD, may be effective in reducing neurodegenerative events (Uddin et al., 2019).

Anthracyanins and their metabolites are being evaluated as novel therapeutic agents for the treatment of NDs (Winter & Bickford, 2019). Prenylate flavonoids and aryl benzofurans; compounds from morus species have shown neuroprotective effects in the management of AD (Hegazi et al., 2024).

Compounds such as alkaloids, polyphenols, terpenoids and flavonoids offer potential benefits in alleviating the symptoms and slowing the progression of these diseases. The multifaceted effects of these metabolites may offer an effective strategy against the complex pathophysiology of NDs. These metabolites are natural products derived from plants and can affect some neurodegenerative processes through their diverse biological activities. Some of them are as follows;

These metabolites are being evaluated as compounds with clinical utility in protein folding disorders and central nervous system conditions (Joyner & Cichewicz, 2011). Antioxidant and anti-inflammatory effects may slow the progression of neurodegenerative processes by targeting these processes in diseases such as Alzheimer's and Parkinson's, where oxidative stress and inflammation play an important role (Cassano et al., 2020; Fakhri et al., 2021). Biflavonoids and tau/amyloid beta aggregation, such as tau and amyloid beta aggregation in AD, can form non-toxic aggregates by altering tau and amyloid beta aggregation and thus reduce neurodegenerative events (Uddin et al., 2019).

Metabolites are common to NDs and metabolites: Metabolites that are commonly altered in NDs such as AD, PD, and ALS include uric acid, choline, creatine, L-glutamine, alanine, creatinine, and N-acetyl-L-aspartate. These metabolites provide important clues in understanding the pathophysiology of diseases (Kori et al., 2016; Kori et al., 2016; Lanznaster et al., 2022). Alanine, aspartate, glutamate, and purine metabolism may function as alternative pathways to overcome insufficient glucose supply and energy crises during neurodegeneration (Kori et al., 2016; Kori et al., 2016). These metabolites may exert neuroprotective effects by targeting signaling pathways such as PI3K/Akt/mTOR (Fakhri et al., 2021; Scotti & Scotti, 2015). It can modulate neurodegenerative processes such as oxidative stress and protein aggregation (Winter & Bickford, 2019; Joyner & Cichewicz, 2011; Scotti & Scotti, 2015). Metabolomics research plays an essential part in understanding the pathophysiology of these diseases and in the development of new therapeutic strategies. Terpenoids, polyphenols, and other secondary metabolites found in plant-based foods may exert preventive effects on the development of NDs (Gomes et al., 2024).

3.2. Which Signaling Pathways Secondary Metabolites Target in Neurodegenerative Diseases

In NDs, secondary metabolites can slow or halt disease progression by targeting various signaling pathways. These signaling pathways play an important role in the pathogenesis of NDs and regulation of these pathways offers potential therapeutic strategies. Some targeted signaling pathways are as follows;

- ***PI3K/Akt/mTOR Pathway:*** Plant-derived secondary metabolites can be used in the treatment of NDs such as Alzheimer's and Parkinson's by targeting the PI3K/Akt/mTOR signaling pathway. Suppression of this pathway is considered a protective step in NDs (Fakhri et al., 2021).
- ***Toll-Like Receptor (TLR) Pathway:*** Polyphenols can reduce neuroinflammation by targeting the TLR signaling pathway, especially the TLR4/NF-κB/STAT pathway, and can be used in the treatment of NDs (Azam et al., 2019).
- ***Nrf2-ARE Pathway:*** Nrf2 has emerged as an important target in the treatment of NDs by reducing oxidative stress and neuroinflammation. Activation of this pathway strengthens cellular defense mechanisms (Buendía et al., 2016).
- ***Wnt/β-catenin pathway:*** This pathway is critical for nervous system development and homeostasis. Regulation of the Wnt/β-catenin pathway may offer new opportunities in the treatment of NDs (Serafino et al., 2020).

- ***Kynurenine Pathway:*** This pathway, which is part of tryptophan metabolism, is involved in the pathogenesis of NDs by producing neuroactive metabolites. Regulation of this pathway may alleviate disease symptoms (Maddison & Giorgini, 2015).
- ***JNK Pathway:*** The JNK signaling pathway contributes to neurodegeneration by leading to synaptic dysfunction and neuronal apoptosis. Targeting this pathway may offer potential therapeutic strategies (Zhao et al., 2022).
- ***Ubiquitin Signaling Pathway:*** The ubiquitin-proteasome system and the autophagy-lysosomal pathway may prevent the progression of NDs by maintaining protein homeostasis (Schmidt et al., 2021). Secondary metabolites may slow or stop the progression of NDs by targeting various signaling pathways. Regulation of these pathways offers potential therapeutic strategies by providing neuroprotective effects. Each pathway may interfere with neurodegenerative processes through different mechanisms, affecting the diversity of therapeutic approaches.

3.3. Challenges in the Treatment of Neurodegenerative Diseases with Secondary Metabolites

These challenges include the bioavailability of metabolites, the need for multiple targeting, and barriers to transition to clinical applications.

- ***Bioavailability and efficacy; the*** bioavailability of secondary metabolites, especially anthocyanins, is low, resulting in limited therapeutic effects (Winter & Bickford, 2019). Therefore, new formulations need to be developed to increase the efficacy of metabolites or their derivatives.
- ***The need for multiple targeting*** involves multiple pathological mechanisms such as NDs, oxidative stress, inflammation, and protein aggregation. Therefore, an effective therapeutic strategy should target more than one of these mechanisms (Winter & Bickford, 2019; Cassano et al., 2020; Fakhri et al., 2022). However, developing such multi-targeting strategies is complex and challenging.
- ***Transition to clinical applications;*** further research is required to prove the efficacy and safety of metabolites at the clinical level (Lanznaster et al., 2022; Joyner & Cichewicz, 2011). Furthermore, it is crucial to fully understand the pharmacokinetic and pharmacodynamic properties of metabolites (Alghamdi et al., 2022).
- ***Microbiota and brain interaction;*** The interaction between the gut microbiota and the brain plays an important part in the development of NDs. Effective use of metabolites targeting this interaction requires metabolomic analyses and microbiota regulatory strategies (Luan et al., 2019; Missiego-Beltrán & Beltrán-Velasco, 2024). As a result, the use of secondary metabolites in the treatment for NDs faces several obstacles, including bioavailability issues, the need for multiple targeting and difficulties in transitioning to clinical applications. Further research and innovative strategies are required to overcome these challenges. Polyphenols may slow the progression of NDs by influencing the gut microbiota. This effect is mediated through the microbiota-brain axis (Zhang et al., 2022).

4. SYNERGISTIC USE OF NANOPARTICLES AND SECONDARY METABOLITES IN NEURODEGENERATIVE DISEASES AND CURRENT CASE STUDIES

This combination could be particularly effective in brain-targeted drug delivery and treatment of diseases at the molecular level. Lipid NPs enable the transmission of innovative molecules to the brain for the treatment of NDs. Lipid NPs and cell-penetrating peptides are used to deliver innovative molecules to the brain for the treatment of NDs (Wu & Angelova, 2023). Iron oxide NPs are used as magnetic resonance imaging (MRI) contrast agents and help in the early diagnosis of diseases such as Alzheimer's (Luo et al., 2020). Cerium oxide NPs (CeONPs) have attracted attention as a potential nanomedicine in the treatment of NDs thanks to their ability to scavenge reactive oxygen species (Naz et al., 2017). Different nanoparticle types offer novel capabilities in the diagnosis and treatment of NDs (Vissers et al., 2019). However, the biological efficacy of NPs has been demonstrated in various preclinical animal models and studies in this field are progressing rapidly (Padmanabhan et al., 2020). In AD, small interfering RNAs (siRNA) and NPs can be used as a synergistic treatment approach in AD. This combination can reduce neurological damage by providing controlled siRNA release in target cells (Li et al., 2023).

In PD, synaptic vesicle-like NPs may offer a synergistic treatment by clearing toxic protein aggregates in PD (Ji et al., 2021). This combination could be particularly effective in brain-targeted drug delivery and treatment of diseases at the molecular level.

Secondary metabolites such as terpenoids, polyphenols, glucosinolates and phytosterols may be involved in the primary prevention of NDs thanks to their antioxidant and anti-Alzheimer's properties (Gomes et al., 2024). Polydatin may have a protective effect against diseases such as Alzheimer's and Parkinson's by modulating mechanisms such as neuroinflammation and oxidative stress (Fakhri et al., 2021). Cannabidiol (CBD) may slow the progression of Alzheimer's and PDs with its anti-inflammatory and antioxidant properties (Cassano et al., 2020).

Gene therapies, especially viral vector-based gene therapies, enable the widespread delivery of genes to the central nervous system (Sun & Roy, 2021; Sudhakar & Richardson, 2019; Chen et al., 2020). Antioxidant molecules are used in the treatment of NDs by reducing oxidative stress. Antioxidants such as polyphenols, carotenes, and vitamins are considered neuroprotective agents (Morén et al., 2022). Immunotherapies provide neuroprotective effects by targeting protein aggregates. These therapies aim to clear protein aggregates in Alzheimer's and other age-related NDs (Kwon et al., 2020). Passive and active vaccines made using monoclonal antibodies and specific antigens are innovative immunotherapeutic strategies being developed to fight neuroinflammation-induced neurodegeneration (Mortada et al., 2021). Transdifferentiation is the process of reprogramming somatic cells into another cell lineage and offers an alternative strategy for NDs. These methods hold promise for obtaining patient-specific neuronal cells and personalized medicine applications (Mollinari et al., 2018).

5. UTILIZATION AND ROLE OF EXTRACELLULAR VESICLES (EXOSOMES) IN NEURODEGENERATIVE DISEASES

A class of membrane particles released from cells, including exosomes, microvesicles, apoptotic bodies and other fragments (Mathieu et al., 2019; Van Niel, et al., 2018). It has a lipid bilayer structure and is known to be 50-150 nm in diameter. They are produced initially as intraluminal vesicles in multivesicular body during endosome maturation and released by fusion with plasma membranes. Recent studies suggest that secretion and transmission of exosomes are implicated in the initiation and advancement of various NDs (You et al., 2019). Disease-causing proteins such as tau and α-Syn have been proposed to be secreted and transmitted between cells via exosomes and may explain the widespread spread of disease protein accumulations in the brains of neurodegenerative disease patients. On the other hand, preventing the abnormal accumulation of disease-causing proteins, because they are thought to suppress the formation of intracellular aggregation or promote the extracellular purification of these proteins, leading to the maintenance of protein homeostasis at the level of the multicellular organism. They are currently thought to have two aspects, pathogenic and protective. Due to these two facets of exosome function, inhibition of their secretion leads not only to the anticipated therapeutic effects, but also to negative effects on cell survival. Exosomes play a critical function in intercellular communication and protein transport and are implicated in the transportation of proteins implicated in NDs (Howitt & Hill, 2016; Vella et al., 2016; Beatriz et al., 2021). When we examine the role and functions of exosomes;

Protein transport and disease spread may contribute to disease spread by facilitating the intercellular transport of pathological protein aggregates (e.g., beta-amyloid and alpha-synuclein) in diseases such as Alzheimer's and Parkinson's (Howitt & Hill, 2016; Vella et al., 2016; Beatriz et al., 2021).

Biomarker potential; these vesicles carry proteins and genetic material that reflect pathological changes in the disease (Alidoust & Jafari, 2020; Rastogi et al., 2021). These vesicles carry proteins that contribute to the spread of disease and can be used as biomarkers.

Due to their ability to cross the blood-brain barrier, exosomes can transport molecules targeted to the central nervous system (CNS) and are therefore seen as a promising tool in the treatment of NDs and can be used to transport therapeutic molecules targeted to the CNS (Gao et al, 2021; Jarmalavičiūtė & Pivoriūnas, 2016; Sarko & McKinney, 2017; He et al., 2023). can be found in cerebrospinal fluid, plasma, serum, saliva, and urine and can be used as valuable biomarkers for diagnosis. These biomarkers can reflect the pathological states of the disease and monitor disease progression (Luo et al., 2020; Wu et al., 2017; Liu et al., 2019). In Alzheimer's and PDs, exosomes transport biomolecules that contribute to disease progression (He et al., 2023; Zhang et al., 2021). Their role in pathogenesis is that exosomes can mediate the spread of misfolded and accumulated alpha-synuclein proteins throughout the brain. This is considered an important mechanism in the pathological progression of PD (Pinnell et al., 2020; Wu et al., 2017; Vella et al., 2016; Porro et al., 2019). Exosomes released from glial cells contribute to the propagation of the neuroinflammatory response, and glia-neuron transfer of alpha-synuclein-containing exosomes can cause neurodegeneration (Izco et al., 2022; Porro et al., 2019).

In therapeutic applications, it provides effective delivery of drugs to the brain, especially in PD (Haney et al., 2015; Luo et al., 2020; Izco et al., 2022). Exosomes have shown neuroprotective effects by being used to transport antioxidants such as catalase. This suggests the potential of exosome-based strategies in the treatment of PD (Haney et al., 2015; Wu et al., 2017; Sun et al., 2020). While alpha-synuclein can mediate the spread and increase of neuroinflammation, they can also be used as biomarkers and drug transporters.

Exosomes (MSC-Exo), especially obtained from mesenchymal stem cells, are nano-sized extracellular vesicles that draw attention with their neuroprotective and immunomodulatory effects.

These exosomes have properties such as reducing neuroinflammation, promoting neurogenesis, and reducing apoptotic loss of neural cells (Gorabi et al., 2019). MSC-Exo shows promise in the treatment of NDs by reducing neuroinflammation and promoting neurogenesis (Gorabi et al., 2019). MSC-Exo may contribute to the recovery of cognitive functions by reducing neuroinflammation and preventing apoptotic loss of neural cells (Liu et al., 2019). MSC- Exosomes can influence pathological processes by enabling intracellular communication and transporting biological information over long distances and can therefore be used as biomarkers and drug carriers (Gorabi et al., 2019; Joo et al., 2020).

It has been shown that MSC-Exo can be made more effective by genetic modification, thereby enhancing its neuroprotective effects (Zhao et al., 2019). In conclusion, exosomes derived from mesenchymal stem cells are promising in the treatment of NDs and neuroinflammation with their neuroprotective and immunomodulatory effects. These exosomes have the potential to be used as biomarkers and drug carriers, and research is ongoing for their more widespread use in clinical applications.

CONCLUSION

NDs represent a major global health burden characterized by progressive neuronal loss, cognitive decline, and limited therapeutic options. The convergence of nanotechnology and bioactive secondary metabolites provides a transformative avenue for addressing the multifaceted mechanisms underlying these disorders. Nanoparticle-based systems offer significant advantages in overcoming biological barriers such as the blood–brain barrier, enabling targeted and controlled delivery of therapeutic agents to affected neuronal tissues. Concurrently, secondary metabolites derived from natural sources exhibit potent antioxidant, anti-inflammatory, and neuroprotective properties, contributing to the modulation of key molecular pathways implicated in neurodegeneration.

The synergistic integration of these two approaches holds substantial promise for enhancing therapeutic efficacy, minimizing systemic toxicity, and improving bioavailability. Nevertheless, several challenges remain, including the optimization of nanoparticle biocompatibility, long-term safety, regulatory standardization, and the limited translational success from preclinical to clinical settings. Future directions should emphasize the development of multifunctional, biodegradable nanocarriers capable of co-delivering neuroprotective compounds and genetic materials, while incorporating precision-targeting mechanisms guided by molecular diagnostics and biomarker profiling.

Ultimately, advancing an integrative nanomedicine framework that harmonizes nanotechnology, molecular neuroscience, and natural product pharmacology may redefine current therapeutic paradigms and contribute to the realization of personalized, mechanism-based interventions for NDs.

REFERENCES

Al-Chalabi, A., Berg, L., & Veldink, J. (2017). Gene discovery in amyotrophic lateral sclerosis: Implications for clinical management. *Nature Reviews. Neurology*, *13*(2), 96–104. DOI: 10.1038/nrneurol.2016.182 PMID: 27982040

Alghamdi, S., Suliman, R., Aljammaz, N., Kahtani, K., Aljatli, D., & Albadrani, G. (2022). Natural products as novel neuroprotective agents: Computational predictions of the molecular targets, ADME properties, and safety profile. *Plants*, *11*(4), 549. DOI: 10.3390/plants11040549 PMID: 35214883

Alidoust, L., & Jafari, A. (2020). Exosomes: Future perspective in neurodegenerative diseases. *Caspian Journal of Neurological Sciences*, *6*(23), 251–258. DOI: 10.32598/cjns.6.23.4

Amiri, H., Saeidi, K., Borhani, P., Manafirad, A., Ghavami, M., & Zerbi, V. (2013). Alzheimer's disease: Pathophysiology and applications of magnetic nanoparticles as MRI theranostic agents. *ACS Chemical Neuroscience*, *4*(11), 1417–1429. DOI: 10.1021/cn4001582 PMID: 24024702

Amor, S., Puentes, F., Baker, D., & Van der Valk, P. (2010). Inflammation in neurodegenerative diseases. *Immunology*, *129*(2), 154–169. DOI: 10.1111/j.1365-2567.2009.03225.x PMID: 20561356

Andrade, S., Nunes, D., Dabur, M., Ramalho, M., Pereira, M., & Loureiro, J. (2023). Therapeutic potential of natural compounds in neurodegenerative diseases: Insights from clinical trials. *Pharmaceutics*, *15*(1), 212. DOI: 10.3390/pharmaceutics15010212 PMID: 36678841

Anselmo, A. C., & Mitragotri, S. (2016). Nanoparticles in the clinic. *Bioengineering & Translational Medicine*, *1*(1), 10–29. DOI: 10.1002/btm2.10003 PMID: 29313004

Asefy, Z., Hoseinnejhad, S., & Ceferov, Z. (2021). Nanoparticles approaches in neurodegenerative diseases diagnosis and treatment. *Neurological Sciences*, *42*, 2653–2660. DOI: 10.1007/s10072-021-05234-x PMID: 33846881

Asil, S., Ahlawat, J., Barroso, G., & Narayan, M. (2020). Nanomaterial based drug delivery systems for the treatment of neurodegenerative diseases. *Biomaterials Science*. Advance online publication. DOI: 10.1039/d0bm00809e PMID: 32638706

Balestrino, R., & Schapira, A. (2019). Parkinson disease. *European Journal of Neurology*, *27*. Advance online publication. DOI: 10.1111/ene.14108 PMID: 31631455

Ball, N., Teo, W., Chandra, S., & Chapman, J. (2019). Parkinson's disease and the environment. *Frontiers in Neurology*, *10*. Advance online publication. DOI: 10.3389/fneur.2019.00218 PMID: 30941085

Ballard, C., Gauthier, S., Corbett, A., Brayne, C., Aarsland, D., & Jones, E. (2011). Alzheimer's disease. *Lancet*, *377*, 1019–1031. DOI: 10.1016/S0140-6736(10)61349-9 PMID: 21371747

Barua, S., & Mitragotri, S. (2014). Challenges associated with penetration of nanoparticles across cell and tissue barriers: A review of current status and future prospects. *Nano Today*, *9*(2), 223–243. DOI: 10.1016/j.nantod.2014.04.008 PMID: 25132862

Baskin, J., Jeon, J., & Lewis, S. (2020). Nanoparticles for drug delivery in Parkinson's disease. *Journal of Neurology, 268*, 1981–1994. DOI: 10.1007/s00415-020-10291-x PMID: 33141248

Beatriz, M., Vilaça, R., & Lopes, C. (2021). Exosomes: Innocent bystanders or critical culprits in neurodegenerative diseases. *Frontiers in Cell and Developmental Biology, 9*. Advance online publication. DOI: 10.3389/fcell.2021.635104 PMID: 34055771

Begum, A., Aguilar, J., Elias, L., & Hong, Y. (2016). Silver nanoparticles exhibit coating and dose-dependent neurotoxicity in glutamatergic neurons derived from human embryonic stem cells. *Neurotoxicology, 57*, 45–53. DOI: 10.1016/j.neuro.2016.08.015 PMID: 27593553

Benfenati, F., & Lanzani, G. (2020). Clinical translation of nanoparticles for neural stimulation. *Nature Reviews. Materials, 6*, 1–4. DOI: 10.1038/s41578-020-00267-8

Bhat, S., Acharya, U., Hagiwara, Y., Dadmehr, N., & Adeli, H. (2018). Parkinson's disease: Cause factors, measurable indicators, and early diagnosis. *Computers in Biology and Medicine, 102*, 234–241. DOI: 10.1016/j.compbiomed.2018.09.008 PMID: 30253869

Bilal, M., Barani, M., Sabir, F., Rahdar, A., & Kyzas, G. (2020). Nanomaterials for the treatment and diagnosis of Alzheimer's disease: An overview. *NanoImpact, 20*, 100251. DOI: 10.1016/j.impact.2020.100251

Bills, G., & Gloer, J. (2016). Biologically active secondary metabolites from the fungi. *Microbiology Spectrum, 4*. Advance online publication. DOI: 10.1128/microbiolspec.funk-0009-2016 PMID: 27809954

Blennow, K., Leon, M., & Zetterberg, H. (2006). Alzheimer's disease. *Lancet, 368*, 387–403. DOI: 10.1016/S0140-6736(06)69113-7 PMID: 16876668

Bloem, B., Okun, M., & Klein, C. (2021). Parkinson's disease. *Lancet, 397*, 2284–2303. DOI: 10.1201/b12952-20 PMID: 33848468

Bonaventura, G., Munafò, A., Bellanca, C., La Cognata, V., Iemmolo, R., Attaguile, G., Di Mauro, R., Di Benedetto, G., Cantarella, G., Barcellona, M., Cavallaro, S., & Bernardini, R. (2021). Stem cells: Innovative therapeutic options for neurodegenerative diseases? *Cells, 10*. Advance online publication. DOI: 10.3390/cells10081992 PMID: 34440761

Buendía, I., Michalska, P., Navarro, E., Gameiro, I., Egea, J., & León, R. (2016). Nrf2–ARE pathway: An emerging target against oxidative stress and neuroinflammation in neurodegenerative diseases. *Pharmacology & Therapeutics, 157*, 84–104. DOI: 10.1016/j.pharmthera.2015.11.003 PMID: 26617217

Cano, A., Ettcheto, M., Chang, J., Barroso, E., Espina, M., Kühne, B., Barenys, M., Auladell, C., Folch, J., Souto, E., Camins, A., Turowski, P., & García, M. (2019). Dual-drug loaded nanoparticles of epigallocatechin-3-gallate (EGCG)/ascorbic acid enhance therapeutic efficacy of EGCG in an APPswe/PS1dE9 Alzheimer's disease mice model. *Journal of Controlled Release, 301*, 62–75. DOI: 10.1016/j.jconrel.2019.03.010 PMID: 30876953

Cano, A., Sánchez-López, E., Ettcheto, M., López-Machado, A., Espina, M., Souto, E., Galindo, R., Camins, A., García, M., & Turowski, P. (2020). Current advances in the development of novel polymeric nanoparticles for the treatment of neurodegenerative diseases. *Nanomedicine (London)*. Advance online publication. DOI: 10.2217/nnm-2019-0443 PMID: 32370600

Cao, Y., & Zhang, R. (2022). The application of nanotechnology in treatment of Alzheimer's disease. *Frontiers in Bioengineering and Biotechnology, 10*. Advance online publication. DOI: 10.3389/fbioe.2022.1042986 PMID: 36466349

Cassano, T., Villani, R., Pace, L., Carbone, A., Bukke, V., Orkisz, S., Avolio, C., & Serviddio, G. (2020). From Cannabis sativa to cannabidiol: Promising therapeutic candidate for the treatment of neurodegenerative diseases. *Frontiers in Pharmacology, 11*. Advance online publication. DOI: 10.3389/fphar.2020.00124 PMID: 32210795

Chakraborty, S., Vishwas, S., Harish, V., Gupta, G., Paudel, K., Dhanasekaran, M., Goh, B., Zacconi, F., De Jesus Andreoli Pinto, T., Kumbhar, P., Disouza, J., Dua, K., & Singh, S. (2024). Exploring nanoparticular platform in delivery of repurposed drug for Alzheimer's disease: Current approaches and future perspectives. *Expert Opinion on Drug Delivery, 21*(1), 1–22. DOI: 10.1080/17425247.2024.2414768 PMID: 39397403

Chen, T., Liu, W., Xiong, S., Li, D., Fang, S., Wu, Z., Wang, Q., & Chen, X. (2019). Nanoparticles mediating the sustained puerarin release facilitate improved brain delivery to treat Parkinson's disease. *ACS Applied Materials & Interfaces, 11*(45), 41828–41840. DOI: 10.1021/acsami.9b16047 PMID: 31638771

Chen, W., Hu, Y., & Ju, D. (2020). Gene therapy for neurodegenerative disorders: Advances, insights and prospects. *Acta Pharmaceutica Sinica. B, 10*(8), 1347–1359. DOI: 10.1016/j.apsb.2020.01.015 PMID: 32963936

Chiò, A., Mazzini, L., & Mora, G. (2020). Disease-modifying therapies in amyotrophic lateral sclerosis. *Neuropharmacology, 167*, 107986. Advance online publication. DOI: 10.1016/j.neuropharm.2020.107986 PMID: 32062193

Cimini, A., D'Angelo, B., Das, S., Gentile, R., Benedetti, E., Singh, V., Monaco, A., Santucci, S., & Seal, S. (2012). Antibody-conjugated PEGylated cerium oxide nanoparticles for specific targeting of Aβ aggregates modulate neuronal survival pathways. *Acta Biomaterialia, 8*(6), 2056–2067. DOI: 10.1016/j.actbio.2012.01.035 PMID: 22343002

Ciurea, A., Mohan, A., Covache-Busuioc, R., Costin, H., Glavan, L., Corlatescu, A., & Saceleanu, V. (2023). Unraveling molecular and genetic insights into neurodegenerative diseases: Advances in understanding Alzheimer's, Parkinson's, and Huntington's diseases and amyotrophic lateral sclerosis. *International Journal of Molecular Sciences, 24*(13), 10809. DOI: 10.3390/ijms241310809 PMID: 37445986

Cunha, A., Gaubert, A., Latxague, L., & Dehay, B. (2021). PLGA-based nanoparticles for neuroprotective drug delivery in neurodegenerative diseases. *Pharmaceutics, 13*(7), 1042. DOI: 10.3390/pharmaceutics13071042 PMID: 34371733

De Gioia, R., Biella, F., Citterio, G., Rizzo, F., Abati, E., Nizzardo, M., Bresolin, N., Comi, G., & Corti, S. (2020). Neural stem cell transplantation for neurodegenerative diseases. *International Journal of Molecular Sciences, 21*(9), 3103. DOI: 10.3390/ijms21093103 PMID: 32354178

De Virgilio, A., Greco, A., Fabbrini, G., Inghilleri, M., Rizzo, M., Gallo, A., Conte, M., Rosato, C., Appiani, C., & De Vincentiis, M. (2016). Parkinson's disease: Autoimmunity and neuroinflammation. *Autoimmunity Reviews, 15*(10), 1005–1011. DOI: 10.1016/j.autrev.2016.07.022 PMID: 27497913

Desai, N. (2012). Challenges in development of nanoparticle-based therapeutics. *The AAPS Journal, 14*(2), 282–295. DOI: 10.1208/s12248-012-9339-4 PMID: 22407288

DeTure, M., & Dickson, D. (2019). The neuropathological diagnosis of Alzheimer's disease. *Molecular Neurodegeneration, 14*(1), 32. DOI: 10.1186/s13024-019-0333-5 PMID: 31375134

Dugger, B., & Dickson, D. (2017). Pathology of neurodegenerative diseases. *Cold Spring Harbor Perspectives in Biology, 9*(7), a028035. DOI: 10.1101/cshperspect.a028035 PMID: 28062563

Durães, F., Pinto, M., & Sousa, E. (2018). Old drugs as new treatments for neurodegenerative diseases. *Pharmaceuticals, 11*(2), 44. DOI: 10.3390/ph11020044 PMID: 29751602

Ealias, A., & Saravanakumar, M. (2017). A review on the classification, characterization, synthesis of nanoparticles and their application. *IOP Conference Series. Materials Science and Engineering, 263*(3), 032019. DOI: 10.1088/1757-899X/263/3/032019

Emerich, D., & Thanos, C. (2003). Nanotechnology and medicine. *Expert Opinion on Biological Therapy, 3*(4), 655–663. DOI: 10.1517/14712598.3.4.655 PMID: 12831370

Erkkinen, M., Kim, M., & Geschwind, M. (2018). Clinical neurology and epidemiology of the major neurodegenerative diseases. *Cold Spring Harbor Perspectives in Biology, 10*(4), a033118. DOI: 10.1101/cshperspect.a033118 PMID: 28716886

Fahn, S. (2003). Description of Parkinson's disease as a clinical syndrome. *Annals of the New York Academy of Sciences, 991*(1), 1–14. DOI: 10.1111/j.1749-6632.2003.tb07458.x PMID: 12846969

Fakhri, S., Abbaszadeh, F., Moradi, S., Cao, H., Khan, H., & Xiao, J. (2022). Effects of polyphenols on oxidative stress, inflammation, and interconnected pathways during spinal cord injury. *Oxidative Medicine and Cellular Longevity, 8100195*. Advance online publication. DOI: 10.1155/2022/8100195 PMID: 35035667

Fakhri, S., Iranpanah, A., Gravandi, M., Moradi, S., Ranjbari, M., Majnooni, M., Echeverría, J., Qi, Y., Wang, M., Liao, P., Farzaei, M., & Xiao, J. (2021). Natural products attenuate PI3K/Akt/mTOR signaling pathway: A promising strategy in regulating neurodegeneration. *Phytomedicine, 91*, 153664. DOI: 10.1016/j.phymed.2021.153664 PMID: 34391082

Fan, L., Mao, C., Hu, X., Zhang, S., Yang, Z., Hu, Z., Sun, H., Fan, Y., Dong, Y., Yang, J., Shi, C., & Xu, Y. (2020). New insights into the pathogenesis of Alzheimer's disease. *Frontiers in Neurology, 10*, 1312. DOI: 10.3389/fneur.2019.01312 PMID: 31998208

Feldman, E., Goutman, S., Petri, S., Mazzini, L., Savelieff, M., Shaw, P., & Sobue, G. (2018). Amyotrophic lateral sclerosis. *Nature Reviews. Disease Primers*, *4*, 3. DOI: 10.1038/nrdp.2017.71

Ferrari, M., & Downing, G. (2005). Medical nanotechnology. *BioDrugs*, *19*(4), 203–210. DOI: 10.2165/00063030-200519040-00001 PMID: 16128604

Finkbeiner, S. (2011). Huntington's disease. *Cold Spring Harbor Perspectives in Biology*, *3*(6), a007476. DOI: 10.1101/cshperspect.a007476 PMID: 21441583

Flores, A., Ye, J., Jarr, K., Hosseini-Nassab, N., Smith, B., & Leeper, N. (2019). Nanoparticle therapy for vascular diseases. *Arteriosclerosis, Thrombosis, and Vascular Biology*, *39*(4), 635–646. DOI: 10.1161/ATVBAHA.118.311569 PMID: 30786744

Fonseca-Santos, B., Gremião, M., & Chorilli, M. (2015). Nanotechnology-based drug delivery systems for the treatment of Alzheimer's disease. *International Journal of Nanomedicine*, *10*, 4981–5003. DOI: 10.2147/IJN.S87148 PMID: 26345528

Gan, L., Cookson, M., Petrucelli, L., & La Spada, A. (2018). Converging pathways in neurodegeneration, from genetics to mechanisms. *Nature Neuroscience*, *21*(10), 1300–1309. DOI: 10.1038/s41593-018-0237-7 PMID: 30258237

Gao, H., & Hong, J. (2008). Why neurodegenerative diseases are progressive: Uncontrolled inflammation drives disease progression. *Trends in Immunology*, *29*(8), 357–365. DOI: 10.1016/j.it.2008.05.002 PMID: 18599350

Gao, P., Li, X., Du, X., Liu, S., & Xu, Y. (2021). Diagnostic and therapeutic potential of exosomes in neurodegenerative diseases. *Frontiers in Aging Neuroscience*, *13*, 790863. DOI: 10.3389/fnagi.2021.790863 PMID: 34975460

Gavas, S., Quazi, S., & Karpiński, T. (2021). Nanoparticles for cancer therapy: Current progress and challenges. *Nanoscale Research Letters*, *16*, 173. DOI: 10.1186/s11671-021-03628-6 PMID: 34866166

Ghasemi, M., & Brown, R. (2018). Genetics of amyotrophic lateral sclerosis. *Cold Spring Harbor Perspectives in Medicine*, *8*(5), a024125. DOI: 10.1101/cshperspect.a024125 PMID: 28270533

Gomes, T., Sousa, P., Campos, C., Perestrelo, R., & Câmara, J. (2024). Secondary bioactive metabolites from foods of plant origin as theravention agents against neurodegenerative disorders. *Foods*, *13*(14), 2289. DOI: 10.3390/foods13142289 PMID: 39063373

Gorabi, A., Kiaie, N., Barreto, G., Read, M., Tafti, H., & Sahebkar, A. (2019). The therapeutic potential of mesenchymal stem cell–derived exosomes in treatment of neurodegenerative diseases. *Molecular Neurobiology*, *56*(12), 8157–8167. DOI: 10.1007/s12035-019-01663-0 PMID: 31197655

Grad, L., Rouleau, G., Ravits, J., & Cashman, N. (2017). Clinical spectrum of amyotrophic lateral sclerosis (ALS). *Cold Spring Harbor Perspectives in Medicine*, *7*(8). Advance online publication. DOI: 10.1101/cshperspect.a024117 PMID: 28003278

Gujrati, M., Malamas, A., Shin, T., Jin, E., Sun, Y., & Lu, Z. R. (2014). Multifunctional cationic lipid-based nanoparticles facilitate endosomal escape and reduction-triggered cytosolic siRNA release. *Molecular Pharmaceutics*, *11*(8), 2734–2744. DOI: 10.1021/mp500171h PMID: 25020033

Guo, S., & Yi, C. (2023). Cell type-targeting nanoparticles in treating central nervous system diseases: Challenges and hopes. *Nanotechnology Reviews*, *12*(1), 20230051. DOI: 10.1515/ntrev-2023-0158

Guo, T., Zhang, D., Zeng, Y., Huang, T., Xu, H., & Zhao, Y. (2020). Molecular and cellular mechanisms underlying the pathogenesis of Alzheimer's disease. *Molecular Neurodegeneration*, *15*, 40. DOI: 10.1186/s13024-020-00391-7 PMID: 32677986

Gupta, J., Fatima, M., Islam, Z., Khan, R., Uversky, V., & Salahuddin, P. (2019). Nanoparticle formulations in the diagnosis and therapy of Alzheimer's disease. *International Journal of Biological Macromolecules*, *130*, 515–526. DOI: 10.1016/j.ijbiomac.2019.02.156 PMID: 30826404

Gusella, J., MacDonald, M., & Lee, J. (2014). Genetic modifiers of Huntington's disease. *Movement Disorders*, *29*(11), 1359–1365. DOI: 10.1002/mds.26001 PMID: 25154728

Hadacek, F. (2002). Secondary metabolites as plant traits: Current assessment and future perspectives. *Critical Reviews in Plant Sciences*, *21*(4), 273–322. DOI: 10.1080/0735-260291044269

Han, Q., Cai, S., Yang, L., Wang, X., Qi, C., Yang, R., & Wang, C. (2017). Molybdenum disulfide nanoparticles as multifunctional inhibitors against Alzheimer's disease. *ACS Applied Materials & Interfaces*, *9*(25), 21116–21123. DOI: 10.1021/acsami.7b03816 PMID: 28613069

Haney, M., Klyachko, N., Zhao, Y., Gupta, R., Plotnikova, E., He, Z., Patel, T., Piroyan, A., Sokolsky, M., Kabanov, A., & Batrakova, E. (2015). Exosomes as drug delivery vehicles for Parkinson's disease therapy. *Journal of Controlled Release*, *207*, 18–30. DOI: 10.1016/j.jconrel.2015.03.033 PMID: 25836593

Hao, C., Qu, A., Xu, L., Sun, M., Zhang, H., Xu, C., & Kuang, H. (2019). Chiral molecule-mediated porous CuxO nanoparticle clusters with antioxidation activity for ameliorating Parkinson's disease. *Journal of the American Chemical Society*, *141*(2), 1091–1099. DOI: 10.1021/jacs.8b11856 PMID: 30540450

Hardiman, O., Berg, L., & Kiernan, M. (2011). Clinical diagnosis and management of amyotrophic lateral sclerosis. *Nature Reviews. Neurology*, *7*(11), 639–649. DOI: 10.1038/nrneurol.2011.153 PMID: 21989247

Harilal, S., Jose, J., Parambi, D., Kumar, R., Mathew, G., Uddin, M., Kim, H., & Mathew, B. (2019). Advancements in nanotherapeutics for Alzheimer's disease: Current perspectives. *The Journal of Pharmacy and Pharmacology*, *71*(9), 1370–1393. DOI: 10.1111/jphp.13132 PMID: 31304982

Harish, V., Tewari, D., Gaur, M., Yadav, A., Swaroop, S., Bechelany, M., & Barhoum, A. (2022). Review on nanoparticles and nanostructured materials: Bioimaging, biosensing, drug delivery, tissue engineering, antimicrobial, and agro-food applications. *Nanomaterials (Basel, Switzerland)*, *12*(3), 457. DOI: 10.3390/nano12030457 PMID: 35159802

Hasan, S. (2015). A review on nanoparticles: Their synthesis and types. *Research Journal of Recent Sciences*, *4*, 1–3.

He, A., Wang, M., Li, X., Chen, H., Lim, K., Lu, L., & Zhang, C. (2023). Role of exosomes in the pathogenesis and theranostic of Alzheimer's disease and Parkinson's disease. *International Journal of Molecular Sciences*, *24*(13), 11054. DOI: 10.3390/ijms241311054 PMID: 37446231

Heemels, M. (2016). Neurodegenerative diseases. *Nature*, *539*(7628), 179. DOI: 10.1038/539179a PMID: 27830810

Hegazi, N., Mohamed, T., Salama, A., Hamed, A., Saad, H., Saleh, I., Reda, E., Elsayed, A., Ibrahim, M., Paré, P., Efferth, T., & Hegazy, M. (2024). Molecular networking-guided investigation of the secondary metabolome of four Morus species and their in vivo neuroprotective potential for the mitigation of Alzheimer's disease. *Food & Function*, *15*(10), 5108–5127. DOI: 10.1039/d3fo05711a PMID: 38533683

Hou, K., Zhao, J., Wang, H., Li, B., Li, K., Shi, X., Wan, K., Ai, J., Lv, J., Wang, D., Huang, Q., Wang, H., Cao, Q., Liu, S., & Tang, Z. (2020). Chiral gold nanoparticles enantioselectively rescue memory deficits in a mouse model of Alzheimer's disease. *Nature Communications*, *11*, 4790. DOI: 10.1038/s41467-020-18525-2 PMID: 32963242

Hou, Y., Dan, X., Babbar, M., Wei, Y., Hasselbalch, S., Croteau, D., & Bohr, V. (2019). Ageing as a risk factor for neurodegenerative disease. *Nature Reviews. Neurology*, *15*(10), 565–581. DOI: 10.1038/s41582-019-0244-7 PMID: 31501588

Howitt, J., & Hill, A. (2016). Exosomes in the pathology of neurodegenerative diseases. *The Journal of Biological Chemistry*, *291*(50), 26589–26597. DOI: 10.1074/jbc.R116.757955 PMID: 27852825

Huang, M., Hu, M., Song, Q., Song, H., Huang, J., Gu, X., Wang, X., Chen, J., Kang, T., Feng, X., Jiang, D., Zheng, G., Chen, H., & Gao, X. (2015). GM1-modified lipoprotein-like nanoparticle: Multifunctional nanoplatform for the combination therapy of Alzheimer's disease. *ACS Nano*, *9*(11), 10801–10816. DOI: 10.1021/acsnano.5b03124 PMID: 26440073

Hulla, J., Sahu, S., & Hayes, A. (2015). Nanotechnology: History and future. *Human and Experimental Toxicology*, *34*(12), 1318–1321. DOI: 10.1177/0960327115603588 PMID: 26614822

Iranpanah, A., Kooshki, L., Moradi, S., Saso, L., Fakhri, S., & Khan, H. (2023). The exosome-mediated PI3K/Akt/mTOR signaling pathway in neurological diseases. *Pharmaceutics*, *15*(3), 1006. DOI: 10.3390/pharmaceutics15031006 PMID: 36986865

Izco, M., Carlos, E., & Álvarez-Erviti, L. (2022). The two faces of exosomes in Parkinson's disease: From pathology to therapy. *The Neuroscientist*, *28*(2), 180–193. DOI: 10.1177/1073858421990001 PMID: 33530851

Jarmalavičiūtė, A., & Pivoriūnas, A. (2016). Exosomes as a potential novel therapeutic tool against neurodegenerative diseases. *Pharmacological Research*, 113(Pt B), 816–822. https://doi.org/DOI: 10.1016/j.phrs.2016.02.002

Ji, W., Li, Y., Liu, R., Lu, Z., Liu, L., Shi, Z., Shen, J., & Zhang, X. (2021). Synaptic vesicle-inspired nanoparticles with spatiotemporally controlled release ability as a "nanoguard" for synergistic treatment of synucleinopathies. *Materials Horizons*, *8*(4), 1199–1206. DOI: 10.1039/D0MH01542C PMID: 34821912

Jimenez-Sanchez, M., Licitra, F., Underwood, B., & Rubinsztein, D. (2017). Huntington's disease: Mechanisms of pathogenesis and therapeutic strategies. *Cold Spring Harbor Perspectives in Medicine*, *7*(7), a024240. DOI: 10.1101/cshperspect.a024240 PMID: 27940602

Joseph, T., Mahapatra, D., Esmaeili, A., Piszczyk, Ł., Hasanin, M., Kattali, M., Haponiuk, J., & Thomas, S. (2023). Nanoparticles: Taking a unique position in medicine. *Nanomaterials (Basel, Switzerland)*, *13*(3), 574. DOI: 10.3390/nano13030574 PMID: 36770535

Joyner, M., & Cichewicz, R. (2011). Bringing natural products into the fold: Exploring the therapeutic lead potential of secondary metabolites for the treatment of protein-misfolding-related neurodegenerative diseases. *Natural Product Reports*, *28*(1), 26–47. DOI: 10.1039/c0np00017e PMID: 20927454

Juźwik, C., Drake, S., Zhang, Y., Paradis-Isler, N., Sylvester, A., Amar-Zifkin, A., Douglas, C., Morquette, B., Moore, C., & Fournier, A. (2019). microRNA dysregulation in neurodegenerative diseases: A systematic review. *Progress in Neurobiology*, *182*, 101664. DOI: 10.1016/j.pneurobio.2019.101664 PMID: 31356849

Kalia, L., & Lang, A. (2015). Parkinson's disease. *Lancet*, *386*(9996), 896–912. DOI: 10.1016/S0140-6736(14)61393-3 PMID: 25904081

Karthivashan, G., Ganesan, P., Park, S., Kim, J., & Choi, D. (2018). Therapeutic strategies and nano-drug delivery applications in management of ageing Alzheimer's disease. *Drug Delivery*, *25*(1), 307–320. DOI: 10.1080/10717544.2018.1428243 PMID: 29350055

Khan, I., Saeed, K., & Khan, I. (2017). Nanoparticles: Properties, applications and toxicities. *Arabian Journal of Chemistry*, *12*(7), 908–931. DOI: 10.1016/j.arabjc.2017.05.01

Kim, B. Y., Rutka, J. T., & Chan, W. C. W. (2010). Nanomedicine. *The New England Journal of Medicine*, *363*(25), 2434–2443. DOI: 10.1056/NEJMra0912273 PMID: 21158659

Kim, M., Réhman, S., Amin, F., & Kim, M. (2017). Enhanced neuroprotection of anthocyanin-loaded PEG-gold nanoparticles against Aβ1-42-induced neuroinflammation and neurodegeneration via the NF-κB/JNK/GSK3β signaling pathway. *Nanomedicine; Nanotechnology, Biology, and Medicine*, *13*(8), 2533–2544. DOI: 10.1016/j.nano.2017.06.022 PMID: 28736294

Knopman, D., Amièva, H., Petersen, R., Chételat, G., Holtzman, D., Hyman, B., Nixon, R., & Jones, D. (2021). Alzheimer disease. *Nature Reviews. Disease Primers*, *7*, 33. DOI: 10.1038/s41572-021-00269-y PMID: 33986301

Kong, Y., Tay, K., Su, Y., Wong, C., Tan, W., & Khaw, K. (2021). Potential of naturally derived alkaloids as multi-targeted therapeutic agents for neurodegenerative diseases. *Molecules (Basel, Switzerland)*, *26*(3), 728. DOI: 10.3390/molecules26030728 PMID: 33573300

Kori, M., Aydin, B., Unal, S., Arğa, K., & Kazan, D. (2016). Metabolic biomarkers and neurodegeneration: A pathway enrichment analysis of Alzheimer's disease, Parkinson's disease, and amyotrophic lateral sclerosis. *OMICS: A Journal of Integrative Biology*, *20*(11), 645–661. DOI: 10.1089/omi.2016.0106 PMID: 27828769

Kovacs, G. (2016). Molecular pathological classification of neurodegenerative diseases: Turning towards precision medicine. *International Journal of Molecular Sciences*, *17*(2), 189. DOI: 10.3390/ijms17020189 PMID: 26848654

Kovacs, G. (2019). Molecular pathology of neurodegenerative diseases: Principles and practice. *Journal of Clinical Pathology*, *72*(11), 725–735. DOI: 10.1136/jclinpath-2019-205952 PMID: 31395625

Kwon, S., Iba, M., Kim, C., & Masliah, E. (2020). Immunotherapies for aging-related neurodegenerative diseases—Emerging perspectives and new targets. *Neurotherapeutics; the Journal of the American Society for Experimental NeuroTherapeutics*, *17*(3), 935–954. DOI: 10.1007/s13311-020-00853-2 PMID: 32347461

Lamptey, R., Chaulagain, B., Trivedi, R., Gothwal, A., Layek, B., & Singh, J. (2022). A review of the common neurodegenerative disorders: Current therapeutic approaches and the potential role of nanotherapeutics. *International Journal of Molecular Sciences*, *23*(3), 1640. DOI: 10.3390/ijms23031851 PMID: 35163773

Lanznaster, D., Dingeo, G., Samey, R., Emond, P., & Blasco, H. (2022). Metabolomics as a crucial tool to develop new therapeutic strategies for neurodegenerative diseases. *Metabolites*, *12*(9), 859. DOI: 10.3390/metabo12090864 PMID: 36144268

Leyva-Gómez, G., Cortés, H., Magaña, J., Leyva-García, N., Quintanar-Guerrero, D., & Florán, B. (2015). Nanoparticle technology for treatment of Parkinson's disease: The role of surface phenomena in reaching the brain. *Drug Discovery Today*, *20*(7), 824–837. DOI: 10.1016/j.drudis.2015.02.009 PMID: 25701281

Li, B., Li, N., Chen, L., Ren, S., Gao, D., Geng, H., Fu, J., Zhou, M., & Xing, C. (2022). Alleviating neuroinflammation through photothermal conjugated polymer nanoparticles by regulating reactive oxygen species and Ca2+ signaling. *ACS Applied Materials & Interfaces*, *14*(46), 51433–51445. DOI: 10.1021/acsami.2c13322 PMID: 36268893

Li, J., Peng, H., Zhang, W., Li, M., Wang, N., Peng, C., Zhang, X., & Li, Y. (2023). Enhanced nose-to-brain delivery of combined small interfering RNAs using lesion-recognizing nanoparticles for the synergistic therapy of Alzheimer's disease. *ACS Applied Materials & Interfaces*, *15*(45), 51916–51929. DOI: 10.1021/acsami.3c08756 PMID: 37939350

Liu, H., Han, Y., Wang, T., Zhang, H., Xu, Q., Yuan, J., & Li, Z. (2020). Targeting microglia for therapy of Parkinson's disease by using biomimetic ultrasmall nanoparticles. *Journal of the American Chemical Society*, *142*(52), 21730–21742. DOI: 10.1021/jacs.0c09390 PMID: 33315369

Liu, H., Yang, H., Fang, Y., Li, K., Tian, L., Liu, X., Zhang, W., Tan, Y., Lai, W., Bian, L., Lin, B., & Xi, Z. (2019). Neurotoxicity and biomarkers of zinc oxide nanoparticles in main functional brain regions and dopaminergic neurons. *The Science of the Total Environment*, *689*, 141–152. DOI: 10.1016/j.scitotenv.2019.06.388 PMID: 31829301

Liu, H., Zhang, W., Fang, Y., Yang, H., Tian, L., Li, K., Lai, W., Bian, L., Lin, B., Liu, X., & Xi, Z. (2020). Neurotoxicity of aluminum oxide nanoparticles and their mechanistic role in dopaminergic neuron injury involving p53-related pathways. *Journal of Hazardous Materials*, *392*, 122312. DOI: 10.1016/j.jhazmat.2020.122312 PMID: 32105957

Liu, W. (2006). Nanoparticles and their biological and environmental applications. *Journal of Bioscience and Bioengineering*, *102*(1), 1–7. DOI: 10.1263/jbb.102.1 PMID: 16952829

Liu, W., Bai, X., Zhang, A., Huang, J., Xu, S., & Zhang, J. (2019). Role of exosomes in central nervous system diseases. *Frontiers in Molecular Neuroscience*, *12*, 240. DOI: 10.3389/fnmol.2019.00240 PMID: 31636538

Liu, Y., & Shen, Y. (2023). Applications of nanoparticles in Alzheimer's disease. *Journal of Alzheimer's Disease*, *92*(3), 1005–1028. DOI: 10.3233/JAD-230098 PMID: 37807779

Liu, Z., Liu, Q., Zhang, B., Liu, Q., Fang, L., & Gou, S. (2021). Blood–brain barrier permeable and NO-releasing multifunctional nanoparticles for Alzheimer's disease treatment: Targeting NO/cGMP/CREB signaling pathways. *Journal of Medicinal Chemistry*, *64*(24), 18110–18126. DOI: 10.1021/acs.jmedchem.1c01240 PMID: 34517696

Liu, Z., Zhou, T., Ziegler, A., Dimitrion, P., & Zuo, L. (2017). Oxidative stress in neurodegenerative diseases: From molecular mechanisms to clinical applications. *Oxidative Medicine and Cellular Longevity*, *2525967*. Advance online publication. DOI: 10.1155/2017/2525967 PMID: 28785371

Lojk, J., Repas, J., Veranič, P., Bregar, V., & Pavlin, M. (2020). Toxicity mechanisms of selected engineered nanoparticles on human neural cells in vitro. *Toxicology*, *434*, 152364. DOI: 10.1016/j.tox.2020.152364 PMID: 31927068

Luan, H., Wang, X., & Cai, Z. (2019). Mass spectrometry-based metabolomics: Targeting the crosstalk between gut microbiota and brain in neurodegenerative disorders. *Mass Spectrometry Reviews*, *38*(1), 22–33. DOI: 10.1002/mas.21553 PMID: 29130504

Luo, S., Cui, C., Zhu, M., Ju, W., Yang, Y., & Wang, X. (2020). Application of iron oxide nanoparticles in the diagnosis and treatment of neurodegenerative diseases with emphasis on Alzheimer's disease. *Frontiers in Cellular Neuroscience*, *14*, 21. DOI: 10.3389/fncel.2020.00021 PMID: 32184709

Macheleidt, J., Mattern, D., Fischer, J., Netzker, T., Weber, J., Schroeckh, V., Valiante, V., & Brakhage, A. (2016). Regulation and role of fungal secondary metabolites. *Annual Review of Genetics*, *50*, 371–392. DOI: 10.1146/annurev-genet-120215-035203 PMID: 27732794

Madariaga-Mazón, A., Hernandez-Alvarado, R., Noriega-Colima, K., Osnaya-Hernández, A., & Martínez-Mayorga, K. (2019). Toxicity of secondary metabolites. *Physical Sciences Reviews*, *4*(7), 20180116. DOI: 10.1515/psr-2018-0116

Maddison, D., & Giorgini, F. (2015). The kynurenine pathway and neurodegenerative disease. *Seminars in Cell & Developmental Biology*, *40*, 134–141. DOI: 10.1016/j.semcdb.2015.03.002 PMID: 25773161

Malik, S., Muhammad, K., & Waheed, Y. (2023). Nanotechnology: A revolution in modern industry. *Molecules (Basel, Switzerland)*, *28*(2), 661. DOI: 10.3390/molecules28020661 PMID: 36677717

Martin, J. (1984). Huntington's disease. *Neurology*, *34*(8), 1059. DOI: 10.1212/WNL.34.8.1059 PMID: 6146951

Masrori, P., & Van Damme, P. (2020). Amyotrophic lateral sclerosis: A clinical review. *European Journal of Neurology*, *27*(10), 1918–1929. DOI: 10.1111/ene.14393 PMID: 32526057

Masters, C., Bateman, R., Blennow, K., Rowe, C., Sperling, R., & Cummings, J. (2015). Alzheimer's disease. *Nature Reviews. Disease Primers*, *1*, 15056. DOI: 10.1038/nrdp.2015.56 PMID: 27188934

Mathieu, M., Martin-Jaular, L., Lavieu, G., & Théry, C. (2019). Specificities of secretion and uptake of exosomes and other extracellular vesicles for cell-to-cell communication. *Nature Cell Biology*, *21*(1), 9–17. DOI: 10.1038/s41556-018-0250-9 PMID: 30602770

McColgan, P., & Tabrizi, S. (2018). Huntington's disease: A clinical review. *European Journal of Neurology*, *25*(1), 24–34. DOI: 10.1111/ene.13413 PMID: 28817209

McNamara, K., & Tofail, S. (2017). Nanoparticles in biomedical applications. *Advances in Physics: X*, *2*(1), 54–88. DOI: 10.1080/23746149.2016.1254570

Missiego-Beltrán, J., & Beltrán-Velasco, A. (2024). The role of microbial metabolites in the progression of neurodegenerative diseases—Therapeutic approaches: A comprehensive review. *International Journal of Molecular Sciences*, *25*(18), 10041. DOI: 10.3390/ijms251810041 PMID: 39337526

Mollinari, C., Zhao, J., Lupacchini, L., Garaci, E., Merlo, D., & Pei, G. (2018). Transdifferentiation: A new promise for neurodegenerative diseases. *Cell Death & Disease*, *9*, 757. DOI: 10.1038/s41419-018-0891-4 PMID: 30082779

Morén, C., deSouza, R., Giraldo, D., & Uff, C. (2022). Antioxidant therapeutic strategies in neurodegenerative diseases. *International Journal of Molecular Sciences*, *23*(16), 9328. DOI: 10.3390/ijms23169328 PMID: 36012599

Morgan, S., & Orrell, R. (2016). Pathogenesis of amyotrophic lateral sclerosis. *British Medical Bulletin*, *119*(1), 87–98. DOI: 10.1093/bmb/ldw026 PMID: 27450455

Mortada, I., Farah, R., Nabha, S., Ojcius, D., Fares, Y., Almawi, W., & Sadier, N. (2021). Immunotherapies for neurodegenerative diseases. *Frontiers in Neurology*, *12*, 654739. DOI: 10.3389/fneur.2021.65473 PMID: 34163421

Mukherjee, S., Madamsetty, V., Bhattacharya, D., Chowdhury, S., Paul, M., & Mukherjee, A. (2020). Recent advancements of nanomedicine in neurodegenerative disorders theranostics. *Advanced Functional Materials*, *30*(40), 2003054. DOI: 10.1002/adfm.202003054

Naz, S., Beach, J., Heckert, B., Tummala, T., Pashchenko, O., Banerjee, T., & Santra, S. (2017). Cerium oxide nanoparticles: A "radical" approach to neurodegenerative disease treatment. *Nanomedicine (London)*, *12*(5), 545–553. DOI: 10.2217/nnm-2016-0399 PMID: 28181459

O'Connor, D., & Boulis, N. (2015). Gene therapy for neurodegenerative diseases. *Trends in Molecular Medicine*, *21*(8), 504–512. DOI: 10.1016/j.molmed.2015.06.001 PMID: 26122838

O'Connor, S. (2015). Engineering of secondary metabolism. *Annual Review of Genetics*, *49*, 71–94. DOI: 10.1146/annurev-genet-120213-092053 PMID: 26393965

Oskarsson, B., Gendron, T., & Staff, N. (2018). Amyotrophic lateral sclerosis: An update for 2018. *Mayo Clinic Proceedings*, *93*(11), 1617–1628. DOI: 10.1016/j.mayocp.2018.04.007 PMID: 30401437

Padmanabhan, P., Palanivel, M., Kumar, A., Máthé, D., Radda, G., Lim, K., & Gulyás, B. (2020). Nanotheranostic agents for neurodegenerative diseases. *Emerging Topics in Life Sciences*, *4*(6), 773–787. DOI: 10.1042/ETLS20190141 PMID: 33320185

Pagare, S., Bhatia, M., Tripathi, N., Pagare, S., & Bansal, Y. (2015). Secondary metabolites of plants and their role: Overview. *Current Trends in Biotechnology and Pharmacy*, *9*(3), 293–304.

Paul, A., & Yadav, K. (2020). Parkinson's disease: Current drug therapy and unraveling the prospects of nanoparticles. *Journal of Drug Delivery Science and Technology*, *58*, 101790. DOI: 10.1016/j.jddst.2020.10179

Peng, C., Trojanowski, J., & Lee, V. (2020). Protein transmission in neurodegenerative disease. *Nature Reviews. Neurology*, *16*(4), 199–212. DOI: 10.1038/s41582-020-0333-7 PMID: 32203399

Pinnell, J., Cui, M., & Tieu, K. (2020). Exosomes in Parkinson disease. *Journal of Neurochemistry*, *157*(4), 413–428. DOI: 10.1111/jnc.15288 PMID: 33372290

Pirhaghi, M., Mamashli, F., Moosavi-Movahedi, F., Arghavani, P., Amiri, A., Davaeil, B., Mohammad-Zaheri, M., Mousavi-Jarrahi, Z., Sharma, D., Langel, Ü., Otzen, D., & Saboury, A. (2024). Cell-penetrating peptides: Promising therapeutics and drug-delivery systems for neurodegenerative diseases. *Molecular Pharmaceutics*, *21*(4), 1375–1398. DOI: 10.1021/acs.molpharmaceut.3c01167 PMID: 38440998

Poovaiah, N., Davoudi, Z., Peng, H., Schlichtmann, B., Mallapragada, S., Narasimhan, B., & Wang, Q. (2018). Treatment of neurodegenerative disorders through the blood–brain barrier using nanocarriers. *Nanoscale*, *10*(36), 16962–16983. DOI: 10.1039/c8nr04073g PMID: 30182106

Porro, C., Panaro, M., Lofrumento, D., Hasalla, E., & Trotta, T. (2019). The multiple roles of exosomes in Parkinson's disease: An overview. *Immunopharmacology and Immunotoxicology*, *41*(4), 469–476. DOI: 10.1080/08923973.2019.1650371 PMID: 31405314

Puri, A., Loomis, K., Smith, B., Lee, J. H., Yavlovich, A., Heldman, E., & Blumenthal, R. (2009). Lipid-based nanoparticles as pharmaceutical drug carriers: From concepts to clinic. *Critical Reviews™ in Therapeutic Drug Carrier Systems*, 26(6), 523–580. https://doi.org/DOI: 10.1615/CritRevTherDrugCarrierSyst.v26.i6.10

Pusztahelyi, T., Holb, I., & Pócsi, I. (2015). Secondary metabolites in fungus–plant interactions. *Frontiers in Plant Science*, *6*, 573. DOI: 10.3389/fpls.2015.00573 PMID: 26300892

Rafe, M. (2024). Drug delivery for neurodegenerative diseases is a problem, but lipid nanocarriers could provide the answer. *Nanotheranostics*, *8*(2), 90–99. DOI: 10.7150/ntno.88849 PMID: 38164504

Raj, R., Wairkar, S., Sridhar, V., & Gaud, R. (2018). Pramipexole dihydrochloride loaded chitosan nanoparticles for nose to brain delivery: Development, characterization and in vivo anti-Parkinson activity. *International Journal of Biological Macromolecules*, *109*, 27–35. DOI: 10.1016/j.ijbiomac.2017.12.056 PMID: 29247729

Ramanathan, S., Archunan, G., Sivakumar, M., Selvan, T., Fred, A., Kumar, S., Gulyás, B., & Padmanabhan, P. (2018). Theranostic applications of nanoparticles in neurodegenerative disorders. *International Journal of Nanomedicine*, *13*, 5561–5576. DOI: 10.2147/IJN.S149022 PMID: 30271147

Rastogi, S., Sharma, V., Bharti, P., Rani, K., Modi, G., Nikolajeff, F., & Kumar, S. (2021). The evolving landscape of exosomes in neurodegenerative diseases: Exosomes characteristics and a promising role in early diagnosis. *International Journal of Molecular Sciences*, *22*(1), 440. DOI: 10.3390/ijms22010440 PMID: 33406804

Rawat, M. K., Jain, A., & Singh, S. (2011). Studies on binary lipid matrix-based solid lipid nanoparticles of repaglinide: In vitro and in vivo evaluation. *Journal of Pharmaceutical Sciences*, *100*(6), 2366–2378. DOI: 10.1002/jps.22443 PMID: 21491449

Reith, W. (2018). Neurodegenerative diseases. *Der Radiologe*, *58*(3), 241–258. DOI: 10.1007/s00117-018-0363 PMID: 29511772

Rekatsina, M., Paladini, A., Piroli, A., Zis, P., Pergolizzi, J., & Varrassi, G. (2020). Pathophysiology and therapeutic perspectives of oxidative stress and neurodegenerative diseases: A narrative review. *Advances in Therapy*, *37*(1), 113–139. DOI: 10.1007/s12325-019-01148-5 PMID: 31782132

Rizwan, M., Shoukat, A., Ayub, A., Razzaq, B., & Tahir, M. (2021). Types and classification of nanomaterials. In Nanomaterials: Synthesis, Characterization, Hazards and Safety (pp. 1–26). *Elsevier*. https://doi.org/DOI: 10.1016/B978-0-12-823823-3.00001-X

Roos, R. (2010). Huntington's disease: A clinical review. *Orphanet Journal of Rare Diseases*, *5*, 40. DOI: 10.1186/1750-1172-5-40 PMID: 21171977

Ross, C., & Poirier, M. (2004). Protein aggregation and neurodegenerative disease. *Nature Medicine*, *10*(Suppl 2), S10–S17. DOI: 10.1038/nm1066 PMID: 15272267

Ross, C., & Tabrizi, S. (2011). Huntington's disease: From molecular pathogenesis to clinical treatment. *Lancet Neurology*, *10*(1), 83–98. DOI: 10.1016/S1474-4422(10)70245-3 PMID: 21163446

Rüb, U., Seidel, K., Heinsen, H., Vonsattel, J., den Dunnen, W., & Korf, H. (2016). Huntington's disease (HD): The neuropathology of a multisystem neurodegenerative disorder of the human brain. *Brain Pathology (Zurich, Switzerland)*, *26*(6), 726–740. DOI: 10.1111/bpa.12426 PMID: 27529157

Ruffini, N., Klingenberg, S., Heese, R., Schweiger, S., & Gerber, S. (2022). The big picture of neurodegeneration: A meta study to extract the essential evidence on neurodegenerative diseases in a network-based approach. *Frontiers in Aging Neuroscience, 14,* 866886. DOI: 10.3389/fnagi.2022.866886 PMID: 35832065

Samii, A., Nutt, J., & Ransom, B. (2004). Parkinson's disease. *Lancet, 363*(9423), 1783–1793. DOI: 10.1016/S0140-6736(04)16305-8 PMID: 15172778

Sanati, M., Khodagholi, F., Aminyavari, S., Ghasemi, F., Gholami, M., Kebriaeezadeh, A., Sabzevari, O., Hajipour, M., Imani, M., Mahmoudi, M., & Sharifzadeh, M. (2019). Impact of gold nanoparticles on amyloid β-induced Alzheimer's disease in a rat animal model: Involvement of STIM proteins. *ACS Chemical Neuroscience, 10*(5), 2299–2309. DOI: 10.1021/acschemneuro.8b00622 PMID: 30933476

Sanghai, N., & Tranmer, G. (2023). Biochemical and molecular pathways in neurodegenerative diseases: An integrated view. *Cells, 12*(18), 2318. DOI: 10.3390/cells12182318 PMID: 37759540

Santiago, J., Bottero, V., & Potashkin, J. (2017). Dissecting the molecular mechanisms of neurodegenerative diseases through network biology. *Frontiers in Aging Neuroscience, 9,* 166. DOI: 10.3389/fnagi.2017.00166 PMID: 28611656

Santos, A., Morais, F., Simões, A., Pereira, I., Sequeira, J., Pereira-Silva, M., Veiga, F., & Ribeiro, A. (2019). Nanotechnology for the development of new cosmetic formulations. *Expert Opinion on Drug Delivery, 16*(4), 313–330. DOI: 10.1080/17425247.2019.1585426 PMID: 30793641

Sanvicens, N., & Marco, M. (2008). Multifunctional nanoparticles—Properties and prospects for their use in human medicine. *Trends in Biotechnology, 26*(8), 425–433. DOI: 10.1016/j.tibtech.2008.04.005 PMID: 18514941

Saraiva, C., Praça, C., Ferreira, R., Santos, T., Ferreira, L., & Bernardino, L. (2016). Nanoparticle-mediated brain drug delivery: Overcoming blood–brain barrier to treat neurodegenerative diseases. *Journal of Controlled Release, 235,* 34–47. DOI: 10.1016/j.jconrel.2016.05.044 PMID: 27208862

Sarko, D., & McKinney, C. (2017). Exosomes: Origins and therapeutic potential for neurodegenerative disease. *Frontiers in Neuroscience, 11,* 82. DOI: 10.3389/fnins.2017.00082 PMID: 28289371

Scharf, D., Heinekamp, T., & Brakhage, A. (2014). Human and plant fungal pathogens: The role of secondary metabolites. *PLoS Pathogens, 10*(1), e1003859. DOI: 10.1371/journal.ppat.1003859 PMID: 24497825

Scheltens, P., De Strooper, B., Kivipelto, M., Holstege, H., Chételat, G., Teunissen, C., Cummings, J., & van der Flier, W. (2021). Alzheimer's disease. *Lancet, 397*(10284), 1577–1590. DOI: 10.1016/S0140-6736(20)32205-4 PMID: 33667416

Schmidt, M., Gan, Z., Komander, D., & Dewson, G. (2021). Ubiquitin signalling in neurodegeneration: Mechanisms and therapeutic opportunities. *Cell Death and Differentiation, 28*(2), 570–590. DOI: 10.1038/s41418-020-00706-7 PMID: 33414510

Schubert, D., Dargusch, R., Raitano, J., & Chan, S. (2006). Cerium and yttrium oxide nanoparticles are neuroprotective. *Biochemical and Biophysical Research Communications, 342*(1), 86–91. DOI: 10.1016/j.bbrc.2006.01.129 PMID: 16480682

Scotti, L., & Scotti, M. (2015). Computer-aided drug design studies in the discovery of secondary metabolites targeted against age-related neurodegenerative diseases. *Current Topics in Medicinal Chemistry, 15*(21), 2239–2252. DOI: 10.2174/1568026615666150610143510 PMID: 26059353

Serafino, A., Giovannini, D., Rossi, S., & Cozzolino, M. (2020). Targeting the Wnt/β-catenin pathway in neurodegenerative diseases: Recent approaches and current challenges. *Expert Opinion on Drug Discovery, 15*(7), 803–822. DOI: 10.1080/17460441.2020.1746266 PMID: 32281421

Singh, N. (2017). Nanotechnology innovations, industrial applications and patents. *Environmental Chemistry Letters, 15*(2), 185–191. DOI: 10.1007/s10311-017-0612-8

Slanzi, A., Iannoto, G., Rossi, B., Zenaro, E., & Constantin, G. (2020). In vitro models of neurodegenerative diseases. *Frontiers in Cell and Developmental Biology, 8*, 328. DOI: 10.3389/fcell.2020.00328 PMID: 32528949

Stephenson, J., Nutma, E., van der Valk, P., & Amor, S. (2018). Inflammation in CNS neurodegenerative diseases. *Immunology, 154*(2), 204–219. DOI: 10.1111/imm.12922 PMID: 29513402

Sudhakar, V., & Richardson, R. (2019). Gene therapy for neurodegenerative diseases. *Neurotherapeutics; the Journal of the American Society for Experimental NeuroTherapeutics, 16*(1), 166–175. DOI: 10.1007/s13311-018-00694-0 PMID: 30542906

Sun, D., Li, N., Zhang, W., Zhao, Z., Mou, Z., Huang, D., Liu, J., & Wang, W. (2016). Design of PLGA-functionalized quercetin nanoparticles for potential use in Alzheimer's disease. *Colloids and Surfaces. B, Biointerfaces, 148*, 116–129. DOI: 10.1016/j.colsurfb.2016.08.052 PMID: 27591943

Sun, J., & Roy, S. (2021). Gene-based therapies for neurodegenerative diseases. *Nature Neuroscience, 24*(3), 297–311. DOI: 10.1038/s41593-020-00778-1 PMID: 33526943

Sun, T., Ding, Z., Luo, X., Liu, Q., & Cheng, Y. (2020). Blood exosomes have neuroprotective effects in a mouse model of Parkinson's disease. *Oxidative Medicine and Cellular Longevity, 3807476*. Advance online publication. DOI: 10.1155/2020/3807476 PMID: 33294121

Takeuchi, T. (2021). Pathogenic and protective roles of extracellular vesicles in neurodegenerative diseases. *Journal of Biochemistry, 169*(2), 181–186. DOI: 10.1093/jb/mvaa131 PMID: 33196835

Tan, S., Karri, V., Tay, N., Chang, K., Ah, H., Ng, P., Ho, H., Keh, H., & Candasamy, M. (2019). Emerging pathways to neurodegeneration: Dissecting the critical molecular mechanisms in Alzheimer's disease, Parkinson's disease. *Biomedicine and Pharmacotherapy, 111*, 765–777. DOI: 10.1016/j.biopha.2018.12.101 PMID: 30612001

Tanaka, M., Toldi, J., & Vécsei, L. (2020). Exploring the etiological links behind neurodegenerative diseases: Inflammatory cytokines and bioactive kynurenines. *International Journal of Molecular Sciences, 21*(7), 2431. DOI: 10.3390/ijms21072431 PMID: 32244523

Thomas, O., & Weber, W. (2019). Overcoming physiological barriers to nanoparticle delivery—Are we there yet? *Frontiers in Bioengineering and Biotechnology*, *7*, 415. DOI: 10.3389/fbioe.2019.00415 PMID: 31921819

Tiwari, S., Agarwal, S., Seth, B., Yadav, A., Nair, S., Bhatnagar, P., Karmakar, M., Kumari, M., Chauhan, L., Patel, D., Srivastava, V., Singh, D., Gupta, S., Tripathi, A., Chaturvedi, R., & Gupta, K. (2014). Curcumin-loaded nanoparticles potently induce adult neurogenesis and reverse cognitive deficits in Alzheimer's disease model via canonical Wnt/β-catenin pathway. *ACS Nano*, *8*(1), 76–103. DOI: 10.1021/nn405077y PMID: 24467380

Torres-Ortega, P., Saludas, L., Hanafy, A., Garbayo, E., & Blanco-Prieto, M. (2019). Micro- and nanotechnology approaches to improve Parkinson's disease therapy. *Journal of Controlled Release*, *295*, 201–213. DOI: 10.1016/j.jconrel.2018.12.036 PMID: 30579984

Uddin, M., Kabir, M., Tewari, D., Mathew, B., & Aleya, L. (2020). Emerging signal-regulating potential of small-molecule biflavonoids to combat neuropathological insults of Alzheimer's disease. *The Science of the Total Environment*, *700*, 134836. DOI: 10.1016/j.scitotenv.2019.134836 PMID: 31704512

van Niel, G., d'Angelo, G., & Raposo, G. (2018). Shedding light on the cell biology of extracellular vesicles. *Nature Reviews. Molecular Cell Biology*, *19*(4), 213–228. DOI: 10.1038/nrm.2017.125 PMID: 29339798

Vella, L., Hill, A., & Cheng, L. (2016). Focus on extracellular vesicles: Exosomes and their role in protein trafficking and biomarker potential in Alzheimer's and Parkinson's disease. *International Journal of Molecular Sciences*, *17*(2), 173. DOI: 10.3390/ijms17020173 PMID: 26861304

Vicente-Zurdo, D., Gómez-Mejía, E., Rosales-Conrado, N., & León-González, M. (2024). A comprehensive analytical review of polyphenols: Evaluating neuroprotection in Alzheimer's disease. *International Journal of Molecular Sciences*, *25*(11), 5906. DOI: 10.3390/ijms25115906 PMID: 38892094

Villain, N., & Dubois, B. (2019). Alzheimer's disease including focal presentations. *Seminars in Neurology*, *39*(2), 213–226. DOI: 10.1055/s-0039-1681041 PMID: 30925614

Virameteekul, S., Lees, A., & Bhidayasiri, R. (2024). Small particles, big potential: Polymeric nanoparticles for drug delivery in Parkinson's disease. *Movement Disorders*. Advance online publication. DOI: 10.1002/mds.29939 PMID: 39077831

Vissers, C., Ming, G., & Song, H. (2019). Nanoparticle technology and stem cell therapy team up against neurodegenerative disorders. *Advanced Drug Delivery Reviews*, *148*, 239–251. DOI: 10.1016/j.addr.2019.02.00 PMID: 30797953

Walker, F. (2007). Huntington's disease. *Lancet*, *369*(9557), 218–228. DOI: 10.1016/S0140-6736(07)60111-1 PMID: 17240289

Wang, E., & Wang, A. (2014). Nanoparticles and their applications in cell and molecular biology. *Integrative Biology*, *6*(1), 9–26. DOI: 10.1039/c3ib40165k PMID: 24104563

Wang, N., Jin, X., Guo, D., Tong, G., & Zhu, X. (2017). Iron chelation nanoparticles with delayed saturation as an effective therapy for Parkinson disease. *Biomacromolecules, 18*(2), 461–474. DOI: 10.1021/acs.biomac.6b0154 PMID: 27989126

Winter, A., & Bickford, P. (2019). Anthocyanins and their metabolites as therapeutic agents for neurodegenerative disease. *Antioxidants, 8*(9), 333. DOI: 10.3390/antiox8090333 PMID: 31443476

Wu, X., Zheng, T., & Zhang, B. (2017). Exosomes in Parkinson's disease. *Neuroscience Bulletin, 33*(3), 331–338. DOI: 10.1007/s12264-016-0092-z PMID: 28025780

Wu, Y., & Angelova, A. (2023). Recent uses of lipid nanoparticles, cell-penetrating and bioactive peptides for the development of brain-targeted nanomedicines against neurodegenerative disorders. *Nanomaterials (Basel, Switzerland), 13*(23), 4004. DOI: 10.3390/nano13233004 PMID: 38063700

Yetisgin, A., Çetinel, S., Zuvin, M., Koşar, A., & Kutlu, O. (2020). Therapeutic nanoparticles and their targeted delivery applications. *Molecules (Basel, Switzerland), 25*(9), 2193. DOI: 10.3390/molecules2509219 PMID: 32397080

You, Y., & Ikezu, T. (2019). Emerging roles of extracellular vesicles in neurodegenerative disorders. *Neurobiology of Disease, 130*, 104512. DOI: 10.1016/j.nbd.2019.104512 PMID: 31229685

Yuan, Q., Bao, B., Li, M., & Tang, Y. (2022). Bioactive composite nanoparticles for effective microenvironment regulation, neuroprotection, and cell differentiation. *ACS Applied Materials & Interfaces, 14*(14), 16435–16448. DOI: 10.1021/acsami.2c00579 PMID: 35322659

Zeng, H., Qi, Y., Zhang, Z., Liu, C., Peng, W., & Zhang, Y. (2021). Nanomaterials toward the treatment of Alzheimer's disease: Recent advances and future trends. *Chinese Chemical Letters, 32*(12), 3774–3785. DOI: 10.1016/j.cclet.2021.01.014

Zhang, T., Sun, S., Lv, J., Wang, X., Afewerky, H., Li, H., & Lu, Y. (2021). The emerging role of exosomes in Alzheimer's disease. *Ageing Research Reviews, 68*, 101321. DOI: 10.1016/j.arr.2021.101321 PMID: 33727157

Zhang, Y., Yu, W., Zhang, L., Wang, M., & Chang, W. (2022). The interaction of polyphenols and the gut microbiota in neurodegenerative diseases. *Nutrients, 14*(24), 5373. DOI: 10.3390/nu14245373 PMID: 36558531

Zhao, Y., Kuča, K., Wu, W., Wang, X., Nepovimova, E., Musílek, K., & Wu, Q. (2022). Hypothesis: JNK signaling is a therapeutic target of neurodegenerative diseases. *Alzheimer's & Dementia, 18*(1), 152–158. DOI: 10.1002/alz.12370 PMID: 34032377

Zheng, X., Zhang, C., Guo, Q., Wan, X., Shao, X., Liu, Q., & Zhang, Q. (2017). Dual-functional nanoparticles for precise drug delivery to Alzheimer's disease lesions: Targeting mechanisms, pharmacodynamics and safety. *International Journal of Pharmaceutics, 525*(1), 237–248. DOI: 10.1016/j.ijpharm.2017.04.033 PMID: 28432017

Ziental, D., Czarczyńska-Goślińska, B., Mlynarczyk, D., Głowacka-Sobotta, A., Stanisz, B., Gośliński, T., & Sobotta, L. (2020). Titanium dioxide nanoparticles: Prospects and applications in medicine. *Nanomaterials (Basel, Switzerland), 10*(2), 387. DOI: 10.3390/nano10020387 PMID: 32102185

Chapter 14
Nanoparticle-Based Interventions Targeting Dysregulated Neuronal Signaling in Neurodegenerative Disorders

Uzma Bashir
https://orcid.org/0000-0002-7904-1192
The Islamia University of Bahawalpur, Pakistan

Muhammad Shahid
The Islamia University of Bahawalpur, Pakistan

ABSTRACT

The rise in life expectancy has led to increased neurodegenerative diseases (NDDs) like Alzheimer's, Parkinson's, and Huntington's, driven by inflammation, oxidative stress, and protein misfolding. Current treatments fail due to the blood-brain barrier (BBB). Nanotechnology offers a solution by enhancing drug delivery and stability. Nano-phytomedicine combines nanoparticles with plant-based compounds, improving efficacy. Metal-based nanoparticles also show promise, paving the way for precision-driven NDD treatments.

INTRODUCTION

The global increase in average life expectancy has been accompanied by a significant rise in the incidence of age-related neurodegenerative diseases (NDs). These disorders are characterized by a progressive decline in neuronal function, often culminating in neuronal death. Mechanistically, dysregulated signaling pathways such as inflammation, apoptosis, oxidative stress, and protein misfolding play pivotal roles in driving neuronal damage. Additionally, in families harboring

DOI: 10.4018/979-8-3373-1365-8.ch014

genetic abnormalities, mutations and duplications in genes encoding α-synuclein and amyloid proteins contribute to the development of specific NDs. Overproduction and aberrant aggregation of these proteins have been extensively demonstrated in various models of neuronal dysfunction and degeneration. NDs constitute a diverse group of age-associated disorders characterized by progressive deterioration of motor and/or cognitive functions, frequently accompanied by psychiatric manifestations. These conditions, often referred to as "protein misfolding" diseases or proteinopathies, include lzheimer's disease (AD), Parkinson's disease (PD), and Huntington's disease (HD). Shared histopathological features among these disorders include insoluble protein deposition, glial cell activation, neuronal loss, and disrupted synaptic connectivity. Currently, no curative or disease-modifying treatments are available for these NDDs. Many therapeutic compounds have failed in clinical trials, primarily due to their inability to traverse the blood-brain barrier (BBB) a highly selective barrier formed by tightly joined endothelial cells that restricts the entry of most therapeutic agents into the brain. Nanotechnology offers a promising solution to overcome the limitations of the BBB by enabling the development of nanoparticles (NPs) engineered at nanoscale dimensions (1–100 nm). These particles exhibit unique physicochemical properties that allow enhanced efficacy, cellular uptake, and targeted delivery of therapeutic agents. Nanotechnology-driven innovations in bioinformatics further enhance the potential of precision medicine for neurological disorders. Advanced computational approaches enable early diagnosis and tailored therapeutic strategies, aligning with the complex requirements of treating NDs. Functionalized NPs, capable of crossing the BBB, can deliver active molecules directly to brain target sites, minimizing off-target effects. The advent and progression of nanotechnology have also enabled the integration of NPs with medicinal plant derivatives, a field termed "nano-phytomedicine." These conjugates combine the natural abundance and therapeutic potential of phytochemicals with advanced nanocarriers, facilitating targeted brain delivery and reducing the likelihood of adverse effects. Phytochemicals, as multi-target therapeutic agents, exhibit neuroprotective effects through diverse mechanisms. However, their application is limited by the BBB and rapid degradation in systemic circulation, necessitating innovative nanoformulations to enhance their stability, bioavailability, and pharmacokinetics. Various nanostructures, including polymeric NPs, lipid-based NPs, nanoliposomes, nano-micelles, and carbon nanotubes (CNTs), alongside diverse vehicular systems such as poly (lactic-co-glycolic acid), lactoferrin, and polybutylcyanoacrylate, have been shown to markedly enhance therapeutic efficacy, reduce adverse effects, improve stability, and optimize pharmacokinetic profiles of numerous pharmaceuticals. NDs, which afflict millions globally, represent a group of debilitating conditions that pose significant clinical challenges. Extensive research has demonstrated that several nanoformulations derived from natural compounds—such as curcumin (Cur), quercetin (QC), resveratrol (RSV), piperine (PIP), Ginkgo biloba, and Nigella sativa—have notably ameliorated the clinical manifestations in patients suffering from NDs. However, the delivery of therapeutics to the central nervous system (CNS) is hindered by multiple barriers, with the BBB emerging as the principal obstacle in the treatment of NDDs. Natural antioxidants, including Ginkgo biloba, curcumin, and tea extracts, have shown promise in mitigating oxidative stress and inflammation associated with NDDs. Nanocarriers enhance the therapeutic efficacy of these compounds by improving controlled release and targeting brain tissues effectively. Complex natural extracts act synergistically on multiple biochemical pathways, offering potential for multi-therapeutic effects in conditions such as AD, PD, HD, and amyotrophic lateral sclerosis (ALS).

Silver nanoparticles (AgNPs), with their size- and shape-dependent optical, antimicrobial, and electrical properties, are gaining attention for their role in biomedical applications. These NPs are synthesized through various methodologies, including laser ablation, chemical reduction, and plant-mediated techniques, which align with the principles of green chemistry. Similarly, zinc oxide nanoparticles (ZnO NPs) have shown significant potential in optoelectronic, pharmaceutical, and cosmetic applications due to their antimicrobial, UV-blocking, and anti-inflammatory properties. ZnO NPs demonstrate the ability to inhibit inflammatory pathways, making them relevant for treating conditions involving neuroinflammation. In conclusion, this study highlights the synergistic potential of nanotechnology and phytomedicine for managing NDs. By integrating advanced nanocarriers with therapeutic agents, innovative strategies can address challenges such as the BBB, enhance therapeutic efficacy, and pave the way for precision-driven neurotherapeutics.

I. DRUG DELIVERY SYSTEM

The drug delivery model is a kind of system that demonstrates how the therapeutic effect of a medicine is achieved by administering the drug differently to people or animals. For this goal, a number of medication delivery methods have been created and refined. The development of a novel medicinal molecule is both expensive and time-consuming. Nonetheless, we want to increase the safety-to-effectiveness ratio of diverse old medications by using a variety of methods throughout the drug delivery system. NDDS is the most efficient and advanced form of medication delivery (Sun, Ma, & Hu, 2021).

1. Novel Drug Delivery System

The Novel Drug Delivery Systems (NDDS) is a revolutionary program that aims to maximize medication delivery. Recent advancements in medication pharmacokinetic and pharmacodynamics effects have resulted in a greater consideration and creation of a unique or improved drug delivery method (Rajput, Tandel, Borse, & Butani, 2021). Large pharmaceutical businesses that have lost market share for their older treatments are developing novel drug delivery systems for their current medications in order to restore their market position, since developing a novel drug is more expensive than developing a new drug delivery system. The medication molecule is bonded to a nano carrier and transferred to the target organ in NDDS (Tiwari & Talreja, 2020). All of the following features should be present in the vehicle or nano-carrier that is chosen.

- Nano-carrier should ensure the controlled-release of the drug from the dosage form
- It will move through the biological barrier including BBB.
- If anticancer drugs are enclosed then it should cross tumor vasculature.
- It must be chemically and physically stable in both in-vivo and in-vitro conditions.
- This should not to be toxic to the body.
- It must be biodegradable.
- It must be biocompatible.
- The Nano carrier should be non-reactive and should not chemically react with active

o The carrier should be no immunogenic so it shouldn't induce any type of immune response

1.1. Polymeric Nanoparticle

Polymeric nano-particles are small solid-particle dispersions or tiny colloidal particles with diameters ranging from 10 nm to 1000 nm (Fissan, Ristig, Kaminski, Asbach, & Epple, 2014). These might be added to a nanoparticle matrix to encapsulate or trap medications or items. Different preparation procedures produce NPs, Nano capsules, or nanospheres of varied sizes. Polymeric nano-particles are now manufactured by employing several kinds of hydrophobic and hydrophilic polymers that are employed for the trapping of proteins, genes, DNA, and other active drugs.

The main features or advantages of polymeric Nano-particles are:

- The active drug can be entrapped into Nano-particles without any chemical reactions.
- These polymeric Nano-particles provide local action with sustained-drug release.
- Because of their small size and diverse surface properties, they can be used to target various organs in both an active and passive manner.
- For active target ion, active ligands are attached to the surface of NPs.

1.2. Metallic-Nanoparticles

Metallic NPs have attracted scientists for more than hundred years and are currently mostly employed in biological and engineering research. They are a source of worry due to their enormous potential in nanotechnology. NPs such as silver NPs, nano shells, and magnetic NPs (iron oxide) have been employed and modified with different chemical functional groups during the course of the year, enabling them to be mixed with antibodies, ligands, and medicines. This opens up a plethora of possible applications in biotechnology, including targeted medication delivery, vehicles for gene and drug transport, and, most significantly, diagnostic imaging (Akhtar & Khan, 2016).

1.3. Gold-Nanoparticles

Colloid gold, also known as gold NPs, is a solution of nanometer-sized gold particles. The colloidal solution is either a bright red (particles smaller than 100 nm) or a dull yellow color (bigger NPs). They have exceptional optical properties and could be used in photo thermal therapy (PTT) (Amaral *et al.,* 2021).

2. Advantages of NDDS

NDDS offer several significant advantages that enhance the overall effectiveness of pharmaceutical treatments. They help reduce the required concentrations of active medicinal ingredients while improving the water solubility of hydrophobic drugs. NDDS also enable the development of sustained-release formulations, extending the duration of action for drugs with short half-lives (Zaman *et al.*, 2019). By delivering medications in a controlled and targeted manner, these systems minimize side effects and enhance therapeutic efficiency. Additionally, NDDS can prevent drug metabolism in the gastrointestinal tract during the first-pass effect, further improving bioavailability. In many cases, the use of NDDS lowers the overall cost of therapy by reducing the patient's hospital stay and, consequently, increases patient compliance.

3. Need for Novel Drug Delivery System

A Nano carrier is a nanomaterial that serves as a vehicle for another chemical, such as a medicine. Nano carriers are being used in the transport of drugs, and their good properties imply that they will be used in chemotherapy in the future. Many different types of nanomaterials utilized in Nano carriers enable the transport of hydrophobic and hydrophilic medications throughout the body. Because the human body is largely composed of water, the ability of Nano carriers to successfully deliver hydrophobic medications to humans is a substantial therapeutic advantage (Mi *et al.*, 2012). Herbal formulations were also selected as a powerful drug candidate for shipping via a nano delivery system because patient resistance may occur due to large doses and lower potency of available formulations, adverse effects, and newly marketed formulations lacking target specificity for many chronic diseases. Bulk drugs are included in nano formulations, implying dose reduction and, finally, effectiveness.

4. Controlled Released Drug Delivery System (CRDDS)

A controlled release drug delivery system is one that maintains consistent drug release from particular devices, resulting in a steady concentration of plasma medications throughout time(Kamaly, Yameen, Wu, & Farokhzad, 2016). The controlled release drug delivery system (CRDDS) employs a variety of mechanisms (Figure 1).

Figure 1. Controlled release drug delivery system

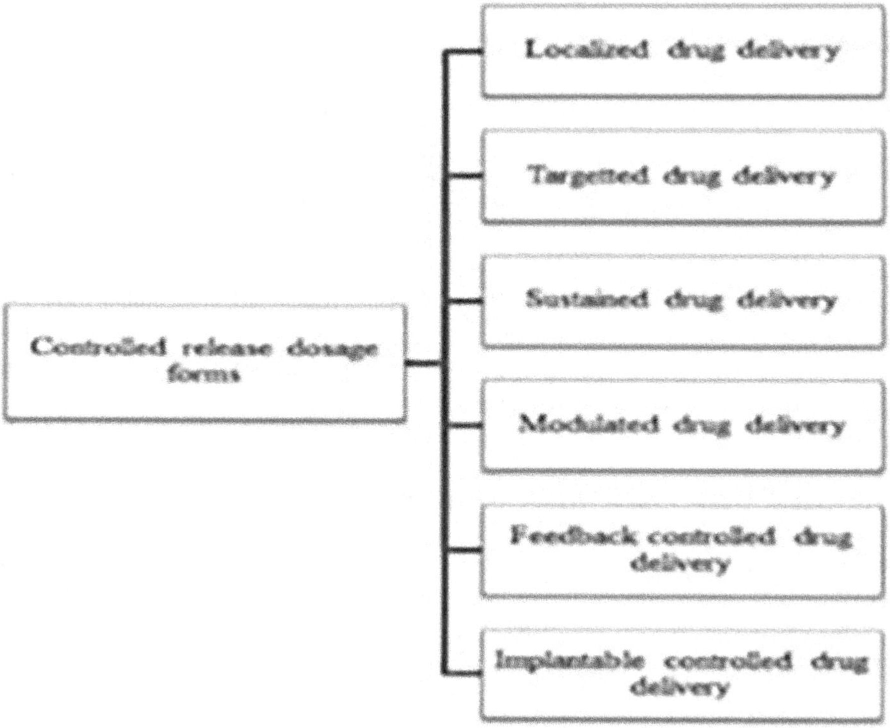

4.1. Localized Drug Delivery System

The medication release in this drug delivery mechanism occurs directly in the affected region or organ. Because the drug is delivered to a particular location in the organ, this drug delivery technology decreases the number of adverse effects associated with many medications. Localized drug delivery systems provide several benefits, including lower medication doses and higher drug concentrations at the location of the illness. Furthermore, drug release, especially in GIT areas, will limit active ingredient degradation and boost medication absorption in the GIT (Florek, Caillard, & Kleitz, 2017).

4.2. SRDDS (Sustained Release Drug Delivery System)

This technique of drug-release necessitates a procedure that allows for time-independent drug release over a lengthy period of time. The continuous release of pharmaceuticals ensures that the serum concentration of active medication stays within its therapeutic window for an extended period of time (Patra *et al.,* 2018).

4.3. MDDS (Modulated Drug Delivery System)

This technology allows drugs to be released based on physiological necessity. Externally, these structures self-modulate. In the case of an externally regulated system, drug release may be adjusted by applying external stimuli. However, drug release may be influenced by changes in the surrounding environment (Lin, 2019).

4.4. Feedback-Controlled Drug Delivery System

This system is an excellent medication delivery system in which patient dosage is based on automated patient sample analysis. This automated sampling may be accomplished by the examination of blood glucose levels or coagulation levels, as well as direct analysis. The sample set has a bi-directional tube that transports the material to the analyzer, and the controller monitors the flow of medicinal fluid to the patient's body based on the analyzer's computation. Feedback drug delivery systems may be provided through a variety of routes, the most prevalent of which being nasal, transdermal, and transmucosal systems (VASUMATHI & Neelamegam, 2017).

4.5. Targeting Drug Delivery System

This system is an excellent medication delivery system in which patient dosage is based on automated patient sample analysis. This automated sampling may be accomplished by the examination of blood glucose levels or coagulation levels, as well as direct analysis. The sample set has a bi-directional tube that transports the material to the analyzer, and the controller monitors the flow of medicinal fluid to the patient's body based on the analyzer's computation. Feedback drug delivery systems may be provided through a variety of routes, the most prevalent of which being nasal, transdermal, and transmucosal systems (Hawkins, Casolaro, Brown, Edwards, & Wikswo, 2020).

4.5.1. Passive Targeting

A targeted drug delivery system is one in which a medicine is released to a specified or target organ or tissue, increasing the drug's therapeutic index. At similar plasma drug levels, TDDS reduces medication doses, minimizes toxicity, and improves therapeutic efficacy. The medicine is delivered to the targeted tissue by passive or active targeting. In active targeting, the drug is conjugated with a carrier, which includes site-specific ligands and is responsible for delivering the drug to a particular portion of the body. The medicine is encased in a carrier in passive targeting, and this carrier uses the regular bodily pathway to reach the target organ shown in Figure 2 (Jain *et al.,* 2018).

It is a novel method of medication delivery. In this procedure, medicines are delivered in order to increase or localize their concentration in the proper place (cell / tissue / organ). The negative effects of the medicine are decreased, while its effectiveness is increased. A targeted drug delivery system assists the active agent in reaching the target organ. This medication delivery technology offers several benefits over other conventional drug administration systems due to its precision and concentration. It limits drug penetration in normal tissue in order to increase drug concentration in sick tissue or organs. This is why the system concentrates mostly on anti-tumor treatments, which may induce cytotoxicity in human cells (Dunuweera *et al.,* 2019).

Figure 2. Passive targeting

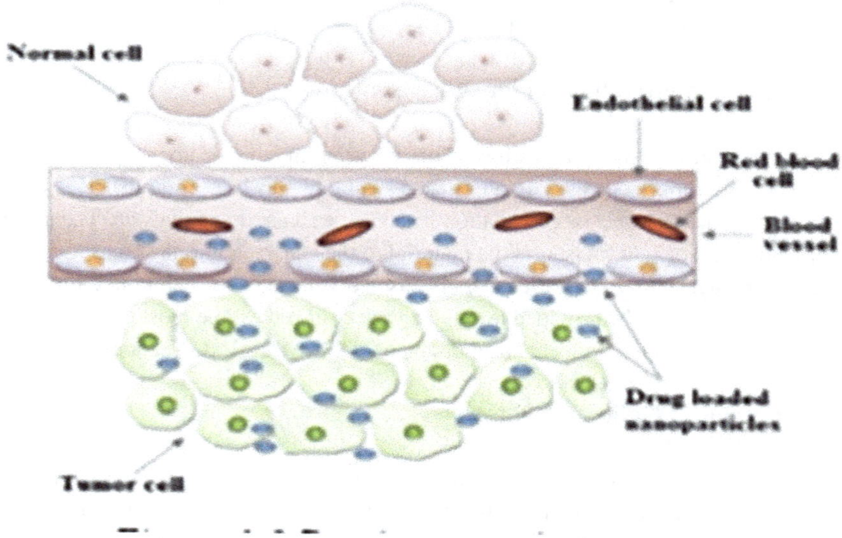

4.5.2. Active Targeting

Active-targeting is a kind of ligand-based targeting in which targeting moieties (ligands attached to drug carriers or vehicles) bind to their target location, which is a tumor cell. Active-targeting ligands should bind to tumor cell receptors but not to normal cells. The primary goal of active-targeting is drug molecule penetration into the intracellular regions of malignant cells shown in Figure 3. Size, shape, surface charge, and surface hydrophobicity are all important aspects impacting the efficiency of NPs. Ligands utilized in active targeting include nucleic acid, peptides, protein, sugar, antibodies, and vitamins. Sugar, proteins, and lipids may be the receptors or targets to which these ligands bind (Kamaly, Xiao, Valencia, Radovic-Moreno, & Farokhzad, 2012).

Figure 3. Active targeting

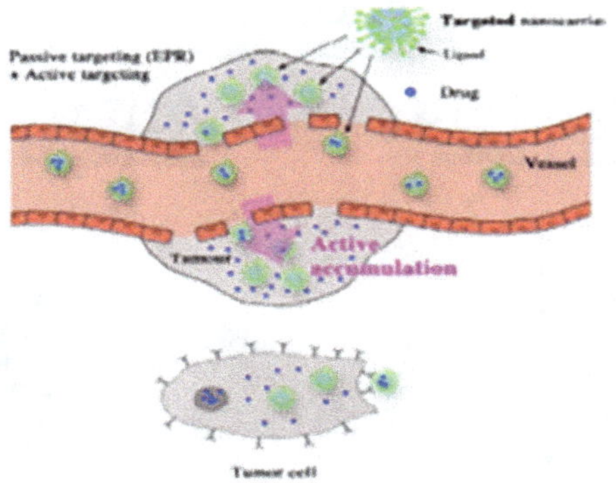

II. NANOTECHNOLOGY

Nano-science and nano-biotechnology have unexpected effects on nanomaterials, making it an advanced study field with unique applications in biotechnology, catalysis, biomedical, and life science. The nano-size of materials offers exceptional features because it connects bulk-sized materials to atomic-sized materials (Pillai & Ganguly).

1. Methods for Synthesis of Nanoparticles

Researchers have been paying close attention to NP synthesis research initiatives throughout the last decade. It is difficult to comprehend the mechanism of the nanoparticle production process, specialized functionality, fine-tuning the physicochemical activities of nano-sized particles, and implementations (Long, Ryan, & Padrela, 2019). The two major methods for the generation of NPs are top-down and bottom-up procedures shown in Figure 4. Both of these methodologies are related to NP and are used to calculate the size, shape, and activities of NPs for definitive implementation.

Figure 4. Nanoparticle's preparation methods

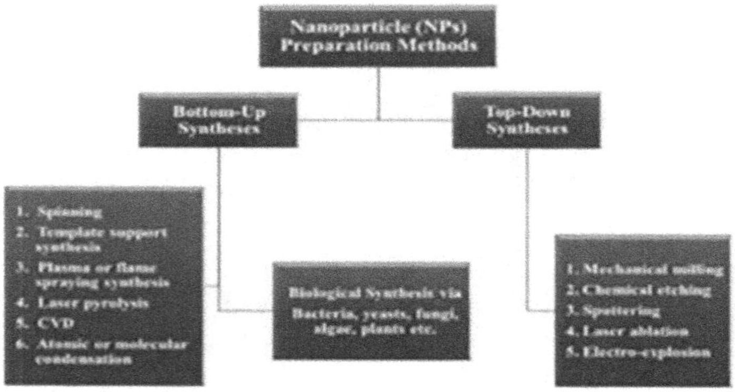

1.1. Top-Down Approach

The top-down method is also called the destructive method because in this process, large materials are reduced into nano-sized materials. This method involves further processes that are used for the synthesis of NPs. Laser ablation, nanolithography, thermal decomposition, mechanical milling, and sputtering are mainly selected for NPs synthesis (Ealia & Saravanakumar, 2017).

1.2. Bottom-Up Approach

The bottom-up strategy is the most appropriate and operational way to approach NP generation. NPs are generated using this approach by the self-association of smaller particles such as atoms and molecules. By varying the reaction conditions, NPs of various sizes and morphologies may be produced. In the bottom-up technique, reaction variables such as temperature, pH, and the concentration of solvent and plant extract may regulate particle size. As a result, NPs are manufactured with excellent precision using this method (Indiarto, Indriana, Andoyo, Subroto, & Nurhadi, 2021).

1.3. Chemical Method

The chemical approach is most often employed to create silver NPs. This method uses chemicals to convert a silver ion (Ag+) into a silver atom (Ag0). This approach produces stable Ag-NPs that disperse as colloidal particles in solvents (water and organic). Sodium borohydride, elemental hydrogen, citrate, and ascorbate are some of the chemicals utilized in reduction. The following procedures are examples of chemical approaches for the creation of NPs (Figure 5): • Chemical /Electrochemical precipitation • Vapor deposition • Molecular condensation • Sol-Gel process • Pyrolysis (Singh et al., 2014).

Typically, the chemical approach is inexpensive and yields a large number of NPs. Lethal chemical solvents and other chemical contaminants are employed in this procedure. As a result, one downside of chemical techniques is that they generate a hazardous by-product that pollutes the environment.

Bottom-up methods have several benefits over top-down methods. This approach saves time when synthesizing a large quantity of NPs. This procedure also has certain drawbacks since it employs hazardous chemicals, which have detrimental impacts on the environment. As a result, researchers are developing green ways for the synthesis of NPs. For extraction in the green approach, a non-hazardous solvent such as water is employed. Hydrothermal, microwave, biological, and low-temperature settings are examples of greener NP synthesis processes. Plant extracts and other natural items are used in the green synthesis of NPs (Remya, Abitha, Rajput, Rane, & Dutta, 2017).

Figure 5. Chemical method for the synthesis of NPs

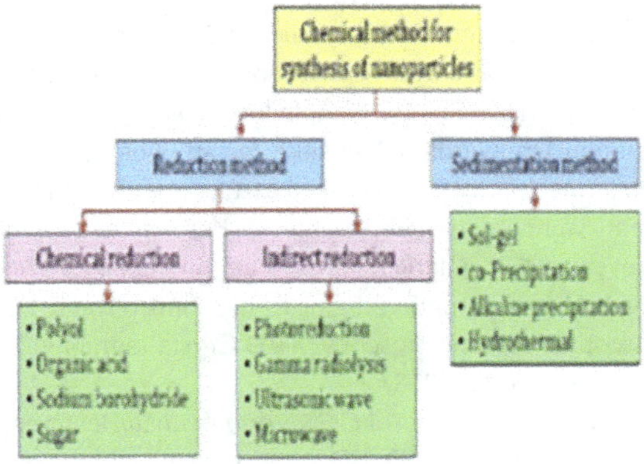

1.4. Biological Method

Besides from numerous chemical approaches, the biological method is the finest replacement method for the synthesis of NPs. Nanoparticle preparation has been a major focus for researchers due to its unique qualities such as particular shape, size, and antibacterial activity. Bacteria, fungus, and plant extracts produce NPs. The mechanism of green nanoparticle production comprises a redox reaction. The typical size of NPs generated via green synthesis ranges from 10 to 100 nm. Adjusting reaction parameters such as reaction temperature, pH, precursor solution concentration, and sample concentration can change the size of NPs. The biological synthesis of NPs is an ecologically acceptable procedure that takes less time to complete than the chemical approach, which is dangerous to the environment (Parveen, Banse, & Ledwani, 2016).

Biologically generated NPs may be used to benefit humans in the food, cosmetics, and pharmaceutical sectors. During the literature search, various plant species of the Apiaceae family were discovered and reported to be employed in the synthesis of Ag-NPs and ZnONPs.

Green synthesis of various sized Ag-NPs from *Trachysperm umammi* and *Papaver somniferum* aqueous extracts was described (Vijayaraghavan, Nalini, Prakash, & Madhankumar, 2012). According to this study, the primary components of *T.ammi* are thymol, p-cymene, and -terpinene, whereas the main components of *P. somniferum* are morphine and codeine. The extract was employed at the same concentrations for the production of AgNPs. These findings demonstrated that the essential oils of T.ammi extract were a more effective reducing agent *than P. somniferum. T. ammi's* Ag-NPs were characterized as triangular-shaped NPs with a size of 87nm.

Biological method for the manufacture of silver NPs was described utilizing the aqueous extract of *Foeniculum vulgare* Mill seeds. *F. vulgare* seed extract reduced silver to form Ag-NPs. The UV-vis spectrum revealed a peak at 475 nm, confirming the production of Ag-NPs. The FTIR spectrum revealed that amides and amine groups were involved in the capping and reduction of AgNO3 into Ag-NPs. EDS was utilized to validate the vibrational analysis, and the XRD method was employed to determine the structure of the NPs. The antibacterial activity of Ag-NPs was investigated using *Pseudomonas aeruginosa, Escherichia coli, Proteus mirabilis, Shigella flexineri,* and *Klebsiella pneumonia*. The findings demonstrated substantial antibacterial activity of Ag-NPs (Vijayaraghavan *et al.,* 2012).

AgNPs of various sizes (30-50 nm) were synthesized utilizing a hydro extract of *Asiatica Centella L.* leaves. This reaction was carried out at room temperature. The antibacterial activity of produced Ag-NPs against human infections, Staphylococcus aureus, was shown (MTCC 7405). This pathogen causes skin inflammation. UV-visible spectrophotometer (Khatoon et al., 2022).

At room temperature, *Pimpinella anisum* L. seed extract was also employed in the production of Ag-NPs. Toxic solvents were not used in the process. For seed extraction, distilled water was employed. This study revealed that spherical Ag-NPs are tiny in size.

TEM examination revealed wavelengths ranging from 3.2nm to 16nm. For the characterization of Ag-NPs, UV spectroscopy, FTIR spectroscopy, XRD Crystallography, TEM Microscopy, and EDX were utilized (Devanesan *et al.,* 2018).

AgNPs derived from the aqueous extract of *Coriandrum sativum* leaf were said to be anti-breast cancer, anti-dandruff. UV–vis, FTIR, XRD, EDS, and Brunauer-Emmett-Teller spectroscopies were utilized to characterize AgNPs. The results showed that green synthesized AgNPs with a size of 37 nm from *C. sativum* have a wide range of biomedical applications.

Green produced AgNPs have been shown to have antimicrobial potential against the gram-negative bacterium E. coli. Hydro leaf extract of Hydro cotyle rotundifolia was used to create AgNPs. UV-Vis Spectrophotometer, TEM, EDX, Selected Area Electron Diffraction (SAED), and FTIR spectroscopy were used to examine these produced NPs (Dhineshbabu, Rajendran, Nithyavathy, & Vetumperumal, 2016).

Green synthesis is an environment friendly method that involves the use of plants, fungi, and bacteria for the synthesis of NPs. Green synthesis is cheap, free of toxic chemical method and it gives high yield of NPs(Yadi *et al.,* 2018). Many plants were used in the synthesis of Ag and ZnONPs. *Polypodium vulgare* is used for the first time in this study for the green synthesis of silver and zinc oxide nanoparticle (Bellanthudawa et al., 2023).

III. MECHANISMS BY WHICH NANOFORMULATIONS MODULATE NEURONAL OXIDATIVE STRESS, INFLAMMATION, APOPTOSIS, AND PROTEIN AGGREGATION IN NDS

NPs, defined as nanoscale structures with dimensions ranging from 1 to 100 nm, can be categorized based on their shape, size, and physicochemical properties. Key classes of NPs include carbon-based materials (e.g., carbon nanotubes and fullerenes), ceramics, metals, semiconductors, lipid-based systems, polymeric NPs, peptide-engineered systems, DNA nanocages, and nanoenzymes. These nanostructures possess unique physical and chemical properties attributed to their small size and large surface area-to-volume ratio, enabling them to exhibit behaviors distinct from their bulk counterparts (S. Khan & Hossain, 2022).

Nanotechnology offers advanced drug delivery platforms capable of improving the diagnosis, monitoring, control, and treatment of neuronal disorders at the molecular level. NPs have demonstrated numerous therapeutic advantages, including superior biodegradability and biocompatibility, enhanced pharmacokinetics and therapeutic efficacy, and reduced adverse drug effects (Mittal et al., 2022). These systems have shown the potential to modulate critical dysregulated pathways implicated in NDDs, such as oxidative stress, inflammation, apoptosis, and protein aggregation. The anti-inflammatory properties of NPs are particularly noteworthy, stemming from their high surface area-to-volume ratio, which enhances their ability to suppress key mediators of inflammation, including enzymes and cytokines involved in inflammatory signaling cascades (Nayal, Mejjo, & Abajy, 2024).

1. Neuronal Oxidative Stress in NDDs: Modulation by Nanoparticles

Oxidative stress refers to an imbalance between the production of reactive oxygen species (ROS) and the activity of antioxidants, leading to damage of cellular macromolecules, including DNA, proteins, carbohydrates, and lipids (Sharifi-Rad et al., 2020). This imbalance plays a central role in the initiation and progression of diseases such as cancer, cardiovascular disorders, chronic obstructive pulmonary disease, asthma, and NDDs, including AD, PD, HD, ALS, and stroke (Figure 6). Oxidative stress-induced macromolecular damage is a hallmark of NDDs, accelerating neuronal degeneration through toxic signaling cascades and neuronal apoptosis (Lara-Reyna et al., 2018). Protein misfolding and aggregation further exacerbate ROS generation, forming a vicious cycle of oxidative damage and neurodegeneration (Gandhi et al., 2019).

Figure 6. Signaling pathway in NDDs

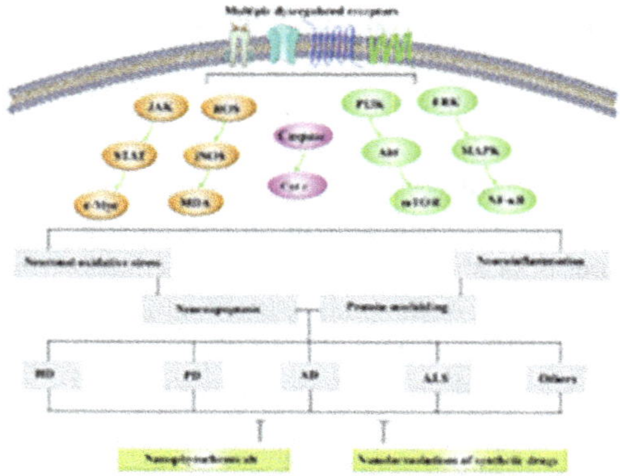

Studies have demonstrated the superior antioxidant activity of CeO2 NPs compared to traditional antioxidants, such as vitamin E analogs and N-acetyl-cysteine. For instance, Caputo et al. reported that CeO2 NPs significantly reduced oxidative signals induced by irradiated titanium dioxide (TiO2) NPs. Similarly, manganese oxide NPs have shown intrinsic SOD-like activity, enhancing magnetic resonance imaging (MRI) contrast while simultaneously providing therapeutic benefits in targeting cancer cells (FLORES, Espino, & PARIENTE, 2023).

Preclinical studies have highlighted the therapeutic potential of CeO2 NPs in various disease models. For example, treatment of transgenic mice with CeO2 NPs suppressed progressive left ventricular dysfunction and reduced myocardial oxidative stress. Additionally, CeO2 NPs attenuated pro-inflammatory cytokine levels, including tumor necrosis factor-alpha (TNF-α), interleukin (IL)-6, and IL-1β, further demonstrating their anti-inflammatory properties (Brandão Da Silva Assis, Nestal De Moraes, & De Souza, 2024).

To enhance their therapeutic potential, NPs can be functionalized with natural or synthetic antioxidant compounds, enzymes, or phytochemicals. Key enzymatic antioxidants such as glutathione peroxidase, SOD, and CAT, as well as non-enzymatic antioxidants like glutathione (GSH), vitamins E, C, and A, and various phytochemicals, have been incorporated into nanoparticle systems. These nanoformulations protect neurons from oxidative stress, delay the progression of NDDs, and provide targeted antioxidant delivery to the central nervous system (CNS) (Ezema, Eze, Ayoka, & Nnadi, 2024).

The development of advanced nanoformulations represents a promising approach to mitigate oxidative stress and its associated pathological consequences in NDDs. By targeting dysregulated pathways and enabling precise delivery of antioxidant compounds to the CNS, NPs hold significant potential for improving therapeutic outcomes and slowing the progression of NDs (Fakhri et al., 2022).

Platinum (Pt) NPs have emerged as potent agents in nanomedicine due to their catalytic capabilities to convert oxygen (O_2) into hydrogen peroxide (H_2O_2) and subsequently facilitate the decomposition of H_2O_2 into water (H_2O) and oxygen (O_2) (Manoharan, Wang, Chen, Li, & Yeh, 2024). This enzymatic mimicry of catalase (CAT) and superoxide dismutase (SOD) positions

Pt NPs as promising candidates for mitigating oxidative stress. Studies have demonstrated that CAT-loaded NPs enhance in vitro neuroprotection against oxidative damage induced by H_2O_2, effectively attenuating mitochondrial membrane depolarization, DNA damage, protein oxidation, and cell membrane degradation. Furthermore, Martín et al. engineered Pt and gold NPs supported on Fenton-treated diamond NPs (Pt/HO-DNP and Au/HO-DNP), which displayed significant biocompatibility, cell membrane permeability, and antioxidant efficacy against reactive oxygen species (ROS)-induced oxidative stress in hepatoma cells (Salatin et al., 2023).

2. Nanoparticles and Neuroinflammation in Neurodegenerative Disorders

Neuroinflammation involves the activation of the brain's innate immune system, often triggered by inflammatory conditions that elicit intricate molecular and cellular responses. This activation promotes glial cell stimulation and increases the production of pro-inflammatory mediators, including chemokines (e.g., CXCL1, CCL5, CCL2), cytokines (e.g., TNF-α, IL-6, IL-1β), ROS, and reactive nitrogen species (RNS) (Theophanous, Sargiannidou, & Kleopa, 2024). Concomitant pathological events such as edema, breakdown of the BBB, infiltration of peripheral immune cells, and elevated BBB permeability exacerbate neuroinflammation. Additionally, mast cells contribute to inflammatory processes by releasing histamines, leukotrienes, and cytokines, while promoting IgE synthesis via B lymphocytes. The nuclear factor kappa-light-chain-enhancer of activated B cells (NF-κB) signaling pathway plays a pivotal role in regulating over 500 genes associated with neuroinflammation, further driving the pathogenesis of neurodegenerative and psychiatric disorders such as AD, PD, ALS, HD, stroke, and multiple sclerosis (MS) (Ryan, 2019).

NPs have shown remarkable anti-inflammatory potential, attributed to their high surface area-to-volume ratio, which enhances their ability to modulate enzymes, cytokines, and mediators involved in inflammation. Several studies have confirmed the in vitro and in vivo anti-inflammatory activities of NPs such as gold, copper, silver, selenium, zinc oxide, magnesium oxide, iron oxide, and cerium oxide (CeO) (Nayal et al., 2024). Zinc oxide NPs, for example, suppress pro-inflammatory cytokines (e.g., TNF-α, IL-1β) by inhibiting NF-κB signaling, caspase-1 activation, and lipopolysaccharide-induced pathways, as well as reducing oxidative stress markers like malondialdehyde. Similarly, gold NPs diminish ROS production, downregulate pro-inflammatory cytokines (e.g., TNF-α, IL-17, IL-12, IL-1β), and modulate pathways such as phosphoinositide 3-kinase (PI3K) and mitogen-activated protein kinase (MAPK). These effects extend to suppressing amyloid-beta (Aβ) aggregation in AD (Summer et al., 2024). Silver NPs exhibit anti-inflammatory properties by reducing hypoxia-inducible factor 1-alpha (HIF-1α) expression and suppressing pro-inflammatory cytokines, including TNF-α, IL-4, IL-5, and IL-13. They also mitigate cyclooxygenase-2 (COX-2) expression, reduce vascular endothelial growth factor (VEGF) levels, and inhibit mucin hypersecretion and T-helper type-2 mediated inflammation. Selenium NPs demonstrate anti-inflammatory activity by inhibiting NF-κB release and suppressing the expression of COX-2 and inducible nitric oxide synthase (iNOS) (Fakhri et al., 2022).

Additionally, titanium dioxide (TiO_2) NPs suppress thrombin-related inflammatory responses by reducing thrombin activity and interfering with protease-activated receptor signaling. CeO NPs, on the other hand, reduce neuronal death, microglial activation, oxidative stress, and pro-inflammatory cytokines (e.g., TNF-α) in models of retinal neurodegeneration (Bjelić & Finšgar, 2021).

2.1. Neuroapoptosis in Neurodegenerative Disorders (NDrs): The Role of Nanoparticles

Apoptosis, a form of programmed cell death, is characterized by distinct morphological changes, including chromatin condensation, nuclear membrane breakdown, cellular shrinkage, and apoptotic body formation. The apoptotic process is mediated through two primary pathways: the extrinsic pathway, also referred to as the death receptor pathway, and the intrinsic or mitochondrial pathway (Mishra et al., 2018). The intrinsic pathway is initiated intracellularly, leading to the upregulation and activation of BH3-only proteins, which subsequently activate Bax. In certain cases, Bak is also elevated, promoting the formation of mitochondrial membrane pores and facilitating the release of cytochrome c, which binds apoptotic peptidase activating factor-1 (APAF-1). This cascade activates caspases, culminating in downstream caspase cleavage and the degradation of cellular proteins. Conversely, the extrinsic pathway is triggered externally via death receptor activation by ligands, leading to caspase-8 activation and the subsequent activation of downstream caspases and BH3-only protein Bid (Roufayel, Younes, Al-Sabi, & Murshid, 2022).

Inhibition of key mediators within both intrinsic and extrinsic apoptotic pathways holds promise as a therapeutic strategy for mitigating neuronal apoptosis. NPs have demonstrated significant potential in enhancing the neuroprotective properties of various compounds in both in vitro and in vivo models. These technologies have been effective in mitigating neuronal inflammation, oxidative stress, and apoptosis (Tian et al., 2023). For instance, Wang et al. developed nanoparticle-based formulations to improve drug delivery, effectively reducing neuronal apoptosis and microglial neurotoxicity by modulating pro-inflammatory factors, inflammatory cell activity, and apoptosis after cerebral ischemia/reperfusion injury (Chu et al., 2024).

In another study, selenium NPs exhibited substantial neuroprotective effects by attenuating oxidative, inflammatory, and apoptotic cascades, reversing neurochemical alterations, and preventing neuronal loss in a mouse model (Yuan et al., 2020). Additionally, antioxidant enzyme-loaded NPs demonstrated protective effects against apoptosis in a rat model of spinal cord injury by improving mitochondrial function, reducing cytochrome c activity, and inhibiting caspase-3 activation, ROS production, and neuronal apoptosis. Moreover, the combination of gold NPs with other therapies significantly reduced neuronal apoptosis, decreased brain infarct volume, and increased neurotrophic factors in ischemic stroke models (Chiang, Yang, Nicol, & Wang, 2024).

2.2. Protein Aggregation in Neurodegenerative Disorders: The Role of Nanoparticles

The aggregation of proteins and peptides into amyloid fibrils is a central pathogenic mechanism in various neurodegenerative disorders, including AD, PD, HD, ALS. These aggregates are often formed from misfolded proteins such as amyloid-beta (Aβ), tau, and alpha-synuclein, which disrupt normal cellular processes and lead to neuronal dysfunction and death (Pathak et al., 2021).

In AD, amyloid plaques—extracellular aggregates composed primarily of Aβ peptides—and neurofibrillary tangles—intracellular accumulations of hyperphosphorylated tau protein—are hallmark features. Aβ aggregation and fibril formation are often initiated by metal-induced aggregation of the Aβ1-42 peptide, culminating in plaque deposition. Tau hyperphosphorylation disrupts microtubule stability, contributing to synaptic dysfunction and neuronal transport impairments (Muralidar, Ambi, Sekaran, Thirumalai, & Palaniappan, 2020).

Similarly, PD is characterized by the accumulation of mutant alpha-synuclein, which forms Lewy bodies within neurons. Mutations such as E46K, A30P, and A53T enhance the propensity of alpha-synuclein to misfold and aggregate, disrupting synaptic signaling and mitochondrial function. In ALS, mutations in genes such as SOD1 and other proteins like TDP-43, FUS, and UBQLN2 lead to the formation of toxic protein aggregates that impair cellular homeostasis and accelerate neuronal degeneration (Saramowicz et al., 2023).

2.3. Dual Functions of Nanoparticles: Neuroprotective and Neurotoxic Potential

NPs have shown considerable promise in enhancing therapeutic outcomes for NDrs by improving drug bioavailability, stability, and half-life, while reducing dosages and side effects. These advantages are particularly valuable in central nervous system (CNS)-targeted treatments. However, NPs also present potential risks, as they can induce inflammation, apoptosis, necrosis, autophagy, and oxidative stress, which may contribute to acute and chronic health complications. Despite these concerns, systematic clinical studies and comprehensive documentation of nanoparticle-induced toxicity in humans remain limited. Most findings regarding their toxicological effects are derived from animal models and cell culture studies (Mohammadinejad et al., 2019).

Understanding the mechanisms underlying nanoparticle-induced toxicity is essential for designing safer nanotechnologies. Reported toxicities include neurotoxicity, pulmonary toxicity, genotoxicity, and immunotoxicity, all of which depend on factors such as dose, frequency of exposure, nanoparticle concentration, route of administration, and physicochemical characteristics such as size, shape, surface charge, roughness, coating, and composition (Siafaka et al., 2024).

One prominent mechanism of nanoparticle-induced toxicity involves the generation of reactive oxygen species (ROS), including hydroxyl radicals, singlet oxygen, superoxide anions, and hydrogen peroxide (H_2O_2). ROS can disrupt cellular homeostasis, causing oxidative stress, membrane damage, DNA fragmentation, cytoskeletal dysfunction, protein misfolding, and enzyme inactivation. The oxidative reaction of NPs' surface groups or transition metals and the enhanced surface area of NPs amplify chemical reactivity and ROS production. Mitochondrial damage, resulting from nanoparticle interactions, further exacerbates ROS release and oxidative imbalance, leading to DNA damage, ion channel disruption, and cellular organelle dysfunction (Kessler, Hedberg, Blomberg, & Odnevall, 2022).

2.4. Nanoparticle-Induced Neurotoxicity: Evidence from Studies

Numerous studies have evaluated the neurotoxicity of NPs:

- **Polysorbate 80-modified chitosan nanoparticles:** Intravenous administration resulted in nanoparticle accumulation in the cerebellum and frontal cortex of rats, inducing oxidative stress, inflammation, and neuronal apoptosis (Ahmad et al., 2024).
- **Liposomal cisplatin formulations:** Direct brain administration caused severe neurotoxic effects due to the combined neurotoxicity of liposomes and cisplatin.

- **Polyamidoamine dendrimers:** These NPs exhibited charge-dependent neurotoxicity, inducing DNA damage, oxidative stress, apoptosis, and mitochondrial dysfunction in neural cells while suppressing neuronal differentiation.
- **Iron oxide nanoparticles:** Treatment led to oxidative stress, excessive iron accumulation, and protein aggregation in neural cells.
- **Silver nanoparticles:** Primary neurotoxicity mechanisms involved ROS generation and oxidative stress in cerebral astrocytes.
- **Silica nanoparticles:** In vivo and in vitro studies demonstrated neuropathology, MAPK activation, behavioral alterations, and neuronal damage.
- **Titanium dioxide (TiO2) nanoparticles:** Their neurotoxicity mechanisms included oxidative stress, genotoxicity, apoptosis, inflammatory responses, neurotransmitter dysregulation, DNA methylation, and autophagy disruption.

Cadmium telluride quantum dots: These NPs induced motor neuron toxicity via oxidative stress, autophagy acceleration, synaptic transmission impairments, and increased cytoplasmic calcium levels (H. Yang, Niu, Guo, & Xue, 2024).

2.5. Nanoparticles as a Strategy to Overcome the Blood-Brain Barrier

The limited ability of therapeutic compounds to traverse the BBB presents a significant obstacle to developing effective treatments for NDrs. The BBB is a highly specialized structure derived from the brain's microvascular system, functioning as a selective barrier between the circulating blood and the brain's extracellular environment in most vertebrates. It comprises a basal lamina, which contains extracellular matrix proteins such as collagen, heparan sulfate, and laminin that surround pericytes and endothelial cells. Astrocytic end feet and interneurons further contribute to the architecture of the BBB (Benz & Liebner, 2020).

Structurally, the BBB is characterized by tight junctions, gap junctions, and adherents' junctions, with tight junctions being the most crucial due to their role in establishing high electrical resistance between endothelial cells. The BBB's primary function is to protect the brain from foreign agents, regulate the ionic composition of the cerebrospinal fluid, metabolize chemical compounds, and dispose of waste products. Alterations in the extracellular matrix proteins can increase BBB permeability, thereby enhancing drug transport (Wu et al., 2023).

NPs are emerging as effective tools for crossing the BBB through passive and active transport mechanisms. Passive transfer involves small lipophilic molecules (<400 Da) and gold NPs, which can diffuse across the BBB (Kou et al., 2020). Active transport, mediated by receptor, carrier, and adsorption-based endocytosis, involves receptors such as low-density lipoprotein (LDL), transferrin, lactoferrin, and insulin. For instance, liposomes and PLGA-based NPs exploit receptor-mediated endocytosis to penetrate the BBB. Furthermore, positively charged NPs are more likely to undergo adsorption-mediated transport due to the negative charge of endothelial cell membranes. Carrier-mediated transport systems, including glucose transporter 1 (GLUT1) and amino acid transporter ASCT2, also facilitate nanoparticle delivery across the BBB (Kou et al., 2020).

IV. NANOFORMULATIONS IN COMBATING NEURODEGENERATION: PRECLINICAL ADVANCES

Recent pharmaceutical advancements have focused on leveraging nanotechnology in drug delivery, particularly for treating NDDs. NPs enhance drug delivery efficiency, improve bioavailability, enable co-delivery of multiple therapeutic agents, target intracellular delivery, and reduce systemic toxicity. These benefits address critical challenges in NDD treatment, particularly the inability of conventional drugs to achieve sufficient concentrations in the brain due to the BBB (Lamptey et al., 2022).

1. Nanomedicine for Alzheimer's Disease

AD, the leading cause of dementia, is characterized by progressive cognitive decline, impairments in language, visuospatial abilities, executive functions, and behavioral changes. AD pathology is marked by extracellular amyloid-beta (Aβ) plaques and intraneuronal tau-containing neurofibrillary tangles, which interact with inflammatory, oxidative stress, and apoptotic pathways. Despite the availability of pharmacological treatments, the BBB limits the brain's drug delivery, necessitating innovative approaches such as nanomedicine (Sohrabi & Weinborn, 2019). Nanoparticle-based therapies hold significant potential for addressing these challenges. They enhance drug biodistribution, improve pharmacokinetics, and facilitate the delivery of therapeutic agents across the BBB. These nanoformulations provide a promising avenue for advancing the treatment of AD and other neurodegenerative conditions by overcoming barriers to CNS drug delivery while minimizing systemic side effects (Anwarkhan, Koilpillai, & Narayanasamy, 2024).

2. Phytochemical-Based Nanoformulations for Alzheimer's Disease

The exploration of phytochemicals from the plant kingdom has unveiled significant potential in mitigating AD pathology. Resveratrol, a natural polyphenolic compound found in grapes, mulberries, peanuts, rhubarb, and wines, has demonstrated the capacity to depolymerize amyloid-beta (Aβ) peptides through proteasome-mediated mechanisms, albeit without direct effects on β- or γ-secretase enzymes or transthyretin (Fonseca-Santos & Chorilli, 2020). Studies indicate that resveratrol exhibits anti-tauopathy activity by reducing tau levels and suppressing tau hyperphosphorylation and accumulation in animal models. However, pharmacokinetic limitations, such as rapid systemic clearance, hinder its therapeutic application. To address these limitations, solid lipid NPs (SLNPs) have been developed for encapsulating resveratrol, enhancing its delivery to brain regions implicated in amyloid fibril formation. SLNs effectively inhibit Aβ(1–42) aggregate production, potentially slowing AD progression. Additionally, nanostructured lipid carriers (NLCs), administered intranasally, have shown superior efficacy compared to oral resveratrol solutions. These NLC-based formulations improved memory function and enhanced nasal mucosal permeability, presenting a viable strategy for AD treatment by leveraging lipid-oil mixtures to enhance particle stability (Grinán-Ferré et al., 2021).

3. Curcumin Nanoformulations for Neuroprotection

Curcumin, derived from *Curcuma longa,* exhibits substantial neurotherapeutic potential in treating neurodegenerative disorders, including AD. Despite its efficacy in reducing Aβ aggregation, oxidative damage, and neuroinflammation, curcumin's therapeutic utility is constrained by low bioavailability, rapid metabolism, and rapid elimination. Nanocurcumin formulations have been shown to overcome these limitations, preventing tau hyperphosphorylation, amyloid misfolding, and neurotoxicity in AD models.

Nanocurcumin enhances curcumin's solubility, stability, and bioavailability, allowing for targeted delivery across the BBB. PLGA (poly(lactic-co-glycolic acid)) NPs encapsulating curcumin protect neuronal cells from oxidative damage with minimal cytotoxicity while reducing Aβ aggregates. PLGA NPs conjugated with ligands for BBB translocation, such as PLGA-PEG-B6, significantly enhance curcumin delivery, improve spatial learning, and reduce hippocampal Aβ deposition and tau hyperphosphorylation in AD models (Lv, Wang, Yang, Ling, & Zhang, 2023).

4. Naringenin and Quercetin Nanoformulations

Naringenin, a flavonoid with antioxidative and anti-inflammatory properties, demonstrates neuroprotective effects but suffers from poor membrane permeability. Nanoemulsion formulations of naringenin significantly reduce Aβ-induced cytotoxicity and tau phosphorylation in neuronal cell models by attenuating amyloidogenic pathways, suggesting their applicability in AD therapeutics (Prasanna & Upadhyay, 2021).

Quercetin, another potent phytochemical, has demonstrated therapeutic efficacy against neurodegenerative disorders. Nanoencapsulation of quercetin into zein NPs enhances its bioavailability and oral absorption, improving cognitive and memory deficits in experimental AD models. Moreover, quercetin encapsulated in cyclodextrin-dodecyl carbonate NPs amplifies its anti-inflammatory activity by modulating toll-like receptor 4 (TLR4) and cyclooxygenase-2 (COX-2) pathways (Ramalho, Andrade, Loureiro, & do Carmo Pereira, 2020).

5. Nanoformulations of Synthetic Drugs for Alzheimer's Disease

Synthetic drugs have long been utilized in the treatment of AD, yet their therapeutic efficacy has been significantly enhanced through advanced nanoformulation approaches.

5.1. Erythropoietin (EPO) as a Neuroprotective Agent

Erythropoietin (EPO), a hematopoietic factor with neuroprotective potential in AD, promotes neuronal survival and regulates neurogenesis. However, its therapeutic application is limited due to poor central nervous system (CNS) penetration, attributed to its high molecular weight, hydrophilic nature, and rapid clearance from circulation. To overcome these limitations, Dara et al. developed EPO-loaded SLNPs. In vivo studies demonstrated that EPO-SLNs provided significant neuroprotection in an intra-hippocampal injection model, with notable recovery in spatial recognition memory compared to free EPO treatments (Dara et al., 2019).

5.2. Nicotinamide Nanoformulations

Nicotinamide, a histone deacetylase inhibitor shown to prevent AD progression in preclinical studies, suffers from poor absorption due to its hydrophilic nature. SLN formulations functionalized with polysorbate 80, phosphatidylserine, or phosphatidic acid significantly improved its brain-targeting capabilities. Among these, phosphatidylserine-functionalized SLNs demonstrated enhanced cognitive restoration, increased neuronal survival in hippocampal subregions, and reduced tau hyperphosphorylation in an AD rat model (Vakilinezhad et al., 2018).

5.3. Acetylcholinesterase Inhibitors: Rivastigmine, Donepezil, and Galantamine

Nanoformulations of acetylcholinesterase inhibitors (AChEIs) have shown substantial benefits in AD management. Rivastigmine-loaded liposomes facilitated faster memory recovery and metabolic stabilization in $AlCl_3$-treated rats, outperforming conventional drug solutions with sustained release and enhanced patient compliance. Rivastigmine-SLNs demonstrated superior drug diffusion without inducing nasociliary damage or necrosis, making them viable for nasal delivery. Similarly, donepezil-loaded liposomes exhibited robust stability, sustained release, and improved brain bioavailability via intranasal administration, allowing painless and self-administered treatment options. Galantamine hydrobromide, known for its limited BBB permeability, achieved substantial memory restoration and enhanced bioavailability through SLN encapsulation in cognitive deficit models (Rompicherla, Arumugam, Bojja, Kumar, & Rao, 2021).

5.4. Memantine Nanoformulations

Memantine, an NMDA receptor antagonist approved for AD treatment, was incorporated into PEGylated biodegradable PLGA NPs using a double-emulsion technique to optimize BBB targeting. Both in vitro and in vivo studies confirmed the NPs' ability to sustain brain drug levels and reduce Aβ plaques, demonstrating their therapeutic potential in AD management (Sánchez-López et al., 2018).

5.5. Estradiol Nanoformulations in Postmenopausal AD

Estrogen deficiency post-menopause has been implicated in increased AD risk in women. Tween 80-coated PLGA NPs loaded with estradiol significantly elevated brain estradiol levels in an ovariectomized rat model after 24 hours, outperforming unmodified formulations. These findings suggest a promising approach for hormone-based AD therapies (Fakhri et al., 2022).

5.6. Tarenflurbil: γ-Secretase Modulator

Tarenflurbil, an Aβ42-selective reducing agent and γ-secretase modulator, showed promising outcomes in early preclinical and Phase II clinical studies. However, limited brain penetration contributed to its failure in Phase III trials. Nanocarriers such as PLGA NPs and SLNs improved its brain biodistribution and pharmacokinetics, providing an avenue for optimizing its clinical efficacy (Duan et al., 2024).

5.7. Clioquinol and Basic Fibroblast Growth Factor (bFGF) Nanoformulations

Clioquinol, a quinolone derivative capable of solubilizing extracellular Aβ plaques in the neocortex, demonstrated improved BBB transport when encapsulated in n-butyl cyanoacrylate NPs. Similarly, basic fibroblast growth factor (bFGF), known to prevent neuronal degeneration and enhance learning in AD models, was successfully delivered via Solanum tuberosum lectin-conjugated PEG-PLGA NPs. This intranasal delivery approach enabled efficient brain targeting with reduced peripheral side effects (Poovaiah et al., 2018).

6. Nanoformulations in Parkinson's Disease

PD is a neurodegenerative disorder characterized by the progressive degeneration of dopaminergic neurons in the substantia nigra pars compacta of the basal ganglia. The primary pathological hallmarks of PD include oxidative stress, neuroinflammation, and apoptosis. Genetic factors, particularly mutations in the α-synuclein gene, are implicated in the pathogenesis, with α-synuclein-encoded protein aggregation playing a pivotal role in disease onset (Guo, Zhao, Li, Li, & Liu, 2018).

6.1. Phytochemical Nanoformulations for Parkinson's Disease

Phytochemicals have garnered attention as potential therapeutic agents for PD, with nanoformulations offering enhanced bioavailability and pharmacological efficacy.

6.1.1. Resveratrol Nanoformulations

Resveratrol, known for its antioxidative and neuroprotective properties, has demonstrated limited therapeutic potential due to rapid clearance and poor CNS penetration. Nanoparticle-based formulations of resveratrol have overcome these barriers, sustaining blood concentrations and enhancing pharmacological outcomes. In experimental models, resveratrol NPs attenuated rotenone-induced behavioral deficits, oxidative stress, and mitochondrial dysfunction, preserving complex-I and tricarboxylic acid enzyme activities (Nowacka et al., 2024).

Magnetically targeted Fe_3O_4-modified resveratrol liposomes enabled sustained and delayed drug release, crossing the BBB efficiently under external magnetic fields. Additionally, resveratrol-loaded nanoemulsions showed high nasal mucosa penetration and reduced oxidative stress and histopathological damage in brain tissues. Polysorbate 80-coated PLA NPs further enhanced brain resveratrol concentrations, demonstrating potential for neuroprotection in NDrs such as PD (Fonseca-Santos & Chorilli, 2020).

6.1.2. Curcumin and Piperine Nanoformulations

Curcumin-loaded polybutyl cyanoacrylate NPs exhibited neuroprotective effects, crossing the BBB and mitigating PD symptoms. Co-encapsulation of curcumin and piperine in glyceryl monooleate NPs enhanced therapeutic efficacy by inhibiting protein oligomerization, reducing rotenone-induced oxidative stress, and initiating autophagic pathways. These NPs reversed motor impairments and prevented dopaminergic neuronal degeneration in in vivo PD models (Fatima & Siddique, 2024).

6.1.3. Naringenin Nanoemulsions

Naringenin, a flavonoid with antioxidant properties, demonstrated synergistic effects with conventional levodopa therapy when administered via vitamin E-loaded nanoemulsions. These nanoemulsions restored motor coordination, grip strength, and swimming activity in 6-hydroxydopamine-treated rats while improving brain bioavailability. Biochemical analysis revealed increased glutathione (GSH) and superoxide dismutase (SOD) levels, with reduced malondialdehyde (MDA) concentrations, highlighting their antioxidative efficacy (Bhia et al., 2021).

6.1.4. Quercetin Nanocrystals

Quercetin, a bioflavonoid with potent mitochondrial and anti-inflammatory effects, exhibits limited bioavailability and solubility, which constrain its therapeutic applications. Nanocrystal formulations of quercetin significantly improved its bioavailability, enhancing mitochondrial quality control, ROS scavenging, and antioxidant enzyme activity in PD-like models. Treated animals exhibited elevated hippocampal GSH levels and reduced lipid peroxidation, signifying neuroprotection (Batiha et al., 2020).

6.1.5. Mechanistic Insights and Therapeutic Potential

Nanoformulations of phytochemicals such as resveratrol, curcumin, piperine, naringenin, and quercetin have demonstrated multifaceted therapeutic effects in PD by enhancing bioavailability, alleviating rotenone-induced toxicity, and mitigating oxidative stress and inflammation. Mechanistically, these nanophytochemicals promote neurogenesis, inhibit apoptotic signaling, and suppress neuroinflammatory pathways, offering a promising avenue for disease management (Fakhri et al., 2022).

6.2. Parkinson's Disease and Nanoformulations of Synthetic Drugs

Several synthetic pharmaceuticals are currently prescribed for the symptomatic management of PD, but their clinical efficacy remains limited by significant pharmacokinetic constraints. Levodopa, a precursor of dopamine, is the most widely used and therapeutically beneficial drug in PD treatment. However, its clinical effectiveness is unpredictable due to inconsistent oral absorption and fluctuations in plasma concentrations, resulting in variations in motor function and the development of drug-induced involuntary movements, complicating PD management (di Biase, Pecoraro, Carbone, Caminiti, & Di Lazzaro, 2023). In rodent models, levodopa administration has been associated with the onset of dyskinesia. Notably, the application of levodopa methyl ester/benserazide-loaded NPs in dyskinetic rats reduced dyskinesia and avoided the accumulation of molecules strongly associated with dyskinesia development (Nakmode, Day, Song, & Garg, 2023).

Bromocriptine, an ergot alkaloid with dopamine receptor agonist activity, is commonly employed to delay the onset of motor fluctuations associated with chronic levodopa therapy. Bromocriptine exerts neuroprotective effects by directly and indirectly stimulating dopamine receptors, safeguarding dopaminergic cells (Carrera et al., 2018). Recent research has suggested that solid lipid NPs (SLNs) composed of a tristearin/tricaprin blend may provide a novel method for controlled delivery of bromocriptine, thereby enhancing therapeutic efficacy through sustained drug release and reducing long-term adverse effects. Intranasal administration of bromocriptine-loaded chitosan NPs further demonstrated improved brain absorption, mitigating symptoms such as catalepsy and akinesia in PD mouse models (Zeeshan, Mishra, & Kesharwani, 2021).

A new delivery system for ropinirole, a non-ergoline dopamine receptor agonist, was developed using PLGA NPs. In animal models, this formulation reversed PD-like symptoms. Ropinirole targets D2-dopamine receptors in the striatum and substantia nigra, mimicking dopamine's action to increase neuronal firing rates (Yergök, Derman, Alakbarli, & Eskhıta, 2024). Nanoparticle-based delivery systems offer an exciting alternative to traditional formulations, potentially reducing dosing frequency, optimizing the therapeutic index, and minimizing peak-to-valley fluctuations while reducing side effects (Fakhri et al., 2022).

Pramipexole, another non-ergot dopamine agonist, has been shown to delay PD progression when used in combination with levodopa or monoamine oxidase-B inhibitors. Intranasal administration of pramipexole-loaded chitosan NPs demonstrated superior control over motor deficits in a rotenone-induced PD model compared to oral or solution formulations. This approach also enhanced antioxidant activity, increasing the levels of superoxide dismutase (SOD) and catalase (CAT) while elevating brain dopamine concentrations (Jing, Yang, Taximaimaiti, & Wang, 2023).

Apomorphine, a dopamine receptor agonist, suffers from poor oral bioavailability (1.7%) due to rapid gastrointestinal degradation and the first-pass effect. To overcome these limitations, SLNs were employed as carriers for apomorphine, improving its bioavailability and brain delivery. This formulation reduced the frequency and dosage required for effective treatment while ensuring targeted delivery to the brain striatum, a region severely affected by PD (Kassubek, Jost, & Schwarz, 2024).

Selegiline, a monoamine oxidase inhibitor, encapsulated in nanoemulsions, demonstrated enhanced antioxidant activity, including elevated GSH and SOD levels, and reduced thiobarbituric acid reactive substances (TBARS) compared to free selegiline. Additionally, in a haloperidol-induced rat model of PD, selegiline-loaded nanoemulsions increased drug bioavailability and brain uptake, leading to decreased dopamine depletion. These findings suggest that nanoformulated selegiline may offer an effective intranasal delivery strategy, mitigating oxidative stress and metabolic alterations in PD (Silva, Almeida, & Vale, 2021).

Pentamidine, a potent inhibitor of S100B, a key mediator of neuroinflammation in PD, has limited ability to cross the BBB. Researchers have developed chitosan-coated niosomes containing pentamidine, which successfully mitigated glial-derived S100B activity, restoring dopaminergic function and significantly improving parkinsonian motor deficits (Rinaldi et al., 2019).

6.2.1. Amyotrophic Lateral Sclerosis and Nanoformulations

Amyotrophic Lateral Sclerosis (ALS), also known as motor neuron disease or Lou Gehrig's disease, is a rapidly progressing neurodegenerative disorder that leads to the failure of motor neurons. ALS specifically targets motor neurons in the brain, brainstem, and spinal cord, resulting in progressive degeneration, muscular atrophy, and ultimately paralysis and death due to respiratory failure. The pathogenesis of ALS is linked to gene mutations that impair normal protein function, oxidative stress, protein misfolding and aggregation, mitochondrial dysfunction, and neuroinflammation (Sacks, Bashford, Wijesekera, Leigh, & Sreedharan, 2022).

6.2.2. Amyotrophic Lateral Sclerosis and Nanoformulations of Phytochemicals

In 2018, Ahmadi et al. conducted a randomized clinical trial evaluating the safety and efficacy of oral capsules containing 80 mg of nanomicelles curcuminoids combined with riluzole in ALS patients. The findings indicated that nanocurcumin was safe, with no significant adverse events, and it enhanced the likelihood of survival, particularly in patients with irritable bladder symptoms (Ahmadi et al., 2018). Furthermore, pathological inclusions, such as superoxide dismutase 1 (SOD1), which are characteristic of ALS, resemble amyloid fibrils formed in vitro. Fibrillar aggregates are prominent in neural tissues of mouse models expressing human SOD1 mutations. Inhibiting the development of SOD1 amyloids presents a potential therapeutic strategy for ALS. Curcumin has been shown to prevent SOD1 fibrillation and promote the formation of smaller, less structured aggregates. The use of curcumin NPs, with enhanced water solubility, yielded similar effects in modulating protein aggregation as bulk curcumin (Wright, Antonyuk, & Hasnain, 2019).

Additionally, Tripodo et al. introduced an innovative drug delivery system to treat NDs, particularly ALS. This system employed a "carrier-in-carrier" approach, loading curcumin and mesenchymal stromal cells into curcumin-loaded micelles. These NPs achieved optimal loading in micelle-loaded mesenchymal stromal cells within minutes, and the loading was found to be concentration-dependent. While free curcumin caused apparent cytotoxicity in mesenchymal stromal cells, the micellar formulation shielded the cells from this adverse effect, providing a safer alternative (Kakkar, Kumari, Kaur, & Chholta, 2024).

6.2.3. Amyotrophic Lateral Sclerosis and Nanoformulations of Synthetic Drugs

Currently, the FDA has approved only two drugs for ALS management: Riluzole and Edaravone, both of which act by scavenging reactive oxygen species (ROS) to protect neurons. Riluzole, being a substrate for P-glycoprotein (P-gp), faces restricted distribution across the BBB, limiting its effectiveness in treating ALS. To overcome this, verapamil, a P-gp inhibitor, was utilized in a liposomal co-delivery system, which successfully reduced P-gp expression and enhanced riluzole delivery to the brain by mitigating efflux transporter activity (T. Yang et al., 2018). This strategy could potentially overcome pharmacoresistance to riluzole, improving ALS therapy outcomes. SLNPs containing riluzole demonstrated superior drug loading and enhanced efficacy compared to free riluzole, showing improved brain delivery with reduced off-target biodistribution in rat models (Correia et al., 2022).

Dimethyl fumarate, another drug used in ALS treatment, is known to cause side effects such as flushing and gastrointestinal discomfort. However, SLNs coated with polysorbate 80 and cationized with dimethyldioctadecylammonium chloride were evaluated as intranasal carriers. In vitro studies using a mouse brain microvascular endothelial cell model revealed that cationic SLNs exhibited improved brain penetration. Fluorescence imaging following intraperitoneal or intranasal administration showed that polysorbate 80-coated SLNs could enter the brain via both routes, although they were predominantly distributed in the liver and spleen when administered intraperitoneally (Lombardo, Musumeci, Carbone, & Pignatello, 2021).

Neurodegeneration in ALS models is often associated with the proliferative repair efforts of ependymal stem progenitor cells, which are typically quiescent in the spinal cord. Modulating the proliferation of these progenitor cells may represent a viable strategy to prevent further neurodegeneration. In the G93A-SOD1 transgenic mouse model of ALS, a mutant human SOD1 gene is overexpressed, leading to accelerated disease progression. The administration of FM19G11, a hypoxia-inducible factor (HIF) modulator loaded into gold NPs, activated the PI3K/Akt signaling pathway and mitochondrial uncoupling protein (UCP2) pathway, both of which regulate stem cell properties such as self-renewal and proliferation. This approach may provide new avenues for slowing disease progression in ALS patients (Cecerska-Heryć et al., 2023).

6.3. Stroke and Nanoformulations

Cerebrovascular diseases represent the primary etiology of physical disabilities, the second leading cause of mortality, and a predominant factor contributing to hospitalization. Stroke, the most prevalent cerebrovascular disorder, ranks as the third leading cause of death in developed nations, resulting in approximately 15 million incidents and 5 million fatalities annually. Ischemic stroke, constituting 80% of all strokes, arises from factors such as hypoperfusion, thrombosis, or embolism (Ezennaka & Dodiyi-Manuel, 2021). The pathophysiology of ischemic stroke is influenced by multiple mechanisms, including oxidative stress, apoptosis, ionic imbalance-induced edema, and inflammation. Clinical studies have demonstrated that many herbal formulations derived from natural products exhibit biological activities that are attributed to their antioxidant properties, which are thought to be central to their therapeutic effects in ischemic stroke management. However, the clinical application of these phytochemicals is hindered by their limited bioavailability, which reduces their efficacy (Parham et al., 2020).

6.3.1. Stroke and Phytochemical Nanoformulations

In an experimental rat model of transient middle cerebral artery occlusion, intra-arterial injection of resveratrol NPs provided enhanced protection against cerebral ischemia/reperfusion (I/R) injury. These NPs alleviated oxidative stress induced by ischemia-reperfusion by reducing malondialdehyde (MDA) levels, mitigated brain edema, and protected neurons from apoptosis by lowering Bax expression and cleaved caspase-3 activity. Additionally, resveratrol NPs promoted neurogenesis through the upregulation of brain-derived neurotrophic factor (BDNF) (Lu et al., 2020).

Curcumin has demonstrated vascular protective effects in individuals at risk for stroke. Curcumin incorporated within SLNPs resists enzymatic degradation during absorption, prolongs circulation time, and reduces systemic elimination. Treatment with curcumin-SLNs enhanced antioxidant enzyme activity, including superoxide dismutase (SOD), catalase (CAT), glutathione (GSH), and mitochondrial complex enzymes, while simultaneously reducing lipid peroxidation, nitrite levels, and acetylcholinesterase activity (Parvez et al., 2022).

In another study, oral administration of nanoencapsulated quercetin demonstrated significant neuroprotection by reducing oxidative damage. Quercetin NPs decreased the expression of inducible nitric oxide synthase (iNOS) and attenuated caspase-3 activity, thus preventing the loss of pyramidal neurons from the hippocampal CA1 and CA3 subfields in both young and aged rats subjected to I/R injury. Early intervention with these nanodrugs improved survival rates by minimizing neuronal loss (Zhang et al., 2023). Furthermore, quercetin-loaded polymeric nanocapsules, which delivered quercetin to mitochondria, exhibited enhanced brain uptake and mitochondrial localization. This mitochondrial-targeted delivery system reduced histopathological damage by preserving mitochondrial integrity and function, decreasing reactive oxygen species (ROS), and increasing antioxidant enzyme activities (SOD, CAT) to regulate mitochondrial ROS-induced apoptotic cell death (T. Khan et al., 2022).

Panax notoginseng's active component, Panax notoginsenoside, has shown potential therapeutic effects when delivered through a core-shell hybrid liposomal vesicle (HLV) system, which enhances the bioactivity and bioavailability of the free drug. This formulation resulted in a significant reduction in brain water content and infarct volume in a model of cerebral ischemia/reperfusion injury. Treatment with Panax notoginsenoside-HLV also restored enzymatic levels and improved physicochemical characteristics compared to the control group, highlighting its potential in ischemic stroke management (Kim et al., 2018).

Nanoformulations of naringenin, another promising phytochemical, demonstrated prolonged release profiles and neuroprotective effects in human mesenchymal stem cells. Naringenin-loaded gelatin-coated polycaprolactone NPs reduced inflammatory responses by modulating the NF-κB signaling pathway, effectively mitigating oxygen-glucose deprivation-induced inflammation (e.g., TNF-α, IL-1β, COX-2, iNOS) (Fakhri et al., 2022).

Retinoic acid NPs have also been shown to modulate microglial activation in vitro, inhibiting lipopolysaccharide (LPS)-induced nitric oxide (NO) release and iNOS expression while promoting arginase-1 and IL-4 production, further contributing to neuroprotection (Machado-Pereira, Santos, Ferreira, Bernardino, & Ferreira, 2018).

6.3.2. Stroke and Nanoformulations of Synthetic Drugs

Research has shown that recombinant tissue plasminogen activator (rtPA) encapsulated in PEG-PCL NPs (at 10% of the standard dose) effectively dissolves fibrin clots while demonstrating significant neuroprotective effects following localized cerebral ischemia. These NPs exhibited a half-life approximately 18 times greater than free rtPA, resulting in reduced infarct volumes and improved neurological function (Zenych, Fournier, & Chauvierre, 2020).

Liposomal fasudil, administered prior to rtPA, has been shown to decrease the incidence of rtPA-induced cerebral hemorrhage and extend the therapeutic window for thrombolytic treatment, presenting a promising combined therapeutic strategy for ischemic stroke (Ma, Jiang, Xu, Liu, & Guo, 2021).

Cerium oxide (CeO) NPs possess antioxidant properties that aid in cellular viability and substantially decrease free radical generation, offering protective effects in an animal model of stroke by reducing cellular death. Additionally, platinum (Pt) NPs act as ROS scavengers, both in vitro and in vivo, mitigating oxidative stress caused by ischemia and improving motor function while reducing infarct volume in the brain (Brandão Da Silva Assis et al., 2024).

Single-walled carbon nanotubes have exhibited anti-inflammatory effects in a rat stroke model, preventing neuronal injury induced by ischemia. Moreover, local administration of superoxide dismutase (SOD)-loaded NPs has shown prolonged neuroprotective effects, with a 65% reduction in infarct volume compared to SOD solution treatment, which increased infarct volume by 25%. The extended lifespan observed in rats treated with SOD-loaded NPs underscores their superior protective efficacy (Kong, Chu, & Wang, 2022).

Vascular endothelial growth factor (VEGF) has been shown to promote neurogenesis and cerebral angiogenesis, reducing ischemic brain damage. VEGF-encoding plasmids encapsulated in liposomes functionalized with transferrin were able to cross the BBB upon intravenous administration, enhancing neuroprotection and stimulating neovascularization (Fakhri et al., 2022).

Similarly, the combined delivery of VEGF and angiopoietin-1 (Ang1) through hyaluronic acid-based biodegradable hydrogels mixed with PLGA (poly(lactic-co-glycolic acid)) has demonstrated potential in inducing angiogenesis and facilitating neural repair in the ischemic brain. These hydrogels support vascularization, axonal development, and favorable niches for tissue repair post-stroke (Pal, Sharma, Bani, & Gupta, 2024).

6.4. Other Neurodegenerative Diseases (NDDs) and Nanoformulations

Nanoformulations have demonstrated significant therapeutic potential in combating various NDs, including Huntington's disease (HD) and multiple sclerosis (MS). HD is a hereditary neurodegenerative disorder inherited in an autosomal dominant manner, characterized by progressive motor dysfunction, cognitive decline, and psychiatric symptoms. The disease is driven by an abnormal trinucleotide repeat expansion (cytosine-adenine-guanine, CAG) within the Huntingtin gene, which leads to an increased polyglutamine sequence at the N-terminal of the huntingtin protein (HTT) (Vishwas et al., 2021). While the precise mechanisms of neuronal death in HD remain elusive, key contributing factors include glutamate excitotoxicity, mitochondrial dysfunction, impaired protein degradation, protein misfolding, caspase activation, transcriptional dysregulation, and defects in proteasome activity and proteolysis (Lontay, Kiss, Virág, & Tar, 2020).

Recent studies have shown the efficacy of curcumin-loaded SLNPs in ameliorating HD-like neuropathology induced by 3-nitropropionic acid (3-NP) in rodents. The enhanced bioavailability of curcumin encapsulated in SLNs offers notable advantages over free curcumin, particularly in improving mitochondrial function and increasing cytochrome enzyme activity. Curcumin-SLN treatment significantly restored glutathione (GSH) levels and superoxide dismutase (SOD) activity, leading to improved neuromotor coordination in 3-NP-treated rats. Additionally, curcumin-loaded NPs, based on amphiphilic hyaluronan conjugates, exhibited neuroprotective effects in an in vitro HD model (Jacob et al., 2024).

Similarly, rosmarinic acid-loaded SLNs demonstrated a marked reduction in 3-NP-induced oxidative stress and neurobehavioral impairments, including deficits in body weight, locomotor activity, and motor coordination. Treatment with rosmarinic acid-loaded SLNs restored endogenous antioxidant enzyme activities (e.g., CAT and GSH) and showed superior results when administered via the nasal route, compared to intravenous delivery, by optimizing brain drug concentration (Islam, Mishra, Ahsan, & Fareha, 2023).

6.4.1. Multiple Sclerosis (MS) and Nanoformulations

Multiple sclerosis (MS) is a chronic, immune-mediated demyelinating disorder of the central nervous system (CNS), which can present as relapsing-remitting or progressive forms, leading to axonal damage and paralysis. Typically manifesting between the ages of 20 and 50, MS affects more women than men and results in a diverse spectrum of physical, neurological, and psychological symptoms, including muscle weakness, spasticity, coordination deficits, and impaired movement. Nanotechnology has emerged as a promising approach for the diagnosis and treatment of MS and other CNS-related disorders (Silveira, Guedes, Maia, Curral, & Coelho, 2019).

Dimethyl fumarate, a drug under investigation for MS treatment, is associated with issues such as frequent dosing, inadequate brain penetration, and gastrointestinal side effects. Recent studies have demonstrated that nanoencapsulation of dimethyl fumarate in lipid-based nanocarriers allows for a once-daily dosing regimen, significantly improving critical clinical parameters such as motor coordination, grip strength, weight maintenance, and locomotor activity. Furthermore, SLN-based formulations of dimethyl fumarate enhanced bioavailability, prolonged systemic residence time, and improved brain absorption, suggesting that lipid nanocarriers hold substantial promise for optimizing MS treatment (Edwards et al., 2021).

6.4.2. Exosomes and Other Nanoparticles in MS Treatment

Exosomes, nanoscale vesicles derived from dendritic cells, have gained attention as an alternative to liposomes for drug delivery in MS. These exosomes, rich in proteins and RNA, can cross the BBB efficiently and without inducing toxicity. They have been shown to promote myelination by stimulating oligodendrocyte precursor cell differentiation into myelin-producing cells, offering a promising strategy for enhancing remyelination following CNS injury (Sidoryk-Węgrzynowicz, Dąbrowska-Bouta, Sulkowski, & Strużyńska, 2021). Furthermore, the nasal administration of serum-derived exosomes in naïve rats improved myelin content, increased oligodendrocyte precursor cells and neural stem cell populations, and reduced oxidative damage in hippocampal slice cultures (Domingues, Falcão, Mendes-Pinto, Salgado, & Teixeira, 2020).

Poly(lactic-co-glycolic acid) (PLGA) NPs conjugated with myelin peptides have demonstrated anti-inflammatory properties and modulated effector T cell function, providing a potential therapeutic approach for MS. Additionally, cerium oxide (CeO) NPs, stabilized with citrate/EDTA, have shown antioxidant properties in reducing reactive oxygen species (ROS) levels in a mouse model of experimental autoimmune encephalomyelitis (EAE), a model of MS. Treatment with CeO NPs significantly reduced clinical symptoms, motor deficits, and CNS damage caused by oxidative stress (Heckman et al., 2020).

6.4.3. Artificial Neurotrophins and Nanoformulations in Neurodegeneration

Artificial neurotrophins, as promising therapeutic agents for NDs, have also been highlighted in recent studies. These compounds are designed to target specific mechanisms involved in neuronal degeneration and repair. Together with other phytochemical and synthetic nanoformulations, they offer potential therapeutic strategies for combating neurodegeneration across a broad spectrum of NDDs (Bhattacharya et al., 2022).

CONCLUSION

In conclusion, the integration of nanotechnology and phytomedicine holds immense potential for revolutionizing NDD therapy. By leveraging advanced nanocarriers, researchers can overcome the formidable challenge posed by the BBB, optimize drug bioavailability, and achieve targeted neuroprotection. The development of multifunctional NPs, capable of simultaneously delivering anti-inflammatory, antioxidant, and neurogenic agents, represents a promising direction for future neurotherapeutics. Furthermore, continued exploration of nano-enabled drug delivery systems, combined with computational modeling and precision medicine, will pave the way for next-generation therapies that not only mitigate disease progression but also restore neuronal function. Ultimately, the convergence of nanotechnology, bioinformatics, and phytomedicine offers a paradigm shift in the treatment of NDDs, bringing us closer to effective, patient-specific, and curative strategies for these debilitating disorders.

REFERENCES

Ahmad, N., Khan, M. F., Ullah, Z., Chaudhary, A. A., Alawam, A. S., Khalid, M. S., & Ali, M. T. (2024). Development and evaluation of polysorbate-80 coated Mangiferin PLGA nanoparticles used in the treatment of cerebral ischemia. *Polymer Bulletin*, *81*(8), 7035–7069.

Ahmadi, M., Agah, E., Nafissi, S., Jaafari, M. R., Harirchian, M. H., Sarraf, P., & Aghamollaii, V. (2018). Safety and efficacy of nanocurcumin as add-on therapy to riluzole in patients with amyotrophic lateral sclerosis: A pilot randomized clinical trial. *Neurotherapeutics; the Journal of the American Society for Experimental NeuroTherapeutics*, *15*(2), 430–438. PMID: 29352425

Akhtar, N., & Khan, R. A. (2016). Liposomal systems as viable drug delivery technology for skin cancer sites with an outlook on lipid-based delivery vehicles and diagnostic imaging inputs for skin conditions'. *Progress in Lipid Research*, *64*, 192–230. PMID: 27697511

Amaral, M., Cruz, N., Rosa, A., Nogueira, B., Costa, D., Santos, F., & Kobozev, Y. (2021). An update of advanced nanoplatforms for Glioblastoma Multiforme Management. *EXCLI Journal*, *20*, 1544. PMID: 34924904

Anwarkhan, S., Koilpillai, J., & Narayanasamy, D. (2024). Utilizing multifaceted approaches to target drug delivery in the brain: From nanoparticles to biological therapies. *Cureus*, *16*(9), e68419. PMID: 39360065

Batiha, G. E.-S., Beshbishy, A. M., Ikram, M., Mulla, Z. S., El-Hack, M. E. A., Taha, A. E., & Elewa, Y. H. A. (2020). The pharmacological activity, biochemical properties, and pharmacokinetics of the major natural polyphenolic flavonoid: Quercetin. *Foods*, *9*(3), 374. PMID: 32210182

Bellanthudawa, B., Nawalage, N., Handapangoda, H., Suvendran, S., Wijayasenarathne, K., Rathnasuriya, M., & Perera, I. (2023). A perspective on biodegradable and non-biodegradable nanoparticles in industrial sectors: Applications, challenges, and future prospects. *Nanotechnology for Environmental Engineering*, *8*(4), 975–1013.

Benz, F., & Liebner, S. (2020). *Structure and function of the blood–brain barrier (BBB) Physiology, Pharmacology and Pathology of the Blood-Brain Barrier*. Springer.

Bhattacharya, T., Soares, G. A. B., Chopra, H., Rahman, M. M., Hasan, Z., Swain, S. S., & Cavalu, S. (2022). Applications of phyto-nanotechnology for the treatment of neurodegenerative disorders. *Materials (Basel)*, *15*(3), 804. PMID: 35160749

Bhia, M., Motallebi, M., Abadi, B., Zarepour, A., Pereira-Silva, M., Saremnejad, F., & Jafari, S. M. (2021). Naringenin nano-delivery systems and their therapeutic applications. *Pharmaceutics*, *13*(2), 291. PMID: 33672366

Bjelić, D., & Finšgar, M. (2021). The role of growth factors in bioactive coatings. *Pharmaceutics*, *13*(7), 1083. PMID: 34371775

Brandão Da Silva Assis, M., Nestal De Moraes, G., & De Souza, K. R. (2024). Cerium oxide nanoparticles: Chemical properties, biological effects and potential therapeutic opportunities. *Biomedical Reports*, *20*(3), 48. PMID: 38357238

Carrera, I., Fernandez-Novoa, L., Sampedro, C., Tarasov, V. V., Aliev, G., & Cacabelos, R. (2018). Dopaminergic Neuroprotection with Atremorine in Parkinson´ s Disease. *Current Medicinal Chemistry, 25*(39), 5372–5388. PMID: 29637853

Cecerska-Heryć, E., Pękała, M., Serwin, N., Gliźniewicz, M., Grygorcewicz, B., Michalczyk, A., & Dołęgowska, B. (2023). The use of stem cells as a potential treatment method for selected neurodegenerative diseases. *Cellular and Molecular Neurobiology, 43*(6), 2643–2673. PMID: 37027074

Chiang, M.-C., Yang, Y.-P., Nicol, C. J., & Wang, C.-J. (2024). Gold nanoparticles in neurological diseases: A review of neuroprotection. *International Journal of Molecular Sciences, 25*(4), 2360. PMID: 38397037

Chu, J., Zhang, W., Liu, Y., Gong, B., Ji, W., Yin, T., & Chen, C. (2024). Biomaterials-based anti-inflammatory treatment strategies for Alzheimer's disease. *Neural Regeneration Research, 19*(1), 100–115. PMID: 37488851

Correia, A., Monteiro, A., Silva, R., Moreira, J., Lobo, J. S., & Silva, A. (2022). Lipid nanoparticles strategies to modify pharmacokinetics of central nervous system targeting drugs: Crossing or circumventing the blood–brain barrier (BBB) to manage neurological disorders. *Advanced Drug Delivery Reviews, 189*, 114485. PMID: 35970274

Dara, T., Vatanara, A., Sharifzadeh, M., Khani, S., Vakilinezhad, M. A., Vakhshiteh, F., & Mosaddegh, M. H. (2019). Improvement of memory deficits in the rat model of Alzheimer's disease by erythropoietin-loaded solid lipid nanoparticles. *Neurobiology of Learning and Memory, 166*, 107082. PMID: 31493483

Dhineshbabu, N., Rajendran, V., Nithyavathy, N., & Vetumperumal, R. (2016). Study of structural and optical properties of cupric oxide nanoparticles. *Applied Nanoscience, 6*(6), 933–939.

di Biase, L., Pecoraro, P. M., Carbone, S. P., Caminiti, M. L., & Di Lazzaro, V. (2023). Levodopa-induced dyskinesias in Parkinson's disease: An overview on pathophysiology, clinical manifestations, therapy management strategies and future directions. *Journal of Clinical Medicine, 12*(13), 4427. PMID: 37445461

Domingues, H. S., Falcão, A. M., Mendes-Pinto, I., Salgado, A. J., & Teixeira, F. G. (2020). Exosome circuitry during (De)(Re) myelination of the central nervous system. *Frontiers in Cell and Developmental Biology, 8*, 483. PMID: 32612996

Duan, M., Cao, R., Yang, Y., Chen, X., Liu, L., Ren, B., & Goh, B.-C. (2024). Blood–brain barrier conquest in glioblastoma nanomedicine: Strategies, clinical advances, and emerging challenges. *Cancers (Basel), 16*(19), 3300. PMID: 39409919

Dunuweera, S. P., Rajapakse, R. M. S. I., Rajapakshe, R. B. S. D., Wijekoon, S. H. D. P., Nirodha Thilakarathna, M. G. G. S., & Rajapakse, R. M. (2019). Review on targeted drug delivery carriers used in nanobiomedical applications. *Current Nanoscience, 15*(4), 382–397.

Ealia, S. A. M., & Saravanakumar, M. (2017). *A review on the classification, characterisation, synthesis of nanoparticles and their application.* Paper presented at the IOP Conference Series: Materials Science and Engineering.

Edwards, K. R., Kamath, A., Button, J., Kamath, V., Mendoza, J. P., Zhu, B., & Penner, N. (2021). A pharmacokinetic and biomarker study of delayed-release dimethyl fumarate in subjects with secondary progressive multiple sclerosis: Evaluation of cerebrospinal fluid penetration and the effects on exploratory biomarkers. *Multiple Sclerosis and Related Disorders*, *51*, 102861. PMID: 33773271

Ezema, B. O., Eze, C. N., Ayoka, T. O., & Nnadi, C. O. (2024). Antioxidant-enzyme Interaction in Non-communicable Diseases. *Journal of Exploratory Research in Pharmacology*, *9*(4), 262–275.

Ezennaka, R., & Dodiyi-Manuel, S. (2021). Stroke; A Major Cause of Cardiovascular Mortality-A Review. *Saudi J Med*, *6*(9), 283–294.

Fakhri, S., Abdian, S., Zarneshan, S. N., Moradi, S. Z., Farzaei, M. H., & Abdollahi, M. (2022). Nanoparticles in combating neuronal dysregulated signaling pathways: Recent approaches to the nanoformulations of phytochemicals and synthetic drugs against neurodegenerative diseases. *International Journal of Nanomedicine*, 299–331. PMID: 35095273

Fatima, J., & Siddique, Y. H. (2024). Application of Nanocomposites and Nanoparticles in Treating Neurodegenerative Disorders. *CNS & Neurological Disorders-Drug Targets-CNS & Neurological Disorders*, *23*(10), 1217–1233. PMID: 38288843

Fissan, H., Ristig, S., Kaminski, H., Asbach, C., & Epple, M. (2014). Comparison of different characterization methods for nanoparticle dispersions before and after aerosolization. *Analytical Methods*, *6*(18), 7324–7334. PMID: 40883942

Florek, J., Caillard, R., & Kleitz, F. (2017). Evaluation of mesoporous silica nanoparticles for oral drug delivery–current status and perspective of MSNs drug carriers. *Nanoscale*, *9*(40), 15252–15277. PMID: 28984885

FLORES, D. G., Espino, J., & PARIENTE, J. A. (2023). Antioxidant potential of nanomaterials. *Turkish Journal of Biology = Türk Biyoloji Dergisi*, *47*(4), 218–235. PMID: 38152621

Fonseca-Santos, B., & Chorilli, M. (2020). The uses of resveratrol for neurological diseases treatment and insights for nanotechnology based-drug delivery systems. *International Journal of Pharmaceutics*, *589*, 119832. PMID: 32877730

Gandhi, J., Antonelli, A. C., Afridi, A., Vatsia, S., Joshi, G., Romanov, V., & Khan, S. A. (2019). Protein misfolding and aggregation in neurodegenerative diseases: A review of pathogeneses, novel detection strategies, and potential therapeutics. *Reviews in the Neurosciences*, *30*(4), 339–358. PMID: 30742586

Grinán-Ferré, C., Bellver-Sanchis, A., Izquierdo, V., Corpas, R., Roig-Soriano, J., Chillón, M., & Sanfeliu, C. (2021). The pleiotropic neuroprotective effects of resveratrol in cognitive decline and Alzheimer's disease pathology: From antioxidant to epigenetic therapy. *Ageing Research Reviews*, *67*, 101271. PMID: 33571701

Guo, J. D., Zhao, X., Li, Y., Li, G. R., & Liu, X. L. (2018). Damage to dopaminergic neurons by oxidative stress in Parkinson's disease. *International Journal of Molecular Medicine*, *41*(4), 1817–1825. PMID: 29393357

Hawkins, K. G., Casolaro, C., Brown, J. A., Edwards, D. A., & Wikswo, J. P. (2020). The Microbiome and the Gut-Liver-Brain Axis for Central Nervous System Clinical Pharmacology: Challenges in Specifying and Integrating In Vitro and In Silico Models. *Clinical Pharmacology and Therapeutics*, *108*(5), 929–948. PMID: 32347548

Heckman, K. L., Estevez, A. Y., DeCoteau, W., Vangellow, S., Ribeiro, S., Chiarenzelli, J., & Erlichman, J. S. (2020). Variable in vivo and in vitro biological effects of cerium oxide nanoparticle formulations. *Frontiers in Pharmacology*, *10*, 1599. PMID: 32047435

Indiarto, R., Indriana, L. P. A., Andoyo, R., Subroto, E., & Nurhadi, B. (2021). Bottom–up nanoparticle synthesis: A review of techniques, polyphenol-based core materials, and their properties. *European Food Research and Technology*, •••, 1–24.

Islam, A., Mishra, A., Ahsan, R., & Fareha, S. (2023). Phytopharmaceuticals and Herbal Approaches to Target Neurodegenerative Disorders. *Drug Research*, *73*(07), 388–407. PMID: 37308092

Jacob, S., Kather, F. S., Morsy, M. A., Boddu, S. H., Attimarad, M., Shah, J., & Nair, A. B. (2024). Advances in Nanocarrier Systems for Overcoming Formulation Challenges of Curcumin: Current Insights. *Nanomaterials (Basel, Switzerland)*, *14*(8), 672. PMID: 38668166

Jain, A., Jain, A., Parajuli, P., Mishra, V., Ghoshal, G., Singh, B., & Kesharwani, P. (2018). Recent advances in galactose-engineered nanocarriers for the site-specific delivery of siRNA and anticancer drugs. *Drug Discovery Today*, *23*(5), 960–973. PMID: 29129804

Jing, X.-Z., Yang, H.-J., Taximaimaiti, R., & Wang, X.-P. (2023). Advances in the therapeutic use of non-ergot dopamine agonists in the treatment of motor and non-motor symptoms of Parkinson's disease. *Current Neuropharmacology*, *21*(5), 1224. PMID: 36111769

Kakkar, V., Kumari, P., Kaur, J., & Chholta, S. (2024). *Curcumin Nanoformulations in Neurodegenerative Diseases Curcumin and Neurodegenerative Diseases: From Traditional to Translational Medicines*. Springer.

Kamaly, N., Xiao, Z., Valencia, P. M., Radovic-Moreno, A. F., & Farokhzad, O. C. (2012). Targeted polymeric therapeutic nanoparticles: Design, development and clinical translation. *Chemical Society Reviews*, *41*(7), 2971–3010. PMID: 22388185

Kamaly, N., Yameen, B., Wu, J., & Farokhzad, O. C. (2016). Degradable controlled-release polymers and polymeric nanoparticles: Mechanisms of controlling drug release. *Chemical Reviews*, *116*(4), 2602–2663. PMID: 26854975

Kassubek, J., Jost, W. H., & Schwarz, J. (2024). Sublingual apomorphine in the treatment of Parkinson's disease. *Journal of Neural Transmission (Vienna, Austria)*, 1–8. PMID: 38743091

Kessler, A., Hedberg, J., Blomberg, E., & Odnevall, I. (2022). Reactive oxygen species formed by metal and metal oxide nanoparticles in physiological media—A review of reactions of importance to nanotoxicity and proposal for categorization. *Nanomaterials (Basel, Switzerland)*, *12*(11), 1922. PMID: 35683777

Khan, S., & Hossain, M. K. (2022). *Classification and properties of nanoparticles Nanoparticle-based polymer composites*. Elsevier.

Khan, T., Waseem, R., Zehra, Z., Aiman, A., Bhardwaj, P., Ansari, J., & Islam, A. (2022). Mitochondrial dysfunction: Pathophysiology and mitochondria-targeted drug delivery approaches. *Pharmaceutics, 14*(12), 2657. PMID: 36559149

Khatoon, S., Kalam, N., Shaikh, M., Hasnain, M. S., Hafiz, A. K., & Ansari, M. T. (2022). Nanoencapsulation of Polyphenols as Drugs and Supplements for Enhancing Therapeutic Profile-A Review. *Current Molecular Pharmacology, 15*(1), 77–107. PMID: 34551693

Kim, H., Lee, J. H., Kim, J. E., Kim, Y. S., Ryu, C. H., Lee, H. J., & Lee, J.-Y. (2018). Micro-/nano-sized delivery systems of ginsenosides for improved systemic bioavailability. *Journal of Ginseng Research, 42*(3), 361–369. PMID: 29983618

Kong, J., Chu, R., & Wang, Y. (2022). Neuroprotective treatments for ischemic stroke: Opportunities for nanotechnology. *Advanced Functional Materials, 32*(52), 2209405.

Kou, L., Yao, Q., Zhang, H., Chu, M., Bhutia, Y. D., Chen, R., & Ganapathy, V. (2020). Transporter-targeted nano-sized vehicles for enhanced and site-specific drug delivery. *Cancers (Basel), 12*(10), 2837. PMID: 33019627

Lamptey, R. N., Chaulagain, B., Trivedi, R., Gothwal, A., Layek, B., & Singh, J. (2022). A review of the common neurodegenerative disorders: Current therapeutic approaches and the potential role of nanotherapeutics. *International Journal of Molecular Sciences, 23*(3), 1851. PMID: 35163773

Lara-Reyna, S., Scambler, T., Holbrook, J., Jarosz-Griffiths, H. H., Peckham, D., & McDermott, M. F. (2018). Regulation of the unfolded protein response in disease: Cellular stress and microRNAs. *Current Immunology Reviews, 14*(1), 3–14.

Lin, M.-Y. (2019). Engineering self-regulating T cells to prevent cytokine release syndrome in CAR-T cell.

Lombardo, R., Musumeci, T., Carbone, C., & Pignatello, R. (2021). Nanotechnologies for intranasal drug delivery: An update of literature. *Pharmaceutical Development and Technology, 26*(8), 824–845. PMID: 34218736

Long, B., Ryan, K. M., & Padrela, L. (2019). From batch to continuous—New opportunities for supercritical CO2 technology in pharmaceutical manufacturing. *European Journal of Pharmaceutical Sciences, 137*, 104971. PMID: 31247295

Lontay, B., Kiss, A., Virág, L., & Tar, K. (2020). How do post-translational modifications influence the pathomechanistic landscape of Huntington's disease? A comprehensive review. *International Journal of Molecular Sciences, 21*(12), 4282. PMID: 32560122

Lu, X., Dong, J., Zheng, D., Li, X., Ding, D., & Xu, H. (2020). Reperfusion combined with intraarterial administration of resveratrol-loaded nanoparticles improved cerebral ischemia–reperfusion injury in rats. *Nanomedicine; Nanotechnology, Biology, and Medicine, 28*, 102208. PMID: 32334100

Lv, H., Wang, Y., Yang, X., Ling, G., & Zhang, P. (2023). Application of curcumin nanoformulations in Alzheimer's disease: Prevention, diagnosis and treatment. *Nutritional Neuroscience, 26*(8), 727–742. PMID: 35694842

Ma, H., Jiang, Z., Xu, J., Liu, J., & Guo, Z.-N. (2021). Targeted nano-delivery strategies for facilitating thrombolysis treatment in ischemic stroke. *Drug Delivery*, *28*(1), 357–371. PMID: 33517820

Machado-Pereira, M., Santos, T., Ferreira, L., Bernardino, L., & Ferreira, R. (2018). Intravenous administration of retinoic acid-loaded polymeric nanoparticles prevents ischemic injury in the immature brain. *Neuroscience Letters*, *673*, 116–121. PMID: 29518539

Manoharan, D., Wang, L. C., Chen, Y. C., Li, W. P., & Yeh, C. S. (2024). Catalytic Nanoparticles in Biomedical Applications: Exploiting Advanced Nanozymes for Therapeutics and Diagnostics. *Advanced Healthcare Materials*, 2400746. PMID: 38683107

Mishra, A. P., Salehi, B., Sharifi-Rad, M., Pezzani, R., Kobarfard, F., Sharifi-Rad, J., & Nigam, M. (2018). Programmed cell death, from a cancer perspective: An overview. *Molecular Diagnosis & Therapy*, *22*, 281–295. PMID: 29560608

Mittal, K. R., Pharasi, N., Sarna, B., & Singh, M., Rachana, Haider, S., Dey, A. (2022). Nanotechnology-based drug delivery for the treatment of CNS disorders. *Translational Neuroscience*, *13*(1), 527–546. PMID: 36741545

Mohammadinejad, R., Moosavi, M. A., Tavakol, S., Vardar, D. Ö., Hosseini, A., Rahmati, M., & Klionsky, D. J. (2019). Necrotic, apoptotic and autophagic cell fates triggered by nanoparticles. *Autophagy*, *15*(1), 4–33. PMID: 30160607

Muralidar, S., Ambi, S. V., Sekaran, S., Thirumalai, D., & Palaniappan, B. (2020). Role of tau protein in Alzheimer's disease: The prime pathological player. *International Journal of Biological Macromolecules*, *163*, 1599–1617. PMID: 32784025

Nakmode, D. D., Day, C. M., Song, Y., & Garg, S. (2023). The management of Parkinson's disease: An overview of the current advancements in drug delivery systems. *Pharmaceutics*, *15*(5), 1503. PMID: 37242745

Nayal, R., Mejjo, D., & Abajy, M. Y. (2024). Anti inflammatory properties and safety of green synthesized metal and metal oxidenanoparticles: A review article. *European Journal of Medicinal Chemistry Reports*, •••, 100169.

Nowacka, A., Śniegocka, M., Smuczyński, W., Liss, S., Ziółkowska, E., Bożiłow, D., & Wiciński, M. (2024). The Potential Application of Resveratrol and Its Derivatives in Central Nervous System Tumors. *International Journal of Molecular Sciences*, *25*(24), 13338. PMID: 39769099

Pal, P., Sharma, M., Bani, K. S., & Gupta, S. K. (2024). Advances in neuronal regeneration: Hydrogel-based delivery systems loaded with extracellular vesicles in modulating neural impulses and tissue repair. *European Polymer Journal*, 113457.

Parham, S., Kharazi, A. Z., Bakhsheshi-Rad, H. R., Nur, H., Ismail, A. F., Sharif, S., & Berto, F. (2020). Antioxidant, antimicrobial and antiviral properties of herbal materials. *Antioxidants*, *9*(12), 1309. PMID: 33371338

Parveen, K., Banse, V., & Ledwani, L. (2016). *Green synthesis of nanoparticles: their advantages and disadvantages.* Paper presented at the AIP conference proceedings.

Parvez, S., Kaushik, M., Ali, M., Alam, M. M., Ali, J., Tabassum, H., & Kaushik, P. (2022). Dodging blood brain barrier with "nano" warriors: Novel strategy against ischemic stroke. *Theranostics*, *12*(2), 689. PMID: 34976208

Pathak, N., Vimal, S. K., Tandon, I., Agrawal, L., Hongyi, C., & Bhattacharyya, S. (2021). Neurodegenerative disorders of alzheimer, parkinsonism, amyotrophic lateral sclerosis and multiple sclerosis: An early diagnostic approach for precision treatment. *Metabolic Brain Disease*, 1–38. PMID: 34719771

Patra, J. K., Das, G., Fraceto, L. F., Campos, E. V. R., Rodriguez-Torres, M. P., Acosta-Torres, L. S., & Sharma, S. (2018). Nano based drug delivery systems: Recent developments and future prospects. *Journal of Nanobiotechnology*, *16*(1), 1–33. PMID: 30231877

Pillai, S. C., & Ganguly, P. 2D nanomaterials and composites for energy storage and conversion. *2D Materials for Energy Storage and Conversion.*

Poovaiah, N., Davoudi, Z., Peng, H., Schlichtmann, B., Mallapragada, S., Narasimhan, B., & Wang, Q. (2018). Treatment of neurodegenerative disorders through the blood–brain barrier using nanocarriers. *Nanoscale*, *10*(36), 16962–16983. PMID: 30182106

Prasanna, P., & Upadhyay, A. (2021). Flavonoid-based nanomedicines in Alzheimer's disease therapeutics: Promises made, a long way to go. *ACS Pharmacology & Translational Science*, *4*(1), 74–95. PMID: 33615162

Rajput, A., Tandel, N., Borse, S., & Butani, S. (2021). Potential of naso-brain drug delivery in glioblastoma therapy *Direct Nose-to-Brain* [Elsevier.]. *Drug Delivery*, •••, 389–404.

Ramalho, M. J., Andrade, S., Loureiro, J. A., & do Carmo Pereira, M. (2020). Nanotechnology to improve the Alzheimer's disease therapy with natural compounds. *Drug Delivery and Translational Research*, *10*, 380–402. PMID: 31773421

Remya, V., Abitha, V., Rajput, P. S., Rane, A. V., & Dutta, A. (2017). Silver nanoparticles green synthesis: A mini review. *Chemistry International*, *3*(2), 165–171.

Rinaldi, F., Seguella, L., Gigli, S., Hanieh, P. N., Del Favero, E., Cantù, L., & Esposito, G. (2019). inPentasomes: An innovative nose-to-brain pentamidine delivery blunts MPTP parkinsonism in mice. *Journal of Controlled Release*, *294*, 17–26. PMID: 30529726

Rompicherla, S. K. L., Arumugam, K., Bojja, S. L., Kumar, N., & Rao, C. M. (2021). Pharmacokinetic and pharmacodynamic evaluation of nasal liposome and nanoparticle based rivastigmine formulations in acute and chronic models of Alzheimer's disease. *Naunyn-Schmiedeberg's Archives of Pharmacology*, *394*(8), 1737–1755. PMID: 34086100

Roufayel, R., Younes, K., Al-Sabi, A., & Murshid, N. (2022). BH3-only proteins Noxa and Puma are key regulators of induced apoptosis. *Life (Chicago, Ill.)*, *12*(2), 256. PMID: 35207544

Ryan, S.-L. (2019). *Targeting the nuclear factor kappa-light-chain-enhancer of activated b cells (NF-kb) pathway to overcome cisplatin-resistance in non-small cell lung cancer.* Queensland University of Technology.

Sacks, B., Bashford, J., Wijesekera, L., Leigh, P. N., & Sreedharan, J. (2022). *Motor Neuron Disease: Amyotrophic Lateral Sclerosis Neuroscience in the 21st Century: From Basic to Clinical*. Springer.

Salatin, S., Farhoudi, M., Farjami, A., Maleki Dizaj, S., Sharifi, S., & Shahi, S. (2023). Nanoparticle formulations of antioxidants for the management of oxidative stress in stroke: A review. *Biomedicines*, *11*(11), 3010. PMID: 38002010

Sánchez-López, E., Ettcheto, M., Egea, M. A., Espina, M., Cano, A., Calpena, A. C., & Souto, E. B. (2018). Memantine loaded PLGA PEGylated nanoparticles for Alzheimer's disease: In vitro and in vivo characterization. *Journal of Nanobiotechnology*, *16*, 1–16. PMID: 29587747

Saramowicz, K., Siwecka, N., Galita, G., Kucharska-Lusina, A., Rozpędek-Kamińska, W., & Majsterek, I. (2023). Alpha-synuclein contribution to neuronal and glial damage in Parkinson's disease. *International Journal of Molecular Sciences*, *25*(1), 360. PMID: 38203531

Sharifi-Rad, M., Anil Kumar, N. V., Zucca, P., Varoni, E. M., Dini, L., Panzarini, E., & Peluso, I. (2020). Lifestyle, oxidative stress, and antioxidants: Back and forth in the pathophysiology of chronic diseases. *Frontiers in Physiology*, *11*, 694. PMID: 32714204

Siafaka, P. I., Bülbül, E. Ö., Miliotou, A. N., Karantas, I. D., Okur, M. E., & Okur, N. Ü. (2024). Nano-based carriers for pulmonary drug delivery: A review on the available drug delivery applications and toxicity issues. *Journal of Drug Delivery Science and Technology*, 105381.

Sidoryk-Węgrzynowicz, M., Dąbrowska-Bouta, B., Sulkowski, G., & Strużyńska, L. (2021). Nanosystems and exosomes as future approaches in treating multiple sclerosis. *The European Journal of Neuroscience*, *54*(9), 7377–7404. PMID: 34561918

Silva, S., Almeida, A. J., & Vale, N. (2021). Importance of nanoparticles for the delivery of antiparkinsonian drugs. *Pharmaceutics*, *13*(4), 508. PMID: 33917696

Silveira, C., Guedes, R., Maia, D., Curral, R., & Coelho, R. (2019). Neuropsychiatric symptoms of multiple sclerosis: State of the art. *Psychiatry Investigation*, *16*(12), 877. PMID: 31805761

Singh, L. P., Bhattacharyya, S. K., Kumar, R., Mishra, G., Sharma, U., Singh, G., & Ahalawat, S. (2014). Sol-Gel processing of silica nanoparticles and their applications. *Advances in Colloid and Interface Science*, *214*, 17–37. PMID: 25466691

Sohrabi, H. R., & Weinborn, M. (2019). Cognitive impairments in alzheimer's disease and other neurodegenerative diseases. *Neurodegeneration and Alzheimer's Disease: The Role of Diabetes, Genetics, Hormones, and Lifestyle*, 267-290.

Summer, M., Ashraf, R., Ali, S., Bach, H., Noor, S., Noor, Q., & Khan, R. R. M. (2024). Inflammatory response of nanoparticles: Mechanisms, consequences, and strategies for mitigation. *Chemosphere*, 142826. PMID: 39002651

Sun, Y., Ma, X., & Hu, H. (2021). Marine polysaccharides as a versatile biomass for the construction of nano drug delivery systems. *Marine Drugs*, *19*(6), 345. PMID: 34208540

Theophanous, S., Sargiannidou, I., & Kleopa, K. A. (2024). Glial Cells as Key Regulators in Neuroinflammatory Mechanisms Associated with Multiple Sclerosis. *International Journal of Molecular Sciences*, *25*(17), 9588. PMID: 39273535

Tian, Q., Liu, S., Han, S.-M., Zhang, W., Qin, X.-Y., Chen, J.-H., & Li, M.-C. (2023). The mechanism and relevant mediators associated with neuronal apoptosis and potential therapeutic targets in subarachnoid hemorrhage. *Neural Regeneration Research*, *18*(2), 244–252. PMID: 35900398

Tiwari, S., & Talreja, S. (2020). Nanotube: A New Approach to Novel Drug Delivery System. *Journal of Pharmaceutical Sciences and Research*, *12*(8), 1024–1028.

Vakilinezhad, M. A., Amini, A., Akbari Javar, H., Baha'addini Beigi Zarandi, B. F., Montaseri, H., & Dinarvand, R. (2018). Nicotinamide loaded functionalized solid lipid nanoparticles improves cognition in Alzheimer's disease animal model by reducing Tau hyperphosphorylation. *Daru : Journal of Faculty of Pharmacy, Tehran University of Medical Sciences*, *26*, 165–177. PMID: 30386982

VASUMATHI, R., & Neelamegam, P. (2017). Evaluation of chloride using embedded system based analyser and artificial neural network computation in biological and environmental samples.

Vijayaraghavan, K., Nalini, S. K., Prakash, N. U., & Madhankumar, D. (2012). One step green synthesis of silver nano/microparticles using extracts of Trachyspermum ammi and Papaver somniferum. *Colloids and Surfaces. B, Biointerfaces*, *94*, 114–117. PMID: 22348989

Vishwas, S., Gulati, M., Kapoor, B., Gupta, S., Singh, S. K., Awasthi, A., & Baishnab, S. (2021). Expanding the arsenal against Huntington's disease-Herbal drugs and their nanoformulations. *Current Neuropharmacology*, *19*(7), 957–989. PMID: 33167841

Wright, G. S., Antonyuk, S. V., & Hasnain, S. S. (2019). The biophysics of superoxide dismutase-1 and amyotrophic lateral sclerosis. *Quarterly Reviews of Biophysics*, *52*, e12. PMID: 31760962

Wu, D., Chen, Q., Chen, X., Han, F., Chen, Z., & Wang, Y. (2023). The blood–brain barrier: Structure, regulation, and drug delivery. *Signal Transduction and Targeted Therapy*, *8*(1), 217. PMID: 37231000

Yadi, M., Mostafavi, E., Saleh, B., Davaran, S., Aliyeva, I., Khalilov, R., Panahi, Y. (2018). Current developments in green synthesis of metallic nanoparticles using plant extracts: a review. *Artificial cells, nanomedicine, and biotechnology, 46*(sup3), S336-S343.

Yang, H., Niu, S., Guo, M., & Xue, Y. (2024). Molecular mechanisms of silver nanoparticle-induced neurotoxic injury and new perspectives for its neurotoxicity studies: A critical review. *Environmental Pollution*, 124934. PMID: 39260546

Yang, T., Ferrill, L., Gallant, L., McGillicuddy, S., Fernandes, T., Schields, N., & Bai, S. (2018). Verapamil and riluzole cocktail liposomes overcome pharmacoresistance by inhibiting P-glycoprotein in brain endothelial and astrocyte cells: A potent approach to treat amyotrophic lateral sclerosis. *European Journal of Pharmaceutical Sciences*, *120*, 30–39. PMID: 29704642

Yergök, R. A., Derman, S., Alakbarli, J., & Eskhıta, A. S. (2024). Brain-Targeted Nano-Drug Delivery for the Treatment of Parkinson's Disease. *Current Research in Health Sciences*, *1*(2), 77–92.

Yuan, X., Fu, Z., Ji, P., Guo, L., Al-Ghamdy, A. O., Alkandiri, A., & Kassab, R. B. (2020). Selenium nanoparticles pre-treatment reverse behavioral, oxidative damage, neuronal loss and neurochemical alterations in pentylenetetrazole-induced epileptic seizures in mice. *International Journal of Nanomedicine*, 6339–6353. PMID: 32922005

Zaman, R., Islam, R. A., Ibnat, N., Othman, I., Zaini, A., Lee, C. Y., & Chowdhury, E. H. (2019). Current strategies in extending half-lives of therapeutic proteins. *Journal of Controlled Release, 301*, 176–189. PMID: 30849445

Zeeshan, F., Mishra, D. K., & Kesharwani, P. (2021). *From the nose to the brain, nanomedicine drug delivery Theory and Applications of Nonparenteral Nanomedicines*. Elsevier.

Zenych, A., Fournier, L., & Chauvierre, C. (2020). Nanomedicine progress in thrombolytic therapy. *Biomaterials, 258*, 120297. PMID: 32818824

Zhang, Y., Fu, K., Wang, C., Ma, C., Gong, L., Zhou, H., & Li, Y. (2023). Protective effects of dietary quercetin on cerebral ischemic injury: Pharmacology, pharmacokinetics and bioavailability-enhancing nanoformulations. *Food & Function, 14*(10), 4470–4489. PMID: 37067239

Chapter 15
Advancing Neurodegenerative Disease Research:
Integrating In Vivo, In Vitro, and In Silico Models for Advancing Nanoparticle Applications

Fouad Bouazzaoui
Department of Biology, Cadi Ayyad University, Morocco

Yasmina Jaouhari
https://orcid.org/0009-0007-4795-3051
Department of Biology, Cadi Ayyad University, Morocco

Ouijdane El Hatimy
https://orcid.org/0009-0007-2446-6058
Department of Biology, Cadi Ayyad University, Morocco

Hajar Azraida
https://orcid.org/0000-0002-2208-8264
Department of Biology, Cadi Ayyad University, Morocco

Hamid Kabdy
https://orcid.org/0009-0004-7438-3638
Department of Biology, Cadi Ayyad University, Morocco

Majda Badaoui
Department of Biology, Cadi Ayyad University, Morocco

Souad Moubtakir
Department of Biology, Cadi Ayyad University, Morocco

Loubna Elyazouli
Department of Biology, Cadi Ayyad University, Morocco

Rachida Aboufatima
Faculty of Sciences and Technology, Sultan Moulay Slimane University, Morocco

Mohamed Loukid
Department of Biology, Cadi Ayyad University, Morocco

Abderrahman Chait
Department of Biology, Cadi Ayyad University, Morocco

DOI: 10.4018/979-8-3373-1365-8.ch015

ABSTRACT

Neurodegenerative diseases pose a major global health challenge, demanding advanced models and innovative therapeutic strategies. This chapter overviews in vivo, in vitro, and in silico models used to study these disorders and their role in pharmacological and therapeutic research. Such models clarify mechanisms underlying Alzheimer's, Parkinson's, and Huntington's disease while serving as essential platforms for drug testing, toxicity evaluation, and therapy development. A key focus is the integration of nanoparticles, whose small size, large surface area, and ability to cross the blood–brain barrier enable targeted drug delivery with fewer side effects. Functionalization with diagnostic markers further supports early biomarker detection and personalized medicine. By combining experimental models with nanotechnology, this approach improves diagnostic precision and therapeutic efficacy, offering promising prospects for more effective and less invasive treatments of neurodegenerative disorders.

1. INTRODUCTION

Neurodegenerative diseases, including Alzheimer's disease (AD), Parkinson's disease, Huntington's disease, and amyotrophic lateral sclerosis (ALS), represent a significant and growing global health burden (Gadhave et al., 2024). These disorders are characterized by the progressive loss of structure or function of neurons in the CNS and PNS, ultimately leading to severe cognitive, motor, or functional impairments (Lamptey et al., 2022). Despite decades of research, the exact mechanisms underlying these conditions remain elusive (Carraro et al., 2024a). The primary risk factor is aging, during which the cellular repair mechanisms gradually weaken. With the aging population on the rise, the prevalence of neurodegenerative diseases is expected to increase dramatically, emphasizing the urgent need for innovative strategies to better understand and treat these conditions (Di Luca et al., 2021).

One of the primary challenges in tackling neurodegenerative diseases is their complex and multifaceted etiology and pathology (Palanisamy et al., 2023). These disorders arise from a combination of genetic, environmental, and lifestyle factors, which interact in ways that are not yet fully understood (Ayeni et al., 2022). Moreover, the heterogeneity in symptoms and progression among patients adds an additional layer of complexity (Fasano & Alberio, 2023). Current therapeutic interventions largely focus on symptom management, with limited options available to halt or reverse neuronal degeneration ; Wareham et al., 2022). This lack of definitive cures highlights a critical gap in both our understanding of disease mechanisms and the development of effective treatments (Wareham et al., 2022).

The development of advanced modeling techniques has emerged as a crucial step toward addressing these challenges (Pasko et al., 2022). Traditional in vivo models, such as genetically engineered animal models, have provided valuable insights but are often limited in their ability to fully replicate human pathophysiology (Dawson et al., 2018). Among these, transgenic models (e.g., APP/PS1 or Tau mice) are frequently used to reproduce hallmark features of Alzheimer's disease, such as amyloid plaque deposition and tau hyperphosphorylation, thereby enabling the investigation of pathogenic mechanisms and therapeutic strategies (Kitazawa et al., 2012). In addition, neurotoxin-induced models, including MPTP for Parkinson's disease and 3-nitropropionic acid for Huntington's disease, are valuable tools for mimicking selective neuronal degeneration and for evaluating neuroprotective agents (Ibrahim & Abdel Rasheed, 2022). Furthermore, lesion-based models, such as those induced by 6-hydroxydopamine (6-OHDA), are widely applied to study dopaminergic neurodegeneration and to test novel interventions aimed at restoring motor function (Slézia et al., 2023).

In vitro systems, particularly those utilizing patient-derived induced pluripotent stem cells (iPSCs), have opened new path for personalized research and drug testing (Shi et al., 2017). Furthermore, recent biotechnological innovations, including organ-on-a-chip platforms, CRISPR-Cas9 gene editing, and omics-based approaches, are revolutionizing pharmacological research (Ansori et al., 2023). These technologies enable more accurate modeling of disease processes and facilitate the identification of novel therapeutic targets such as nanoparticles (Krsek & Baticic, 2024).

Nanoparticles, due to their small size and high surface area, have unique properties that make them highly effective for medical applications (Joseph et al., 2023). They can cross biological barriers, such as the blood-brain barrier, which is particularly important for neurodegenerative diseases (Sadat Razavi et al., 2025). Nanoparticles can be engineered for targeted drug delivery, allowing therapeutic agents to be delivered precisely to affected brain regions, minimizing side effects (Yetisgin et al., 2020). They can also be functionalized with diagnostic markers for early detection of disease biomarkers, enhancing the potential for early intervention and treatment (Bodaghi et al., 2023).

These models are also part of a broader exploration of innovative approaches for the use of nanoparticles in both the diagnosis and treatment of neurodegenerative diseases (Kabanov & Gendelman, 2007). Nanoparticles offer unique capabilities for targeted delivery of therapeutics and advanced imaging, making them a promising tool in addressing the challenges of these complex disorders (Cheng et al., 2023).

The aim of this chapter is to provide a comprehensive overview of *in vivo, in vitro,* and computational *in silico* models used in neurodegenerative disorders, which can revolutionize pharmacological research and therapeutic development. Additionally, this chapter highlights nanoparticles as a new potential tool for both treatment and diagnosis, utilizing in vivo, in vitro, and in silico models.

2. *IN VIVO* MODELS

2.1. Recent Advances

In vivo models have been instrumental in advancing our understanding of neurodegenerative diseases, offering crucial insights into disease mechanisms, progression, and therapeutic interventions (Strome & Doudet, 2007).

Over the years, significant progress has been made in the development of transgenic and humanized animal models, which aim to closely mimic human pathologies (Walsh et al., 2017). A wide variety of animals are used to study neurodegenerative diseases, ranging from invertebrates to mammals (Table 1).

Invertebrates, such as *Drosophila melanogaster* (fruit flies) and *Caenorhabditis elegans* (nematode worms), are commonly used for genetic screenings and drug testing due to their simplicity, short lifespans, and well-mapped genomes (N. Sharma et al., 2017). Zebrafish, with their rapid breeding and transparent embryos, have gained popularity for studying the molecular mechanisms of diseases like AD (Newman et al., 2014).

Rodents, particularly mice and rats, remain the most widely used models due to their genetic tractability, availability of transgenic lines, and similarities in neuroanatomy with humans (Harvey et al., 2011). Larger animals, such as pigs, rabbits, and non-human primates, are utilized when a closer approximation of human brain structure and function is required (Taha et al., 2022; Yin et al., 2022).

To overcome the limitations of rodent models, humanized models like transgenic pigs have been developed. With larger brains and human-like physiology, these pigs provide a better platform for studying diseases like Huntington's (Yan et al., 2018). Transgenic pigs expressing mutant HTT (huntingtin) genes bridge the gap between rodent studies and human research by aiding the study of motor deficits and brain atrophy. Non-human primates are also valuable models for studying diseases like Parkinson's and Huntington's due to their brain similarity to humans. (Pan et al., 2024; Emborg, 2017).

Table 1. Overview of experimental models for studying neurodegenerative diseases

Type	Overview	Applications
Toxin-Induced Models	Models use chemical agents to selectively induce neuronal damage and mimic disease pathology.	- 1-méthyl-4-phényl-1,2,3,6-tétrahydropyridine **(MPTP) for Parkinson's Disease (PD)**: Destroys dopaminergic neurons (Narmashiri et al., 2022) - **6-OHDA for PD**: Targets the nigrostriatal pathway (Deumens et al., 2002). - **Kainic Acid**: Used for epilepsy and neurodegeneration studies (Q. Wang et al., 2005).
Protein/Peptide Injection Models	Models involve injecting disease-related proteins or peptides into the brain to replicate key pathological features.	- **Amyloid-β Injection**: Simulates amyloid plaque formation in AD (Poon et al., 2020). - **Tau Protein Injection**: Mimics tauopathies seen in frontotemporal dementia and AD (Robert et al., 2021).
Viral Vector Models	Models rely on viral vectors to introduce specific genetic material into the brain, inducing or modifying disease mechanisms.	- **AAV or Lentivirus Vectors**: Overexpress or knockdown disease-related genes (Mijanović et al., 2020). - Example: **α-Synuclein Overexpression** for Parkinson's Disease (PD) (Ulusoy et al., 2010).
Dietary Models	These models use special diets or dietary compounds to induce features of neurodegeneration.	- **BSSG Diet for ALS**: β-sitosterol β-D-glucoside diet induces motor neuron degeneration (Tabata et al., 2008).
CRISPR-Cas9 Technology	A revolutionary gene-editing tool that allows for precise mutations to be introduced in vivo, creating models that closely mimic human diseases.	- **Familial Alzheimer's Disease**: Mutation of APP or presenilin genes in rodents (Bhardwaj et al., 2021). - **Huntington's Disease**: Introduction of CAG repeats in mice to study progression (Seo et al., 2023). - **Other Neurodegenerative Diseases**: Precise mutations for modeling genetic disease mechanisms (Nouri Nojadeh et al., 2023).
Transgenic Models	Genetically modified animals are engineered to carry human gene mutations associated with neurodegenerative diseases.	- **AD**: Mice with APP or presenilin mutations (Elder et al., 2010). - **Huntington's Disease**: Models with expanded CAG repeats in the huntingtin gene (Kaye et al., 2021). - **ALS**: SOD1-mutant mice (Julien & Kriz, 2006). - **PD**, rodent models with mutations in genes like SNCA (α-synuclein) or LRRK2 have helped elucidate mechanisms of dopaminergic neuron degeneration (G. Liu et al., 2012).

2.2. Importance of Mimicking Human Pathology

Neurodegenerative diseases involve complex interactions between genetic mutations, environmental factors, and cellular processes (Firdaus & Li, 2024). Transgenic animal models, which incorporate disease-related genes, allow researchers to study these interactions in a living organism (Houdebine, 2006). By mimicking key pathological features such as amyloid plaque deposition in AD or dopaminergic neuron loss in Parkinson's disease (PD), these models provide a platform for testing potential treatments and monitoring disease progression (Banerjee et al., 2022). In vivo studies enable researchers to track disease evolution over time and evaluate how treatments affect disease trajectories, offering insights into potential clinical outcomes (Chornyy et al., 2023; Palanisamy et al., 2023).

2.3. Parameters evaluated in the validation of animal models

When validating animal models for neurodegenerative diseases, several parameters are evaluated to ensure they accurately mimic human conditions and are effective for therapeutic testing (Tello et al., 2022). Construct validity examines disease-specific pathological features, such as amyloid plaques in Alzheimer's disease (Tello et al., 2022) and Lewy bodies in Parkinson's disease (Harvey et al., 2023), as well as relevant genetic mutations like SNCA (Jahabardeen et al., 2024), PARKIN (Madsen et al., 2021), and PINK1 (Samaranch et al., 2010). Face validity assesses the replication of motor symptoms, including akinesia and tremors, alongside non-motor symptoms such as sleep disturbances and gastrointestinal issues (Guimarães et al., 2024). Predictive validity evaluates the model's ability to predict treatment efficacy and replicate human treatment-related adverse effects. In addition, biochemical parameters such as oxidative stress markers (e.g., malondialdehyde, superoxide dismutase, catalase, glutathione) and inflammatory mediators (e.g., TNF-α, IL-1β, IL-6, COX-2, NF-κB signaling) are commonly measured to assess the underlying molecular and cellular processes that drive neurodegeneration (Ashok et al., 2022). These parameters help confirm whether the model recapitulates the oxidative and neuroinflammatory environment observed in patients. Behavioral and cognitive changes are also evaluated through memory and learning tests in Alzheimer's models and motor coordination tasks such as the rotarod or pole test in Parkinson's models(Fan et al., 2021). Finally, neurodegeneration parameters include neuronal loss in disease-relevant regions, such as the substantia nigra in Parkinson's, and the effectiveness of neuroprotective therapies (Sarkar et al., 2016).

3. INNOVATIONS IN *IN VITRO* PLATFORMS

3.1. Brain Organoids

Brain organoids are three-dimensional (3D) self-organizing structures derived from human pluripotent stem cells (hPSCs), such as iPSCs or embryonic stem cells (ESCs) (Lee et al., 2017). These organoids closely mimic the architecture and some functional aspects of the developing human brain, making them valuable tools for studying neurodevelopment and neurodegenerative diseases (Xu & Wen, 2021).

Organoids can be generated through a series of carefully controlled steps (X. Yin et al., 2016). The process begins with stem cell cultivation, where stem cells are cultured in a suitable medium to maintain their pluripotency, either with feeder cells or under feeder-free conditions. Next, neural induction is initiated by transferring the embryoid bodies (EBs) into a specialized medium that guides cells toward a neural lineage, often supplemented with components like N2 and non-essential amino acids. During aggregation, stem cells are dissociated into small clumps and cultured in conditions promoting formation of EBs, frequently using specialized media such as AggreWell EB formation medium. Differentiation follows as the EBs are maintained in a neural induction medium until they develop neuroepithelial structures, a process that can take several days to weeks depending on the protocol. Finally, the organoids enter the maturation phase, where they are cultured in a differentiation medium supporting their development into functional brain tissue models (Figure 1). Long-term culture and periodic passaging help maintain organoid health and functionality (Zhao et al., 2022).

In the context of neurodegenerative diseases, brain organoids provide a powerful platform for modeling conditions such as A D (Sreenivasamurthy et al., 2022) and PD (Cui et al., 2024). They allow the investigation of key pathological features, including amyloid-beta and tau protein accumulation in AD, as well as dopaminergic neuron loss in Parkinson's Disease . Researchers can explore biological changes like synaptic dysfunction, neuroinflammation, and cell death (Kumar & Hassan, 2021). Functional assessments, such as measuring neuronal firing rates and network activity, can be conducted using electrophysiological tools (Kang et al., 2024). Furthermore, brain organoids serve as valuable models for drug testing, enabling researchers to evaluate the effects of potential therapeutic compounds on reducing protein aggregation and improving cellular health (Zhou et al., 2023).

Figure 1. The process of brain organoid generation

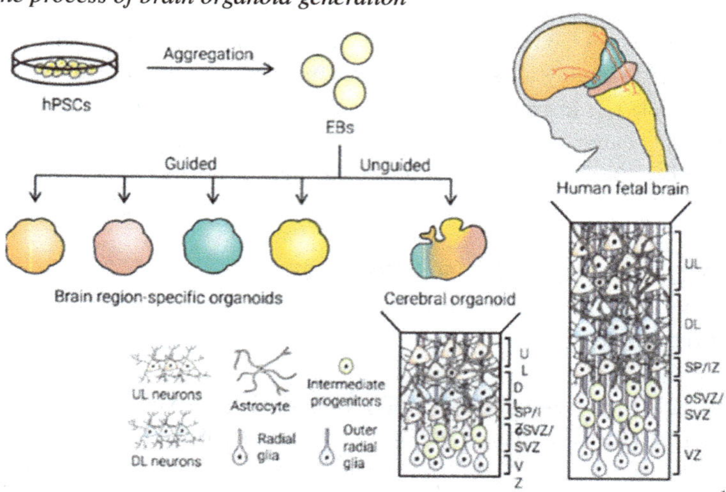

Source: Jang et al. (2022)

3.2. Patient-Derived Induced Pluripotent Stem Cells (iPSCs)

Patient-derived iPSCs are generated by reprogramming somatic cells, such as skin fibroblasts or blood cells, back into a pluripotent state (Al Abbar et al., 2020). This process enables differentiation into various cell types, including neurons and glial cells, making them valuable tools for creating patient-specific models of neurodegenerative diseases (Albert et al., 2021). These personalized models provide critical insights into disease mechanisms and therapeutic responses.

iPSC technology has been instrumental in studying neurodegenerative conditions like Alzheimer's Disease (AD), Parkinson's Disease (PD), and Huntington's Disease (HD). In AD, iPSCs can model key pathological features such as amyloid-beta plaque formation and tau protein phosphorylation, allowing researchers to explore disease progression at the cellular level (Essayan-Perez et al., 2019). For PD, iPSCs facilitate the study of alpha-synuclein aggregation and mitochondrial dysfunction, providing insights into the disease's underlying mechanisms (Spathopoulou et al., 2022). Similarly, in HD, iPSCs help investigate the effects of polyglutamine expansions on cellular function (Kaye et al., 2022). These models can be used to test various biological parameters, including oxidative stress markers, synaptic density, and cellular viability, through techniques such as immunohistochemistry, Western blotting, and live-cell imaging (Guerreiro & Maciel, 2023). Drug responses can also be evaluated by exposing the cells to different compounds and assessing changes in disease-related markers (Figure 2).

iPSC models offer significant advantages, enabling personalized medicine with patient-specific models for tailored insights into disease mechanisms and therapeutic responses (Paik et al., 2020). They also enhance understanding of disease pathology by recapitulating disease-specific phenotypes, aiding in the study of complex cellular interactions in neurodegenerative diseases (Li & Fraenkel, 2021). Additionally, iPSC models support therapeutic development through screening and testing potential compounds, potentially advancing treatments for conditions like AD, PD, and HD (Farkhondeh et al., 2019).

Figure 2. Creation of control and patient-specific iPSCs, followed by their differentiation into neurons and glial cells to evaluate disease-specific changes

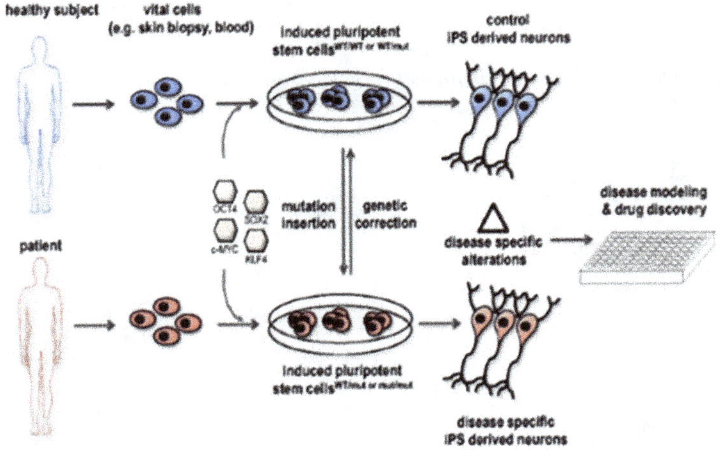

Source: Jungverdorben et al. (2017)

3.3. Microfluidic Technologies

Microfluidic technologies utilize micro-scale fluidic devices that mimic the cellular microenvironment of brain tissues, providing precise control over fluid dynamics (Jahromi et al., 2019). These devices simulate brain microenvironments and facilitate the study of synaptic transmission, neuronal connectivity, and network activity (Habibey et al., 2022).

Microfluidic platforms have proven valuable in neurodegenerative disease research by modeling synaptic activity and connectivity deficits observed in conditions like Alzheimer's Disease and Parkinson's Disease (Miny et al., 2022). They also enable the co-culture of neurons and glial cells, essential for studying neuroinflammation and cell-cell interactions, helping researchers explore how different cell types influence each other in health and disease (Sanapathi et al., 2023). Parameters such as neurotransmitter release, network firing patterns, and calcium flux can be measured using advanced neurophysiological methods. Calcium imaging employs fluorescent indicators to visualize intracellular calcium dynamics, providing insights into synaptic activity, excitotoxicity, and neuronal survival, which are often impaired in disorders like Alzheimer's and Parkinson's (Holloway et al., 2021). Patch-clamp electrophysiology allows the direct recording of ionic currents and action potentials at the single-cell level, enabling the detection of synaptic dysfunction, channelopathies, and altered neuronal excitability characteristic of neurodegeneration(Gabriel et al., 2023). Biosensors, including enzyme-based or genetically encoded sensors, can be used to quantify neurotransmitter release (e.g., dopamine, glutamate, acetylcholine) in real time, shedding light on the deficits in dopaminergic and glutamatergic signaling that underlie cognitive and motor impairments in these diseases (Leopold et al., 2019). Additionally, microfluidic devices support high-throughput drug screening due to their capacity to test multiple conditions in parallel while minimizing reagent use.

Key advantages of microfluidic technologies include precise microenvironment control, allowing regulation of fluid flow, pressure, and chemical gradients to simulate physiological conditions. They also promote efficient use of resources, requiring minimal cells and reagents, making them suitable for studies involving rare or costly biological materials (Figure 3). Furthermore, their flexibility and scalability make them adaptable to diverse experimental needs in neuroscience research (Halldorsson et al., 2015).

Figure 3. Prospective in vitro models for neurodegenerative disease

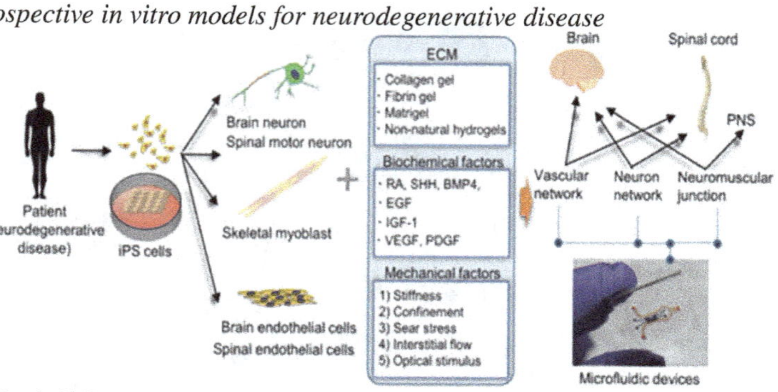

Source: Osaki et al. (2018)

4. BIOTECHNOLOGICAL INNOVATIONS IN DISEASE MODELING

4.1. In Silico Models

In silico models play crucial roles in disease modeling, particularly in neurodegenerative diseases. Single-cell omics technologies, such as single-cell RNA sequencing (scRNA-seq) and single-cell ATAC sequencing (scATAC-seq), along with AI-powered tools, represent in silico models (Choi et al., 2024). These models rely on computational analysis to explore gene expression patterns, detect early cellular changes, and identify biomarkers linked to disease progression (Choi et al., 2024; Erfanian et al., 2023). For instance, AI tools like the single cell generative pretrained transformer (scGPT) can predict gene expression levels and integrate diverse datasets, enhancing our understanding of cellular heterogeneity in neurodegenerative diseases (Wong & carrissa, 2023). These purely computational models are valuable for providing data-driven insights without direct biological experimentation, which is especially useful for studying complex diseases like Alzheimer's and Parkinson's (Geerts et al., 2017).

4.2. Hybrid Models

Neuromodulation techniques such as optogenetics and chemogenetics are considered hybrid models (Vlasov et al., 2018). They combine in silico elements with *in vitro* and in vivo experimentation, allowing for precise computational control of neuronal activity (Won et al., 2020). These techniques require experimental validation in biological systems, making them powerful tools for both theoretical analysis and practical intervention. For example, optogenetics can be used to modulate specific neuronal populations in animal models of neurodegenerative diseases, providing insights into disease mechanisms and potential therapeutic targets (Vann & Xiong, 2016). Both in silico and hybrid models contribute significantly to neurodegenerative disease research. They provide complementary insights into disease mechanisms, facilitate the development of targeted therapies, and enhance our understanding of cellular and molecular changes associated with neurodegeneration (Makhouri & Ghasemi, 2018). Single-cell multiomics technologies, in particular, have been instrumental in identifying specific neuronal subtypes vulnerable to neurodegenerative diseases, which can inform the design of more effective treatments (Carraro et al., 2024). Overall, these models are essential for advancing our understanding and treatment of neurodegenerative diseases.

5. NANOTECHNOLOGY IN NEURODEGENERATIVE DISEASE RESEARCH

5.1. Overview of Nanoparticles in Biomedical Applications

Nanoparticles are tiny structures, typically ranging from 1 to 100 nanometers, with unique physical and chemical properties that make them highly valuable in biomedical applications (Khan et al., 2019). Their small size allows for enhanced interaction with biological systems at the molecular level, while surface modifications can improve their stability, biocompatibility, and targeting capabilities (Altammar, 2023). In the context of neurological diseases, nanoparticles are especially suited due to their ability to cross the blood-brain barrier (BBB), a selective barrier that restricts the passage of most therapeutic agents to the brain (Hersh et al., 2022). Their functional versatility enables the delivery of drugs, imaging agents, or genetic tools directly to affected areas, offering targeted and efficient solutions for neurodegenerative conditions (Sim & Wong, 2021).

5.2. Nanoparticles for Diagnosis and Detection

Nanoparticles are revolutionizing the early diagnosis and detection of neurodegenerative diseases by enhancing imaging techniques such as magnetic resonance imaging (MRI), positron emission tomography (PET), and fluorescence imaging (Joshi et al., 2023). Functionalized nanoparticles, which are nanoscale particles engineered with specific surface modifications to enable targeted binding and improved biocompatibility, can bind selectively to disease-specific biomarkers, such as amyloid-beta plaques and tau tangles in Alzheimer's disease (Chakraborty et al., 2023). This specificity allows for improved visualization of pathological changes at early stages, potentially enabling pre-symptomatic diagnosis. For instance, nanoparticles coated with ligands that recognize and bind to amyloid-beta or alpha-synuclein proteins can provide detailed images of these aggregates, paving the way for more accurate and earlier interventions (Javdani et al., 2019).

5.3. Nanoparticle-Based Therapeutics

The ability of nanoparticles to cross the BBB and target diseased cells has positioned them as a cornerstone in therapeutic development for neurodegenerative diseases (Rodrigues et al., 2024). These tiny carriers are being used to deliver drugs, anti-inflammatory agents, or genetic tools, such as siRNA and CRISPR, directly to affected regions (Chen et al., 2023a). For example, nanoparticles can transport agents that mitigate alpha-synuclein aggregation in Parkinson's disease (PD) or target mutant huntingtin proteins in HD (Zoey et al., 2021;Chánez-Cárdenas & Vázquez-Contreras, 2012; Wang et al., 2022). Their precision reduces systemic side effects while enhancing therapeutic efficacy (Elumalai et al., 2024). Additionally, nanoparticles can be engineered to release their cargo in response to specific environmental triggers, such as pH changes or enzyme activity, ensuring that treatments are activated only where needed (Toader et al., 2024) (Table 2).

Table 2. Nanoparticle applications in neurodegenerative disease treatment: in vitro, in vivo, and in silico models

Model	Type of Nanoparticle	Application	Conducted Methodology	Focus	Reference
In Vitro	Gold Nanoparticles (AuNPs)	AD	Cultured neuronal cells treated with AuNPs to assess neuroprotective effects against amyloid-beta toxicity.	Treatment	(Wang et al., 2023)
	Cerium Oxide Nanoparticles (CeO_2)	PD	Neuronal cell lines exposed to CeO_2 NPs to measure ROS levels and mitochondrial function.	Treatment	(Cicek & Danışman, 2023)
	Polymeric Nanoparticles	General ND treatment	Drug-loaded polymeric NPs tested in cell cultures for drug release profiles and cytotoxicity assessments.	Treatment	(Jesus et al., 2019)
In Vivo	Magnetic Iron Oxide Nanoparticles	AD	Administered to AD model mice, followed by behavioral tests and imaging to evaluate cognitive function.	Treatment	(Sharma & Pervushin, 2020)
	PLGA/Puerarin Nanoparticles	PD	Injected into PD model mice; assessed for behavioral improvements and dopamine levels in the brain.	Treatment	(Zhao et al., 2020)
	Hydroxyapatite Nanoparticles	General ND treatment	Evaluated in animal models for inflammatory response and neurodegeneration markers post-injection.	Treatment	(Mursal et al., 2024)
In Silico	Computational Models	Drug design and efficacy prediction	Simulations conducted to model NP interactions with biological systems, using molecular dynamics approaches.	Diagnosis & Treatment	(Agamah et al., 2019)
	Molecular Dynamics Simulations	Understanding NP behavior in biological environments	Used to predict stability, bioavailability, and therapeutic outcomes of NPs based on simulated conditions.	Diagnosis & Treatment	(Krsek & Baticic, 2024)

5.4. Advanced Functionalization and Theranostics

Theranostics, the combination of therapy and diagnostics in a single platform, is an emerging application of multifunctional nanoparticles. These advanced systems can simultaneously deliver neuroprotective agents and monitor disease progression in real time (Ekhator et al., 2023). For instance, nanoparticles functionalized with both therapeutic molecules and imaging agents allow clinicians to observe the effectiveness of treatment while administering it (Janib et al., 2010). This dual functionality is particularly valuable for diseases like AD and PD, where continuous monitoring of pathological changes can inform treatment adjustments and improve outcomes (Roghani et al., 2024;Van Vliet et al., 2023). The precision of these systems facilitates a more individualized approach to the management of neurodegenerative diseases.

6. CHALLENGES AND FUTURE DIRECTIONS IN NANOTECHNOLOGY FOR NEURODEGENERATION

Despite their promise, nanoparticles face significant challenges in clinical translation (Liu et al., 2024). Biocompatibility and potential toxicity remain primary concerns, as the long-term effects of nanoparticles on human health are not fully understood (Xuan et al., 2023). Furthermore, the scalability of nanoparticle production and the rigorous regulatory requirements for clinical approval pose additional hurdles (Desai, 2012). However, the future of nanotechnology in neurodegeneration is bright, especially when integrated with advanced tools like artificial intelligence and CRISPR gene-editing technologies (Chen et al., 2023). These combinations could enable highly personalized and precise treatments, tailored to the genetic and molecular profiles of individual patients (Ho et al., 2020). Overcoming current limitations will require interdisciplinary collaboration and innovative approaches to ensure the safe and effective application of nanotechnology in treating neurodegenerative diseases (Asil et al., 2020).

7. CONCLUSION

Recent advances in neurodegenerative disease modeling are transforming biomedical research by bridging the gap between traditional models and human pathophysiology. Beyond classical pathological and behavioral assessments, the integration of oxidative, inflammatory, biochemical, and neurophysiological parameters has enhanced the translational relevance of these models. Nanotechnology and theranostic platforms further expand this potential by enabling targeted therapy alongside real-time monitoring of disease progression. Despite these advances, challenges such as safety, reproducibility, and clinical scalability remain major hurdles. Future progress will rely on interdisciplinary strategies that combine neuroscience, nanomedicine, and computational tools like AI and CRISPR. Together, these innovations hold promise for delivering the next generation of personalized therapies for Alzheimer's, Parkinson's, and related disorders.

REFERENCES

Agamah, F. E., Mazandu, G. K., Hassan, R., Bope, C. D., Thomford, N. E., Ghansah, A., & Chimusa, E. R. (2019). Computational/in silico methods in drug target and lead prediction. *Briefings in Bioinformatics, 21*(5), 1663–1675. DOI: 10.1093/bib/bbz103 PMID: 31711157

Al Abbar, A., Ngai, S. C., Nograles, N., Alhaji, S. Y., & Abdullah, S. (2020). Induced Pluripotent Stem Cells: Reprogramming Platforms and Applications in Cell Replacement Therapy. *BioResearch Open Access, 9*(1), 121–136. DOI: 10.1089/biores.2019.0046 PMID: 32368414

Albert, K., Niskanen, J., Kälvälä, S., & Lehtonen, Š. (2021). Utilising Induced Pluripotent Stem Cells in Neurodegenerative Disease Research: Focus on Glia. *International Journal of Molecular Sciences, 22*(9), 4334. DOI: 10.3390/ijms22094334 PMID: 33919317

Altammar, K. A. (2023). A review on nanoparticles: Characteristics, synthesis, applications, and challenges. *Frontiers in Microbiology, 14*, 1155622. DOI: 10.3389/fmicb.2023.1155622 PMID: 37180257

Ansori, A. NM., Antonius, Y., Susilo, R. JK., Hayaza, S., Kharisma, V. D., Parikesit, A. A., Zainul, R., Jakhmola, V., Saklani, T., Rebezov, M., & Ullah, Md. E., Maksimiuk, N., Derkho, M., & Burkov, P. (2023). Application of CRISPR-Cas9 genome editing technology in various fields: A review. *Narra J, 3*(2), e184. DOI: 10.52225/narra.v3i2.184 PMID: 38450259

Ashok, A., Andrabi, S. S., & Labhasetwar, V. (2022). Antioxidant therapy in oxidative stress-induced neurodegenerative diseases: Role of nanoparticle-based drug delivery systems in clinical translation. *Antioxidants, 11*(2), 408. DOI: 10.3390/antiox11020408 PMID: 35204290

Asil, S. M., Ahlawat, J., Barroso, G. G., & Narayan, M. (2020). Nanomaterial Based Drug Delivery Systems for the Treatment of Neurodegenerative Diseases. *Biomaterials Science, 8*(15), 4109–4128. DOI: 10.1039/d0bm00809e PMID: 32638706

Ayeni, E. A., Aldossary, A. M., Ayejoto, D. A., Gbadegesin, L. A., Alshehri, A. A., Alfassam, H. A., Afewerky, H. K., Almughem, F. A., Bello, S. M., & Tawfik, E. A. (2022). Neurodegenerative Diseases: Implications of Environmental and Climatic Influences on Neurotransmitters and Neuronal Hormones Activities. *International Journal of Environmental Research and Public Health, 19*(19), 12495. DOI: 10.3390/ijerph191912495 PMID: 36231792

Banerjee, R., Rai, A., Iyer, S. M., Narwal, S., & Tare, M. (2022). Animal models in the study of Alzheimer's disease and Parkinson's disease: A historical perspective. *Animal Models and Experimental Medicine, 5*(1), 27–37. DOI: 10.1002/ame2.12209 PMID: 35229999

Bhardwaj, S., Kesari, K. K., Rachamalla, M., Mani, S., & Ashraf, G. Md., Jha, S. K., Kumar, P., Ambasta, R. K., Dureja, H., Devkota, H. P., Gupta, G., Chellappan, D. K., Singh, S. K., Dua, K., Ruokolainen, J., Kamal, M. A., Ojha, S., & Jha, N. K. (2021). CRISPR/Cas9 gene editing: New hope for Alzheimer's disease therapeutics. *Journal of Advanced Research, 40*, 207–221. DOI: 10.1016/j.jare.2021.07.001 PMID: 36100328

Carraro, C., Montgomery, J. V., Klimmt, J., Paquet, D., Schultze, J. L., & Beyer, M. D. (2024a). Tackling neurodegeneration in vitro with omics: A path towards new targets and drugs. *Frontiers in Molecular Neuroscience, 17*. Advance online publication. DOI: 10.3389/fnmol.2024.1414886 PMID: 38952421

Carraro, C., Montgomery, J. V., Klimmt, J., Paquet, D., Schultze, J. L., & Beyer, M. D. (2024b). Tackling neurodegeneration in vitro with omics: A path towards new targets and drugs. *Frontiers in Molecular Neuroscience, 17*, 1414886. DOI: 10.3389/fnmol.2024.1414886 PMID: 38952421

Chakraborty, A., Mohapatra, S. S., Barik, S., Roy, I., Gupta, B., & Biswas, A. (2023). Impact of nanoparticles on amyloid β-induced Alzheimer's disease, tuberculosis, leprosy and cancer: A systematic review. *Bioscience Reports, 43*(2), BSR20220324. Advance online publication. DOI: 10.1042/BSR20220324 PMID: 36630532

Chánez-Cárdenas, M. E., & Vázquez-Contreras, E. (2012). The Aggregation of Huntingtin and α-Synuclein. *Journal of Biophysics, 606172*. Advance online publication. DOI: 10.1155/2012/606172 PMID: 22899913

Chen, C., Zhong, W., Du, S., Li, Y., Zeng, Y., Liu, K., Yang, J., Guan, X., & Han, X. (2023a). Intelligent nanotherapeutic strategies for the delivery of CRISPR system. *Acta Pharmaceutica Sinica. B, 13*(6), 2510–2543. DOI: 10.1016/j.apsb.2022.12.013 PMID: 37425051

Chen, C., Zhong, W., Du, S., Li, Y., Zeng, Y., Liu, K., Yang, J., Guan, X., & Han, X. (2023b). Intelligent nanotherapeutic strategies for the delivery of CRISPR system. *Acta Pharmaceutica Sinica. B, 13*(6), 2510–2543. DOI: 10.1016/j.apsb.2022.12.013 PMID: 37425051

Cheng, X., Xie, Q., & Sun, Y. (2023). Advances in nanomaterial-based targeted drug delivery systems. *Frontiers in Bioengineering and Biotechnology, 11*, 1177151. DOI: 10.3389/fbioe.2023.1177151 PMID: 37122851

Choi, H., Kim, H., Chung, H., Lee, D.-S., & Kim, J. (2024). Application of computational algorithms for single-cell RNA-seq and ATAC-seq in neurodegenerative diseases. *Briefings in Functional Genomics*, •••, elae044. Advance online publication. DOI: 10.1093/bfgp/elae044 PMID: 39500613

Chornyy, S., Borovicka, J. A., Patel, D., Shin, M.-K., Vázquez-Rosa, E., Miller, E., Wilson, B., Pieper, A. A., & Dana, H. (2023). Longitudinal in vivo monitoring of axonal degeneration after brain injury. *Cell Reports Methods, 3*(5), 100481. DOI: 10.1016/j.crmeth.2023.100481 PMID: 37323578

Cicek, B., & Danışman, B. (2023). Cerium Oxide Nanoparticles Rescue Dopaminergic Neurons in Parkinson's Disease Model of SH-SY5Y Cells via Modulating Nrf2 Signaling and Ameliorating Apoptotic Cell Death. *Archives of Basic and Clinical Research, 5*(2), 284–290. DOI: 10.5152/ABCR.2023.23150

Cui, X., Li, X., Zheng, H., Su, Y., Zhang, S., Li, M., Hao, X., Zhang, S., Hu, Z., Xia, Z., Shi, C., Xu, Y., & Mao, C. (2024). Human midbrain organoids: A powerful tool for advanced Parkinson's disease modeling and therapy exploration. *NPJ Parkinson's Disease, 10*(1), 1–14. DOI: 10.1038/s41531-024-00799-8 PMID: 39428415

Dawson, T. M., Golde, T. E., & Tourenne, C. L. (2018). Animal Models of Neurodegenerative Diseases. *Nature Neuroscience, 21*(10), 1370–1379. DOI: 10.1038/s41593-018-0236-8 PMID: 30250265

Desai, N. (2012). Challenges in Development of Nanoparticle-Based Therapeutics. *The AAPS Journal, 14*(2), 282–295. DOI: 10.1208/s12248-012-9339-4 PMID: 22407288

Deumens, R., Blokland, A., & Prickaerts, J. (2002). Modeling Parkinson's Disease in Rats: An Evaluation of 6-OHDA Lesions of the Nigrostriatal Pathway. *Experimental Neurology, 175*(2), 303–317. DOI: 10.1006/exnr.2002.7891 PMID: 12061862

Di Luca, M., Destrebecq, F., & Kramer, S. (2021). Future of the aging brain: Bridging the gap between research and policy. *Aging Brain, 1*, 100002. DOI: 10.1016/j.nbas.2020.100002 PMID: 36911502

Ekhator, C., Qureshi, M. Q., Zuberi, A. W., Hussain, M., Sangroula, N., Yerra, S., Devi, M., Naseem, M. A., Bellegarde, S. B., & Pendyala, P. R. (2023). Advances and Opportunities in Nanoparticle Drug Delivery for Central Nervous System Disorders: A Review of Current Advances. *Cureus, 15*(8), e44302. DOI: 10.7759/cureus.44302 PMID: 37649926

Elder, G. A., Sosa, M. A. G., De Gasperi, R., Dickstein, D. L., & Hof, P. R. (2010). Presenilin transgenic mice as models of Alzheimer's disease. *Brain Structure & Function, 214*(0), 127–143. DOI: 10.1007/s00429-009-0227-3 PMID: 19921519

Elumalai, K., Srinivasan, S., & Shanmugam, A. (2024). Review of the efficacy of nanoparticle-based drug delivery systems for cancer treatment. *Biomedical Technology, 5*, 109–122. DOI: 10.1016/j.bmt.2023.09.001

Emborg, M. E. (2017). Nonhuman Primate Models of Neurodegenerative Disorders. *ILAR Journal, 58*(2), 190–201. DOI: 10.1093/ilar/ilx021 PMID: 28985333

Erfanian, N., Heydari, A. A., Feriz, A. M., Iañez, P., Derakhshani, A., Ghasemigol, M., Farahpour, M., Razavi, S. M., Nasseri, S., Safarpour, H., & Sahebkar, A. (2023). Deep learning applications in single-cell genomics and transcriptomics data analysis. *Biomedicine and Pharmacotherapy, 165*, 115077. DOI: 10.1016/j.biopha.2023.115077 PMID: 37393865

Essayan-Perez, S., Zhou, B., Nabet, A. M., Wernig, M., & Huang, Y.-W. A. (2019). Modeling Alzheimer's disease with human iPS cells: Advancements, lessons, and applications. *Neurobiology of Disease, 130*, 104503. DOI: 10.1016/j.nbd.2019.104503 PMID: 31202913

Fan, Y., Han, J., Zhao, L., Wu, C., Wu, P., Huang, Z., Hao, X., Ji, Y., Chen, D., & Zhu, M. (2021). Experimental Models of Cognitive Impairment for Use in Parkinson's Disease Research: The Distance Between Reality and Ideal. *Frontiers in Aging Neuroscience, 13*, 745438. DOI: 10.3389/fnagi.2021.745438 PMID: 34912207

Farkhondeh, A., Li, R., Gorshkov, K., Chen, K. G., Might, M., Rodems, S., Lo, D. C., & Zheng, W. (2019). Induced pluripotent stem cells for neural drug discovery. *Drug Discovery Today, 24*(4), 992–999. DOI: 10.1016/j.drudis.2019.01.007 PMID: 30664937

Fasano, M., & Alberio, T. (2023). Neurodegenerative disorders: From clinicopathology convergence to systems biology divergence. In Espay, A. J. (Ed.), *Handbook of Clinical Neurology* (Vol. 192, pp. 73–86). Elsevier., DOI: 10.1016/B978-0-323-85538-9.00007-9

Firdaus, Z., & Li, X. (2024). Unraveling the Genetic Landscape of Neurological Disorders: Insights into Pathogenesis, Techniques for Variant Identification, and Therapeutic Approaches. *International Journal of Molecular Sciences*, *25*(4), 4. Advance online publication. DOI: 10.3390/ijms25042320 PMID: 38396996

Gabriel, R., Boreland, A., & Pang, Z. P. (2023). Whole cell patch clamp electrophysiology in human neuronal cells. In Pang, Z. P. (Ed.), *Methods in molecular biology* (Vol. 2683, pp. 259–276). Humana Press., DOI: 10.1007/978-1-0716-3195-0_16

Gadhave, D. G., Sugandhi, V. V., Jha, S. K., Nangare, S. N., Gupta, G., Singh, S. K., Dua, K., Cho, H., Hansbro, P. M., & Paudel, K. R. (2024). Neurodegenerative disorders: Mechanisms of degeneration and therapeutic approaches with their clinical relevance. *Ageing Research Reviews*, *99*, 102357. DOI: 10.1016/j.arr.2024.102357 PMID: 38830548

Geerts, H., Hofmann-Apitius, M., & Anastasio, T. J., & Brain Health Modeling Initiative. (2017). Knowledge-driven computational modeling in Alzheimer's disease research: Current state and future trends. *Alzheimer's & Dementia*, *13*(11), 1292–1302. DOI: 10.1016/j.jalz.2017.08.011 PMID: 28917669

Guerreiro, S., & Maciel, P. (2023). Transition from Animal-Based to Human Induced Pluripotent Stem Cells (iPSCs)-Based Models of Neurodevelopmental Disorders: Opportunities and Challenges. *Cells*, *12*(4), 4. Advance online publication. DOI: 10.3390/cells12040538 PMID: 36831205

Guimarães, R. P., de Resende, M. C. S., Tavares, M. M., Belardinelli de Azevedo, C., Ruiz, M. C. M., & Mortari, M. R. (2024). Construct, Face, and Predictive Validity of Parkinson's Disease Rodent Models. *International Journal of Molecular Sciences*, *25*(16), 8971. DOI: 10.3390/ijms25168971 PMID: 39201659

Habibey, R., Rojo Arias, J. E., Striebel, J., & Busskamp, V. (2022). Microfluidics for Neuronal Cell and Circuit Engineering. *Chemical Reviews*, *122*(18), 14842–14880. DOI: 10.1021/acs.chemrev.2c00212 PMID: 36070858

Halldorsson, S., Lucumi, E., Gómez-Sjöberg, R., & Fleming, R. M. T. (2015). Advantages and challenges of microfluidic cell culture in polydimethylsiloxane devices. *Biosensors & Bioelectronics*, *63*, 218–231. DOI: 10.1016/j.bios.2014.07.029 PMID: 25105943

Harvey, B. K., Richie, C. T., Hoffer, B. J., & Airavaara, M. (2011). Transgenic animal models of neurodegeneration based on human genetic studies. *Journal of Neural Transmission (Vienna, Austria : 1996)*, *118*(1), 27–45. https://doi.org/DOI: 10.1007/s00702-010-0476-6

Harvey, J., Pishva, E., Chouliaras, L., & Lunnon, K. (2023). Elucidating distinct molecular signatures of Lewy body dementias. *Neurobiology of Disease*, *188*, 106337. DOI: 10.1016/j.nbd.2023.106337 PMID: 37918758

Hersh, A. M., Alomari, S., & Tyler, B. M. (2022). Crossing the Blood-Brain Barrier: Advances in Nanoparticle Technology for Drug Delivery in Neuro-Oncology. *International Journal of Molecular Sciences, 23*(8), 4153. DOI: 10.3390/ijms23084153 PMID: 35456971

Ho, D., Quake, S. R., McCabe, E. R. B., Chng, W. J., Chow, E. K., Ding, X., Gelb, B. D., Ginsburg, G. S., Hassenstab, J., Ho, C.-M., Mobley, W. C., Nolan, G. P., Rosen, S. T., Tan, P., Yen, Y., & Zarrinpar, A. (2020). Enabling technologies for personalized and precision medicine. *Trends in Biotechnology, 38*(5), 497–518. DOI: 10.1016/j.tibtech.2019.12.021 PMID: 31980301

Holloway, P. M., Willaime-Morawek, S., Siow, R., Barber, M., Owens, R. M., Sharma, A. D., Rowan, W., Hill, E., & Zagnoni, M. (2021). Advances in microfluidic in vitro systems for neurological disease modeling. *Journal of Neuroscience Research, 99*(5), 1276–1307. DOI: 10.1002/jnr.24794 PMID: 33583054

Houdebine, L.-M. (2006). Transgenic Animal Models in Biomedical Research. In S. Mouldy, *Target Discovery and Validation Reviews and Protocols* (Vol. 360, pp. 163–202). Humana Press. https://doi.org/DOI: 10.1385/1-59745-165-7:163

Ibrahim, W., & Abdel Rasheed, N. O. (2022). Diapocynin neuroprotective effects in 3-nitropropionic acid Huntington's disease model in rats: Emphasis on Sirt1/Nrf2 signaling pathway. *Inflammopharmacology, 30*(5), 1745–1758. DOI: 10.1007/s10787-022-01057-0 PMID: 35639233

Jahabardeen, A., S, N., J, N., & V, C. (2024). A Review on the Role of SNCA Gene in Neurodegenerative Diseases. *Cureus, 16*(9), e69450. DOI: 10.7759/cureus.69450 PMID: 39411638

Jahromi, M. A. M., Abdoli, A., Rahmanian, M., Bardania, H., Bayandori, M., Basri, S. M. M., Kalbasi, A., Aref, A. R., Karimi, M., & Hamblin, M. R. (2019). Microfluidic Brain-on-a-Chip: Perspectives for Mimicking Neural System Disorders. *Molecular Neurobiology, 56*(12), 8489–8512. DOI: 10.1007/s12035-019-01653-2 PMID: 31264092

Jang, H., Kim, S., Koh, Y., & Yoon, K.-J. (2022). Engineering Brain Organoids: Toward Mature Neural Circuitry with an Intact Cytoarchitecture. *International Journal of Stem Cells, 15*, 41–59. DOI: 10.15283/ijsc22004 PMID: 35220291

Janib, S. M., Moses, A. S., & MacKay, J. A. (2010). Imaging and drug delivery using theranostic nanoparticles. *Advanced Drug Delivery Reviews, 62*(11), 1052–1063. DOI: 10.1016/j.addr.2010.08.004 PMID: 20709124

Javdani, N., Rahpeyma, S. S., Ghasemi, Y., & Raheb, J. (2019). Effect of superparamagnetic nanoparticles coated with various electric charges on α-synuclein and β-amyloid proteins fibrillation process. *International Journal of Nanomedicine, 14*, 799–808. DOI: 10.2147/IJN.S190354 PMID: 30774334

Jesus, S., Schmutz, M., Som, C., Borchard, G., Wick, P., & Borges, O. (2019). Hazard Assessment of Polymeric Nanobiomaterials for Drug Delivery: What Can We Learn From Literature So Far. *Frontiers in Bioengineering and Biotechnology, 7*. Advance online publication. DOI: 10.3389/fbioe.2019.00261 PMID: 31709243

Joseph, T. M., Kar Mahapatra, D., Esmaeili, A., Piszczyk, Ł., Hasanin, M. S., Kattali, M., Haponiuk, J., & Thomas, S. (2023). Nanoparticles: Taking a Unique Position in Medicine. *Nanomaterials (Basel, Switzerland), 13*(3), 574. DOI: 10.3390/nano13030574 PMID: 36770535

Joshi, R., Missong, H., Mishra, J., Kaur, S., Saini, S., Kandimalla, R., Reddy, P. H., Babu, A., Bhatti, G. K., & Bhatti, J. S. (2023). Nanotheranostics revolutionizing neurodegenerative diseases: From precision diagnosis to targeted therapies. *Journal of Drug Delivery Science and Technology, 89*, 105067. DOI: 10.1016/j.jddst.2023.105067

Julien, J.-P., & Kriz, J. (2006). Transgenic mouse models of amyotrophic lateral sclerosis. *Biochimica et Biophysica Acta (BBA) -. Molecular Basis of Disease, 1762*(11–12), 1013–1024. DOI: 10.1016/j.bbadis.2006.03.006 PMID: 16675207

Jungverdorben, J., Till, A., & Brüstle, O. (2017). Induced pluripotent stem cell-based modeling of neurodegenerative diseases: A focus on autophagy. *Journal of Molecular Medicine, 95*(7), 705–718. DOI: 10.1007/s00109-017-1533-5 PMID: 28593578

Kabanov, A. V., & Gendelman, H. E. (2007). Nanomedicine in the diagnosis and therapy of neurodegenerative disorders. *Progress in Polymer Science, 32*(8–9), 1054–1082. DOI: 10.1016/j.progpolymsci.2007.05.014 PMID: 20234846

Kang, R., Park, S., Shin, S., Bak, G., & Park, J.-C. (2024). Electrophysiological insights with brain organoid models: A brief review. *BMB Reports, 57*(7), 311–317. DOI: 10.5483/BMBRep.2024-0077 PMID: 38919012

Kaye, J., Reisine, T., & Finkbeiner, S. (2021). Huntington's disease mouse models: Unraveling the pathology caused by CAG repeat expansion. *Faculty Reviews, 10*, 77. DOI: 10.12703/r/10-77 PMID: 34746930

Kaye, J., Reisine, T., & Finkbeiner, S. (2022). Huntington's disease iPSC models—Using human patient cells to understand the pathology caused by expanded CAG repeats. *Faculty Reviews, 11*, 16. DOI: 10.12703/r/11-16 PMID: 35865413

Khan, I., Saeed, K., & Khan, I. (2019). Nanoparticles: Properties, applications and toxicities. *Arabian Journal of Chemistry, 12*(7), 908–931. DOI: 10.1016/j.arabjc.2017.05.011

Kitazawa, M., Medeiros, R., & LaFerla, F. M. (2012). Transgenic mouse models of Alzheimer disease: Developing a better model as a tool for therapeutic interventions. *Current Pharmaceutical Design, 18*(8), 1131–1147. DOI: 10.2174/138161212799315786 PMID: 22288400

Krsek, A., & Baticic, L. (2024). Nanotechnology-Driven Therapeutic Innovations in Neurodegenerative Disorders: A Focus on Alzheimer's and Parkinson's Disease. *Future Pharmacology, 4*(2), 2. Advance online publication. DOI: 10.3390/futurepharmacol4020020

Kumar, D., & Hassan, Md. I. (2021). Neurodegenerative brain models vs. cell replacement or restoration therapy: A review on promises and pitfalls. *Biochemical and Biophysical Research Communications, 585*, 124–131. DOI: 10.1016/j.bbrc.2021.11.040 PMID: 34801932

Lamptey, R. N. L., Chaulagain, B., Trivedi, R., Gothwal, A., Layek, B., & Singh, J. (2022). A Review of the Common Neurodegenerative Disorders: Current Therapeutic Approaches and the Potential Role of Nanotherapeutics. *International Journal of Molecular Sciences, 23*(3), 3. Advance online publication. DOI: 10.3390/ijms23031851 PMID: 35163773

Lee, C.-T., Bendriem, R. M., Wu, W. W., & Shen, R.-F. (2017). 3D brain Organoids derived from pluripotent stem cells: Promising experimental models for brain development and neurodegenerative disorders. *Journal of Biomedical Science, 24*(1), 59. DOI: 10.1186/s12929-017-0362-8 PMID: 28822354

Leopold, A. V., Shcherbakova, D. M., & Verkhusha, V. V. (2019). Fluorescent biosensors for neurotransmission and neuromodulation: Engineering and applications. *Frontiers in Cellular Neuroscience, 13*, 474. DOI: 10.3389/fncel.2019.00474 PMID: 31708747

Li, J., & Fraenkel, E. (2021). Phenotyping Neurodegeneration in Human iPSCs. *Annual Review of Biomedical Data Science, 4*, 83–100. DOI: 10.1146/annurev-biodatasci-092820-025214 PMID: 34465166

Liu, G., Aliaga, L., & Cai, H. (2012). α-synuclein, LRRK2 and their interplay in Parkinson's disease. *Future Neurology, 7*(2), 145–153. PMID: 22563296

Liu, Y., Zhang, Y., Li, H., & Hu, T. Y. (2024). Recent advances in the bench-to-bedside translation of cancer nanomedicines. *Acta Pharmaceutica Sinica. B, S2211383524004647*. Advance online publication. DOI: 10.1016/j.apsb.2024.12.007 PMID: 40041906

Madsen, D. A., Schmidt, S. I., Blaabjerg, M., & Meyer, M. (2021). Interaction between Parkin and α-Synuclein in PARK2-Mediated Parkinson's Disease. *Cells, 10*(2), 2. Advance online publication. DOI: 10.3390/cells10020283 PMID: 33572534

Makhouri, F. R., & Ghasemi, J. B. (2018). In Silico Studies in Drug Research Against Neurodegenerative Diseases. *Current Neuropharmacology, 16*(6), 664–725. DOI: 10.2174/1570159X15666170823095628 PMID: 28831921

Mijanović, O., Branković, A., Borovjagin, A. V., Butnaru, D. V., Bezrukov, E. A., Sukhanov, R. B., Shpichka, A., Timashev, P., & Ulasov, I. (2020). Battling Neurodegenerative Diseases with Adeno-Associated Virus-Based Approaches. *Viruses, 12*(4), 460. DOI: 10.3390/v12040460 PMID: 32325732

Miny, L., Maisonneuve, B. G. C., Quadrio, I., & Honegger, T. (2022). Modeling Neurodegenerative Diseases Using In Vitro Compartmentalized Microfluidic Devices. *Frontiers in Bioengineering and Biotechnology, 10*, 919646. DOI: 10.3389/fbioe.2022.919646 PMID: 35813998

Mursal, M., Kumar, A., Hasan, S. M., Hussain, S., Singh, K., Kushwaha, S. P., Arif, M., Kumar Singh, R., Singh, D., Mohammad, A., & Fatima, S. K. (2024). Role of natural bioactive compounds in the management of neurodegenerative disorders. *Intelligent Pharmacy, 2*(1), 102–113. DOI: 10.1016/j.ipha.2023.09.006

Narmashiri, A., Abbaszadeh, M., & Ghazizadeh, A. (2022). The effects of 1-methyl-4-phenyl-1,2,3,6-tetrahydropyridine (MPTP) on the cognitive and motor functions in rodents: A systematic review and meta-analysis. *Neuroscience and Biobehavioral Reviews*, *140*, 104792. DOI: 10.1016/j.neubiorev.2022.104792 PMID: 35872230

Newman, M., Ebrahimie, E., & Lardelli, M. (2014). Using the zebrafish model for Alzheimer's disease research. *Frontiers in Genetics*, *5*, 189. DOI: 10.3389/fgene.2014.00189 PMID: 25071820

Nouri Nojadeh, J., Bildiren Eryilmaz, N. S., & Ergüder, B. I. (2023). CRISPR/Cas9 genome editing for neurodegenerative diseases. *EXCLI Journal*, *22*, 567–582. DOI: 10.17179/excli2023-6155 PMID: 37636024

Osaki, T., Shin, Y., Sivathanu, V., Campisi, M., & Kamm, R. D. (2018). In Vitro Microfluidic Models for Neurodegenerative Disorders. *Advanced Healthcare Materials*, *7*(2), 1700489. DOI: 10.1002/adhm.201700489 PMID: 28881425

Paik, D. T., Chandy, M., & Wu, J. C. (2020). Patient and Disease–Specific Induced Pluripotent Stem Cells for Discovery of Personalized Cardiovascular Drugs and Therapeutics. *Pharmacological Reviews*, *72*(1), 320–342. DOI: 10.1124/pr.116.013003 PMID: 31871214

Palanisamy, C. P., Pei, J., Alugoju, P., Anthikapalli, N. V. A., Jayaraman, S., Veeraraghavan, V. P., Gopathy, S., Roy, J. R., Janaki, C. S., Thalamati, D., Mironescu, M., Luo, Q., Miao, Y., Chai, Y., & Long, Q. (2023). New strategies of neurodegenerative disease treatment with extracellular vesicles (EVs) derived from mesenchymal stem cells (MSCs). *Theranostics*, *13*(12), 4138–4165. DOI: 10.7150/thno.83066 PMID: 37554286

Pan, M.-T., Zhang, H., Li, X.-J., & Guo, X.-Y. (2024). Genetically modified non-human primate models for research on neurodegenerative diseases. *Zoological Research*, *45*(2), 263–274. DOI: 10.24272/j.issn.2095-8137.2023.197 PMID: 38287907

Pasko, V. I., Churkina, A. S., Shakhov, A. S., Kotlobay, A. A., & Alieva, I. B. (2022). Modeling of Neurodegenerative Diseases: 'Step by Step' and 'Network' Organization of the Complexes of Model Systems. *International Journal of Molecular Sciences*, *24*(1), 604. DOI: 10.3390/ijms24010604 PMID: 36614047

Poon, C. H., Wang, Y., Fung, M.-L., Zhang, C., & Lim, L. W. (2020). Rodent Models of Amyloid-Beta Feature of Alzheimer's Disease: Development and Potential Treatment Implications. *Aging and Disease*, *11*(5), 1235–1259. DOI: 10.14336/AD.2019.1026 PMID: 33014535

Robert, A., Schöll, M., & Vogels, T. (2021). Tau Seeding Mouse Models with Patient Brain-Derived Aggregates. *International Journal of Molecular Sciences*, *22*(11), 6132. DOI: 10.3390/ijms22116132 PMID: 34200180

Rodrigues, R. O., Shin, S.-R., & Bañobre-López, M. (2024). Brain-on-a-chip: An emerging platform for studying the nanotechnology-biology interface for neurodegenerative disorders. *Journal of Nanobiotechnology*, *22*, 573. DOI: 10.1186/s12951-024-02720-0 PMID: 39294645

Roghani, A. K., Garcia, R. I., Roghani, A., Reddy, A., Khemka, S., Reddy, R. P., Pattoor, V., Jacob, M., Reddy, P. H., & Sehar, U. (2024). Treating Alzheimer's disease using nanoparticle-mediated drug delivery strategies/systems. *Ageing Research Reviews*, *97*, 102291. DOI: 10.1016/j.arr.2024.102291 PMID: 38614367

Sadat Razavi, Z., Sina Alizadeh, S., Sadat Razavi, F., Souri, M., & Soltani, M. (2025). Advancing neurological disorders therapies: Organic nanoparticles as a key to blood-brain barrier penetration. *International Journal of Pharmaceutics*, *670*, 125186. DOI: 10.1016/j.ijpharm.2025.125186 PMID: 39788400

Samaranch, L., Lorenzo-Betancor, O., Arbelo, J. M., Ferrer, I., Lorenzo, E., Irigoyen, J., Pastor, M. A., Marrero, C., Isla, C., Herrera-Henriquez, J., & Pastor, P. (2010). PINK1-linked parkinsonism is associated with Lewy body pathology. *Brain*, *133*(4), 1128–1142. DOI: 10.1093/brain/awq051 PMID: 20356854

Sanapathi, J., Vipparthi, P., Mishra, S., Sosnik, A., & Kumarasamy, M. (2023). Microfluidics for brain endothelial cell-astrocyte interactions. *Organs-on-a-Chip*, *5*, 100033. DOI: 10.1016/j.ooc.2023.100033

Sarkar, S., Raymick, J., & Imam, S. (2016). Neuroprotective and Therapeutic Strategies against Parkinson's Disease: Recent Perspectives. *International Journal of Molecular Sciences*, *17*(6), 904. DOI: 10.3390/ijms17060904 PMID: 27338353

Seo, J. H., Shin, J. H., Lee, J., Kim, D., Hwang, H.-Y., Nam, B.-G., Lee, J., Kim, H. H., & Cho, S.-R. (2023). DNA double-strand break-free CRISPR interference delays Huntington's disease progression in mice. *Communications Biology*, *6*(1), 1–12. DOI: 10.1038/s42003-023-04829-8 PMID: 37117485

Sharma, B., & Pervushin, K. (2020). Magnetic Nanoparticles as In Vivo Tracers for Alzheimer's Disease. *Magnetochemistry (Basel, Switzerland)*, *6*(1), 1. Advance online publication. DOI: 10.3390/magnetochemistry6010013

Sharma, N., Khurana, N., & Muthuraman, A. (2017). Lower vertebrate and invertebrate models of Alzheimer's disease – A review. *European Journal of Pharmacology*, *815*, 312–323. DOI: 10.1016/j.ejphar.2017.09.017 PMID: 28943103

Shi, Y., Inoue, H., Wu, J. C., & Yamanaka, S. (2017). Induced pluripotent stem cell technology: A decade of progress. *Nature Reviews. Drug Discovery*, *16*(2), 115–130. DOI: 10.1038/nrd.2016.245 PMID: 27980341

Sim, S., & Wong, N. K. (2021). Nanotechnology and its use in imaging and drug delivery [Review]. *Biomedical Reports*, *14*(5), 42. DOI: 10.3892/br.2021.1418 PMID: 33728048

Slézia, A., Hegedüs, P., Rusina, E., Lengyel, K., Solari, N., Kaszas, A., Balázsfi, D., Botzanowski, B., Acerbo, E., Missey, F., Williamson, A., & Hangya, B. (2023). Behavioral, neural and ultrastructural alterations in a graded-dose 6-OHDA mouse model of early-stage Parkinson's disease. *Scientific Reports*, *13*(1), 19478. DOI: 10.1038/s41598-023-46501-5 PMID: 37945922

Spathopoulou, A., Edenhofer, F., & Fellner, L. (2022). Targeting α-Synuclein in Parkinson's Disease by Induced Pluripotent Stem Cell Models. *Frontiers in Neurology, 12*, 786835. DOI: 10.3389/fneur.2021.786835 PMID: 35145469

Sreenivasamurthy, S., Laul, M., Zhao, N., Kim, T., & Zhu, D. (2022). Current progress of cerebral organoids for modeling Alzheimer's disease origins and mechanisms. *Bioengineering & Translational Medicine, 8*(2), e10378. DOI: 10.1002/btm2.10378 PMID: 36925717

Strome, E. M., & Doudet, D. J. (2007). Animal Models of Neurodegenerative Disease: Insights from In vivo Imaging Studies. *Molecular Imaging & Biology, 9*(4), 186–195. DOI: 10.1007/s11307-007-0093-4 PMID: 17357857

Tabata, R. C., Wilson, J. M. B., Ly, P., Zwiegers, P., Kwok, D., Van Kampen, J. M., Cashman, N., & Shaw, C. A. (2008). Chronic Exposure to Dietary Sterol Glucosides is Neurotoxic to Motor Neurons and Induces an ALS-PDC Phenotype. *Neuromolecular Medicine, 10*(1), 24. DOI: 10.1007/s12017-007-8020-z PMID: 18196479

Taha, A., Bobi, J., Dammers, R., Dijkhuizen, R., Dreyer, A. Y., van Es, A., Ferrara, F., Gounis, M. J., Nitzsche, B., Platt, S., Stoffel, M. H., Volovici, V., del Zoppo, G. J., Duncker, D. J., Dippel, D. W. J., Boltze, J., & van Beusekom, H. (2022). Comparison of Large Animal Models for Acute Ischemic Stroke: Which model to use? *Stroke, 53*(4), 1411–1422. DOI: 10.1161/STROKEAHA.121.036050 PMID: 35164533

Tello, J. A., Williams, H. E., Eppler, R. M., Steinhilb, M. L., & Khanna, M. (2022). Animal Models of Neurodegenerative Disease: Recent Advances in Fly Highlight Innovative Approaches to Drug Discovery. *Frontiers in Molecular Neuroscience, 15*, 883358. DOI: 10.3389/fnmol.2022.883358 PMID: 35514431

Toader, C., Dumitru, A. V., Eva, L., Serban, M., Covache-Busuioc, R.-A., & Ciurea, A. V. (2024). Nanoparticle Strategies for Treating CNS Disorders: A Comprehensive Review of Drug Delivery and Theranostic Applications. *International Journal of Molecular Sciences, 25*(24), 13302. DOI: 10.3390/ijms252413302 PMID: 39769066

Ulusoy, A., Decressac, M., Kirik, D., & Björklund, A. (2010). Viral vector-mediated overexpression of α-synuclein as a progressive model of Parkinson's disease. In Vol. 184, pp. 89–111). Progress in Brain Research. Elsevier., DOI: 10.1016/S0079-6123(10)84005-1

Van Vliet, E. F., Knol, M. J., Schiffelers, R. M., Caiazzo, M., & Fens, M. H. A. M. (2023). Levodopa-loaded nanoparticles for the treatment of Parkinson's disease. *Journal of Controlled Release, 360*, 212–224. DOI: 10.1016/j.jconrel.2023.06.026 PMID: 37343725

Vann, K. T., & Xiong, Z.-G. (2016). Optogenetics for neurodegenerative diseases. *International Journal of Physiology, Pathophysiology and Pharmacology, 8*(1), 1–8. PMID: 27186317

Vlasov, K., Van Dort, C. J., & Solt, K. (2018). Optogenetics and Chemogenetics. *Methods in Enzymology, 603*, 181–196. DOI: 10.1016/bs.mie.2018.01.022 PMID: 29673525

Walsh, N., Kenney, L., Jangalwe, S., Aryee, K.-E., Greiner, D. L., Brehm, M. A., & Shultz, L. D. (2017). Humanized mouse models of clinical disease. *Annual Review of Pathology, 12*, 187–215. DOI: 10.1146/annurev-pathol-052016-100332 PMID: 27959627

Wang, G., Shen, X., Song, X., Wang, N., Wo, X., & Gao, Y. (2023). Protective mechanism of gold nanoparticles on human neural stem cells injured by β-amyloid protein through miR-21–5p/SOCS6 pathway. *Neurotoxicology, 95*, 12–22. DOI: 10.1016/j.neuro.2022.12.011 PMID: 36623431

Wang, Q., Yu, S., Simonyi, A., Sun, G. Y., & Sun, A. Y. (2005). Kainic Acid-Mediated Excitotoxicity as a Model for Neurodegeneration. *Molecular Neurobiology, 31*(1–3), 003–016. https://doi.org/DOI: 10.1385/MN:31:1-3:003

Wang, S., Zheng, J., Ma, L., Petersen, R. B., Xu, L., & Huang, K. (2022). Inhibiting protein aggregation with nanomaterials: The underlying mechanisms and impact factors. *Biochimica et Biophysica Acta (BBA) -. General Subjects, 1866*(2), 130061. DOI: 10.1016/j.bbagen.2021.130061 PMID: 34822925

Wareham, L. K., Liddelow, S. A., Temple, S., Benowitz, L. I., Di Polo, A., Wellington, C., Goldberg, J. L., He, Z., Duan, X., Bu, G., Davis, A. A., Shekhar, K., Torre, A. L., Chan, D. C., Canto-Soler, M. V., Flanagan, J. G., Subramanian, P., Rossi, S., Brunner, T., & Calkins, D. J. (2022). Solving neurodegeneration: Common mechanisms and strategies for new treatments. *Molecular Neurodegeneration, 17*, 23. DOI: 10.1186/s13024-022-00524-0 PMID: 35313950

Won, S. M., Song, E., Reeder, J. T., & Rogers, J. A. (2020). Emerging Modalities and Implantable Technologies for Neuromodulation. *Cell, 181*(1), 115–135. DOI: 10.1016/j.cell.2020.02.054 PMID: 32220309

wong, carrissa. (2023, August). *A New AI Tool Predicts Gene Expression in a Single Cell.* The Scientist Magazine®. https://www.the-scientist.com/a-new-ai-tool-predicts-gene-expression-in-a-single-cell-71295

Xu, J., & Wen, Z. (2021). Brain Organoids: Studying Human Brain Development and Diseases in a Dish. *Stem Cells International, 5902824*. Advance online publication. DOI: 10.1155/2021/5902824 PMID: 34539790

Xuan, L., Ju, Z., Skonieczna, M., Zhou, P., & Huang, R. (2023). Nanoparticles-induced potential toxicity on human health: Applications, toxicity mechanisms, and evaluation models. *MedComm, 4*(4), e327. DOI: 10.1002/mco2.327 PMID: 37457660

Yan, S., Tu, Z., Liu, Z., Fan, N., Yang, H., Yang, S., Yang, W., Zhao, Y., Ouyang, Z., Lai, C., Yang, H., Li, L., Liu, Q., Shi, H., Xu, G., Zhao, H., Wei, H., Pei, Z., Li, S., & Li, X.-J. (2018). A huntingtin knock-in pig model recapitulates features of selective neurodegeneration in Huntington's disease. *Cell, 173*(4), 989–1002.e13. DOI: 10.1016/j.cell.2018.03.005 PMID: 29606351

Yetisgin, A. A., Cetinel, S., Zuvin, M., Kosar, A., & Kutlu, O. (2020). Therapeutic Nanoparticles and Their Targeted Delivery Applications. *Molecules (Basel, Switzerland), 25*(9), 2193. DOI: 10.3390/molecules25092193 PMID: 32397080

Yin, P., Li, S., Li, X.-J., & Yang, W. (2022). New pathogenic insights from large animal models of neurodegenerative diseases. *Protein & Cell, 13*(10), 707–720. DOI: 10.1007/s13238-022-00912-8 PMID: 35334073

Yin, X., Mead, B. E., Safaee, H., Langer, R., Karp, J. M., & Levy, O. (2016). Stem Cell Organoid Engineering. *Cell Stem Cell, 18*(1), 25–38. DOI: 10.1016/j.stem.2015.12.005 PMID: 26748754

Zhao, Y., Xiong, S., Liu, P., Liu, W., Wang, Q., Liu, Y., Tan, H., Chen, X., Shi, X., Wang, Q., & Chen, T. (2020). Polymeric Nanoparticles-Based Brain Delivery with Improved Therapeutic Efficacy of Ginkgolide B in Parkinson's Disease. *International Journal of Nanomedicine*, *15*, 10453–10467. DOI: 10.2147/IJN.S272831 PMID: 33380795

Zhao, Z., Chen, X., Dowbaj, A. M., Sljukic, A., Bratlie, K., Lin, L., Fong, E. L. S., Balachander, G. M., Chen, Z., Soragni, A., Huch, M., Zeng, Y. A., Wang, Q., & Yu, H. (2022). Organoids. *Nature Reviews. Methods Primers*, *2*(1), 1–21. DOI: 10.1038/s43586-022-00174-y PMID: 37325195

Zhou, J.-Q., Zeng, L.-H., Li, C.-T., He, D.-H., Zhao, H.-D., Xu, Y.-N., Jin, Z.-T., & Gao, C. (2023). Brain organoids are new tool for drug screening of neurological diseases. *Neural Regeneration Research*, *18*(9), 1884–1889. DOI: 10.4103/1673-5374.367983 PMID: 36926704

Zoey, F. L., Palanivel, M., Padmanabhan, P., & Gulyás, B. (2021). Parkinson's Disease: A Nanotheranostic Approach Targeting Alpha-Synuclein Aggregation. *Frontiers in Cell and Developmental Biology*, *9*, 707441. DOI: 10.3389/fcell.2021.707441 PMID: 34490255

Chapter 16
Emerging Trends in the Treatment of Neurodegenerative Diseases

Punet Kumar
https://orcid.org/0000-0001-6047-6615
Shri Gopichand College of Pharmacy, India

Deepak Kumar
https://orcid.org/0009-0002-7598-5484
Shri Gopichand College of Pharmacy, India

Sangam Singh
https://orcid.org/0000-0002-1341-4451
Oxford College of Pharmacy, India

Mazid Ali Khan
https://orcid.org/0009-0002-7300-5219
Oxford College of Pharmacy, India

Jagjeet
https://orcid.org/0009-0004-9934-4811
Oxford College of Pharmacy, India

ABSTRACT

Alzheimer's, Parkinson's, and Huntington's are devastating neurodegenerative diseases posing major challenges to global health and economies. Current therapies often fail to halt progression, lack targeting, and provide limited efficacy. This chapter highlights innovations poised to transform ND care, including personalized medicine tailored to genetic, molecular, and environmental factors, enabling precise, low-dose treatments. Nanotechnology advances drug delivery, with nanoparticles loaded with secondary metabolites showing promising neuroprotective effects in preclinical and clinical studies. Natural products, particularly secondary metabolites, offer neuroprotection via antioxidation, anti-inflammation, and apoptosis modulation. Together, these

DOI: 10.4018/979-8-3373-1365-8.ch016

strategies present a synergistic roadmap for ND therapy. While ethical, regulatory, and scalability challenges remain, this chapter envisions a future of transformative treatments, offering guidance for clinicians and researchers in revolutionizing ND management.

INTRODUCTION

Neurodegenerative diseases comprise a heterogeneous assemblage of chronic illnesses. Through time, these diseases progressively break down the neurons leading to cell death. Central and peripheral nervous systems affect the distinct thus making these diseases the leading causes of disability and mortality all over the world. Alzheimer's, Parkinson's, Huntington's, ALS, and Multiple sclerosis are the most common neurodegenerative diseases known to man. One of the main characteristics of those diseases is that abnormal proteins in the brain are deposited causing disturbances in proper cellular functions and leading to the death of nerve cells (Gao, & Hong, 2008). Although they appear to be having wide variations in their symptoms, all of them indeed have some similarities. These include proteins misfolded that accumulate and interfere with the functions of nerve cells. Ultimately, this causes cellular demise. Moreover, mutations in the huntingtin protein lie at the core of Huntington's disease pathology. Furthermore, scientists link frontotemporal dementia and amyotrophic lateral sclerosis to clumped TDP-43 or FUS wearing a disguise (Mackenzie, et al., 2010). Cellular malfunctions cause proteinopathies to weaken essential processes, genomic housekeeping, and quality control pathways, including protein degradation, mitochondrial function, and synaptic activity initiation. Certain illnesses predominantly stem from solitary gene mutations, such as Huntington's disease deriving from autosomal dominant alterations to HTT (Ajitkumar et al., 2023). Alternatively, sporadic Parkinson's and Alzheimer's commonly emerge from an interplay of genetic susceptibilities and outside exposures over extended lifespans. Prolonged aging deteriorates the maintenance and repair of neurons, exacerbating oxidative stress and cellular fatigue. As protective mechanisms fade, microscopic trauma accumulates, priming the conditions for neurodegeneration (Mackenzie, et al., 2010). Symptoms of neurodegenerative diseases depend on which areas of the brain or spinal cord are affected. Alzheimer's affects cognitive functions generally, causing memory loss and impaired mental processes. Parkinson's disease implies motor difficulties, such as tremors, rigidity, and slowness of movements. Motor dysfunction and mood disturbance constitute Huntington's disease and lead to dementia overlap (Kumar et al., 2024). Motor neuron diseases are neurodegenerative diseases that initially manifest with muscle weakness caused by motor neuron degeneration, i.e., ALS the slowly progressive form. Multiple sclerosis is characterized by discrete attacks of neurological dysfunction caused by inflammation and damage to the myelin sheaths (Rojas, et al., 2020).

Neurodegenerative disease symptoms differ depending on whether regions of the brain or spinal cord are targeted. Alzheimer's disease is characterized by the cognitive system, which means forgetfulness and lack of mental acumen are mainly affected by this illness. Parkinson's disease is more a disorder of movement (tremors, rigidity, and slowness) whereas Huntington's disease affects the brain leading to motor dysfunction, mood disturbances, and eventually dementia (Lamptey et al., 2022). The diagnosis of neurodegenerative disorders is difficult and usually entails a tangle between clinical examination, advanced imaging, and biomarker analysis MRI and PET scans are the imaging methods that serve as a window to structural or functional brain abnormalities. Novel diagnostic modalities continuously bring new dimensions for early detection, increasing diagnostic accuracy (Aljthalin, et al. 2024).

The treatment of neurodegenerative diseases is centered on symptom management and quality of life improvement. Meds that are part and parcel of Alzheimer's include cholinesterase inhibitors and, more recently, NMDA receptor antagonists. The primary treatments for Parkinson's are dopamine replacement (levodopa). Many treatments are symptom-orientated, although newer strategies involving gene therapy and manipulation, neuroprotectants, and immunotherapies are beginning to show promise in modifying disease course (Mathur, et al., 2023). Novel methods are also now being investigated, such as CRISPR-Cas9 gene editing, stem cell therapy, and precision medicine. In addition, behavior changes (e.g., exercise, diet, cognitive and lifestyle interventions to prevent the onset/progression of these) are increasingly seen as valuable tools for delaying the onset and evolution of this pathology (Caobi, et al., 2020). Neurodegenerative diseases are a major global health problem debilitating individuals and societies. Although our understanding of their underlying mechanisms has grown significantly over the years; there still exists an urgent requirement for therapies that can impact these diseases. Successfully reconciling the challenge will necessitate coordinated research with cross-talk between genetics, biochemistry, neuroimaging, and clinical cohort studies to unravel these intricacies better and lead to future therapeutics (Ayeni, et al., 2022).

Epidemiology and Socioeconomic Impact

Neurodegenerative diseases are complex and confound therapeutic development. Mechanisms of protein misfolding, oxidative stress, mitochondrial dysfunction, and neuroinflammation are but partly defined. Traditional ways of treating amyloid-beta plaques and tau tangles seen with the diseases of Alzheimer's have been unfulfilling because of the multi-factorial cause underlying most forms of dementia. Then lastly, it's an extremely early diagnosis which is important but difficult as we don't have good biomarkers. Extensive loss of neurons is observed at the time when symptoms develop, thus treatments are less effective (Sweeney, et al., 2017). Drug delivery to the CNS due to the limited permeability of the blood-brain barrier (BBB), is another limitation concerning treatment. Most current treatments come with substantial side effects. Gene therapy and monoclonal antibodies are advanced strategies, but to expensive, off-limits in slow-access countries, and probably unsafe in the long term. Global health problems such as these require stakeholders to work together for solutions that lead to better patient outcomes (Teleanu et al., 2022). The Global Socioeconomic Burden of Neurodegenerative Diseases: Prevalence and Economic Costs by Region Prevalence in millions, costs in billions USD As shown in Figure 1, Asia has the highest prevalence (21.3 million cases) and economic burden ($600B yearly) followed by Europe (14.7 million cases, $400 billion) and North America (10.5 million cases, $320 billion). South America (7.5 million), Africa (2.2 million), and Oceania (1.6 million) respectively, with even lower prevalence rates of $220 billion, $92 billion, and $28 billion as well (Huang, et al., 2023).

Figure 1. Global socioeconomic burden of neurodegenerative diseases: Prevalence and economic costs by region

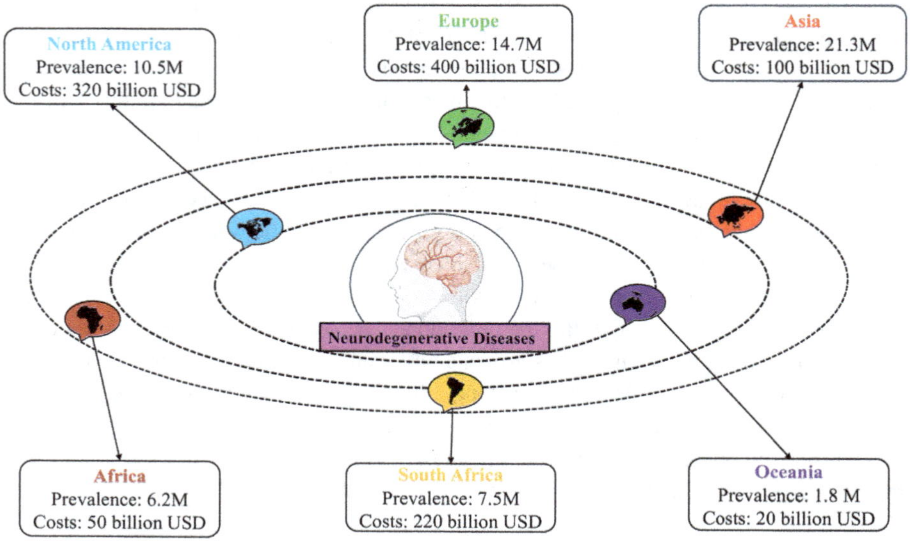

Current Therapeutic Limitations

Understanding that brain diseases are evolving and hard to treat is still a big problem. All existing treatments are symptomatic and provide symptom relief without modulating the root cause. This eventually accelerates disease progression, and patients are decompensating. For instance, despite treatments for Alzheimer's having symptomatic value, the cholinesterase inhibitors (donepezil, rivastigmine) and NMDA receptor antagonists (memantine) do not confer protection from the disease (Passeri, et al., 2022). Likewise, Parkinson's medications, which address motor symptoms with levodopa and dopamine agonists, do not stop neuron loss. Neurodegenerative diseases are ultimately held back by their complexity, and that makes therapeutic development trickier. The mechanisms of protein misfolding, oxidative stress, and mitochondrial stress in brain aging have been somewhat elucidated. Amyloid-beta plaques and tau tangles key targets in Alzheimer's are treated poorly, in large part due to the multifaceted nature of the disease (Murakami, et al., 2023). Moreover, early detection is important but hard to obtain due to the lack of effective biomarkers. By the time symptoms manifest, extensive neuronal loss has already occurred, limiting treatment (Klyucherev, et al., 2022). Current treatments have many side effects. Not surprisingly, however, levodopa therapy causes dyskinesias in the majority of PD patients, and immune-suppressive therapies for MS predispose patients to infections (excesses of cytokines). Gene therapy and monoclonal antibodies are more elaborate approaches that remain expensive and are at the limit of accessibility in low-income countries, with unclear long-term benefits. It is no wonder that identifying new drug candidates and transitioning drugs to patients takes a long time, and most clinical trials end in failure. Next-generation strategies must be based on detailed insights, early diagnostics, bypassing BBB barriers, and access to therapies. Stakeholder collaborations are needed to overcome these challenges and improve patient outcomes worldwide (Thanvi, et al., 2007).

Defeating complex neurodegenerative diseases will not be easy, leading to exhaustive therapeutic solutions. Early detection is essential, but extremely hard because of the lack of reliable biomarkers. Symptoms already result in extensive neuronal injury, so treatment effectiveness is low by the time you see anything. Drug delivery across the blood-brain barrier (BBB) further impedes central nervous system treatment. The sluggish drug development and numerous clinical trial failures highlight the requirement for fresh approaches. The next steps forward will require deeper biological insights into disease etiology, reliable early diagnostics, and time-tested BBB cracks as well as stakeholder solutions to get access therapies. Collaboration among researchers, healthcare providers, health policymakers, and industry stakeholders will be necessary to solve this challenge and drive better patient results worldwide (Kim, et al., 2022).

Need for Innovative Approaches

As the occurrence of neurodegenerative diseases increases in both poignancy and amount, we need immediate progress on research, that is urgent to therapy. Limited current treatments treat symptoms rather than prevent or reverse the conditions underlying diseases like Alzheimer's, Parkinson's' and amyotrophic lateral sclerosis (ALS), Alzheimer´s, etc., making it imperative to develop novel therapeutic strategies (Wareham, et al., 2022). Next-generation therapies must attack the root cause of neurodegeneration such as protein aggregation, oxidative stress, and mitochondrial dysfunction that are fundamental to the pathogenesis of many diseases. Gene editing (e.g., CRISPR/Cas9) and stem cell therapies are innovations that allow one to correct genetic mutations by using as well as restoring dead neurons. Biomarkers and next-generation imaging/genetic profiling may change what we use to predict disease progression, to tailor treatment strategies. Immunotherapy is emerging as a promising field, with monoclonal antibodies successfully targeting amyloid-beta plaques in Alzheimer's disease. Additionally, immunomodulatory therapies aim to mitigate neuroinflammation, a key factor in neurodegenerative diseases (Wang & Wang, 2023). The therapeutic landscape broadens thanks to regenerative medicine and nanomedicine. Nanotechnology-based neurotrophic factors (e.g., brain-derived neurotrophic factor, BDNF) support neuronal repair and growth, nanomedicine improving drug delivery across the BBB. In addition, they may also provide earlier diagnosis thus improving intervention outcomes. It seems that the next level of therapeutics for neurodegenerative disease is ripe for an exciting change. However, their success eventually requires collaborative efforts between researchers and clinicians to translate scientific advances into accessible medicines for patients globally (Wu, et al., 2022).

Challenges in Conventional Treatments

Current conventional treatments for neurodegenerative diseases have only helped to treat symptoms and are constrained by formidable impediments. Most of the time these treatments are symptom-mitigating rather than disease-specific (for example targeting like in Alzheimer's & Parkinson's) thus underlying neurodegenerative processes are left to auto-fulfill. So, disease progression happens that subsequently causes the deterioration of symptoms (Palanisamy, et al., 2023). As an illustration, although cholinesterase inhibitors in Alzheimer's disease may only be of temporary cognition benefit; they do not protect against or delay disease progression. Those same drugs, the Parkinsonian ones e.g. levodopa provide momentary relief from motor symptoms that invariably cause long-term side effects such as dyskinesia. On the other hand, these are mostly not well taken due to adverse events like nausea, dizziness, and also cardiovascular issues which reduce tolerability leading to poor adherence with long-term benefits (Wilkinson, et al., 2004).

There is the blood-brain barrier (BBB) to overcome as well as the very limited ability current drugs have in treating most therapeutic agents that lead to the brain, thus lowering the efficacy of many conventional drugs. On top of the difficulties already, diagnosis of neurodegenerative diseases comes quite late as current tools are unable to reveal early-stage pathology. By the time symptoms appear, considerable neuronal damage has begun and medications already on the market are less efficacious. It is crucial to move away from traditional disease-focused therapeutic strategies towards innovative, disease-modifying, and personalized treatments for more treatment success in neurodegenerative diseases (Hersh, et al., 2016).

Role of Precision Medicine in Neurodegenerative Diseases

Precision medicine offers the chance to transform the management of neurodegenerative diseases by taking an individualized approach toward medication prescribed to them based on their genetic, molecular, and environmental phenotypes. In my opinion, this individualized intervention is very useful for diseases such as Alzheimer's, Parkinson's, Huntington's, and ALS- with inherently heterogeneous course and therapeutic response. Using the latest genetic and molecular profiling, we can find disease-specific mutations and biomarkers (Ex; APP, PSEN1/2 in Alzheimer´s or HTT gene in Huntington´s). This information assists in early detection and individualized interventions. Precision medicine depends on early diagnosis since most neurodegenerative diseases are still undifferentiated in their incipient stage. The result is enhanced diagnostic precision and the capacity to intervene at key windows in disease development (Nuzziello, et al., 2019).

Precision medicine in turn leads the way toward more personalized treatment for patients taking into account their genetics and molecular make-up. Many people with Parkinson's have LRRK2 mutations that could be treatable, others will not derive benefit from any approach. Precision medicine also helps with clinical trials by targeting subsets of patients who are homogeneous genetically or molecularly. For example, Alzheimer's disease amyloid-targeting trials could select biomarker-positive participants, making trials more robust. Precision medicine also accounts for the root cause: of neuroinflammation, protein misfolding, and mitochondrial dysfunction instead of merely targeting symptoms (Arafah, et al., 2023). Precision medicine offers a way to early detection, personalized treatments, and enhanced outcomes simply by harnessing genetic, molecular & environmental information that appears very hopeful regarding the field of neurodegenerative diseases.

PERSONALIZED MEDICINE AND NANOTECHNOLOGY

Understanding Personalized Medicine

Precision medicine, also called personalized medicine changes healthcare by individualizing prevention and treatment to an individual based on a genomic, molecular, environmental, and lifestyle profile. Apart from conventional ways, it ensures that interventions are matched with the personalized characteristics of an individual and results in improved outcomes (Goetz & Schork, 2018). Genomics and molecular diagnostics provide the tools necessary for personalized medicine to be realized with many genetic predispositions shown both in disease risk as well as response to treatment. Nanotechnology boosts personalized medicine with the help of nano diagnostics and targeted drug delivery. Integrating nanotechnology nanoparticles tackle diseased cells which minimizes damage to healthy tissue, and nano-sensors enable early diagnosis of biomarkers like amyloid plaques seen in Alzheimer's disease. In this way, personalized medicine and nanotechnology converge to provide patient-focused care to improve treatment outcomes while minimizing risks through the triangulation of genetic, molecular, and environmental data (Sim & Wong, 2021).

Nanotechnology in Neurodegenerative Diseases

Nanotechnology has been identified as an enabling tool for the management of neurodegenerative disorders (NDs): Alzheimer´s disease, Parkinson´s disease, Huntington´s choreoathetosis, and amyotrophic lateral sclerosis which are some typical examples. Typical of these diseases is the chronic degeneration of neuro cells and current therapeutic approaches have been difficult because of the blood-brain barrier (BBB) and disease complexity. Nanotechnology has the potential to solve these restrictions and would pave the pathway for better drug delivery, diagnostics, and in the future even disease-modifying treatments (Bhattacharya, et al., 2022). Table 1 provides a comparative analysis of various nanoparticle types used in the treatment of neurodegenerative diseases, highlighting their unique properties and applications.

Nanoparticles for Targeted Drug Delivery

Nanotechnology holds great potential for the treatment of neurodegenerative diseases by optimization of targeted drug delivery within the brain. The blood-brain barrier (BBB) a blockade of substances from entering the brain through the blood supply is a major obstacle for existing therapies in neurological diseases. Nanoparticles could be engineered such that these would more easily cross the BBB due to their size and application with a specific targeting agent for the systemically delivered drugs or therapeutic molecules could be targeted to just the site of injury in the brain where they are needed, allowing for higher efficacy and reduced toxicity (Hersh, et al., 2022). In addition, the nanoparticle functionalization with appropriate ligands (i.e. antibodies or peptides epitope against BBB receptors) can increase this targeting properties and also improve drug delivery to neuronal cells. Moreover, in the future, nanoparticles may also be used to administer biologics like siRNA or gene editing tools (CRISPR-Cas9) which could potentially repair the genetic underpinnings of neurodegenerative diseases. Nanoparticles help serve the delivery-overcoming bottleneck for these biologics when delivered straight to brain cells (Duan, et al., 2021). Figure 2 illustrates the mechanism of Nanoparticle-Driven Targeted Drug Delivery, demonstrating its interaction with the blood-brain barrier for effective therapeutic delivery.

Table 1. Comparison of nanoparticle types for treating neurodegenerative diseases

Nanoparticle Type	Characteristics	Applications in Neurodegenerative Diseases	Advantages	Limitations
Liposomes	Spherical vesicles with a phospholipid bilayer. It can encapsulate both hydrophilic and hydrophobic drugs.	Alzheimer's (Rivastigmine-loaded liposomes). Parkinson's (L-DOPA encapsulation).	Biocompatible, capable of crossing BBB, sustained release, low toxicity.	Stability issues, risk of leakage, high production cost. (Allen, & Cullis. 2013).
Dendrimers	Branched, tree-like polymers with functional groups on the surface, enabling drug conjugation or encapsulation.	Gene delivery for Huntington's disease. Alzheimer's therapy targeting tau proteins.	High drug-loading capacity and surface functionalization allow precision targeting.	Potential toxicity due to reactive surface groups, and complex and costly synthesis. (Patri, et. al., 2002).
Polymeric Nanoparticles	Solid colloidal particles are made of biodegradable polymers like PLGA or chitosan.	Delivery of curcumin for Alzheimer's. Dopamine delivery for Parkinson's disease.	Biodegradable, versatile, controlled release, can penetrate BBB with surface modifications.	Risk of solvent toxicity, and stability issues under physiological conditions (Hans, & Lowman, 2002).
Gold Nanoparticles (AuNPs)	Spherical nanoparticles with tunable optical and chemical properties.	Imaging and drug delivery for Parkinson's and Alzheimer's. Theranostic applications.	BBB penetration, minimal immune response, excellent in diagnostics and drug delivery.	Potential toxicity at higher concentrations, and high production costs (Huang, et al, 2007).
Silica Nanoparticles	Porous structures with a high surface area and tunable pore sizes.	Drug delivery for Alzheimer's using donepezil and galantamine.	High drug-loading capacity, tailored for controlled release, stable.	Potential long-term accumulation limited BBB penetration without functionalization (Slowing, 2007).
Magnetic Nanoparticles	Iron oxide-based particles functionalized for targeted delivery.	Parkinson's (levodopa delivery), Alzheimer's imaging.	Targeted delivery using external magnetic fields, dual-use in imaging and therapy.	It requires external magnetic devices, has a risk of aggregation, and has limited safety data (Pankhurst. et al, 2003).

Figure 2. Illustration of nanoparticle-driven targeted drug delivery and interaction with the blood-brain barrier

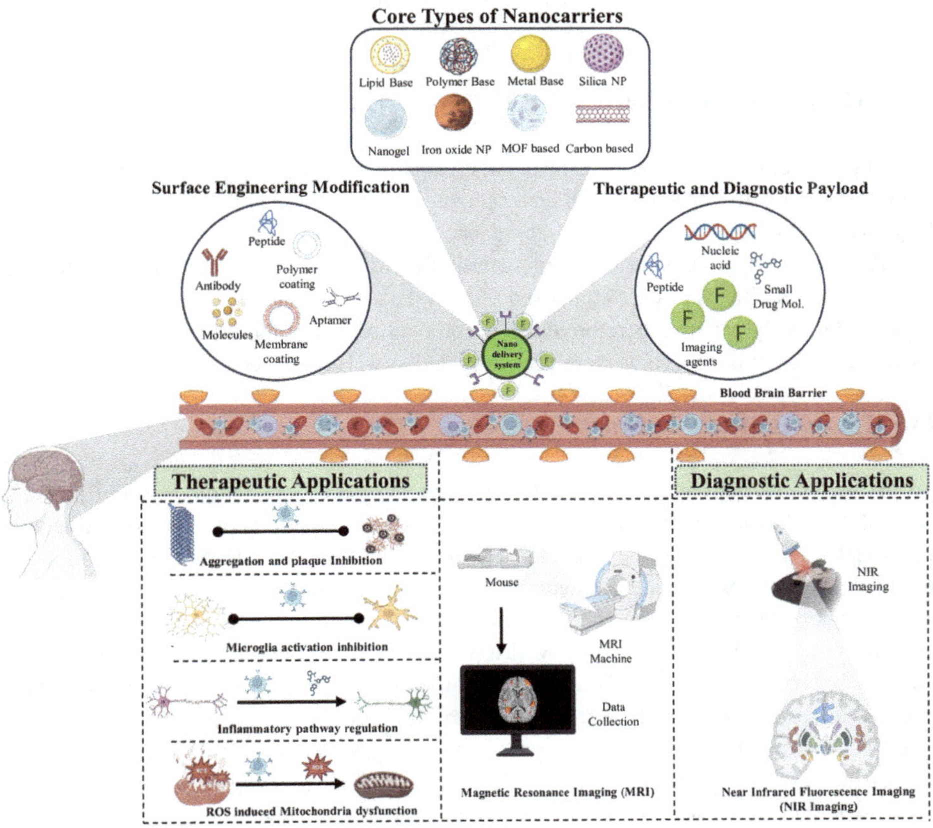

Application of Secondary Metabolites in Nanoparticle Formulations

Plants, fungi, and bacteria produce natural products called secondary metabolites with diverse bioactive properties that could help treat neurodegenerative diseases. The most common types comprise alkaloids, flavonoids, terpenoids, and polyphenols and are often used for their antioxidant, anti-inflammatory, neuroprotective, and transcription factor activities. Secondary metabolites are used in the form of nanoparticles, which have recently attracted research interest in modifying their therapeutic effects and reducing toxicity (Roy, et al., 2022). Encapsulation of these compounds in nanoparticles may protect them from degradation and enhance their solubility. These nanoparticles are capable of targeting specific brain regions, followed by a therapeutic efficacy in preclinical models of neurodegenerative diseases controlled by the release of the nanoparticles (Saraiva, et al., 2016). Similarly, various natural substances including resveratrol, quercetin and epigallocatechin gallate (EGCG) have been successfully nano-encapsulated for the treatment of neurodegenerative disorders. These compounds which are known for their antioxidant and anti-inflammatory action can help to reduce oxidative stress- a common feature of many neurodegenerative diseases through the protection of neurons (Singh, et al., 2016).

ARTIFICIAL INTELLIGENCE IN NANOPARTICLE DELIVERY SYSTEMS

Artificial Intelligence (AI) is focusing on the nanoparticle-based drug delivery systems (DDS) research field for better efficiency and better-targeted administration of these. The marriage of AI into nanotechnology provides novel approaches to enhancing the creation, structuring, and use of drug delivery systems for managing debilitating illnesses such as cancer, and neuro and infective diseases. Processors are composed of high-datasets which AI better than humans can analyse and identify patterns therefore it has become a strong tool for optimizing NP design as well performance of nanoparticles targeting drug delivery (Heydari, et al., 2024).

AI for Drug Delivery Optimization

AI is revolutionizing nanoparticle drug delivery by optimizing interactions between drugs and nanoparticles. Machine learning algorithms analyze extensive datasets of molecular structures, drug properties, and nanoparticle characteristics to predict the most effective combinations for targeted delivery. These models evaluate drug attributes such as size, solubility, charge, and hydrophobicity, alongside nanoparticle features like size, surface charge, and functionalization (Serrano, et al., 2024). By predicting how drugs behave within nanoparticles and interact with biological systems, AI ensures precise targeting with minimal side effects and maximal efficacy. It also facilitates the fabrication of smart nanoparticles that would be responsive to biological triggers such as pH, temperature, or enzyme and for targeted drug delivery. This AI-powered approach increases the specificity of drug delivery, decreases the toxicity, and allows us to design individualized delivery systems for patients. The incorporation of machine learning within drug delivery takes us one big step closer to safer and more effective, personalized treatment approaches (Cheng, et al., 2023). Figure 3 presents a flowchart detailing the AI-driven optimization workflow for nanoparticle drug delivery, encompassing key steps such as data inputs, predictive modeling, and optimization processes.

Figure 3. Flowchart depicting the AI-driven optimization workflow for nanoparticle drug delivery, including data inputs, predictive modelling, and optimization steps

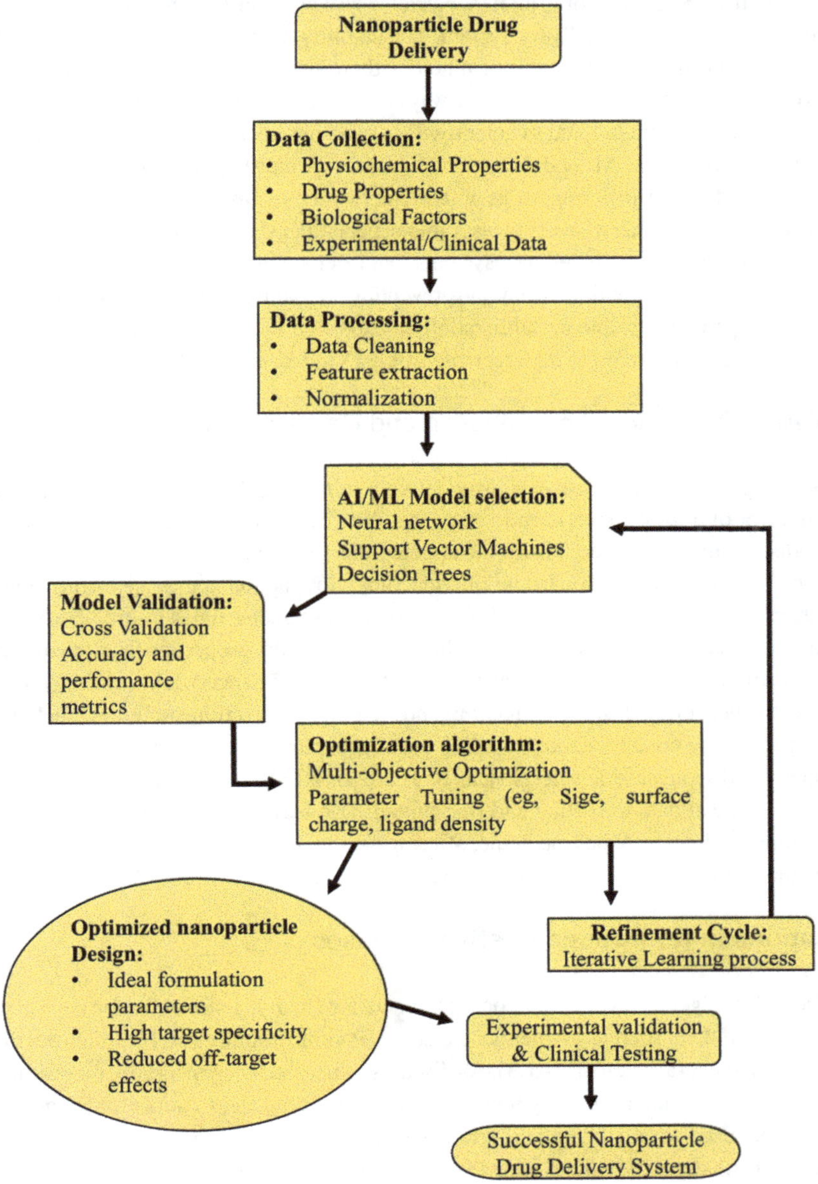

Machine Learning for Predicting Drug-Nanoparticle Interactions

Predicting the interactions of drug nanoparticles will be an important task in the process of developing an effective drug delivery system. The stability, release profile, and biodistribution of therapeutic molecules are largely determined by their interactions with nanoparticle carriers. Before AI and machine learning, this was always an age-old method of trial and error but now we can do this systematically and efficiently. By using historical data, chemical interactions, and biological parameters, AI models (especially deep learning) approaches along with neural networks can predict how differing drugs will interact with these nanoparticles (Singh & Lillard, 2009). AI can predict, otherwise un-measurable aspects like stability of the drug-nanoparticle complex too. Developing drug delivery systems that offer stable ex-vivo and in-vivo releases is critical to preserving therapeutic efficacy and preventing premature drug release. Machine learning algorithms can predict possible destabilization hazards, and therefore improve the formulation so that the drug gets delivered to its target site safely (Vidhya, et al., 2023).

Role of AI in Optimizing Biodistribution and Release Kinetics

AI is changing nanoparticle-based drug delivery to miniaturize biodistribution and release kinetics in favor of improved therapeutic responses. The distribution of nanoparticles within the body (biodistribution) is a prerequisite for targeted delivery and minimizes off-target activity. AI relies on data about the nanoparticle-pharmacology interaction aggregated from biological, physicochemical, and pharmacokinetic information about nanoparticles to predict tissue accumulation (Das & J, 2023). AI skilfully uses these to decide what properties (size, shape, surface charge, and circulation lifetime) nanoparticles must have for enhanced delivery. Machine learning models further refine designs by recommending functionalization strategies, such as using ligands or peptides to enable selective binding to cell receptors. This enhances tissue specificity, and efficacy, and reduces side effects (Das & J, 2023). AI also optimizes release kinetics by analyzing preclinical data to predict and fine-tune drug release profiles. This synergy between AI and nanotechnology reshapes medicine, offering personalized, efficient, and safer therapeutic strategies that significantly improve patient outcomes (Ghayoor & Kohan, 2024).

Applications in Neurodegenerative Diseases

Artificial Intelligence (AI) shows good prospects for changing strategies for neurodegenerative diseases (NDDs), such as Alzheimer's disease, Parkinson's disease, Huntington's disease, and Amyotrophic lateral sclerosis (ALS). As diseases of neuronal degeneration, these conditions are extremely hard to diagnose early regarding drug discovery and treatment management. AI has been widely used within therapeutic strategies, including improved versions of personalized medicine and drug development, as well as optimization of nano-carriers (Kumar, et al., 2023).

Case Examples of AI-Driven Therapeutic Systems

Artificial intelligence (AI) is becoming an invaluable innovation for NDDs- from early diagnosis to adapting treatment plans based on individual patient traits. AI uses machine-learning algorithms to assess neuroimaging data (e.g., MRI, PET) in the case of Alzheimer's disease and catch preclinical biomarkers as well as subtle structural alterations that suggest early onset of the disease. With it, we can catch the disease earlier and better forecast disease progression. AI Also aids in first the discovery of drugs by screening huge chemical datasets to identify potential compounds against tau protein aggregation a strong pathological sign. Wearable devices and sensors connected to waking AI in Parkinson's detect motor impairments like tremors, and bradykinesia - providing data to support tailored symptom management in real time (Kumar, et al., 2024). AI helps discover therapeutic targets through proteomics and genomics data interpretation to find cures for dopaminergic neuron degeneration by AI as well. In Huntington's disease, AI enhances genetics through GWAS-based analysis of the whole genome to identify genetic variants and possible pathways for therapy. It also automates clinical trials by managing patient recruitment, trial selection, and real-time auditing of the trials to shorten the drug development time (Parekh et al., 2023). We apply AI to data from patient-specific genetic, neuroimaging, and clinical datasets in amyotrophic lateral sclerosis (ALS) to identify potential therapeutic targets and predict disease evolution. It improves drug dosing and early treatment response to therapies such as gene therapies and neuroprotective agents in the context of individualized care. For these diseases, the capability of AI to perform systemic analysis and predict therapeutic outcomes challenges the practice of treatment (Pun, et al., 2022).

Addressing Barriers to Scalability and Implementation

AI-powered therapeutic systems present a large opportunity in the modulation of neurodegenerative diseases, but there are huge hurdles for scaling and mitigating operational issues. The biggest setback is the absence of high-quality, comparable data. Everybody works with different neuroimaging, genomic, and clinical datasets from each institution, which does not generalize the AI models very well. Creating shared platforms aimed at collaborative data sharing and standards for datasets need to be established to navigate this challenge. Gaps in the expertise of neuroscientists, computer scientists, and clinicians are a further hurdle to overcome as interdisciplinary collaborative efforts are required (Borchert, et al., 2023. Cross-educational and relatedly practice-partnership is required in the opened AI potential for healthcare. Regulatory and ethical issues, data privacy, transparency, and bias are some of the other problems that come with AI implementation. By proactively working to dissolve these barriers, AI-driven systems can be made to accelerate early detection, precision medicines, and effective therapies that will revolutionize the management of neurodegenerative diseases with improved patient outcomes (Bajwa, et al., 2021).

NATURAL COMPOUNDS AS NEUROPROTECTIVE AGENTS

Therapeutic Potential of Natural Compounds

More and more diverse pharmacological properties in natural compounds have been highlighted as promising neuroprotective agents with few side effects against the background of increasing attention. Coming from plants, fungi, and sea invertebrates, these compounds are justified for having different beneficial effects (antioxidant properties, anti-inflammatory, and anti-apoptotic actions), making them interesting candidates for the treatment of neurodegenerative diseases.

As synthetic drugs have additional limitations in treating diseases (Alzheimer's, Parkinson's, Huntington's), natural compounds provide viable alternatives or adjuvants in neuroprotective (Lim, et al., 224). For example, many traditional medicinal plants like *Ginkgo biloba*, *Curcuma longa* (turmeric), *Withania somnifera* (Ashwagandha), and Panax ginseng have neuroprotective potential in both preclinical and clinical trials to protect against neuronal injury. As evidenced by the highly relevant active constituents (i.e., flavonoids and terpenoids) in *Ginkgo biloba,* these plant combinations are redox-scavengers and antioxidants that will prevent oxidative damage alleviating cognitive functions (Gregory, et al., 2021). Analogously, Curcuma longa has curcumin, a polyphenolic substance that was able to decrease amyloid plaques critical to Alzheimer's disease and boost neurogenesis. In addition to plant-based compounds, natural compounds derived from fungi and marine sources are also being explored for their neuroprotective properties (Genchi, et al., 2024). Marine-derived compounds like omega-3 fatty acids from fish oils are well known for their neuroprotective actions, supporting brain health and mitigating neuroinflammation (Ashraf, et al., 2020). Table 2 summarizes neuroprotective natural compounds, detailing their sources, mechanisms of action, experimental models, and targeted neurodegenerative diseases.

Mechanisms of Neuroprotection

Naturally occurring molecules have potent neuroprotective properties by going straight to the root cause of the pathological processes involved in neurodegenerative diseases like those of Alzheimer's, Parkinson's, and MS. One of their main actions is their antioxidative capabilities, which might help prevent some cell damage by counteracting oxidative stress provoked by reactive oxygen and nitrogen species. Curcumin, resveratrol from grapes, and the polyphenol epigallocatechin gallate in green tea boost antioxidant enzyme function: neurodegeneration is synonymous with neuronal damage, which is largely due to loss (and subsequent dysfunction) of one or more of these in the CNS (Babazadeh, et al., 2023). Also, a central mechanism of action is the anti-inflammatory effect. Curcumin, quercetin, and nobiletin suppress pro-inflammatory cytokines Like TNF-α, IL-1b, and IL-6 (one of the NF-κB pathways) a sample of which blocks chronic inflammation that would be loading on neurons' neurites. Curcumin and KEA-like natural compounds also suppress or disaggregate the aggregated proteins of amyloid-beta and α-synuclein resulting in the alteration of neuronal dysfunction. Although preclinical studies are encouraging warranting their use and potential implementation into treatment strategies; require rigorous testing further afield (Assi, et al., 2023). Figure 4 illustrates the mechanisms through which natural compounds provide neuroprotection in neurodegenerative diseases, highlighting their therapeutic potential.

Table 2. Neuroprotective natural compounds: Sources, mechanisms, models, and neurodegenerative disease targets

Natural Compound	Plant Source	Targeted Diseases	Study Model	Mechanism of Action
Curcumin	*Curcuma longa* (Turmeric)	Alzheimer's, Parkinson's	In vivo/In vitro	Antioxidant, anti-inflammatory, and anti-apoptotic effects (Abdul-Rahman, et al., 2024).
Resveratrol	Grapes, Red wine	Alzheimer's, Huntington's	In vivo/In vitro	Antioxidant, modulates protein aggregation, anti-apoptotic (Sahebkar, A., 2010).
Epigallocatechin Gallate	Green Tea	Parkinson's, Alzheimer's	In vivo/In vitro	Inhibits α-synuclein aggregation, antioxidant, anti-apoptotic (Gadhave, et al., 2024).
Ginsenosides	*Panax ginseng*	Alzheimer's	In vivo/In vitro	Promotes neurogenesis, enhances synaptic plasticity, anti-apoptotic (Zheng, et al., 2018).
Quercetin	Apples, Onions	Alzheimer's, Parkinson's	In vivo/In vitro	Reduces COX-2 and nitric oxide (NO) production, antioxidant (D'Andrea, G. 2015).
Nobiletin	Citrus Fruits	Alzheimer's	In vivo/In vitro	Promotes synaptic plasticity, anti-inflammatory, and antioxidant (Pang, et al., 2023).
Omega-3 Fatty Acids	Fish Oils	Alzheimer's, Multiple Sclerosis	In vivo	Neuroprotective, supports brain health, anti-inflammatory (Bazinet, and Layé, 2014).
Cordycepin	*Cordyceps sinensis*	Alzheimer's, Parkinson's	In vivo/In vitro	Modulates apoptotic pathways, antioxidant, and anti-inflammatory (Paterson, and Anderson, 2005).
Berberine	*Berberis* species	Alzheimer's	In vivo/In vitro	Regulates mitochondrial dysfunction, anti-inflammatory, and antioxidant (Habtemariam, 2020.).
Huperzine A	*Huperzia serrata*	Alzheimer's	In vivo	Inhibits acetylcholinesterase, and enhances cognitive performance (Xing, et al., 2014).
Withanolides	*Withania somnifera*	Parkinson's, Alzheimer's	In vivo/In vitro	Stimulates neurogenesis, anti-apoptotic, and antioxidant (Speers, et al., 2021).
Bacoside A	*Bacopa monnieri*	Alzheimer's	In vivo/In vitro	Improves synaptic plasticity, anti-inflammatory, and antioxidant (Goyal, et al., 2023).
Luteolin	Parsley, Celery	Alzheimer's, Parkinson's	In vivo/In vitro	Reduces TNF-α and IL-6 levels, antioxidant, and anti-inflammatory (Nabavi, et al., 2015).
Baicalein	*Scutellaria baicalensis*	Alzheimer's, Parkinson's	In vitro	Inhibits protein aggregation, antioxidant, and anti-apoptotic (Li, et al., 2017).
Ferulic Acid	Rice Bran, Oats	Alzheimer's, Huntington's	In vivo/In vitro	Enhances mitochondrial function, anti-inflammatory, and antioxidant (Ou, & Kwok, 2004).

Figure 4. Mechanisms of neuroprotection by natural compounds in neurodegenerative diseases

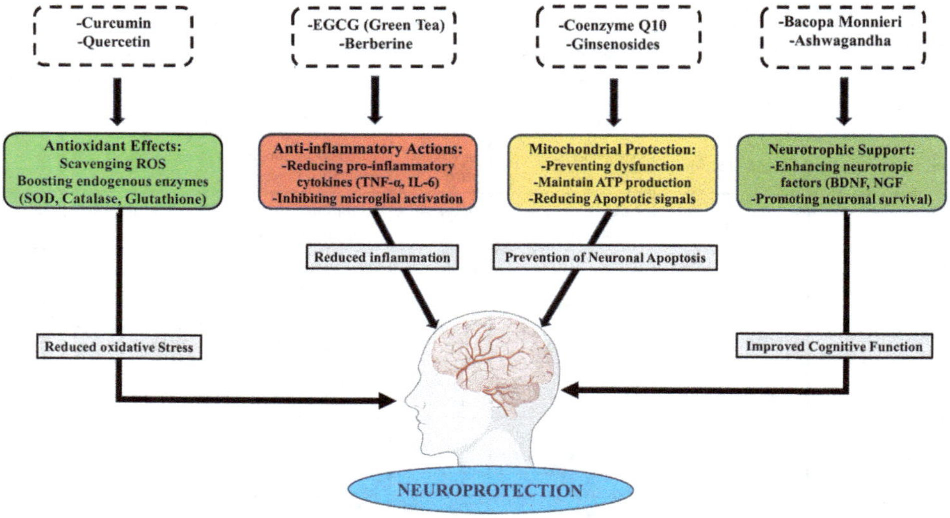

Secondary Metabolites in Focus

Secondary metabolites from plants offer significant neuroprotective potential, targeting diverse pathways involved in neurodegenerative diseases. Alkaloids like galantamine, derived from the snowdrop plant, act as acetylcholinesterase inhibitors, enhancing acetylcholine levels and improving cognition in Alzheimer's patients (Marucci, et al., 2021). Berberine, from *Berberi's* species, exhibits antioxidant and anti-inflammatory properties, making it promising for Alzheimer's and Parkinson's disease. Flavonoids, such as quercetin (from apples and onions) and epigallocatechin gallate from green tea, are potent antioxidants and anti-inflammatory agents (Sehgal, et al., 2012). EGCG reduces amyloid-beta accumulation, a hallmark of Alzheimer's, while quercetin modulates cytokines, protecting neurons from damage. Terpenoids, including ginsenosides from Panax ginseng and curcumin from turmeric, support cognitive function, reduce neuroinflammation, and promote neurogenesis. Curcumin also prevents oxidative damage and inhibits amyloid aggregation, aiding brain health (Hussain, et al., 2018). Phenolic compounds like resveratrol, found in red wine and grapes, enhance mitochondrial function, reduce oxidative stress, and mitigate neuroinflammation. Ferulic acid, from rice and wheat, also demonstrates strong antioxidative and neuroprotective effects. Saponins such as withaferin A from Withania somnifera (ashwagandha) promote neurogenesis and reduce amyloid-beta aggregation, offering potential Alzheimer's therapy (Venkatesan, et al., 2015). Polysaccharides from medicinal mushrooms like Ganoderma lucidum (reishi) exhibit immunomodulatory effects, supporting neuronal survival and cognitive improvement. These metabolites represent promising agents for managing neurodegenerative diseases (Venturini, et al., 2010).

Enhancements in Bioprospecting and Bioavailability:

Bioprospecting quest of natural organisms for novel bioactive compounds has been a primary driver for identifying novel targets in neurodegenerative diseases. Marine bioprospecting such as extraction of eckol from *Ecklonia Cava, lamellarins* from the marine sponge and so on are characterized by scavenging free radicals to ensure their antioxidative & anti-inflammation neuroprotective effects (Sekurova, et al., 2019). Compounds discovered from fungal bioprospecting include cordycepin (promotes neuronal survival and reduces inflammation) from *Cordyceps militaris* for its neuron protection and ergothioneine (antioxidant protection against oxidative stress) from *Ganoderma lucidum* (Reishi mushroom). Dialing up the difficulty, one of the biggest implications of the use of secondary metabolites as a therapeutic agent is profound bioavailability (Ashraf, et al., 2020). Other compounds like quercetin and ginsenosides are also included in drug delivery to cope with concerns about bioavailability. Future breakthroughs in bioprospecting, drug delivery, and pharmacokinetic optimization will address the current problematics, and these otherwise promising compounds may become practical therapeutics. The grape and berry-derived resveratrol is a potent antioxidant/anti-inflammatory that also supports cardiovascular health and cognitive function. Ginsenosides, in ginseng which is related to fatigue reduction immune enhancement, and neuroprotection, Epigallocatechin Gallate from Green tea (EGCG) inhibits cancer cell line proliferation Therefore, preventing carcinogenesis and cardioprotective effects. Silibin (extracted from the milk thistle) protects the liver and seems to be a liver (hepatitis C) specific intervention. *Bacopa Monnieri* - improves performance in mental tasks and reduces anxiety, and *Ginkgo Biloba* supports cognitive function, and circulation (particularly in Alzheimer's & Dementia). All of the natural compounds, working by many different mechanisms indicate their potential as personalized medicine and disease management, so innovative therapeutic formulation can come (Mikulska, et al., 2023).

SYNERGISTIC APPROACHES

Representations that marry nanotechnology, artificial intelligence (AI), and natural-based therapies hold the promise of new innovative treatments to be applied to neurodegenerative disease problems. Nanotechnology aids their effective delivery and bioavailability by facilitating natural compounds to cross the blood-brain barrier. Remediation of therapeutic outcomes is achieved through the use of AI in optimizing drug formulations and designing patient-tailored therapy based on specific data reported. This integrated approach has the potential to transform the management of neurodegenerative diseases and systematic personalized therapies (Kulkarni, et al., 2024).

Combining Nanotechnology, AI, and Natural Compounds

Nanotechnology and natural compounds, together with artificial intelligence (AI), are scientifically sound solutions that address the hurdles that present-day neurodegeneration therapies encounter. Together, these technologies allow for the enhancement of bioavailability, therapeutic targeting, and overall efficacy of treatment (Adir, et al., 2020). Nanoparticles may be fabricated to penetrate the blood-brain barrier, which remains a significant challenge for drug delivery. The natural drugs curcumin and resveratrol, as well as the targeted delivery of ginsenosides to the brain, which are known to have neuroprotective effects, can be incorporated into nanoparticles like liposomes, dendrimers, or polymeric nanoparticles. This makes the particles capable of targeting particular cells or tissues, thereby minimizing unwanted side effects and improving therapeutic benefits. (Pinheiro, et al., 2021).

AI, especially machine learning, can improve treatment strategies by forecasting natural compound interactions with nanoparticles and tuning formulations of drugs. AI can process large data sets and derive the most potent nanoparticle combinations, drug dose plans, and predictions of biodistribution patterns. It can also optimize drug release kinetics to ensure the drugs are released at their desired time and place for optimum efficacy. In neurodegenerative diseases, for instance, AI will facilitate the creation of genetically and molecularly tailored treatment protocols specific to different patients. (Serrano, et al., 2024). Shown in Figure 5, the integration of artificial intelligence (AI), nanotechnology and natural compounds to constitute a single therapeutic model and emphasizing their combinatory capacity within current strategies in clinical medicine.

Figure 5. Depiction of the integration of AI, nanotechnology, and natural compounds into a unified therapeutic approach

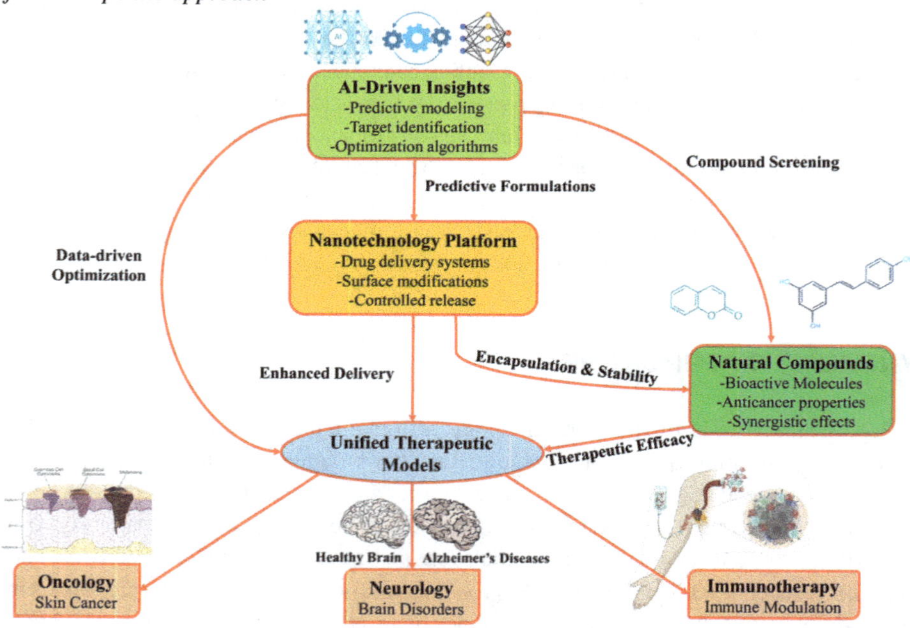

Naturally occurring compounds, such as curcumin, resveratrol, and ginseng, demonstrate neuroprotective actions through antioxidation, anti-inflammation, and neurotrophic factor regulation. However, their low bioavailability and BBB specificity negatively influence their therapeutic effects. In combination with nanotechnology and AI, these compounds represent an opportunity to finally realize their full therapeutic potential. To elaborate, curcumin nanoparticles might specifically bind to and clear amyloid-beta aggregates in Alzheimer's disease patients, while resveratrol nanoparticles could alleviate oxidative stress in models of Parkinson's disease (Cole, et al., 2007).

AI-Designed Optimization of Curcumin Nanoparticles for Alzheimer's Disease:

A recent study incubated curcumin into nanoparticles to allow better brain-targeting delivery for Alzheimer's disease treatment. The optimization of the nanoparticle formulation using AI (size, surface charge, and coating) so that it might enhance curcumin stability and enhancement in brain delivery was done by researchers. The peptide-conjugated nanoparticles were functionalized for targeting amyloid-beta plaques, the hallmark of Alzheimer's disease (Cheng, et al., 2013). The AI model predicted that this would lead to more drug-taking targets, and an optimized release profile. The curcumin nanoparticles decreased the accumulation of amyloid β and improved cognition in a preclinical amyloid mouse model. This tells of how nanotechnology and AI can be ideally employed to fortify the nano-delivery of natural products, thereby advancing clinical practice (Cheng, et al., 2013).

Ginsenoside Nanoparticles from Parkinson's Disease

Ginsenosides, the major bioactive constituents of Panax ginseng, exhibit neuroprotective effects against models of PD. Unfortunately, their clinical utility has been hindered by low bioavailability and rapid metabolism. When delivered as nanoparticles, they were able to cross the BBB in animal models and deliver ginsenosides directly to dopaminergic neurons where they demonstrated antioxidant, and anti-inflammatory effects. It leads to marked amelioration in motor performance and neuroprotection. AI was also employed in this study to forecast the most suitable size and surface changes on nanoparticles for better targeting as well as therapeutic results (Van, et al., 2014).

Combination Therapy of Resveratrol and Curcumin Using Nanoparticles and AI

Both resveratrol and curcumin are established natural compounds with neuroprotective effects. Together, they are additive in vitro and in vivo and demonstrate a synergistic activity, thereby increasing the therapeutic possibility for neurodegenerative diseases (Zaky, et al., 2017). In the study on Alzheimer's and Parkinson's diseases, both resveratrol and curcumin were delivered through nanoparticles, along with an artificial intelligence optimization intended to co-deliver them. In the dual-drug-loaded nanoparticles, the aim was to increase the solubilization, stability, and bioavailability of both drugs. The recommended dose for the neuroprotective effects, while minimizing toxicity, was determined by an AI-driven optimization algorithm of curcumin: resveratrol ratios. When compared with single-drug therapies, this combination therapy generated improved cognitive results and reduced oxidative stress along with neuroprotection in preclinical models (Zaky, et al., 2017).

Bacopa Monnieri and Nanoparticles for Cognitive Enhancement

A Traditional Herb for Cognitive Enhancement-*Bacopa monnieri* has so far become popular in treating cognitive decline of neurodegenerative diseases. However, poor bioavailability of Bacopa inhibited its clinical translation. To address this, researchers developed Bacopa-loaded nanoparticles and used AI to model the best formulation for targeted delivery to the brain. The AI-driven design incorporated surface modifications to increase the nanoparticle affinity for brain cells (Abdul, et al., 2019). Preclinical studies demonstrated that the Bacopa-loaded nanoparticles enhanced cognitive function and promoted neurogenesis in animal models of Alzheimer's disease. This integrated approach of combining natural compounds with nanotechnology and AI optimization resulted in significantly improved outcomes compared to traditional formulations of Bacopa (Chaudhari, et al., 2017).

AI-Driven Nanoparticle Design for Targeted Drug Delivery in Huntington's Disease

A genetic neurodegenerative disorder that causes motor and cognitive decline is known as Huntington's disease. As much as current pharmacotherapy can be good at hitting the region of the brain that is damaged by the disease. The integrated single nanoparticle with an AI-driven design for both AI-guided natural derivative combinations to enhance targeted delivery of therapeutic agents was conceptualized in a study (Ganesh, et al., 2023). AI, size; charge, and surface properties of the nanoparticles were employed to channel them specifically into the striatum the most hypodense region. Preclinical studies demonstrated improved motor performance and decreased neuronal loss in animal models. By combining the AI, nanotech, and natural compound approach, we increase the application of your medicines: precision delivery to target tissues and superior therapeutic efficacy. The emergence of these technologies bodes well for safer, personalized therapies in the treatment of neurodegenerative diseases (Adir, et al., 2020).

CHALLENGES AND FUTURE DIRECTIONS

Integration of Personalized medicine, artificial intelligence (AI), and nanotechnology has the panacea potential to revolutionize our approach to neurodegenerative disease treatment. Nonetheless, several roadblocks inhibit the uptake of these ground-cutting technologies. Issues in implementing effective, safe, and equitable treatments are: Ethical issues Regulatory obstacles technical constraints Personalized medicine helps direct therapies to an individual's genetic/environmental profile, for precision on conditions such as Alzheimer's or Parkinson's. Yet there are great hurdles in data privacy, genetic discrimination, and access to expensive meds. Protecting patient privacy, combating misuse and affordability are key factors to help enable personalized medicine in a very diverse population (Chen, et al., 2024). Besides regulatory and technical restrictions, AI and nanotechnology also face challenges. Current regulatory frameworks for assessing the safety as well as the effectiveness of AI-based systems and nanoparticles are not well established, which creates a bottleneck for process approval. Nanoparticles may be a very useful method of targeted drug delivery, but on the other hand, they have toxicity and bio-persistency in tissues. Creating a precise, standardized set of guidelines for AI and nanotechnology applications is crucial to enable safe use in clinical practice (Rajaraman, et al., 2024, Nasarian, et al., 2024). However, the future of treatment of neurodegenerative disease appears positive, all things considered. Nanomaterials, biocompatible to AI-fuelled drug discovery, are likely to parallel the advancement of relevant effective therapies. However, collaboration between researchers, clinicians, ethicists, and regulators is needed for the successful navigation of these gateways. Mentoring search for ways to overcome the current limits will lead to a paradigm shift in neurodegenerative care by applying personalized medicine, nanotechnologies, and AI and make treatment safer for patients and more effective as well as individualized to reach significantly better outcomes in quality of life (Fehr, et al., 2024).

CONCLUSION

The removal of neurodegenerative disease is undergoing a disruption to its normal progress due to precision medicine, artificial intelligence, nanotechnology, and biotechnology. Within the past ten years, greater insights into the molecular mechanisms of aeromedical conditions have helped advance more precise and individualized management approaches. Genetic markers help to perform quick diagnoses and measures, on the other hand, various intervention strategies are constructed by AI instruments. Nanotechnology allows drug delivery across the blood-brain barrier. In addition to this, natural organic substances from the plant's world are also reported to have the effect of neuroprotection, which can be used as additional medications. These advancements are already being incorporated in preclinical and clinical environments which would subsequently offer solutions that are effective and meet the medical gaps. The intersection of personalized medicine, AI, and nano-technology has never been so precise in the formulation and administration of any medication. In the subsequent period, there are prospects of greater progress in the form of rehabilitative treatments and stem cell therapy aimed at replacing or mending damaged neural tissues. Affected individuals will receive treatment based around their genes and molecules as well and the development of drugs would progress at a faster rate due to Artificial Intelligence. Nonetheless, difficulties remain that include ethics, regulatory restrictions, and developers' capability to expand. Eliminating these challenges will be important to ensure that these improvements are available and cheap everywhere in the world. On the other hand, the outlook for treatment of neurodegenerative disorders is optimistic. It is expected that, as the integration of these technologies progresses, it will not only stop diseases from progressing but also help reverse disease effects and improve the overall life of billions of individuals across the globe.

REFERENCES

Abdul-Rahman, T., Awuah, W. A., Mikhailova, T., Kalmanovich, J., Mehta, A., Ng, J. C., & Ashraf, G. M. (2024). Antioxidant, anti-inflammatory, and epigenetic potential of curcumin in Alzheimer's disease. *BioFactors (Oxford, England)*. Advance online publication. DOI: 10.1002/biof.2039 PMID: 38226733

Ajitkumar, A., & De Jesus, O. (2024 Jan). Huntington Disease. [Updated 2023 Aug 23] In *StatPearls* [Internet]. StatPearls Publishing., Available from www.ncbi.nlm.nih.gov/books/NBK559166/

Allen, T. M., & Cullis, P. R. (2013). Liposomal drug delivery systems: From concept to clinical applications. *Advanced Drug Delivery Reviews*, *65*(1), 36–48. DOI: 10.1016/j.addr.2012.09.037 PMID: 23036225

Ashraf, S. A., Elkhalifa, A. E. O., Siddiqui, A. J., Patel, M., Awadelkareem, A. M., Snoussi, M., & Hadi, S. (2020). Cordycepin for health and wellbeing: A potent bioactive metabolite of an entomopathogenic medicinal fungus Cordyceps with its nutraceutical and therapeutic potential. *Molecules (Basel, Switzerland)*, *25*(12), 2735. DOI: 10.3390/molecules25122735 PMID: 32545666

Assi, A. A., Farrag, M. M., Badary, D. M., Allam, E. A., & Nicola, M. A. (2023). Protective effects of curcumin and Ginkgo biloba extract combination on a new model of Alzheimer's disease. *Inflammopharmacology*, *31*(3), 1449–1464. DOI: 10.1007/s10787-023-01164-6 PMID: 36856916

Ayeni, E. A., Aldossary, A. M., Ayejoto, D. A., Gbadegesin, L. A., Alshehri, A. A., Alfassam, H. A., & Tawfik, E. A. (2022). Neurodegenerative diseases: Implications of environmental and climatic influences on neurotransmitters and neuronal hormones activities. *International Journal of Environmental Research and Public Health*, *19*(19), 12495. DOI: 10.3390/ijerph191912495 PMID: 36231792

Babazadeh, A., Vahed, F. M., Liu, Q., Siddiqui, S. A., Kharazmi, M. S., & Jafari, S. M. (2023). Natural bioactive molecules as neuro medicines for the treatment/prevention of neurodegenerative diseases. *ACS Omega*, *8*(4), 3667–3683. DOI: 10.1021/acsomega.2c06098 PMID: 36743024

Bazinet, R. P., & Layé, S. (2014). Polyunsaturated fatty acids and their metabolites in brain function and disease. *Nature Reviews. Neuroscience*, *15*(12), 771–785. DOI: 10.1038/nrn3820 PMID: 25387473

Borchert, R. J., Azevedo, T., Badhwar, A., Bernal, J., Betts, M., Bruffaerts, R., & Rittman, T. (2023). Artificial intelligence for diagnostic and prognostic neuroimaging in dementia: A systematic review. *Alzheimer's & Dementia*, *19*(12), 5885–5904. DOI: 10.1002/alz.13412 PMID: 37563912

Chaudhari, K. S., Tiwari, N. R., Tiwari, R. R., & Sharma, R. S. (2017). Neurocognitive effect of nootropic drug Brahmi (Bacopa monnieri) in Alzheimer's disease. *Annals of Neurosciences*, *24*(2), 111–122. PMID: 28588366

Chen, C. (2024). AI-Aided Diagnosis For Neurodegenerative Diseases: Prospects And Challenges. Transactions on Materials. *Biotechnology and Life Sciences*, *3*, 66–72. DOI: 10.62051/72thpj80

Cheng, K. K., Yeung, C. F., Ho, S. W., Chow, S. F., Chow, A. H., & Baum, L. (2013). Highly stabilized curcumin nanoparticles tested in an in vitro blood–brain barrier model and in Alzheimer's disease Tg2576 mice. *The AAPS Journal, 15*, 324–336. DOI: 10.1208/s12248-012-9444-4 PMID: 23229335

Cheng, X., Xie, Q., & Sun, Y. (2023). Advances in nanomaterial-based targeted drug delivery systems. *Frontiers in Bioengineering and Biotechnology, 11*, 1177151. DOI: 10.3389/fbioe.2023.1177151 PMID: 37122851

Cole, G. M., Teter, B., & Frautschy, S. A. (2007). *Neuroprotective effects of curcumin.* The molecular targets and therapeutic uses of curcumin in health and disease, 197-212. doi:DOI: 10.1007/978-0-387-46401-5_8

D'Andrea, G. (2015). Quercetin: A flavonol with multifaceted therapeutic applications. *Fitoterapia, 106*, 256–271. DOI: 10.1016/j.fitote.2015.09.018 PMID: 26393898

Das, K. P. (2023). Nanoparticles and convergence of artificial intelligence for targeted drug delivery for cancer therapy: Current progress and challenges. *Frontiers in Medical Technology, 4*, 1067144. DOI: 10.3389/fmedt.2022.1067144 PMID: 36688144

Duan, L., Ouyang, K., Xu, X., Xu, L., Wen, C., Zhou, X., & Liang, Y. (2021). Nanoparticle delivery of CRISPR/Cas9 for genome editing. *Frontiers in Genetics, 12*, 673286. DOI: 10.3389/fgene.2021.673286 PMID: 34054927

Fehr, J., Citro, B., Malpani, R., Lippert, C., & Madai, V. I. (2024). A trustworthy AI reality-check: The lack of transparency of artificial intelligence products in healthcare. *Frontiers in Digital Health, 6*, 1267290. DOI: 10.3389/fdgth.2024.1267290 PMID: 38455991

Gadhave, D. G., Sugandhi, V. V., Jha, S. K., Nangare, S. N., Gupta, G., Singh, S. K., & Paudel, K. R. (2024). Neurodegenerative Disorders: Mechanisms of Degeneration and Therapeutic Approaches with Their Clinical Relevance. *Ageing Research Reviews, 102357.* Advance online publication. DOI: 10.1016/j.arr.2024.102357 PMID: 38830548

Ganesh, S., Chithambaram, T., Krishnan, N. R., Vincent, D. R., Kaliappan, J., & Srinivasan, K. (2023). Exploring Huntington's disease diagnosis via artificial intelligence models: A comprehensive review. *Diagnostics (Basel), 13*(23), 3592. DOI: 10.3390/diagnostics13233592 PMID: 38066833

Gao, H. M., & Hong, J. S. (2008). Why neurodegenerative diseases are progressive: Uncontrolled inflammation drives disease progression. *Trends in Immunology, 29*(8), 357–365. DOI: 10.1016/j.it.2008.05.002 PMID: 18599350

Goetz, L. H., & Schork, N. J. (2018). Personalized medicine: Motivation, challenges, and progress. *Fertility and Sterility, 109*(6), 952–963. DOI: 10.1016/j.fertnstert.2018.05.006 PMID: 29935653

Goyal, A., Solanki, A., & Verma, A. (2023). Preclinical evidence-based review on therapeutic potential of eugenol for the treatment of brain disorders. *Current Molecular Medicine, 23*(5), 390–400. DOI: 10.2174/1566524022666220525145521 PMID: 35619280

Gregory, J., Vengalasetti, Y. V., Bredesen, D. E., & Rao, R. V. (2021). Neuroprotective herbs for the management of Alzheimer's disease. *Biomolecules, 11*(4), 543. DOI: 10.3390/biom11040543 PMID: 33917843

Habtemariam, S. (2020). Berberine pharmacology and the gut microbiota: A hidden therapeutic link. *Pharmacological Research, 155*, 104722. DOI: 10.1016/j.phrs.2020.104722 PMID: 32105754

Hans, M. L., & Lowman, A. M. (2002). Biodegradable nanoparticles for drug delivery and targeting. *Current Opinion in Solid State and Materials Science, 6*(4), 319–327. DOI: 10.1016/S1359-0286(02)00117-1

Hersh, A. M., Alomari, S., & Tyler, B. M. (2022). Crossing the blood-brain barrier: Advances in nanoparticle technology for drug delivery in neuro-oncology. *International Journal of Molecular Sciences, 23*(8), 4153. DOI: 10.3390/ijms23084153 PMID: 35456971

Hersh, S, D., S Wadajkar, A., B Roberts, N., G Perez, J., P Connolly, N., Frenkel, V., ... & J Kim, A. (2016). Evolving drug delivery strategies to overcome the blood brain barrier. *Current Pharmaceutical Design, 22*(9), 1177–1193. DOI: 10.2174/1381612822666151221150733 PMID: 26685681

Heydari, S., Masoumi, N., Esmaeeli, E., Ayyoubzadeh, S. M., Ghorbani-Bidkorpeh, F., & Ahmadi, M. (2024). Artificial intelligence in nanotechnology for the treatment of diseases. *Journal of Drug Targeting, 32*(10), 1247–1266. DOI: 10.1080/1061186X.2024.2393417 PMID: 39155708

Huang, X., Jain, P. K., El-Sayed, I. H., & El-Sayed, M. A. (2007). Gold nanoparticles: Interesting optical properties and recent applications in cancer diagnostics and therapy. *Nanomedicine (London), 2*(5), 681–693. DOI: 10.2217/17435889.2.5.681 PMID: 17976030

Huang, Y., Li, Y., Pan, H., & Han, L. (2023). Global, regional, and national burden of neurological disorders in 204 countries and territories worldwide. *Journal of Global Health, 13*. Advance online publication. DOI: 10.7189/jogh.13.04160 PMID: 38018250

Kim, C. K., Lee, Y. R., Ong, L., Gold, M., Kalali, A., & Sarkar, J. (2022). Alzheimer's disease: Key insights from two decades of clinical trial failures. *Journal of Alzheimer's Disease, 87*(1), 83–100. DOI: 10.3233/JAD-215699 PMID: 35342092

Klyucherev, T. O., Olszewski, P., Shalimova, A. A., Chubarev, V. N., Tarasov, V. V., Attwood, M. M., & Schiöth, H. B. (2022). Advances in the development of new biomarkers for Alzheimer's disease. *Translational Neurodegeneration, 11*(1), 25. DOI: 10.1186/s40035-022-00296-z PMID: 35449079

Kulkarni, M., Patel, K., Patel, A., Patel, S., Desai, J., Patel, M., & Solanki, N. (2024). Nanomaterials as drug delivery agents for overcoming the blood-brain barrier: A comprehensive review. *ADMET & DMPK, 12*(1), 63–105. DOI: 10.5599/admet.2043 PMID: 38560713

Kumar, D., Kumar, P., Ahmed, I., & Singh, S. (2023). Integrating artificial intelligence in disease diagnosis, treatment, and formulation development: A review. *Asian Journal of Pharmaceutical and Clinical Research, 16*(3), 1–8.

Kumar, D., Kumar, P., Pal, S., & Singh, S. (2025). AI-Enhanced Neurophysiological Assessment. In Bansal, R., Maqableh, T., Shuklaa, G., Rabby, F., & Lathabhavan, R. (Eds.), *Transforming Neuropsychology and Cognitive Psychology with AI and Machine Learning* (pp. 33–64). IGI Global Scientific Publishing., DOI: 10.4018/979-8-3693-9341-3.ch002

Kumar, R., Waisberg, E., Ong, J., Paladugu, P., Amiri, D., Saintyl, J., & Tavakkoli, A. (2024). Artificial Intelligence-Based Methodologies for Early Diagnostic Precision and Personalized Therapeutic Strategies in Neuro-Ophthalmic and Neurodegenerative Pathologies. *Brain Sciences*, *14*(12), 1266. DOI: 10.3390/brainsci14121266 PMID: 39766465

Lamptey, R. N., Chaulagain, B., Trivedi, R., Gothwal, A., Layek, B., & Singh, J. (2022). A review of the common neurodegenerative disorders: Current therapeutic approaches and the potential role of nanotherapeutics. *International Journal of Molecular Sciences*, *23*(3), 1851. DOI: 10.3390/ijms23031851 PMID: 35163773

Li, Y., Zhao, J., & Hölscher, C. (2017). Therapeutic potential of baicalein in Alzheimer's disease and Parkinson's disease. *CNS Drugs*, *31*, 639–652. DOI: 10.1007/s40263-017-0451-y PMID: 28634902

Lim, D. W., Lee, J. E., Lee, C., & Kim, Y. T. (2024). Natural Products and Their Neuroprotective Effects in Degenerative Brain Diseases: A Comprehensive Review. *International Journal of Molecular Sciences*, *25*(20), 11223. DOI: 10.3390/ijms252011223 PMID: 39457003

Lu, D., Popuri, K., Ding, G. W., Balachandar, R., & Beg, M. F. (2018). Multimodal and multiscale deep neural networks for the early diagnosis of Alzheimer's disease using structural MR and FDG-PET images. *Scientific Reports*, *8*(1), 5697. DOI: 10.1038/s41598-018-22871-z PMID: 29632364

Mackenzie, I. R., Rademakers, R., & Neumann, M. (2010). TDP-43 and FUS in amyotrophic lateral sclerosis and frontotemporal dementia. *Lancet Neurology*, *9*(10), 995–1007. DOI: 10.1016/S1474-4422(10)70195-2 PMID: 20864052

Marucci, G., Buccioni, M., Dal Ben, D., Lambertucci, C., Volpini, R., & Amenta, F. (2021). Efficacy of acetylcholinesterase inhibitors in Alzheimer's disease. *Neuropharmacology*, *190*, 108352. DOI: 10.1016/j.neuropharm.2020.108352 PMID: 33035532

Mathur, S., Gawas, C., Ahmad, I. Z., Wani, M., & Tabassum, H. (2023). Neurodegenerative disorders: Assessing the impact of natural vs drug-induced treatment options. *Aging Medicine (Milton, N.S.W.)*, *6*(1), 82–97. DOI: 10.1002/agm2.12243 PMID: 36911087

Mikulska, P., Malinowska, M., Ignacyk, M., Szustowski, P., Nowak, J., Pesta, K., & Cielecka-Piontek, J. (2023). Ashwagandha (Withania somnifera)-Current research on the health-promoting activities: A narrative review. *Pharmaceutics*, *15*(4), 1057. DOI: 10.3390/pharmaceutics15041057 PMID: 37111543

Murakami, H., Shiraishi, T., Umehara, T., Omoto, S., & Iguchi, Y. (2023). Recent advances in drug therapy for parkinson's disease. *Internal Medicine (Tokyo, Japan)*, *62*(1), 33–42. DOI: 10.2169/internalmedicine.8940-21 PMID: 35110492

Nabavi, S. F., Braidy, N., Gortzi, O., Sobarzo-Sanchez, E., Daglia, M., Skalicka-Woźniak, K., & Nabavi, S. M. (2015). Luteolin as an anti-inflammatory and neuroprotective agent: A brief review. *Brain Research Bulletin*, *119*, 1–11. DOI: 10.1016/j.brainresbull.2015.09.002 PMID: 26361743

Nuzziello, N., Ciaccia, L., & Liguori, M. (2019). Precision medicine in neurodegenerative diseases: Some promising tips coming from the microRNAs' world. *Cells*, *9*(1), 75. DOI: 10.3390/cells9010075 PMID: 31892254

Ou, S., & Kwok, K. C. (2004). Ferulic acid: Pharmaceutical functions, preparation and applications in foods. *Journal of the Science of Food and Agriculture*, *84*(11), 1261–1269. DOI: 10.1002/jsfa.1873

Palanisamy, C. P., Pei, J., Alugoju, P., Anthikapalli, N. V. A., Jayaraman, S., Veeraraghavan, V. P., & Long, Q. (2023). New strategies of neurodegenerative disease treatment with extracellular vehicles (EVs) derived from mesenchymal stem cells (MSCs). *Theranostic*, *13*(12), 4138. DOI: 10.7150/thno.83066 PMID: 37554286

Pang, Y., Xiong, J., Wu, Y., & Ding, W. (2023). A review on recent advances on nobiletin in central and peripheral nervous system diseases. *European Journal of Medical Research*, *28*(1), 485. DOI: 10.1186/s40001-023-01450-7 PMID: 37932838

Pankhurst, Q. A., Connolly, J., Jones, S. K., & Dobson, J. (2003). Applications of magnetic nanoparticles in biomedicine. *Journal of Physics. D, Applied Physics*, *36*(13), R167. DOI: 10.1088/0022-3727/36/13/201

Passeri, E., Elkhoury, K., Morsink, M., Broersen, K., Linder, M., Tamayol, A., & Arab-Tehrany, E. (2022). Alzheimer's disease: Treatment strategies and their limitations. *International Journal of Molecular Sciences*, *23*(22), 13954. DOI: 10.3390/ijms232213954 PMID: 36430432

Paterson, I., & Anderson, E. A. (2005). The renaissance of natural products as drug candidates. *Science*, *310*(5747), 451–453. DOI: 10.1126/science.1116364 PMID: 16239465

Patri, A. K., Majoros, I. J., & Baker, J. R.Jr. (2002). Dendritic polymer macromolecular carriers for drug delivery. *Current Opinion in Chemical Biology*, *6*(4), 466–471. DOI: 10.1016/S1367-5931(02)00347-2 PMID: 12133722

Pinheiro, R. G., Coutinho, A. J., Pinheiro, M., & Neves, A. R. (2021). Nanoparticles for targeted brain drug delivery: What do we know. *International Journal of Molecular Sciences*, *22*(21), 11654. DOI: 10.3390/ijms222111654 PMID: 34769082

Pun, F. W., Liu, B. H. M., Long, X., Leung, H. W., Leung, G. H. D., Mewborne, Q. T., & Zhavoronkov, A. (2022). Identification of therapeutic targets for amyotrophic lateral sclerosis using PandaOmics–an AI-enabled biological target discovery platform. *Frontiers in Aging Neuroscience*, *14*, 914017. DOI: 10.3389/fnagi.2022.914017 PMID: 35837482

Rajaraman, S. (2024). Artificial Intelligence in Neuro Degenerative Diseases: Opportunities and Challenges. *AI and Neuro-Degenerative Diseases: Insights and Solutions*, 133-153. doi:DOI: 10.1007/978-3-031-53148-4_8

Roy, A., Khan, A., Ahmad, I., Alghamdi, S., Rajab, B. S., Babalghith, A. O., & Islam, M. R. (2022). Flavonoids a bioactive compound from medicinal plants and its therapeutic applications. *BioMed Research International, 2022*(1), 5445291. DOI: 10.1155/2022/5445291 PMID: 35707379

Sahebkar, A. (2010). Neuroprotective effects of resveratrol: Potential mechanisms. *Neurochemistry International, 57*(6), 621–622. DOI: 10.1016/j.neuint.2010.06.014 PMID: 20600434

Saraiva, C., Praça, C., Ferreira, R., Santos, T., Ferreira, L., & Bernardino, L. (2016). Nanoparticle-mediated brain drug delivery: Overcoming blood-brain barrier to treat neurodegenerative diseases. *Journal of Controlled Release, 235*, 34–47. DOI: 10.1016/j.jconrel.2016.05.044 PMID: 27208862

Sehgal, N., Gupta, A., Valli, R. K., Joshi, S. D., Mills, J. T., Hamel, E., & Ravindranath, V. (2012). Withania somnifera reverses Alzheimer's disease pathology by enhancing low-density lipoprotein receptor-related protein in liver. *Proceedings of the National Academy of Sciences of the United States of America, 109*(9), 3510–3515. DOI: 10.1073/pnas.1112209109 PMID: 22308347

Sekurova, O. N., Schneider, O., & Zotchev, S. B. (2019). Novel bioactive natural products from bacteria via bioprospecting, genome mining and metabolic engineering. *Microbial Biotechnology, 12*(5), 828–844. DOI: 10.1111/1751-7915.13398 PMID: 30834674

Serrano, D. R., Luciano, F. C., Anaya, B. J., Ongoren, B., Kara, A., Molina, G., & Lalatsa, A. (2024). Artificial intelligence (AI) applications in drug discovery and drug delivery: Revolutionizing personalized medicine. *Pharmaceutics, 16*(10), 1328. DOI: 10.3390/pharmaceutics16101328 PMID: 39458657

Sim, S., & Wong, N. K. (2021). Nanotechnology and its use in imaging and drug delivery. *Biomedical Reports, 14*(5), 42. DOI: 10.3892/br.2021.1418 PMID: 33728048

Singh, N. A., Mandal, A. K. A., & Khan, Z. A. (2015). Potential neuroprotective properties of epigallocatechin-3-gallate (EGCG). *Nutrition Journal, 15*, 1–17. DOI: 10.1186/s12937-016-0179-4 PMID: 27268025

Singh, R., & Lillard, J. W.Jr. (2009). Nanoparticle-based targeted drug delivery. *Experimental and Molecular Pathology, 86*(3), 215–223. DOI: 10.1016/j.yexmp.2008.12.004 PMID: 19186176

Slowing, I. I., Trewyn, B. G., & Lin, V. S. Y. (2007). Mesoporous silica nanoparticles for intracellular delivery of membrane-impermeable proteins. *Journal of the American Chemical Society, 129*(28), 8845–8849. DOI: 10.1021/ja0719780 PMID: 17589996

Speers, A. B., Cabey, K. A., Soumyanath, A., & Wright, K. M. (2021). Effects of Withania somnifera (Ashwagandha) on stress and the stress-related neuropsychiatric disorders anxiety, depression, and insomnia. *Current Neuropharmacology, 19*(9), 1468. DOI: 10.2174/1570159X19666210712151556 PMID: 34254920

Sweeney, P., Park, H., Baumann, M., Dunlop, J., Frydman, J., Kopito, R., & Hodgson, R. (2017). Protein misfolding in neurodegenerative diseases: Implications and strategies. *Translational Neurodegeneration, 6*, 1–13. DOI: 10.1186/s40035-017-0077-5 PMID: 28293421

Teleanu, R. I., Preda, M. D., Niculescu, A. G., Vladâcenco, O., Radu, C. I., Grumezescu, A. M., & Teleanu, D. M. (2022). Current strategies to enhance delivery of drugs across the blood-brain barrier. *Pharmaceutics*, *14*(5), 987. DOI: 10.3390/pharmaceutics14050987 PMID: 35631573

Thanvi, B., Lo, N., & Robinson, T. (2007). Levodopa-induced dyskinesia in Parkinson's disease: Clinical features, pathogenesis, prevention and treatment. *Postgraduate Medical Journal*, *83*(980), 384–388. DOI: 10.1136/pgmj.2006.054759 PMID: 17551069

Venkatesan, R., Ji, E., & Kim, S. Y. (2015). Phytochemicals that regulate neurodegenerative disease by targeting neurotrophins: A comprehensive review. *BioMed Research International*, *2015*(1), 814068. DOI: 10.1155/2015/814068 PMID: 26075266

Venturini, C. D., Merlo, S., Souto, A. A., Fernandes, M. D. C., Gomez, R., & Rhoden, C. R. (2010). Resveratrol and red wine function as antioxidants in the nervous system without cellular proliferative effects during experimental diabetes. *Oxidative Medicine and Cellular Longevity*, *3*(6), 434–441. DOI: 10.4161/oxim.3.6.14741 PMID: 21307644

Vidhya, K. S., Sultana, A., Kumar, N., & Rangareddy, H. (2023). Artificial intelligence's impact on drug discovery and development from bench to bedside. *Cureus*, *15*(10). Advance online publication. DOI: 10.7759/cureus.47486 PMID: 37881323

Wareham, L. K., Liddelow, S. A., Temple, S., Benowitz, L. I., Di Polo, A., Wellington, C., & Calkins, D. J. (2022). Solving neurodegeneration: Common mechanisms and strategies for new treatments. *Molecular Neurodegeneration*, *17*(1), 23. DOI: 10.1186/s13024-022-00524-0 PMID: 35313950

Wu, Y., Rakotoarisoa, M., Angelov, B., Deng, Y., & Angelova, A. (2022). Self-assembled nanoscale materials for neuronal regeneration: A focus on BDNF protein and nucleic acid biotherapeutic delivery. *Nanomaterials (Basel, Switzerland)*, *12*(13), 2267. DOI: 10.3390/nano12132267 PMID: 35808102

Xing, S. H., Zhu, C. X., Zhang, R., & An, L. (2014). Huperzine A in the Treatment of Alzheimer′ s Disease and Vascular Dementia: A Meta-Analysis. *Evidence-Based Complementary and Alternative Medicine : eCAM*, *2014*(1), 363985. DOI: 10.1155/2014/363985 PMID: 24639880

Zaky, A., Bassiouny, A., Farghaly, M., & El-Sabaa, B. M. (2017). A combination of resveratrol and curcumin is effective against aluminum chloride-induced neuroinflammation in rats. *Journal of Alzheimer's Disease*, *60*(s1), S221–S235. PMID: 28222524

Zhavoronkov, A., Ivanenkov, Y. A., Aliper, A., Veselov, M. S., Aladinskiy, V. A., Aladinskaya, A. V., & Aspuru-Guzik, A. (2019). Deep learning enables rapid identification of potent DDR1 kinase inhibitors. *Nature Biotechnology*, *37*(9), 1038–1040. DOI: 10.1038/s41587-019-0224-x PMID: 31477924

Zheng, M., Xin, Y., Li, Y., Xu, F., Xi, X., Guo, H., & Han, C. (2018). Ginsenosides: A potential neuroprotective agent. *BioMed Research International*, *2018*(1), 8174345. DOI: 10.1155/2018/8174345 PMID: 29854792

About the Contributors

Jawad Laadraoui is a professor of neurophysiology and neuroscience at the Department of Biology, Faculty of Sciences Ain Chock, Hassan II University of Casablanca, Morocco. His research focuses on neurodegenerative diseases and the effects of natural compounds on brain health, with particular emphasis on the interplay between neuroinflammation, seizure activity, and the therapeutic potential of natural compounds, aiming to develop innovative treatment strategies.

Bilal El-Mansoury is a Professor at the Department of Biology, Faculty of Sciences, Chouaib Doukkali University. He served as a Postdoc Researcher in Neuroscience and Neurology at the Royal College of Surgeons in Ireland, University of Medicine and Health Sciences. His work centers on understanding how glial cells dysfunction contributes to the pathophysiology of hepatic encephalopathy and various neurodegenerative disorders. Dr. El-Mansoury also investigated the role of miRNAs in CDKL5 deficiency disorder, with the aim of identifying novel therapeutic targets for this rare epileptic encephalopathy. In addition to his research, Dr. El-Mansoury is Editor of the Book *Physiology and Function of Glial Cells in Health and Disease.*

***

Jagjeet is an Assistant Professor of Pharmaceutics at Oxford College of Pharmacy, holds an M.Pharm from Dr. A.P.J. Abdul Kalam Technical University. With two years of teaching experience and two publications, he specializes in pharmaceutical research and innovative drug delivery systems, contributing to academia and the pharmaceutical industry.

Imade Armadi is currently pursuing a PhD in Biology and Biomedicine at the Center for Regenerative Medicine. His research focuses on the biomedical applications of chitosan, particularly in the context of clinical therapy. His work aims to explore the potential of chitosan in drug delivery systems, tissue engineering, and other therapeutic applications, contributing to advancements in regenerative medicine and clinical treatments.

Tejal Bele is an accomplished academic and research scholar with a deep passion for pharmacology, currently pursuing her Ph.D. at Savitribai Phule University, Pune, with an M.Pharm in Pharmacology, where she was awarded a Gold Medal for her academic excellence. With over six years of experience in both research and teaching, she has contributed significantly to advancing the field of pharmacology. Her work is focused on preclinical evaluation of potential neuropharmacological agents and utilizing cutting-edge computational methods such as network pharmacology, molecular docking and simulation to predict the interaction between therapeutic molecules and biological targets. As a researcher, she had the privilege of presenting her work at an international & national conference and received the Best Oral Presentation Award. She has published several research papers in peer-reviewed journals and is actively involved in research activities. Also, she guides students on their research projects, encouraging them to pursue innovative research and contributing to their academic growth.

Rathi Priya Bollu Yashwant is a research scholar currently pursuing a Ph.D. in Medical Science at Taylor's University. With a strong academic foundation that includes a Master's in Neuroscience and another Master's in Counselling Psychology, Rathi Priya bridges the gap between brain and mind. The main research focuses on phytochemicals and neurodegenerative disease, e.g., exploring the potential role of phytochemicals underlying neuroprotective mechanisms in neurodegenerative disease like Alzheimer's and Parkinson's disease. Rathi Priya is experienced in scientific writing with more than 10 years of experience in crafting articles related to therapeutic areas like stroke, depression, and neurodegeneration. Rathi Priya combines expertise in neuroscience and psychology to uncover insights that contribute to theoretical knowledge and practical applications in mental health and wellness.

Cansu Çadırci received her Master's degree in Histology and Embryology from Aydın Adnan Menderes University, Health Sciences Institute. Her academic interests include cellular development and reproductive biology.

Hicham Chatoui holds a PhD in Neurosciences, Physiology, and Eco-Health from the Sultan Moulay Slimane University of Beni Mellal. He began his teaching career in 2009 at the Faculty of Science Semlalia of Cadi Ayyad University and served as Director of the Health and Agro-Industry Engineering Department at the Private University of Marrakech (UPM). Currently, he is a Professor-Researcher at ISPITS Marrakech under the Ministry of Health and Social Protection in Morocco. An accomplished author and organizer of international scientific events, his research spans telemedicine, e-health, neuroscience, neurophysiology, neuro-immunology, biomedical engineering, nutrition, food safety, and environmental health.

Shraddha Devarshi is a Doctor of Pharmacy graduate from BVDU's Poona College of Pharmacy in 2016. During this tenure, she has worked on a project titled quality of life of Beta-thalassemia patients. She has been awarded Gold medal for her academic excellence. She is currently working as Assistant Professor in SCES's Indira College of Pharmacy, Pune. She has 9 years of teaching experience. She has 5 research articles, 2 case reports, and 1 book chapter published in national and international journals with total citation of 10 and hi-index of 2. She has also presented her work in national and international conferences. She guides Pharm.D. students in their project work, demonstrating her keen interest in research. She had delivered guest lecture on Writing SOAP notes, how to teach Pharmacotherapeutics effectively?, Do's & Don't of Project & Clerkship. She was also a panelist at Clinical Pharmacy Summit 2024 focused on bridging gap between academia and clinical practice. Her areas of research include drug utilisation pattern studies, quality of life studies, and health services.

Sreeram E. is currently based in Chennai and pursuing my final year of B.Tech in Biotechnology. Coming from a middle-class family, my father is an ex-serviceman (Indian Army), and my mother is a homemaker. As a biotechnology student, I am passionate about research and innovation, especially in the fields of food preservation and wastewater diagnosis. My current projects focus on: Food Preservation: Exploring natural and bio-based methods to enhance shelf life and maintain food quality. Wastewater Diagnosis Kit: Developing an efficient diagnostic kit to detect contaminants and improve water quality monitoring. With a keen interest in applying biotechnology for real-world challenges, I am eager to contribute to advancements in environmental sustainability and health sciences. My academic journey, combined with a strong work ethic, drives me to explore innovative solutions that can make a difference.

Hanane El Fatimi is a PhD student in neuroscience, Laboratory of Neuroscience and Cellular Physiology, Mohammed VI University of Health and Sciences, Morocco.

El Hiba Omar is professor of physiopathology, neuroscience and histology, Faculty of Sciences, Chouaib Doukkali University, He was born in 12 September 1985 in Marrakech-Safi state (Morocco) and was graduated in 2007 from Cadi Ayyad University (Morocco) with a bachelor degree in Biology. In 2009, hewas graduated from the same university with Master's degree in Life and Health Sciences. He obtained his PhD in 2015 on the "Physiopathology of Hepatic Encephalopathy" from Cadi Ayyad University. He is, since 2014, member of the International Parkinson and Movement Disorders Society (MDS), USA and the vice-President of the Moroccan Society of Ethics and Animal Research (MoSEAR).

Yasmina Jaouhari is a second-year PhD researcher at the Laboratory of Pharmacology, Neurobiology, Anthropobiology, and Environment at Cadi Ayyad University, Morocco. Her research focuses on the pharmacological evaluation of medicinal plants. She has contributed to scientific publications and conference presentations and is actively involved in academic mentorships.

Rutuja Kamble is a dynamic and dedicated professional with 15 years of experience. Her contributions encompass teaching, with a deep passion for inspiring and mentoring young minds, as well as fostering the growth and development of the institution. She has published several papers in national and international journals, presented numerous posters at various national and international platforms, and received Special Appreciation, secured First Rank, and been awarded Best Presenter at both national and international levels. Actively engaged in continuous professional development, she regularly participates in Faculty Development Programmes to further enhance her expertise and teaching practices. Her research interests include the synthesis, purification, and characterization of organic compounds, as well as online in silico studies and docking studies

Mazid Ali Khan is an Assistant Professor of Pharmaceutics at Oxford College of Pharmacy, holds an M.Pharm from Dr. A.P.J. Abdul Kalam Technical University. With two years of teaching experience and two publications, he specializes in pharmaceutical research and innovative drug delivery systems, contributing to academia and industry.

Deepak Kumar is a passionate pharmacologist, researcher, and educator with a commitment to innovation and excellence in pharmaceutical sciences. He earned his Bachelor of Pharmacy from Shri Gopichand College of Pharmacy, Baghpat, in 2021, and his Master of Pharmacy in Pharmacology from I.T.S College of Pharmacy, Ghaziabad, in 2023, both with first-class honors. Deepak's academic journey includes groundbreaking research on Coleus forskohlii, where he developed an emulgel for anti-psoriatic activity, demonstrating his expertise in pharmacology and drug delivery systems. Currently an Assistant Professor, he is dedicated to mentoring future pharmacists and advancing knowledge in areas like pharmacovigilance, clinical research, and drug safety. An accomplished author, Deepak has published research on psoriasis, artificial intelligence in pharmaceuticals, and other critical topics. His credentials include certifications in drug safety, clinical practices, and quality assurance from esteemed institutions. With aspirations to establish a pharmaceutical company, Deepak aims to focus on innovative, patient-centered products. He is driven to shape research policies, foster collaborations, and make a lasting impact on healthcare and society.

Puneet Kumar is an Associate Professor of Pharmaceutical Chemistry at Shri Gopichand College of Pharmacy, affiliated with Dr. APJ Abdul Kalam Technical University, Lucknow. He holds a Bachelor's degree in Pharmacy (2015) and a Master's degree in Pharmacy (2017) from IFTM University, Moradabad. With over 7 years of teaching experience, he has published 8 research articles and 19 review articles in peer-reviewed journals. His research interests focus on Synthetic Organic Chemistry, Synthetic Medicinal Chemistry, Medicinal and Pharmaceutical Chemistry, and Chromatography. He is also actively involved in studying diseases such as Middle East Respiratory Syndrome Coronavirus (MERS-CoV), COVID-19, and Epilepsy, particularly Absence Epilepsy and the development of Antiepileptic Drugs.

Rajasekhar Komarla Kumarachari is a Professor specializing in Pharmaceutical and Medicinal Chemistry. He holds a Ph.D. in Pharmaceutical Sciences from Andhra University. His research interests include Heterocyclic Small Molecule Synthesis, Bioactivity Screening, Antimicrobials, Drug Design, Lead Optimization, Chemical Analysis, Natural Products, and In silico methods. He has contributed to several publications in reputable journals, book chapters, and textbooks, and has presented his work at national and international conferences. Rajasekhar Komarla Kumarachari is dedicated to advancing knowledge in his field and is committed to making meaningful contributions to academia and society.

Madhumitha L is a dedicated B.Tech Biotechnology student with a strong interest in immunotechnology, research and development, and sustainable biotechnological innovations. Passionate about harnessing nature-inspired approaches, she is actively engaged in a project on creating a bio-bandage using bacterial cellulose. Her work focuses on developing sustainable and effective healthcare solutions, envisioning a future where biotechnology plays a pivotal role in improving lives. With a deep commitment to advancing medical biotechnology, her goal is to contribute to cutting-edge innovations that enhance healthcare accessibility and patient outcomes. --- Let me know if

Hanane Moummou is a professor at the Private University of Marrakech, specializing in health and agro-industry engineering since 2016. She holds a PhD in Biochemistry and

Molecular Biology from Université Cadi Ayyad (Morocco) and Université Paul Sabatier (France), a Master's degree in Water and Environment, and a Bachelor's degree in Food Technology, all from Cadi Ayyad University. Her research focuses on biotechnology, bio-process engineering, plant genomics, food safety, nutritional health, and the environmental impacts of agricultural practices, with particular interest in waste bioprocessing and bioactive compound production. She has authored and co-edited several scientific works, including Nutrition and Human Health: Effects and Environmental Impacts (Springer), and published articles in leading journals such as BMC Plant Biology and Journal of Plant Physiology.

Hariharan N.M. currently working as an Assistant Professor at Vel Tech Rangarajan Dr. Sagunthala R&D Institute of Science and Technology and has teaching experience of 11 years. He has published more than 35 research articles in international and national journals to his credit. His area of expertise includes thermoacoustic energy conversion, nanoparticle synthesis and applications, environmental remediation using novel methodologies.

Vivek Pazhamalai is working as Associate Professor at Vels Institute of Science, Technology and Advanced Studies, and has teaching experience of 15 years. Published more than 60 research articles in international and national journals to his credit.

Subha Lakshmi S pursuing undergraduate degree in btech biotechnology at vistas. co - author of 'Emerging trends in multidisciplinary science and technology (volume-2)' in SRR publicizing research.

Meenambiga S S is an Assistant Professor completed 6 years of teaching experience and 4 years of research experience in the field of Biotechnology. Passionate about teaching and have trained many young students in the field of Mycology, Natural Product Research, Nanoparticle synthesis, Drug Discovery from Plants and fungal sources and in-silico analysis of drug targets using modern Bio-informatics tools. Skilled in scientific writing covering the following aspects: Proposal submission, research articles suitable for national and international publications,drafting, proof reading and editing research thesis and expertise in critically reviewing the research manuscript for indexed journals. Member of Board of Studies for B.Tech Biotechnology curriculum in VISTAS. Guided more than 20 undergraduate students projects and served as Principal Investigator in the student's project funded by Tamilnadu State Council for Science and Technology. Published more than 30 research and review articles in national and international indexed journal and has volunteered as reviewer for many reputed journals. Life member of Mycological Society of India, BRSI, IARA, IAENG and many other professional organizations. Completed

Sangam Singh is a passionate Ph.D. Scholar in Pharmaceutical Chemistry at IFTM University, Moradabad, and an Assistant Professor at Oxford College of Pharmacy, Hapur. He has an impressive portfolio and published 13 research and review papers in reputed journals. His research centered on drug design and synthesis of skin cancer drugs. Singh's contributions extend beyond academia, holding potential for practical applications in the pharmaceutical industry.

Pallavi Singh Chouhan is a highly skilled academician and researcher with 8 years 10 months of professional experience. For the past 8 years, she has been serving as an Assistant Professor in the Department of Pharmaceutical Chemistry at Indira College of Pharmacy. Additionally, she has gained valuable industry experience as a QA executive in the Packaging and Manufacturing department. She is a university approved guide and involved in mentoring M. Pharm Quality Assurance students through their research projects and dissertations. Her research interests lie in the application of Quality by Design (QBD), analytical method development, stability testing in pharmaceuticals, and Pharmaceutical Validation. She has filed a patent and has authored one research and 3 review publications. Vidwan: 229124, Google scholar: 1HVSCWUAAAAJ Orchid: 0000-0001-8755-3422

Gokul V. has a strong background in cosmetic and food technology and has worked extensively on natural based sunscreen formulation.

Amruta Walvekar is an academician and researcher with 9 years of professional experience, including 8 years at Indira College of Pharmacy, Pune. She is Passionate about mentoring students, fostering academic excellence, and accomplishing institutional goals. She has contributed extensively to teaching, delivering well-structured and engaging lectures in pharmaceutical chemistry related subjects. She has authored 2 research and 2 review papers in reputed journals with citation index of 7 and h-index of 2. She has contributed to 1 book chapter which is indexed in international publication in the field of pharmaceutical science. She has been actively involved and participated in poster and oral presentation at various conferences to forecast innovative ideas. She has keen interest to attend various seminars and FDPs organized by various organization at national and international levels. She has guided under graduate and post graduate students on academic projects, career planning, personal development, emph

Eng Hwa Wong is a leading academic in medical sciences, currently serving as Professor at the School of Medicine, Faculty of Health and Medical Sciences. She has been the Programme Director for the School's Postgraduate Programme since 2018 and has held key academic leadership roles, including Associate Dean for Postgraduate, Research and Innovation. With a PhD in Medical Sciences from the University of Malaya and a BSc in Microbiology from Universiti Putra Malaysia, her work focuses on advancing postgraduate education, translational research, and interdisciplinary collaboration in health sciences.

Index

Symbols

A

B

C

D

E

O

P

R

S

T

www.ingramcontent.com/pod-product-compliance
Lightning Source LLC
La Vergne TN
LVHW062336300926
849161LV00007B/1715